Computational Probability

INTERNATIONAL SERIES IN OPERATIONS RESEARCH & MANAGEMENT SCIENCE

Frederick S. Hillier, Series Editor
Stanford University

Saigal, R. / *LINEAR PROGRAMMING: A Modern Integrated Analysis*

Nagurney, A. & Zhang, D. / *PROJECTED DYNAMICAL SYSTEMS AND VARIATIONAL INEQUALITIES WITH APPLICATIONS*

Padberg, M. & Rijal, M. / *LOCATION, SCHEDULING, DESIGN AND INTEGER PROGRAMMING*

Vanderbei, R. / *LINEAR PROGRAMMING: Foundations and Extensions*

Jaiswal, N.K. / *MILITARY OPERATIONS RESEARCH: Quantitative Decision Making*

Gal, T. & Greenberg, H. / *ADVANCES IN SENSITIVITY ANALYSIS AND PARAMETRIC PROGRAMMING*

Prabhu, N.U. / *FOUNDATIONS OF QUEUEING THEORY*

Fang, S.-C., Rajasekera, J.R. & Tsao, H.-S.J. / *ENTROPY OPTIMIZATION AND MATHEMATICAL PROGRAMMING*

Yu, G. / *OPERATIONS RESEARCH IN THE AIRLINE INDUSTRY*

Ho, T.-H. & Tang, C. S. / *PRODUCT VARIETY MANAGEMENT*

El-Taha, M. & Stidham, S. / *SAMPLE-PATH ANALYSIS OF QUEUEING SYSTEMS*

Miettinen, K. M. / *NONLINEAR MULTIOBJECTIVE OPTIMIZATION*

Chao, H. & Huntington, H. G. / *DESIGNING COMPETITIVE ELECTRICITY MARKETS*

Weglarz, J. / *PROJECT SCHEDULING: Recent Models, Algorithms & Applications*

Sahin, I. & Polatoglu, H. / *QUALITY, WARRANTY AND PREVENTIVE MAINTENANCE*

Tavares, L. V. / *ADVANCED MODELS FOR PROJECT MANAGEMENT*

Tayur, S., Ganeshan, R. & Magazine, M. / *QUANTITATIVE MODELING FOR SUPPLY CHAIN MANAGEMENT*

Weyant, J./ *ENERGY AND ENVIRONMENTAL POLICY MODELING*

Shanthikumar, J.G. & Sumita, U./*APPLIED PROBABILITY AND STOCHASTIC PROCESSES*

Liu, B. & Esogbue, A.O. / *DECISION CRITERIA AND OPTIMAL INVENTORY PROCESSES*

Gal, Stewart & Hanne/ *MULTICRITERIA DECISION MAKING: Advances in MCDM Models, Algorithms, Theory, and Applications*

Fox, B. L./ *STRATEGIES FOR QUASI-MONTE CARLO*

Hall, R.W. / *HANDBOOK OF TRANSPORTATION SCIENCE*

Computational Probability

Edited by
Winfried K. Grassmann
University of Saskatchewan

Springer Science+Business Media, LLC

Library of Congress Cataloging-in-Publication Data

A C.I.P. Catalogue record for this book is available
from the Library of Congress.

DOI 10.1007/978-1-4757-4828-4

Copyright © 2000 by Springer Science+Business Media New York
Originally published by Kluwer Academic Publishers in 2000
MyCopy version of the original edition 2000

All rights reserved. No part of this publication may be reproduced, stored in a retrieval system or transmitted in any form or by any means, mechanical, photocopying, recording, or otherwise, without the prior written permission of the publisher, Springer Science+Business Media, LLC.

Printed on acid-free paper.
www.springer.com/mycopy

Contents

Preface	vii
1 Computational Probability: Challenges and Limitations *Winfried K. Grassmann*	1
2 Tools for Formulating Markov Models *Gianfranco Ciardo*	11
3 Transient Solutions for Markov Chains *Edmundo de Souza e Silva and H. Richard Gail*	43
4 Numerical Methods for Computing Stationary Distributions of Finite Irreducible Markov Chains *William J. Stewart*	81
5 Stochastic Automata Networks *Brigitte Plateau and William J. Stewart*	113
6 Matrix Analytic Methods *Winfried K. Grassmann and David A. Stanford*	153
7 Use of Characteristic Roots for Solving Infinite State Markov Chains *H. Richard Gail, Sidney L. Hantler and B. Alan Taylor*	205
8 An Introduction to Numerical Transform Inversion and Its Application to Probability Models *Joseph Abate, Gagan L. Choudhury and Ward Whitt*	257

9
Optimal Control of Markov Chains　325
Shaler Stidham, Jr.

10
On Numerical Computations of Some Discrete-Time Queues　365
Mohan L. Chaudhry

11
The Product Form Tool for Queueing Networks　409
Nico M. van Dijk, Winfried K. Grassmann

12
Techniques for System Dependability Evaluation　445
Jogesh K. Muppala, Ricardo M. Fricks, and Kishor S. Trivedi

Index　481

Preface

In recent years, great advances have been in the field of computational probability, in particular in areas related to queueing systems, stochastic Petri-nets and systems dealing with reliability. The objective of this book is to make these topics accessible to researchers, graduate students, and, hopefully, practitioners.

Great care was taken to make the exposition as clear as possible. Every line in this text has been evaluated, and changes have been made whenever it was felt that the initial exposition was not clear enough for the intended readership.

The topics were selected with great care to cover as wide a range as possible, with particular emphasis on Markov modeling and queueing applications. I feel that I was very privileged to obtain the contributions of so many outstanding researchers in this field, and I thank all the contributors for their great effort. It is not easy to explain and summarize the sophisticated techniques used in Markov modeling and queueing, but I am happy to say that all contributers did an outstanding job.

The outline of this book is as follows. The first chapter describes, in non-mathematical terms, the challenges in computational probability. Most of the models used in this book are based on Markov chains, and Chapter 2 describes the methodologies available to obtain the transition matrices for these Markov chains, with particular emphasis on stochastic Petri-nets. Chapter 3 discusses how to find transient probabilities and transient rewards for theses Markov chains. The next two chapters indicate how to find steady-state probabilities for Markov chains with a finite number of states. In this case, the equilibrium probabilities can be found by solving a system of linear equations, with one equation for each state. If the number of states is not too large, one can use methods based on Gaussian elimination, that is, direct methods, to solve these systems. However, as the number of states increases, iterative methods become advantageous. Both direct and iterative methods are described in Chapter 4. In many situations, the transition matrices are similar to Kronecker products, and this can be exploited for reducing the storage requirements and/or the number of operations needed to find equilibrium solutions. Details for these methods are given in Chapter 5. The next two chapters deal with infinite-state Markov chains. Infinite-state Markov chains occur frequently in queueing, because one typically does not want to set a bound for all queues. There are two main methods to analyze problems with infinite queues: matrix analytical methods and methods based on spectral expansion. These are discussed in Chapters 6 and 7, respectively.

Chapter 8 deals with transforms, in particular Laplace transforms. Transform methods have been used extensively in theoretical work, but their practical value was questioned. This has changed, mainly due to the effort of Whitt and his collaborators, who have developed a number of numerical methods for transform inversions. They describe their work in Chapter 8.

Often one is not satisfied with improving the system, but one wants to configure the system in the best possible way, that is, one wants to optimize the system. One way to do the optimization is through *Markov decision making*, a topic described in Chapter 9.

Markov modeling has found applications in many areas, three of which are described in some detail: Chapters 10 analyses discrete-time queues, Chapter 11 describes networks of queues, and Chapter 12 deals with reliability theory.

I would like to thank all the people who have helped me to make the project a success. I would also like to thank the Natural Science and Engineering Council of Canada (NSERC) who provided me with a generous operating grant, part of which was used for this project, The Department of Computer Science of the University of Saskatchewan and the Department of Statistical and Actuarial Sciences of the University of Western Ontario have provided me with computer facilities and technical advice, which was very much appreciated. Last but not least, I would like to thank all the authors for their excellent contributions.

1 COMPUTATIONAL PROBABILITY: CHALLENGES AND LIMITATIONS

Winfried K. Grassmann

Department of Computer Science
University of Saskatchewan
Saskatoon, Sask S7N 5A9
Canada

grassman@cs.usask.ca

1 INTRODUCTION

Computational probability is the science of calculating probabilities and expectations. As this book demonstrates, there are many mathematical challenges in the area of computational probability. To set the stage, we discuss the objectives of computational probability in more detail, and we point out the difficulties one encounters in this area. We also contrast computational probability with other approaches.

2 STOCHASTIC SYSTEMS AND THEIR ANALYSIS

Probability theory deals with uncertainty, that is, with situations admitting different possible outcomes. The set of all outcomes forms the *sample space*. In contrast to other methods, such as fuzzy set theory [Klir and Yuan, 1995], probability theory associates probabilities with outcomes. If the number of outcomes is countable, a probability can be associated with each outcome. However, to deal with continuous sample spaces, one typically associates probabilities with sets of outcomes.

A system is called *stochastic* if it can behave in different ways, and if one can associate a probability with each set of possible behaviors. Most stochastic systems are also *dynamic*, that is, they change with time. This book deals predominantly with stochastic dynamic systems, including queueing systems, communication systems, and systems dealing with reliability.

To analyze a stochastic system mathematically, it must be formulated in mathematical terms which can then be manipulated. In other words, a mathematical model must be built. Ideally, the model should behave like the real system, that is, it should predict the system behavior accurately. This may not be possible for two reasons: firstly, most real systems are too complicated to be dealt with mathematically, and secondly, the working of the real system may be unknown. The realism of the model can be assessed by comparing its behavior with that of the real system, at least as long as the system is deterministic. However, in stochastic systems, different realizations are possible, and the comparison with the real system must be based on probabilities and expectations.

There are three approaches to analyze stochastic systems and predicting their behavior: analytical methods, numerical methods, and methods based on Monte Carlo simulation. All of these methods are now described, and their strengths and weaknesses are outlined.

Analytical methods express dependencies between the variables of a model explicitly, usually by means of a formula. For instance, Little's theorem provides a simple formula relating the average number in a system, the arrival rate to the system, and the time spent in the system. Simple explicit formulas such as this are definitely the ideal. Unfortunately, many explicit formulas are far from simple. In particular, using symbol manipulation programs, such as Maple or Mathematica, one can easily derive analytical results of amazing complexity. It is questionable whether or not complex formulas have any direct benefit because the human mind cannot deal with more than three to four variables at a time. However, complex formulas may have indirect benefits if they allow one to derive simple results, or if they form the basis of an algorithm. This brings us to the second approach, the numerical approach.

In the numerical approach, one fixes all input parameters of the model to certain values, and one explicitly calculates the probabilities and/or expectations of interest. To do this, an algorithm is created, and a program is written and executed on a computer, using the prescribed input parameters. For instance, to find the waiting time in a GI/G/1 queue, one must use numerical methods, such as the ones described in Chapters 8 and 10. However, once these methods are implemented, the user may specify an appropriate distribution, and set appropriate parameters, and the computer delivers the results. This allows the users to experiment without having to know the underlying mathematical theory.

The methods underlying a numerical algorithm need not be simple. Since the user no longer needs to know about the underlying mathematical details, even extremely complex analytical results may became useful once programmed. In fact, some analytical results, which, at the time of their discovery, seemed to be only of theoretical interest, turned out to be very useful as the basis of numerical algorithms. As a case in point, consider Chapter 8, which describes a number of methods for the numerical inversion of Laplace transforms. These methods make it practical to actually use theoretical results involving transforms derived by theoreticians. On the other hand, some analytical results are numerically very inefficient. This is particularly true for most formulas that depend on Cramer's rule for solving linear equations.

The third method for dealing with complex stochastic dynamic system is through Monte Carlo simulation. In Monte Carlo simulation, one randomly selects a large number of outcomes, and for each outcome selected, one evaluates the system response. Since a single outcome is completely specified, no probabilities are needed. Probabilities, and other measures of interest, are obtained statistically by averaging over all selected outcomes. In short, probabilities and other measures are estimated, rather than calculated. Similar to other estimation problems, one must deal with confidence levels and confidence intervals. To reduce the confidence interval by a factor k, one typically has to increase the computational effort by a factor k^2. For instance, to obtain an additional significant digit of a result, one must reduce the confidence interval by a factor of 10, and to achieve this, the computational effort must be increased by a factor of 100. It is thus very expensive to obtain accurate results by simulation. Estimation of events having a very small probability is also very difficult. In this case, one needs a lot of observations, and therefore a lot of computer time, in order to just observe a few of the rare events whose probability must be estimated. On the other hand, the computational complexity increases only slightly as the number of variables needed to specify the state of the system modeled increases. This is in sharp contrast to numerical methods as will be shown later.

Since simulation does not really deal with probabilities, but with estimation, it uses methodologies that are quite different from the ones used in computational probability. For this reason, we will not discuss simulation in this book, but we refer to the literature on this topic [Bratley et al., 1996, Banks et al., 1996]. Nevertheless, the methodology of simulation is very much relevant in computational probability, as will be shown next.

3 DISCRETE EVENT SYSTEMS AND MARKOV CHAINS

One major branch of simulation is *discrete event simulation*, which, as the name suggests, involves the simulation of discrete event systems. In a discrete event system, one has a number of *entities*, which have certain *properties*, and these interact to change the state of the system. The entities, and their properties completely describe the state of the system. In some systems [Schruben, 1995], it is assumed that only the number of entities is of importance. In this case, the state of the system is described by a number of discrete variables, called *state variables*. We will call such systems *discrete state systems*. Anything that changes one of the state variables is called an *event*. In discrete event simulation, events are *scheduled* in advance. The time between the scheduling of an event, and the time the event affects the system will be called *incubation time*. Typically, the incubation times are random variables.

Ideally, computational probability should be able to deal with any discrete state system, and in fact, attempts were made as early as 1966 by Victor

Wallace [Wallace, 1966] to this effect. Unfortunately, major difficulties arise when applying numerical methods to analyze discrete state systems. These difficulties arise from two sources: the continuity of time, and the difficulty to deal with the joint probabilities of many dependent random variables. This section will only deal with the treatment of time. The problem posed by dealing with many dependent random variables will be discussed in the next section.

Clearly, computers are digital, which means that they cannot handle continuous systems exactly. One always has to find some method to convert the discrete system into a continuous one. This conversion is typically done through discretization, a method that gives only approximate results. In special cases, however, there are alternatives.

Even though the state variables in a discrete state system are discrete, time is usually a continuous variable, which means that the incubation times are continuous as well. There are, however, exceptions to this rule. In particular, in communication, messages are often slotted, with one message per slot. This effectively makes time a discrete variable. Such systems are discussed in Chapter 10. However, most other chapters, and indeed, most papers that have appeared in literature deal with models that are continuous in time.

If all incubation times are exponential random variables, the time since the event in question was scheduled does not matter. In other words, only the values of the state variables at the present moment matters, that is, the system is Markovian. If some incubation times are not exponential, the system is not Markovian, and non-Markovian systems are difficult to analyze. In almost all cases, non-Markovian systems must be converted into Markovian systems to make their analysis possible. The most direct approach is to add new variables, so called *supplementary variables*, which allow one to deal with the scheduling of events. The supplementary variables are added to the state space and treated like any other state variable. Hence, the state is described by two types of variables: the proper variables (the variables of the original model), and the supplementary variables, which provide information for scheduling events. Since the supplementary variables are part of the state description, they too will be considered as state variables, even though they are not proper state variables in the sense of simulation.

There are in principle two methods to use supplementary variables: one can use the elapsed incubation times as supplementary variables, or one can use the remaining incubation times as supplementary variables. Of course, in continuous time systems, all these times are continuous.

As an example of how to deal with supplementary variables, consider the GI/G/1 queue. In the GI/G/1 queue, there are two types of events: arrivals and departures. Each event is associated with an incubation time. The elapsed incubation time associated with arrivals is the time since the last arrival, and the elapsed incubation time of a departure is the time since the service started. When using elapsed times as supplementary variables to make the GI/G/1 queue Markovian, one therefore has to add the time since the last arrival and the elapsed service time as state variables. Alternatively, one can use the remaining times as supplementary variables, that is, the time until next arrival, respectively, the remaining service time.

The fact that exponential incubation times do not lead to continuous supplementary variables suggests that one might be able to use exponential random variables as building blocks for constructing continuous distributions. This idea leads to the *method of phases*. In this method, one assumes that there are different phases, which are often, but not always, completed in sequence, and the rate of completing the incubation time depends on the phase one is in. If the sequence of phases is Markovian, then the resulting distributions are said to be *phase type*. If all incubation times are phase-type, then one can use a discrete supplementary state variable for each incubation time to represent its phase. Details of this method will be discussed in Chapter 2. There are other methods besides the method of phases for dealing with continuous incubation times. In particular, transform methods can deal with continuous incubation times, as indicated in Chapter 8.

The imbedded Markov chain technique is another technique to convert non-Markovian systems into Markov chains. Generally, a process is said to have an imbedded Markov chain if there is a sequence of epochs $t_1, t_2, \ldots$ such that the process $\{X(t_n), n = 1, 2, \ldots\}$ forms a Markov chain. The t_n are called *regeneration points*. It is usually not difficult to analyze the problem at the regeneration points t_n, $n > 0$. After that, one is able to find $X(t)$ for arbitrary t.

In summary, to apply the methods of computational probability, one usually converts the given system into a Markov chain before numerical methods are applied. Hence, much of computational probability deals with what is often referred to as *Markov modelling*. The system is described by a number of state variables, which possibly include supplementary variables. The state is defined jointly by all variables, which means that the state space is a subset of the Cartesian product of the ranges of all state variables. Since one typically needs several state variables to describe a model, the state space tends to be huge. If all variables forming the state have a finite range, the state space is finite. If any state variable or supplementary variable is continuous, the state space is continuous.

4 TRANSIENT AND STEADY STATE SOLUTIONS

We now assume that the number of states is countable, and that it is therefore possible to associate a number with each state, say state 1, 2, 3 .... The rates, respectively, the probabilities, of going from state number i to state number j, $i \neq j$ can now be extracted from the model to be analyzed. In the discrete case, the transition matrix $\mathbf{P}$ is given by the p_{ij}, where p_{ij} is the probability of going from i to j. In the continuous case, one defines $\mathbf{Q} = [q_{ij}]$, with q_{ij}, $i \neq j$ being the rate of making a transition form i to j, and $q_{ii} = -\sum_{j=1}^{\infty} q_{ij}$. As described in Chapter 2, the construction of transition matrices is typically done by specially designed programs.

Once the transition matrix is given, transient and steady state results can be obtained. This will now be discussed. For simplicity, we restrict ourselves to continuous-time Markov chains, or CTMCs. Of basic interest is the probability $\pi_i(t)$ of being in state number i at time t. This probability depends on the initial state, and it may change with increasing t. The $\pi_i(t)$ are in this sense *transient probabilities*. In many Markov chains, the $\pi_i(t)$ converge toward certain *steady state* values π_i. Transient probabilities are discussed in Chapter 3. Since the number of states is typically large—problems with 1000 or 10,000 states are not uncommon—just listing all $\pi_i(t)$ for different values of t is unsatisfactory because the human brain cannot handle so many numbers at once. Hence, certain descriptive measures, such as averages, probabilities for sets of states, and so on have to be found. To deal with this, so called *rewards* are useful. Specifically, with each state i, a number r_i is associated. The original idea is that while in state i, there is a reward of r_i per time unit. Since $\pi_i(t)$ is the probability of being in state i, the expected reward rate at time t is $\sum_{i=1}^{\infty} r_i \pi_i(t)$. The mathematical construct of rewards can also be used in situations unrelated to monetary awards. In particular, they can be used to state that the system is in a certain set A of states. To do this, let $\mathcal{I}\{B\}$ be the *indicator function* for predicate B, that is, $\mathcal{I}\{B\} = 1$ if B is true, and $\mathcal{I}\{B\} = 0$ if B is false. If $r_i = \mathcal{I}\{i \in A\}$, then the probability of being in A at t can be expressed as

$$P_t\{A\} = \sum_{i=1}^{\infty} r_i \pi_i(t) = \sum_{i=1}^{\infty} \mathcal{I}\{i \in A\} \pi_i(t)\}$$

In many cases, one is not only interested in the reward rate, but also in the reward over a particular time period, say form 0 to t. To formulate this accumulated reward, let $X(t)$ be the number of the state the system occupies at t, and let $r(t) = r_{X(t)}$. The total reward from 0 to T is then given as

$$R(T) = \int_0^T r(t)dt$$

Since $R(T)$ is a random variable, one may want to find its distribution. This problem will be addressed in Chapter 3. If $r_i = \mathcal{I}\{i \in A\}$ for some set A, then $R(T)$ is the time spent in A during the interval $(0, T)$, and $R(T)/T$ is therefore the proportion of time spent in A.

Most researchers concentrate on steady state analysis, and there are good reasons for this. First of all, steady state analysis tends to be simpler. Secondly, there is typically only one set of steady state probabilities, whereas there is a probability vector $[\pi_i(t)]$ for each t. Moreover, if t is large enough, then most of these vectors are close to steady state, and their calculation is therefore redundant. Also, for long time intervals,

$$E(R(T)) = T \sum_{i=1}^{\infty} r_i \pi_i$$

whereas this equation does not hold if the $\pi_i(t)$ change with t.

5 THE LIMITS OF COMPUTATIONAL PROBABILITY

The number of states in Markov modelling increases exponentially with the number of state variables, and this severely limits the size of the models that can be analyzed by numerical methods. One may believe that these computational problems can be overcome by the advances in computer technology. Unfortunately, this is not true. In fact, the more difficult a problem is to solve, the less it benefits from advances in computer technology. To make things concrete, assume that the number of calculation per time unit a computer can do increases 1000 fold. In this case, it seems reasonable to expect that the size of the problems one can solve in, say, one hour also increases by a factor of 1000. Unfortunately, in the case of Markov chains, this is not the case. The so called direct methods for solving Markov chains take of the order $O(n^3)$ operations, where n is the number of states. Hence, as the computer power increases by a factor of 1000, the size of Markov chains solvable in one hour increases only by a factor of 10, which is two orders of magnitude below what one might naively expect. Indeed, the fact that direct methods are $O(n^3)$ has motivated the development of so called iterative methods, which have a complexity of somewhat less than $O(n^3)$. For large problems, and if computer speed increases, more and more large problems will be analyzed, even a small reduction in the order, say from $O(n^3)$ to $O(n^{2.5})$, makes a huge difference. If we take a medium size problem with 1,000 states, then it is easy to see that n^3 is $\sqrt{n} = 31.62\ldots$ times larger than $n^{2.5}$, and if $n = 10,000$, than n^3 is two orders of magnitude larger than $n^{2.5}$.

The large number of states one typically observes in Markov modelling is due to the fact that the number of states increases exponentially with the number of state variables. This is known as the *curse of dimensionality*. To demonstrate the effect of this curse, suppose a system has m state variables, and that all state variables can have only two values. If all state variables can assume these values in every configuration, then the number of states is 2^m. For many techniques discussed in this book, one needs to enumerate all states, which takes $O(2^m)$ operations. With the equipment of today, one can do this enumeration for m up to about 30. A 1000 fold increase in computer power would mean that the enumeration can proceed up to 40, that is, the problem size can only increase by about 25%. Moreover, the transition matrices corresponding to 2^m states have 2^{2m} entries, which causes major problems for storage. Direct solutions would require 2^{3m} operations. In fact, a problem with $m = 10$ and 2^m states has recently been solved by a direct method [Grassmann and Zhao, 1997]. To solve a problem of size $m = 14$ would require a four-thousand fold increase in computing speed! But 14 is still a small number! Solving problems with, say, $m = 100$ in this fashion will be elusive forever. In cases like this, simulation may be the only effective method.

The huge number of states caused by the curse of dimensionality is felt throughout computational probability. To begin with, generating transition matrices of high dimension by hand is time-consuming and error-prone, and this has led to the design of programs which automatically generate transition matrices from a high-level description. Storing the transition matrices also causes problems, and dense storage methods must be used whenever possible. Modelling methods leading to sparse transition matrices are preferred, and so solution methods will attempt to preserve sparsity. The methods used to find transient and steady-state probabilities should have the lowest possible order in regard to the computational time complexity. One has to be aware in this regard that the computational time complexity of simulation increases only linearly with the number of states, which means that for large problems, simulation is the more efficient method. In fact, really large problems can only be solved by simulation.

References

[Banks et al., 1996] Banks, J., Carson, J. S. I., and Nelson, B. L. (1996). *Discrete-Event System Simulation.* Prentice Hall, Upper Saddle River, NJ.

[Bratley et al., 1996] Bratley, P., Fox, B. L., and Schrage, L. E. (1996). *A Guide to Simulation.* Springer Verlag, New York, NY, 2nd edition.

[Grassmann and Zhao, 1997] Grassmann, W. K. and Zhao, Y. Q. (1997). Heterogeneous multiserver queues with general input. *INFOR*, 35:208–224.

[Klir and Yuan, 1995] Klir, G. J. and Yuan, B. (1995). *Fuzzy sets and fuzzy logic : theory and applications.* Prentice Hall, Upper Saddle River, NJ.

[Schruben, 1995] Schruben, L. W. (1995). *Graphical Simulation Modeling and Analysis Using SIGMA for Window.* Boyd & Fraser, Danvers, MA.

[Wallace, 1966] Wallace, V. L. (1966). RQA-1: The recursive queue analyser. Technical Report 2, Systems Eng. Laboratory, University of Michigan, Ann Arbor, Michigan.

Winfried Grassmann got his education in economics in Zurich, Switzerland. After his Masters, he joined the Operations Research Department of Swissair, the Swiss flag carrier. There, he developed a system for inventory control, and a system for dealing with rotating parts. Both systems were implemented with great success. While at Swissair, he also finished his Ph.D., which he defended in 1968 with summa cum laude. He then joined the Computer Science Department of the University of Saskatchewan, where he taught Operations Research and Computer Science. Dr. Grassmann was on the editorial boards of Naval Research Logistics and Operations Research, and he is presently associate editor of the INFORMS Journal on Computing. He has written a book on stochastic processes and, more recently, a book on logic and discrete mathematics. His main areas of research are queueing theory and Markov modelling, two areas in which he has published widely. His papers have appeared in Operations Research, Journal of Applied Probability, Interfaces, Naval Research Logistics, INFOR and other journals. For his lifetime achievements, he received the 1999 merit award of the Canadian Operational Research Society.

2 TOOLS FOR FORMULATING MARKOV MODELS

Gianfranco Ciardo

Department of Computer Science
College of William and Mary
Williamsburg, VA 23187-8795
USA

ciardo@cs.wm.edu

1 MARKOV CHAIN MODELING

Many man-made systems, especially those in the areas of computer and communication, are so complex that it is essential to study them with simplified mathematical models during their design, prototyping, and deployment.

When the system's behavior or external stimuli are not completely understood, or simply when it is not feasible to describe them in full detail, a probabilistic, or stochastic, representation can be employed. For example, one could think of an Internet Service Provider that must satisfy connection requests, or a distributed database that must grant access to its data. Both types of traffic are highly variable, yet these systems need to provide acceptable levels of service, which might be defined in terms of availability, response time, or other appropriate measure.

We limit our discussion to stochastic models having a discrete state space S. In this case, the model remains in a given state $i \in S$ for a certain random amount of time, then it moves to a state $j \in S$.

The discrete-state stochastic models used in practical application can be classified according to the solution approach used. As indicated in Chapter 1, the most powerful technique is discrete-event simulation. The alternative is the use of analytical/numerical techniques, for which the Markov approach is a prime example. In this chapter, we restrict ourselves to continuous-time Markov chains (CTMCs), where the interval of time between the occurrence of an event to the next is exponentially distributed with a rate that depends exclusively on the current state. Discrete-time Markov chains (DTMCs) can also be defined, where the time between each change of state is instead geometrically distributed. Indeed, CTMCs and DTMCs share many properties, and it is easy to translate one into the other; this is sometimes done even just for solution purposes (e.g., "embedding" for steady-state analysis [Ciardo et al., 1991] and "uniformization" for transient analysis of CTMCs [de Souza e Silva and Gail, 1989, Grassmann, 1991, Jensen, 1953]).

We should stress that there are other analytical techniques, from more specialized ones (e.g., product-form queuing networks [Baskett et al., 1975], which are less powerful in terms of modeling capabilities but have very efficient solution algorithms) to more general ones (such as those based on semi-Markov or Markov-regenerative processes [Choi et al., 1994, Ciardo et al., 1993b, Çinlar, 1975]). From a stochastic point of view, these can be thought of as either restrictions or extensions of Markov chains, and much of the foundation for their analysis is based on Markov concepts.

While the algorithmic steps for the numerical CTMC solution are well understood, formidable practical challenges exists. Ultimately, these are related to the size of state space, which grows combinatorially state variables, as indicated in Chapter 1.

- The state space S must be specified. Since 10^5, 10^6, or even more states are often required to model a system, this cannot be done "manually". High-level formalisms, discussed in Section 2, are used to describe S in a compact way. Of course, the quantities we want to compute from our models are also to be expressed in terms of high-level model entities, using the concept of rewards described in Section 3.

- CTMCs assume exponentially distributed timing. If this is not the case for a particular event in a system, we can use phase-type distributions to represent more accurately its duration. However, this increase in model fidelity comes at a price: the size of the state space grows dramatically. On the other hand, sometimes we can assume that an event is so fast that it can be considered immediate; doing so reduces the size of the state space. In both cases, the one-to-one correspondence between model states and CTMC states is lost. Section 4 addresses these issues.

- Prior to attempting a numerical solution, S must be generated and stored explicitly. This step can be automated in a software package, but the amount memory required is a limiting factor. The same can be said for the infinitesimal generator matrix $\mathbf{Q}$ describing the CTMC, since the number $\eta(\mathbf{Q})$ of nonzero entries in it corresponds to the possible state-to-state transitions and is considerably larger than $|S|$ itself. Section 5 discusses the generation and storage of S and $\mathbf{Q}$.

Before continuing, we briefly review the CTMC terminology and notation.

1.1 Basic Markov Chain Definitions

A CTMC is a discrete-state stochastic process $\{X(t) : t \geq 0\}$ possessing the Markov property: given the present state, knowledge about the past is irrelevant in assessing the future behavior. Formally, for all $t_1 < \cdots < t_n < t$ and $i_1, \ldots i_n, i \in \mathcal{S}$,

$$\Pr\{X(t) = i \mid X(t_n) = i_n \wedge \cdots \wedge X(t_1) = i_1\} = \Pr\{X(t) = i \mid X(t_n) = i_n\}.$$

A CTMC is completely defined by its infinitesimal generator $\mathbf{Q}$ and its initial probability distribution $\boldsymbol{\pi}(0)$. For $i, j \in \mathcal{S}$, $i \neq j$, $q_{i,j}$ is the rate at which the CTMC goes to state j, given that it is in state i. The diagonal of $\mathbf{Q}$ is instead defined as $q_{i,i} = -\sum_{j \neq i} q_{i,j}$. For each $i \in \mathcal{S}$, $\pi_i(0)$ is the probability that the CTMC is in state i initially, that is, at time 0.

For most types of analysis, it is important to classify the states of the CTMC. In general, the state space $\mathcal{S}$ can be partitioned into one set of transient states, $\mathcal{S}^0$, and m recurrent classes, $\mathcal{S}^l$, $1 \leq l \leq m$ (if $\mathcal{S}$ is infinite, m can be infinite as well). For finite state spaces, this partition can be based exclusively on the graph that has the states as nodes and an arc from i to j iff $q_{i,j} > 0$. Then, a state i is transient if there is a path to some state j with no return path to back to state i. A recurrent class $\mathcal{S}^l$ is simply a strongly-connected component without arcs leaving it; in particular, a state with no outgoing arcs is called an absorbing state.

Given a CTMC, we might be interested in its transient or steady-state analysis. Specifically, for a finite time $t \geq 0$, we can compute $\boldsymbol{\pi}(t)$, the vector specifying the probability of being in each state $i \in \mathcal{S}$ at time t. Formally, $\boldsymbol{\pi}(t)$ is the solution of the system of ordinary differential equations with constant coefficients

$$\frac{d\boldsymbol{\pi}(t)}{dt} = \boldsymbol{\pi}(t) \cdot \mathbf{Q} \quad \text{subject to the initial condition } \boldsymbol{\pi}(0).$$

As $t \to \infty$, $\boldsymbol{\pi}(t)$ may converge toward an equilibrium vector $\boldsymbol{\pi} = \lim_{t\to\infty} \boldsymbol{\pi}(t)$. If $\mathcal{S}$ is finite, $\boldsymbol{\pi}$ always exists. If the CTMC contains a single recurrent class, we compute $\boldsymbol{\pi}$ by solving

$$\boldsymbol{\pi} \cdot \mathbf{Q} = 0 \quad \text{subject to } \sum_{i \in \mathcal{S}} \pi_i = 1$$

and, in this case, $\boldsymbol{\pi}$ does not depend on the initial probability vector $\boldsymbol{\pi}(0)$. If the CTMC contains multiple recurrent classes, $\boldsymbol{\pi}(0)$ must instead be used to determine the probability of eventually reaching each of these classes first. Then, the conditional probability of being in each state of a given recurrent class $\mathcal{S}^l$, given that the CTMC has entered $\mathcal{S}^l$, can be computed in isolation, and the results merged. See [Ciardo et al., 1993a] for details. As will be shown in Chapter 3, it is also possible to find the proportion of time the system spends in state i from 0 to t. As t goes to infinity, these probabilities also converge toward $\boldsymbol{\pi}$, that is, the $\boldsymbol{\pi}$ indicate the proportion of time the system is in state i.

2 HIGH-LEVEL FORMALISMS

We now discuss how a particular system can be represented in practice using a high-level model. Since high-level formalisms are quite compact, extremely complex behaviors can be modeled, even if the underlying CTMC is afflicted by the state space explosion. In other words, high-level formalisms help with the *specification* of the process, by making it both faster and more natural (hence less error-prone), but not with its *solution*. We address the translation of the model into the underlying CTMC in Section 5.

Various approaches have been proposed to describe a CTMC in a compact way. Under the assumption that all activities in the model have an exponentially-distributed duration, the purpose of a high-level model is to define the state space $S = \{s_1, s_2, \ldots, s_n\}$, the set of possible state-state transitions $\mathcal{A} \subseteq S \times S$, and their rates $\lambda : \mathcal{A} \to R^+$, in a compact way.

2.1 General Discrete-Event Systems

One high-level formalism is given by discrete-event systems (DESs) [Cassandras, 1993]. In discrete event systems, one deals with entities, which have attributes, and which interact in some way to change the state of the system. Depending on the application, the entities might be jobs in a computer system, messages in a communication system, products in a job shop, or people waiting for elevators or in restaurants. If the entities are jobs in a computer system, their attributes may be the time at which they arrived, their priority, their type, say real time or batch, and so on. The entities and their attributes collectively describe the state of the system, and the state space S is given by all possible states.

Intuitively, an event is anything that changes the state of the system. Typical events include: the creation of an entity, the change of the attribute of an entity, and the destruction of an entity. We assume that all events have predictable effects, that is, given the original state and the event, there is only one possible final state after the event has taken place. This is no real restriction. Events with K possible outcomes are merely considered as K different events. Hence, an event e is a function, say $new(e, s)$, which maps S into S. The events are *discrete* in the sense that there is an $\epsilon > 0$ such that every event either creates or deletes entities, leaves the system unchanged, or changes the attributes of at least one entity by more than ϵ. In this way, one excludes cases where attributes change in a continuous fashion, such as the amount of work completed in the case of modeling jobs in computer systems.

Events are *scheduled*, that is, if the present clock time is t while the scheduling occurs, a time $x \geq 0$ is selected, typically by a random process, and the event will occur at time $t+x$. Events that have been scheduled, but have not occurred yet are in *incubation*, and the length of the time from the moment the event was scheduled until the moment it takes place is the *incubation time*. The incubation time may be zero. In this case, the event occurs immediately.

Certain events can only be scheduled in a subset of states. If the system is in a state s which allows event e to be scheduled, then e is said to be *enabled*. Otherwise, event e is *disabled*. Problems arise when several events take place simultaneously, because the order in which they occur affects the system behavior. If the incubation times are independent continuous random variables, this possibility can be ignored because the probability of realizing such a coincidence is negligible. However, if the incubation times are not continuous, or if they are dependent, then a decision must be made as to which event takes place first. This applies, in particular, to immediate events. The choice as to which event occurs first can either be deterministic, using a set of predetermined priorities, or stochastic, in which case each event is given a certain chance to occur. This will be discussed later in the context of Petri nets.

In some cases, the future of the system depends on the time and state at the moment a certain event was scheduled. This can be done by adding the appropriate information to the entity causing the event, or by creating a new event for this very purpose. For instance, if one needs to find the time an entity waits in a certain queue, one must add the arrival time as an attribute to each entity once it is scheduled to arrive at the queue. Similar techniques can be employed to find the amount of service an entity has received by a certain server, an information needed to represent a preemptive-repeat-identical (PRI) behavior. With these conventions, a full snapshot of the DES includes the current state, s, plus the current value of the timer for each event e in incubation.

2.2 Simplifying Discrete Event Systems

The DESs we just described are very powerful, but, in general, not even simulation can handle such system in their full generality. In fact, to fit the simulation into a finite computer storage, the number of entities in the system, and the number of attributes an event can have must both be kept below a certain limit. Additional restrictions have to be added to make the system Markovian.

In general, the attributes are allowed to be real random variables. However, to make the system treatable, we assume that the attributes are discrete. This excludes recording say, arrival times as attributes, because arrival times are continuous. Similarly, the remaining service time of a job interrupted by a high priority job cannot be recorded. Without loss of generality, we can actually assume that the attributes are non-negative integers. Of course, the number of entities must be below a known fixed number, and the same must be required for the number of attributes an entity can have. In this case, the state of the system can be characterized by a finite number of non-negative integers, that is, there is a number K such that the state is completely described by the vector $(i_1, i_2, \ldots, i_K)$, where i_k, $k = 1, 2, \ldots, n$ are non-negative integers. To make the system Markovian, we assume that the timers are set according to an exponential distribution which only depends on the present state of the system.

Consequently, to completely define the dynamic behavior of the model under the assumption of exponentially-distributed activities, we only need:

16 COMPUTATIONAL PROBABILITY

- A set of possible events, $\mathcal{E}$.

- An initial state, $i^{[0]} = (i_1^{[0]}, \ldots, i_K^{[0]}) \in \mathcal{S}$.

- Rules to define, for each event $e \in \mathcal{E}$:
  its ability to occur in a state $i \in \mathcal{S}$, $active(e, i) \in \{true, false\}$,

- its rate of occurrence when active, $rate(e, i) > 0$, and

- its effect on the state, that is, the state obtained when e occurs in state i, $new(e, i) \in \mathcal{S}$.

In this chapter, we assume that $\mathcal{S}$ is finite. The approaches to deal with infinite state spaces are discussed in Chapters 6 and 7. Infinite state spaces can be managed only if they exhibit certain regularities, or if they are automatically truncated by the solution methods employed (leading to an approximation [Ciardo, 1995, Van Dijk, 1991]). Hence, we assume that if the state is given by the tuple $(i_1, i_2, \ldots, i_K)$, then all i_k are within certain limits, say $0 \leq i_k \leq I_k$. Alternatively, one can bound the states collectively, say by requiring that $i_1 + i_2 + \cdots + i_K \leq r$, or, equivalently, $i_0 + i_1 + \cdots + i_K = r$, where i_0 takes up the slack. This type of bound is for instance applicable in semi-open queuing networks with a single type of customers, where the total number in the system is restricted to r. According to Feller [Feller, 1962], the number of different states in this case is given by

$$\binom{K+r}{r} = \binom{K+r}{K}.$$

For a fixed K, the number of states increases as r^K. Hence, one will typically be confronted with large state spaces.

The number of states can often be reduced by lumping many states together into a single state. This is in particular true if entities are indistinguishable. For instance, in queuing situations, the entitles forming the queue can often be treated as indistinguishable, and all that is needed is the number in the queue. Similarly, if all servers of a service system are indistinguishable, one only needs to know the number of busy servers, but not their identity. If there are entities with different attributes in a queue, one can reduce the number of states by only retaining the number in the queue with a given attribute.

If the state space is restricted by inequalities such as $0 \leq i_k \leq I_k$ or $i_1 + i_2 + \cdots + i_K \leq r$, then the set of all states satisfying these inequalities will be called the *potential state space* and denoted by $\hat{\mathcal{S}}$. Frequently, some states $j \in \hat{\mathcal{S}}$ cannot be reached from the initial state $i^{[0]}$ and, in this sense, such states do not belong to the state space $\mathcal{S}$. In this case, $\mathcal{S}$ is a proper subset of $\hat{\mathcal{S}}$. Moreover, if there are transient states, or if there are several recurrent classes $\mathcal{S}^l$, then the set of states reachable from a different initial state i may be a proper subset of $\mathcal{S}$. Of course, exploring the ergodicity structure of $\mathcal{S}$ is of great importance. However, for simplicity, we assume that there is only one recurrent class, and no transient states, in which case $\mathcal{S}$ is exactly the set of

$BuildRS$(in: $i^{[0]}, \mathcal{E}, active, new$; out: $\mathcal{S}$);
1. $\mathcal{S} \leftarrow \emptyset$; • $\mathcal{S}$: states explored so far
2. $\mathcal{U} \leftarrow \{i^{[0]}\}$; • $\mathcal{U}$: states found but not yet explored
3. while $\mathcal{U} \neq \emptyset$
4.   "remove the first $i \in \mathcal{U}$"; • $\mathcal{U}$ is managed as a FIFO list
5. $\mathcal{S} \leftarrow \mathcal{S} \cup \{i\}$;
6.   for each $e \in \mathcal{E}$ such that $active(e, i)$
7. $j \leftarrow new(e, i)$;
8.   if $j \notin \mathcal{S} \cup \mathcal{U}$ then
9. $\mathcal{U} \leftarrow \mathcal{U} \cup \{j\}$;

Figure 1 Algorithm $BuildRS$

reachable states. Then, we can use the straightforward reachability algorithm shown in Fig. 1, a breadth-first exploration of the graph implicitly defined by the model. The idea is to create a set $\mathcal{U}$ of states found and not yet explored, and recursively find all states reachable from $\mathcal{U}$. Of course, algorithm $BuildRS$ runs out of memory if $\mathcal{S}$ is infinite, but this case was excluded earlier.

Based on ideas similar to the ones above, a number of approaches for formulating Markovian systems have been described in literature. In particular, Wallace used a representation like the one used here for his RQA analyzer [Wallace and Rosenberg, 1994]. He even introduced immediate events, which he called exo-events. Other approaches were given by Stewart [Stewart, 1991], Gross and Miller [Gross and Miller, 1984], Grassmann [Grassmann, 1991] and Berson et al. [Berson et al., 1991]. By far the most popular methodology for describing stochastic systems, and the one we like best, is the Petri net methodology, which will be discussed next.

2.3 Stochastic Petri Nets

According to Ajmone et al. [Ajmone Marsan et al., 1995], Petri nets (PNs) were originally introduced by Carl Adam Petri to describe concurrent systems. Since timing posed considerable mathematical problems, Petri assumed that any change to the system would occur instantaneously. In other words, all events were immediate. Even with this restriction, important result can be obtained. For instance, it is possible to find the set of states reachable form a given starting state, and whether or not the system contains deadlocks.

The popularity of Petri nets encouraged several researchers to add new features to Petri nets. Here, we are mainly interested in stochastic Petri nets (SPNs) as they were introduced by several authors, including Simons, Florin, Natkin, and Molloy (see [Ajmone Marsan et al., 1995] for details). In a SPN, changes are not immediate, but they occur only after a random delay. The delay is always exponentially distributed. Originally, SPNs did not admit immediate events. If immediate events can occur, one speaks of a GSPN, or a Generalized Stochastic Petri Net.

A GSPN is a bipartite graph with two types of nodes: *places* and *transitions*. The places can contain *tokens* which correspond to the entities in discrete event systems. One can also use tokens to represent conditions. The transitions correspond to events. Places are represented by circles, tokens by bullets, and transitions by bars. The distribution of tokens in the different places is called a *marking*. Each possible marking corresponds to a state. There are two types of transitions: *immediate transitions* and *timed transitions*. Immediate transitions are slim solid bars, whereas timed transitions are bars in outline

All arcs are either *input arcs* or *output arcs*. Input arcs go from a place to a transition, and output arcs go from a transition to a place. There may be multiple arcs between a place and a transition. A transition can *fire*, which means that it takes a token from each input arc and places a token into each output arc. The input arcs take the tokens from the places where they originate, and the output arcs place the tokens into the place where they terminate. If there are not enough tokens in the places corresponding to the input arcs, a transition cannot fire, that is, the transition is *disabled*. A transition which is not disabled is said to be *enabled*. If the transition is immediate, it fires as soon as it is enabled. Timed enabled transitions fire at a certain rate which may depend on the marking. For simplicity, we first assume that all transitions are timed, that is, we deal with a SPN rather than a GSPN.

Formally, a SPN is a tuple

$$P = \left(\mathcal{P}, \mathcal{T}, D^-, D^+, \mathbf{m}^{[0]}, \lambda\right)$$

where

- $\mathcal{P} = \{p_1, ..., p_{|\mathcal{P}|}\}$ is a finite set of places. The vector $\mathbf{m} = [\mathbf{m}_p, p \in \mathcal{P}]$, $\mathbf{m}_p \in \mathbb{N}$ is used for describing the number of tokens in each place. This vector represents the marking.

- $\mathcal{T} = \{t_1, ..., t_{|\mathcal{T}|}\}$ is a finite set of transitions ($\mathcal{P} \cap \mathcal{T} = \emptyset$).

- The multiplicities for the input arcs from $p \in \mathcal{P}$ to $t \in \mathcal{T}$ are denoted by $D^-_{p,t} \in \mathbb{N}$, with $D^- = [D^-_{p,t}]$. The matrix D^+ similarly describes the output arcs.

- $\mathbf{m}^{[0]}$ is the initial marking.

- Given $t \in \mathcal{T}$, $\lambda_t(\mathbf{m})$ is the rate of the exponential distribution for the firing time of transition t in marking $\mathbf{m}$. Also, $\lambda = [\lambda_t]$.

A transition $t \in \mathcal{T}$ is *enabled* in $\mathbf{m}$ iff each of its inputs arcs is satisfied:

$$\forall p \in \mathcal{P}(\ \mathbf{m}_p \geq D^-_{p,t}).$$

An enabled transition t in $\mathbf{m}$ can *fire*, leading to a new marking $\mathbf{n}$, we write $\mathbf{m} \xrightarrow{t} \mathbf{n}$, satisfying

$$\forall p \in \mathcal{P}(\ \mathbf{n}_p = \mathbf{m}_p - D^-_{p,t} + D^+_{p,t}).$$

While the transition is enabled, firing is Markovian with a rate $\lambda(\mathbf{m})$.

The set S of markings reachable from $\mathbf{m}^{[0]}$ is called the *reachability set*, while the reachability graph has nodes S and an arc from $\mathbf{m}$ to $\mathbf{n}$ labeled with t iff $\mathbf{m} \xrightarrow{t} \mathbf{n}$. If in a marking, no transition is enabled, the system is said to be *deadlocked*.

In the generalized SPN (GSPN) [Ajmone Marsan et al., 1995], some of the transitions can be immediate, that is, they have automatic priority over the (timed) exponentially-distributed transitions, and fire in zero time. Markings enabling immediate transitions are called *vanishing*. Any marking that is not vanishing is called *tangible*. Hence, in a vanishing marking, only immediate transitions can fire. If there are several immediate transitions, one has to decide which one fires first. This can be done in a deterministic or stochastic way. If the firing is deterministic, firing priorities are set in the model, and the transitions fire according to their priority. Alternatively, one can assign each transition a *firing weight* $w_t(\mathbf{m})$, which may depend on $\mathbf{m}$. If $\mathcal{E}(\mathbf{m})$ is the set of immediate transitions enabled in marking $\mathbf{m}$, the probability that $t \in \mathcal{E}$ fires is then calculated as

$$w_t(\mathbf{m}) \left(\sum_{u \in \mathcal{E}(\mathbf{m})} w_u(\mathbf{m}) \right)^{-1}$$

In conclusion, a GSPN is given by the tuple

$$P = \left(\mathcal{P}, \mathcal{T}, D^-, D^+, \mathbf{m}^{[0]}, \succ, w, \lambda \right)$$

Here, $\succ$ represents the firing priorities of the transitions, and w the weights.

To allow even more modeling power and flexibility, several extensions have been defined:

- Inhibitor arcs can be used to prevent the enabling of a transition due to the presence of tokens in a place (unlike ordinary input arcs, which disable a transition only due to a lack of tokens). Formally, t is not enabled if there is a $p \in P$ such that $\mathbf{m}_p \geq D^\circ_{p,t}$, where $D^\circ_{p,t} \in \mathbb{N} \cup \{\infty\}$.

- To make modeling more convenient, we can define boolean functions of the marking that act as transition guards. Then, t can be enabled in $\mathbf{m}$ iff $g_t(\mathbf{m}) = true$. One disadvantage of this flexibility is that guards are not represented graphically.

- Another type of extension is to allow the definition of the input, output, and inhibitor arcs to depend on the marking [Ciardo, 1994]. These self-modifying GSPNs are very convenient, for example, when the firing of a transition must empty a place, possibly moving its contents to other places.

- To model distinguishable entities, we can make use of colored tokens [Jensen, 1987], but the resulting colored GSPNs have a much larger state

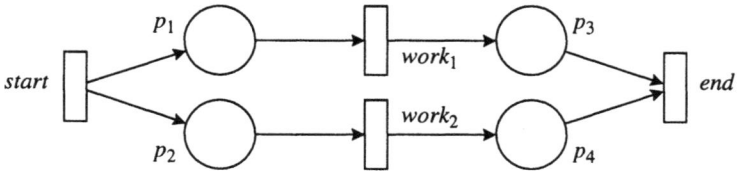

Figure 2 Modeling queuing and overtaking in SPNs.

space. Of particular interest is the case where symmetries due to colored tokens can be automatically detected and used to generate a symbolic reachability graph where the colors are not explicitly unfolded [Chiola et al., 1993].

Inhibitor or transition priorities arcs make the GSPN formalism Turing-equivalent, which implies that there is no general way to prove that the resulting models have a finite reachability set. For the feasibility of other types of analysis, see [Murata, 1989]. For this reason, inhibitor arcs are sometimes avoided. Of course, once a machine is Turing equivalent, it is powerful enough to model any discrete-state system. Hence, further extensions do not add to the power of the formalism, even though they make the formalism much more convenient to use.

There are a number of generalizations of GSPNs. We mention here, in particular, stochastic reward nets [Ciardo et al., 1993a] and stochastic activity networks [Sanders and Meyer, 1991], which stress the reward aspects of the model (Sect. 3 discusses rewards). A CTMC is obtained also in the case of GSPNs with phase-type (not just exponential) distributions [Cumani, 1985]; these models require careful specification of the execution policies [Ajmone Marsan et al., 1989] and increase the state space considerably, as discussed in Sect. 4.1. Discrete-time Markov GSPNs have also been defined. If the firing times have geometric [Molloy, 1985] or, more generally, discrete phase [Ciardo, 1995] distributions, the underlying process is a DTMC.

2.4 Examples of Petri Nets

A number of examples to demonstrate GSPNs and their extensions are now introduced. In all cases, it is assumed that there are no multiple arcs, that is, D^-_{pt}, D^+_{tp}, and D°_{pt} can only assume the values As our first example, consider the SPN in Fig. 2. The transition *start* is always enabled, and it fires at a certain rate $\lambda(\mathbf{m})$. Whenever it fires, a token is placed into the places p_1 and p_2. This may represent a machine that is taken apart, with each token representing a component. If there is a token in p_1, $work_1$ can fire, placing the token in p_3. Similarly, if $work_2$ is enabled, that is, if there is a token in p_2, then it fires at a certain rate, placing one token in p_4. As soon as both p_3 and p_4 contain at least one token, the transition *end* can fire, which means that the two parts are put back together again. Note that p_1 and p_2 may contain several tokens.

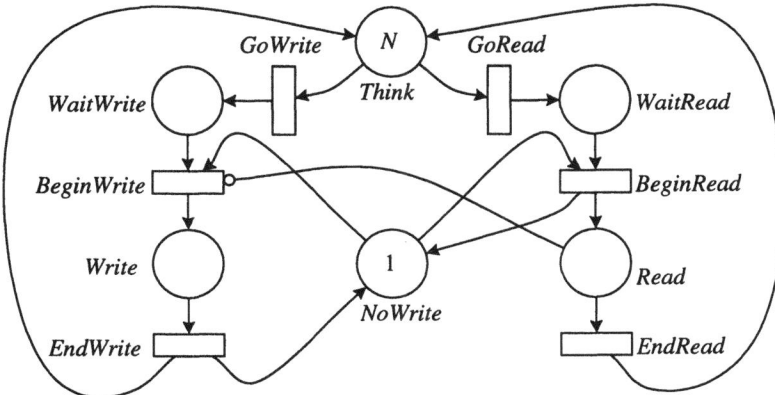

Figure 3 An example of SPN.

We now consider the classical "multiple-readers single-writer mutual exclusion" problem as an example containing immediate events and inhibitor arcs. In this example, multiple (software) processes need to access data in a "critical section", either in "write-mode" or in "read-mode". To avoid inconsistencies, only one process is allowed to be in write mode, and no process is allowed to read while this happens. On the other hand, any number of processes can read the data at the same time, since the data itself is not modified by the operation.

The GSPN in Fig. 3 models this problem. N processes execute locally (in place *Think*) but occasionally require access to the critical section, either in read or in write mode. The presence of a token in place *NoWrite* signifies that no writer is currently accessing the critical section. If this is the case, and readers (in place *WaitRead*) want to access it, they can proceed (transition *BeginRead*). Further readers can arrive and access the critical section as well, because the token in place *NoWrite* is returned right away: the transition *BeginRead* is immediate. On the other hand, processes that decide to ask for write access must wait (in place *WaitWrite*) as long as any reader is in the critical section, due to the inhibitor arc from *Read* to *BeginWrite*. When all readers finally have left the critical section (place *Read* is empty), one of the waiting writers is allowed to enter it, but this removes the token from *NoWrite*, thus no other writer or reader is allowed to enter the critical section until the writer completes (transition *EndWrite*).

No token selection policy is explicitly represented by the model. This is reflected in two aspects: first, the identity of the writers waiting to access the critical section is not modeled, thus there is no concept of "selecting the first one" second, and most important, the model does not dictate how to choose between the waiting writers and readers, after a writer exits the critical section.

This second aspect is modeled by an appropriate definition of the enabling and the relative value of the weights for *BeginWrite* and *BeginRead*.

- To always choose a writer (or a reader), set the priority $BeginWrite \succ BeginRead$ (or $BeginRead \succ BeginWrite$).

- To choose a writer or a reader according to which group has the largest waiting population, set the guards as follows

$$g_{BeginWrite}(\mathbf{m}) = (\mathbf{m}_{WaitWrite} \geq \mathbf{m}_{WaitRead})$$
$$g_{BeginRead}(\mathbf{m}) = (\mathbf{m}_{WaitWrite} \leq \mathbf{m}_{WaitRead} \vee \mathbf{m}_{Read} > 0) \, .$$

- To choose a writer or a reader with equal probability, set weights $w_{BeginWrite}(\mathbf{m}) = w_{BeginRead}(\mathbf{m})$; to choose with a probability proportional to the waiting populations, set the weight $w_{BeginWrite}(\mathbf{m})$ proportional to the number of tokens in $WaitWrite$, and use a similar method for finding $w_{BeginRead}(\mathbf{m})$. Clearly, this is an example where marking-dependent rates and weights are essential for a compact model definition. Note that weights should also be used with the previous policy, since both guards are true whenever the number of tokens in the places $BeginWrite$ and $BeginRead$ are equal.

As an example of a GSPN requiring colored tokens, consider once more the Petri net of Figure 2. However, assume now that instead of a product in a shop, we are dealing with a a job in a computer system which is divided into two threads to be processed by two parallel processors. In contrast to the earlier model, the threads joined at the end must belong to the same job. In the SPN of Figure 2, there is no assurance that this will happen, unless it is changed to make this explicit. To this end, we must make the tokens distinct by using colored tokens, using enough colors to provide for the maximum number of jobs to be processed. This is much easier than to include in the GSPN an unwieldy sequence of transitions and places large enough to accommodate the maximum number of tokens to be matched. Alternatively, a built-in primitive such as queuing places [Bause et al., 1995] can be used. Of course, the underlying state space increases drastically regardless of how this behavior is modeled in the high-level formalism.

2.5 A Prolog-Like Formalism

The transitions of a Petri net can be seen as a way to provide a set of rules that we can use to generate new states from the current ones. Using an artificial intelligence approach, we can carry this idea further, until the model becomes simply a set of assertions about the behavior of the system. Some of these assertions are statements about the initial state (e.g., there are N working units of type a) while others describe the preconditions and postconditions for applying an event rule.

This is the approach taken by TANGRAM [Berson et al., 1991, Page et al., 1989]. TANGRAM follows an object-oriented paradigm: objects can generate events and evolve in isolation, but they can also interact by means of messages. However, we stress a different aspect of TANGRAM: the use of Prolog [Clocksin and Mellish, 1984] for the definition of complex interactions among objects. By listing a set of rules about the behavior of localized portions of the system, the overall logic is implicitly defined as well. Since Prolog is a programming language in itself, the authors of TANGRAM could make use of the compilers and interpreters already available for it. Furthermore, the nature of Prolog is such that experimentation with model variants and structural what-if questions can be easily specified and investigated.

The flexibility of Prolog has a price, however: the generation of the state space S and of the infinitesimal generator $\mathbf{Q}$ is likely to be slower than if it were performed by a specialized and optimized tool written in a procedural language such as C. This can become a problem in practice, especially if the generation time becomes the dominant factor in the overall solution process.

2.6 A Domain-Oriented Formalism

If one is willing to restrict the formalism, and the modeling tool associated with it, to a specific application domain, generality can be traded off for ease of specification.

Petri nets with any of the extensions we described (inhibitor arcs, priorities, or guards) are Turing-equivalent, hence, in principle, they can model any discrete behavior. However, it is easy to argue that, for example, distributed software is best modeled using a pseudo-code-like language with built-in primitives such as local and shared variables, tasks, synchronizations, and if-then-else and while-loop statements; communication protocols are easier to describe in terms of packets, messages, acknowledgments, routers, and buffers; and so on.

By focusing on a specific field, we might even have at our disposal a library of predefined parameterized submodels for well-known types of systems (e.g., a mailbox-type primitive for inter-task data exchange, or the standard Ethernet protocol) which can be simply used as plug-ins when building more complex models.

A particularly successful example of this type of formalisms and tools is SAVE, the System Availability Estimator [Blum et al., 1994], a package targeted to the reliability and availability modeling of computer systems. The SAVE language allows the modeler to describe the number and type of components in a system, the rate at which they fail, and how their failures might affect other components (multiple failure modes may be defined for a given component). Then, repair facilities and their rates can be defined for each component type (shared facilities can be defined, together with their scheduling policies to choose which components should be under repair in any state). Finally, statements defining whether the system is considered functioning or not in a given state can be specified.

For example, Fig. 4 shows the declaration of a component named proc. For its correct operation and for its repair when failed, the component power must be operational as well. Failures for component proc occur with an overall rate of lambda, but they can be of two different types. With probability alpha, they require a repair of type reboot while, with the complementary probability, they

```
COMPONENT :                        proc
    OPERATION DEPENDS UPON:        power
    REPAIR DEPENDS UPON:           power
    FAILURE RATE:                  lambda
    FAILURE MODE PROBABILITIES:    alpha, 1-alpha
    REPAIR RATE:                   mu, gamma
    REPAIR CLASS USED:             reboot, maintenance
    COMPONENTS AFFECTED:           NONE, data
        data:                      1-coverage
```

Figure 4   Declaration of a component in SAVE.

require the repair facility called `maintenance`. The first type of failure does not affect any other component, while the second type corrupts the component `data` with probability `1-coverage`. To complete the example, we would have to define the remaining components, the value of parameters such as `lambda` and `alpha`, the criteria that define when the system is up, and the strategy used by the repair facilities.

From these specifications, the SAVE compiler can automatically generate the underlying CTMC and organize the computation of the required outputs.

2.7 Queueing Networks

In many situations, one has M of service facilities, called *centers*. The service times of all centers are Markovian, and so are all arrivals to the system. The service rate of center k may depend on the number of entities within this center, but not on the number within any other center. The probabilities that an entity having completed service in center i moves to center j is p_{ij}, and this *routing probability* is completely independent of the present state of the system and of the past history of the entity in question. Such a queuing net is completely characterized by the arrival rates λ_i of elements coming from outside and joining center i, the routing matrix $P = [p_{ij}]$, the service rates $\lambda_i(n_i)$, where n_i is the number in center i, $i = 1, 2, \ldots M$, and the initial state. Systems of this type are called *Jackson networks*. The description of Jackson networks is thus rather compact. More importantly, Jackson networks have a so called *product form solution* (see Chapter 11) which makes it easy to find their steady-state behavior. The Jackson networks come in two flavors: open and closed. In contrast to open Jackson networks, no arrivals are possible in a closed Jackson network. This makes the numerical solutions of closed networks somewhat more difficult than open networks.

Generally, the product form is destroyed if the routing probabilities depend on the system state, or if the past history of the entities affects their behavior. For instance, the product form is no longer applicable if servers can be blocked, or if the routing may be changed to adapt to the present traffic. Other restrictions apply as shown in Chapter 11. This severely limits the possible applications of Jackson networks.

2.8 Conclusion

The previous sections described a number of formalisms, and we now give some criteria to compare them. The criteria we propose are

- The *power* of the method, that is, what systems can in principle be modeled

- The *conciseness* of the description, that is, how many symbols are needed to describe a system once it is converted to the formalism

- The *convenience* of its use, that is, how easy is it to convert the system to be modeled into the formalism in question

Here, we are dealing with finite state Markov chains, and there are a number of formalisms that allow one to formulate any CTMC with a finite state space. For instance, every CTMC can be considered as a closed Jackson network, with the states represented as centers, and with a single element in the system visiting the centers. However, the number of symbols one must use for this formulation increases in direct proportion with the size of the CTMC, which is impractical for large Markov chains. Hence, such methods loose out on account of conciseness.

In infinite state systems, power and conciseness seem to go hand in hand. When saying that a formalism has the power to formulate a certain system, one implies that the formulation makes use of a finite number of symbols only. Hence, if the number of symbols needed for formulation increases to infinity with the number of states, then the underlying formalism is not powerful enough to formulate infinite state systems. It is very likely that a more powerful system gives a more concise representation of the system even if the system has a finite number of states only. Hence, the best methodologies tend to be the methodologies which are Turing equivalent, such as GSPNs with inhibitor arcs. With these formalisms, one can formulate essentially every possible Markov chains. The price that has to be paid, however, is undecidability. For instance, given an arbitrary GSPN, it is impossible to decide if its reachability graph is finite or infinite.

Whereas in regard to compactness, one would expect the most powerful formalism to be best, the most powerful formalism may not be the most convenient one to use. On the contrary, the least powerful formalism that can model the system in consideration is often the best one. For example, using a GSPN to model a system that can be modeled by a Jackson networks only complicates the solution, with no advantage. In a related but slightly different line, a formalism based on primitive objects resembling those in the system under study is also going to offer greater ease-of-use, of course at the cost of reducing its field of applicability.

3 REWARD STRUCTURE

The problem of specifying a CTMC is solved using a high-level model. However, when we solve a CTMC, the resulting solution vectors are not only too detailed, but also "at the wrong level", since they relate to the state of the CTMC, not the entities in the high-level model. The quantities of interest must then be defined also at the high-level model, using a reward structure consisting of a reward rate vector ρ and a reward impulse matrix Δ. For $i \in S$, ρ_i is the rate at which reward is accumulated while the process is in state i. Hence, a sojourn of length x in state i earns a reward $x \cdot \rho_i$. For $i, j \in S$, $\Delta_{i,j}$ is the reward earned when the process moves from state i to state j in one step (since transitions from i to itself are not defined for a CTMC, $\Delta_{i,i} = 0$).

Then, we can compute the expected instantaneous reward rate at time t for a CTMC as

$$\sum_{i \in S} \rho_i \cdot \pi_i(t) + \sum_{i,j \in S} \Delta_{i,j} \cdot \Phi_{i,j}(t) = \sum_{i \in S} \left(\rho_i + \sum_{j \in S} \Delta_{i,j} \cdot q_{i,j} \right) \cdot \pi_i(t),$$

where $\Phi_{i,j}(t) = \pi_i(t) \cdot q_{i,j}$ is the unconditional rate of going from i to j at time t.

From the above expression, it is clear that, by redefining the reward rate of state i as

$$\rho'_i = \rho_i + \sum_{j \in S} \Delta_{i,j} \cdot q_{i,j},$$

we can restrict ourselves to computing quantities expressed using just reward rates, without loss of generality.

4 NONEXPONENTIAL TIMING

The assumption of exponentially distributed durations for the incubation times in the model is essential if we are to represent the state of the system without having to know how when the incubation times started. This is due to the fact that the remaining incubation time of event e has the same exponential distribution Expo($rate(e, s)$), with the same rate $rate(e, s)$, no matter how much time has passed since the event was enabled. Only if the state changes because of some other event, then $rate$ may change. It follows that by allowing $rate$ to depend on the state, we can model a situation where this distribution varies over time, but only insofar as the state itself changes, not as a direct function of the local time (i.e., from the instant state s has been entered) or global time (i.e., from the instant we start observing the process). However, many real-life activities do not behave this way. If the runtime of a program does not vary much between executions, a program that has been running for a while is more likely to complete before one that just started. On the other hand, tasks submitted to a computer system could have bimodal execution requirements, so that most tasks complete quickly, but some require long run-time; if a task has been running for quite a while, it is then more likely to be a "long" task, hence its remaining life might be, on the average, longer than that of the "typical" task. In reliability theory, a similar need arises when modeling component lifetimes. An old memory unit, for example, might be more likely to fail in the next 1,000 hours than a newer one. This corresponds to an increasing failure rate (IFR), which implies a new-better-than-used (NBU) distribution. However, brand-new components are subject to "infant mortality": most defects are discovered during the initial period of operation; this phase then possesses a decreasing failure rate (DFR), which implies a new-worse-than-used (NWU) distribution.

In modeling these durations, we often focus on the first two moments of the distribution, so we need to model activities whose duration has a coefficient of variation $cv_X = \sigma/\mu$ smaller or greater than one, respectively. While a single exponentially-distributed duration alone cannot accomplish this, several of exponentially-distributed "phases" appropriately connected can. These are called phase-type distributions [Cox, 1955] and are discussed next.

4.1 Phase-Type Distributions

A (continuous) phase-type distribution can be thought of as the time to reach the absorbing state 0 in a CTMC where all other states are transient, from a given initial state distribution.

For example, to model an activity with a duration having a small coefficient of variation, we can use an Erlang distribution with n stages, as shown in Fig. 5 on the top. Note that the initial state is n with probability one, so the distribution describes the time to go through n independent stages, each of them with duration $\text{Expo}(\lambda)$. If $X \sim \text{Erlang}(n, \lambda)$, its expectation and coefficient of variation are, respectively,

$$E[X] = \frac{n}{\lambda} \quad \text{and} \quad cv_X = \frac{1}{\sqrt{n}}.$$

Hence, even a constant distribution, which has coefficient of variation equal zero, can in principle be approximated arbitrarily well by an Erlang distribution with enough stages.

On the other hand, to model a highly-variable duration, we can use a hyperexponential distribution as shown in Fig. 5 on the bottom. Here, we start either in state 1 with probability p_1 or in state 2 with probability $p_2 = 1 - p_1$; in the former case we reach state 0 after a time $\sim \text{Expo}(\lambda)$, in the latter after a time $\sim \text{Expo}(\mu)$. Then,

$$E[X] = \frac{p_1}{\lambda} + \frac{p_2}{\mu} \quad \text{and} \quad cv_X = \sqrt{2\frac{\frac{p_1}{\lambda^2} + \frac{p_2}{\mu^2}}{\left(\frac{p_1}{\lambda} + \frac{p_2}{\mu}\right)^2} - 1}.$$

It is easy to see that cv_X is greater than one (unless $\lambda = \mu$, $p_1 = 0$, or $p_1 = 1$, in which case we obtain the exponential case), and that it can become arbitrarily large by setting the ratio λ/μ and p_1 appropriately.

The use of phase-type distributions has its disadvantages. First, its destroys the one-to-one mapping between states of the high-level model and states of the CTMC. At any given time, there may be several events in incubation, and for each of these events, one must record the phase of the incubation

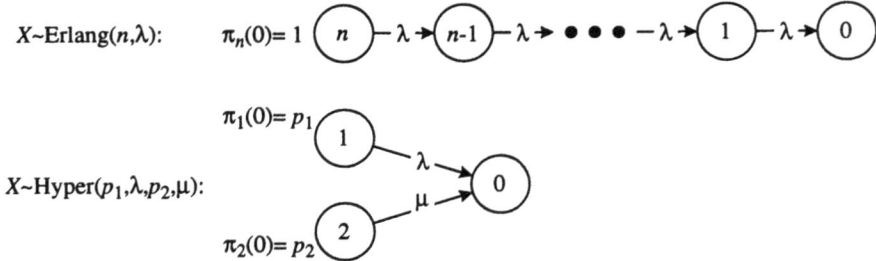

Figure 5 Example of phase-type distributions.

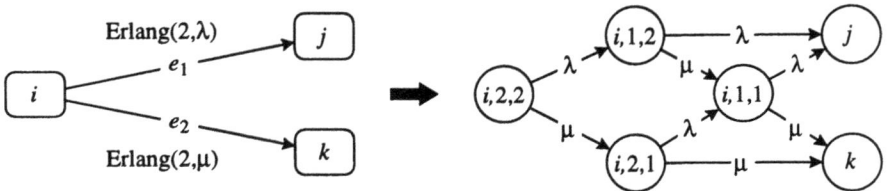

Figure 6 Processing phase-type events.

by using a separate state variable. Hence, if several events are in incubation at the same time, one needs to add many state variables for recording their phases, and this increases the state space drastically. This is true even if the occurrence of one event cancels one or more events still in incubation. Consider for example Fig. 6. In state i, events e_1 and e_2 are in incubation, with duration $\sim$ Erlang$(2, \lambda)$ and $\sim$ Erlang$(2, \mu)$, respectively. The resulting CTMC on the right must explicitly represent the phase of the Erlang distribution in (some of) the states [Bobbio and Cumani, 1984]. Hence, state i must be refined into states $(i, 2, 2)$, $(i, 1, 2)$, $(i, 2, 1)$, and $(i, 1, 1)$. Indeed, this CTMC assumes that the occurrence of one event cancels the other one. Were this not the case, states j and k would have to be duplicated to record the phase of the event that did not occur [Ajmone Marsan et al., 1989].

This refinement process always causes the state space of the CTMC to be larger than that of the original model. Fortunately, all interesting structural properties can still be studied in the original high-level model. For example, the high-level model (or the CTMC obtained assuming exponential distributions) contains a deadlock iff the CTMC obtained assuming phase-type distributions does.

The main problem of using phase-type distributions, however, is that, eventually (certainly before starting a numerical solution) the state-expansion must be performed, and this can dramatically increase the state space of the underlying CTMC. In real modeling applications, deviations from the exponential assumption should then be explicitly represented only when strictly necessary.

We conclude this section by observing that non-exponential behaviors can be represented explicitly in the high-level model [Chen et al., 1989], but doing so

FORMULATING MARKOV MODELS   29

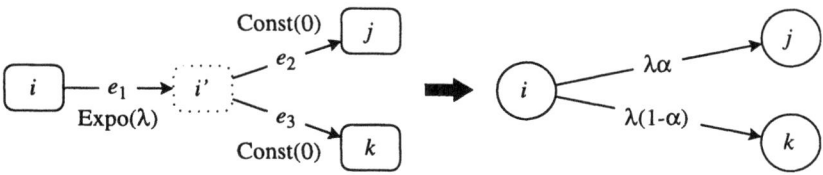

Figure 7 Processing immediate events.

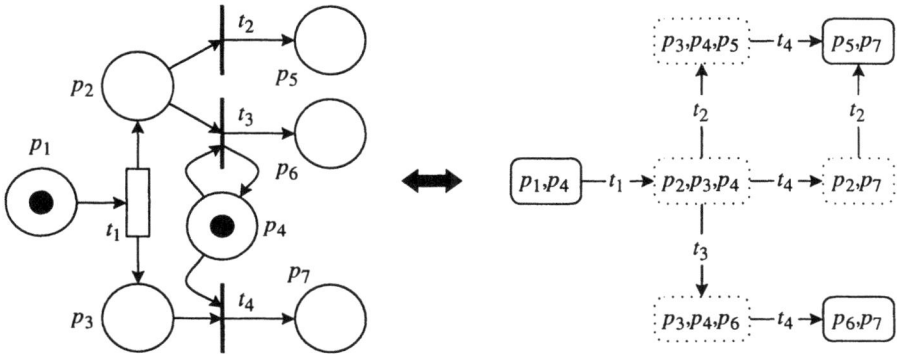

Figure 8 Stochastic confusion.

increases the burden on the modeler and unnecessarily complicates the model by mixing its structural and logical aspects with its timing behavior.

4.2 Immediate Events

Another extension of CTMC-based high-level models is the introduction of instantaneous events, as shown in Fig. 7. Here, the occurrence of event e_1 in state i leads to state i', where instantaneous events e_2 and e_3 can occur. Since they both attempt to occur immediately, a probabilistic choice must be performed. If the high-level model specifies a probability α for e_2 and $1-\alpha$ for e_3, the resulting CTMC is shown on the right (this is analogous to thinning a Poisson process). Hence, the state space of the resulting CTMC on the right does not include the instantaneous state i'. In general, there can be arbitrarily long paths of instantaneous events. Indeed, it is even possible to give a semantic interpretation to a set of instantaneous states from where no transition to timed states is possible [Ciardo, 1989, Grassmann and Wang, 1995], although these situations are rarely needed in practice and are normally treated as an error by modeling software packages.

From a logical modeling point of view, instantaneous events introduce a behavior not otherwise found in continuous-time models: the need to choose or sequentialize simultaneous events, as done by the introduction of the "switch" with probabilities α and $1-\alpha$. Unfortunately, this additional information can be extremely critical, and must be specified with care, if the high-level model is to represent the real system accurately.

For example, consider the portion of GSPN shown in Fig. 8 on the left (transitions t_2, t_3, and t_4 are immediate). From the initial marking $\{p_1, p_4\}$, the firing sequences given on the right side of Fig. 8 can occur. Specifically, one has following possibilities

$$\{p_1,p_4\} \xrightarrow{t_1} \{p_2,p_3,p_4\} \xrightarrow{t_2} \{p_3,p_4,p_5\} \xrightarrow{t_4} \{p_5,p_7\}$$
$$\{p_1,p_4\} \xrightarrow{t_1} \{p_2,p_3,p_4\} \xrightarrow{t_3} \{p_3,p_4,p_6\} \xrightarrow{t_4} \{p_6,p_7\}$$
$$\{p_1,p_4\} \xrightarrow{t_1} \{p_2,p_3,p_4\} \xrightarrow{t_4} \{p_2,p_7\} \xrightarrow{t_2} \{p_5,p_7\}$$

It is apparent that, after the firing of t_1 leads to marking $\{p_2,p_3,p_4\}$, transitions t_2 and t_3 are in symmetric conflict, since firing one disables the other, while t_3 and t_4 are in asymmetric conflict, since the firing of t_4 disables t_3, but not vice versa. However, if we naively assign probabilities α and $1-\alpha$ to t_2 and t_3, the probability of reaching markings $\{p_5,p_7\}$ and $\{p_6,p_7\}$ is not α and $1-\alpha$, respectively. This is because the probabilistic split α and $1-\alpha$ is used in the first and second firing sequence, when t_4 is chosen to fire after the choice between t_2 and t_3 has been made, but, if instead t_4 is fired first, only marking $\{p_5,p_7\}$ remains reachable. In other words, the probabilistic choice of whether to fire t_2 or t_3 disappears, t_2 is chosen deterministically, and the probabilities become 1 and 0, instead of α and $1-\alpha$, respectively.

Of course, specifying a choice between t_3 and t_4 with probabilities β and $1-\beta$ does not solve the problem either. Only by realizing that the entire set of transitions $\{t_2,t_3,t_4\}$ is in "extended conflict", and by assigning firing probabilities to its elements, we can fully specify the behavior of the model.

This and other types of "stochastic confusion" have been considered in the literature, mostly for formalisms based on Petri nets, but they arise in any formalism that allows instantaneous events. It should be stressed that the correct specification of the choices and sequentializations for instantaneous events can affect not only the stochastic behavior, but also the logical behavior. For example, it is easy to build a model where a deadlock can be reached under certain sequentializations decisions but not under others. Fortunately, there exist algorithms to decide whether a model is well-defined (i.e., whether the information provided in the high-level model is sufficient to fully determine the underlying stochastic process) [Ciardo and Zijal, 1996, Qureshi et al., 1995, Sanders, 1988].

Instantaneous events are convenient when modeling complex logical behavior in the high-level model. They are also very appropriate when modeling activities that, while not exactly instantaneous in the real system, are orders of magnitude faster then the other activities being modeled. Using an instantaneous event for them has a two important advantages. First, if we were using ordinary (timed) events with very high rates, the resulting CTMC could be very stiff, hence difficult to solve numerically. Second, by using instead an instantaneous representation for these activities, the corresponding states where they can occur are also instantaneous, hence they can be eliminated.

Thus, again the bijection between the set of reachable states in the high-level model and the set of states of the underlying CTMC is lost, but this time this is an advantage, since the CTMC has fewer states than the high-level model.

In a model with instantaneous events (and no phase-type events), the "logical" state space $\mathcal{S}$ can be partitioned into $\mathcal{S}_T$ (the state space of the underlying CTMC, containing all the timed states) and $\mathcal{S}_I$ (containing all the instantaneous states) and the matrix $\mathbf{Q}$ can be obtained by eliminating the vanishing states. We can perform this elimination "on-the-fly" in procedure $BuildRS$ of Fig. 1 so that, whenever a newly found state j is determined to be instantaneous, the set of timed states reachable from j is explored and appropriately added to $\mathcal{U}$ instead. No instantaneous state is ever added to $\mathcal{U}$, hence the final result is $\mathcal{S}_T$, not $\mathcal{S}$, although all states in $\mathcal{S}_I$ will have been explored by the time $BuildRS$ completes its execution.

An alternative approach is to simply classify the states found as instantaneous or timed, but otherwise treat them in the same way in $BuildRS$. Once $\mathcal{S} = \mathcal{S}_T \cup \mathcal{S}_I$ has been built, we can define

$$\mathbf{A} = \left[\begin{array}{c|c} \mathbf{A}_{I,I} & \mathbf{A}_{I,T} \\ \hline \mathbf{A}_{T,I} & \mathbf{A}_{T,T} \end{array} \right],$$

where the matrices $\mathbf{A}_{I,I}$ and $\mathbf{A}_{I,T}$ contain the one-step *probabilities* of going from each instantaneous state to each instantaneous or timed state, respectively, while $\mathbf{A}_{T,I}$ and $\mathbf{A}_{T,T}$ contain the *rates* at which the model goes from each timed state to each instantaneous or timed state, respectively. Then, we can compute the infinitesimal generator for the underlying CTMC as

$$\mathbf{Q} = \mathbf{A}_{T,T} + \mathbf{A}_{T,I} \cdot (\mathbf{I} - \mathbf{A}_{I,I})^{-1} \cdot \mathbf{A}_{I,T}.$$

Of course, this is only possible if $(\mathbf{I} - \mathbf{A}_{I,I})^{-1}$ exists, which is the case iff the probability of eventually leaving the set of instantaneous states equals one. This elimination "after-the-fact" has both advantages and disadvantages with respect to the one on-the-fly. It correctly manages cycles of instantaneous states such as those shown at the top of Fig. 9, while the on-the-fly algorithm can at best recognize them, if the exploration from an instantaneous state proceeds in depth-first fashion, but it cannot compute the probability of eventually reaching j or k from i, since it would have to follow paths of arbitrary length. Also, if an instantaneous state can be reached through multiple paths, elimination on-the-fly explores it multiple times, but this is usually a small price to pay for not having to store the instantaneous states.

Neither approach, though, performs well if the model contains many "n-to-m switches" as shown in the bottom of Fig. 9. The elimination of a single instantaneous state removes $n + m$ state-to-state transitions, but adds $n \cdot m$ nonzero entries in $\mathbf{Q}$. For this reason, an approach to "preserve" instantaneous states was studied in [Ciardo et al., 1991]. The idea can be described as an embedding of the model at the time any change of state occurs, regardless of whether it is caused by an instantaneous or timed event. Thus, a transition

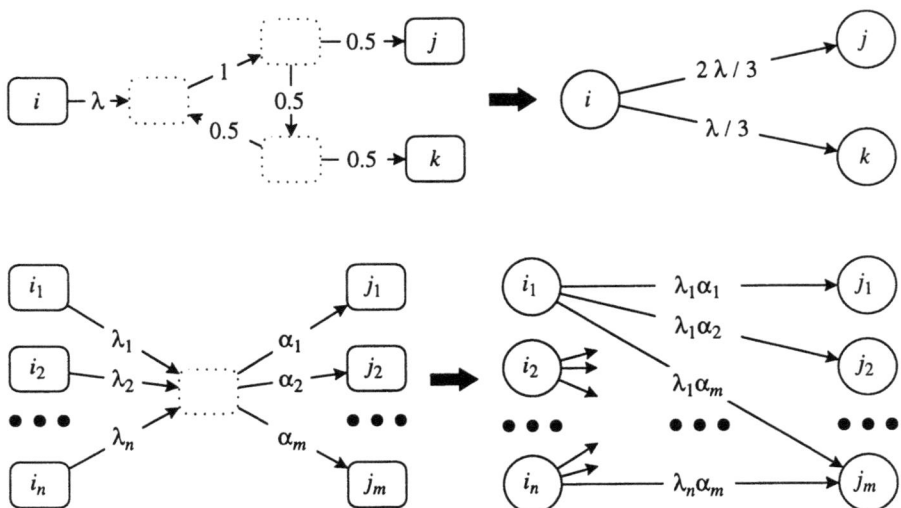

Figure 9 Processing instantaneous cycles and n-to-m switches.

probability matrix

$$\tilde{\mathbf{P}} = \left[\begin{array}{c|c} \mathbf{A}_{I,I} & \mathbf{A}_{I,T} \\ \hline \mathrm{diag}(\mathbf{h}) \cdot \mathbf{A}_{T,I} & \mathrm{diag}(\mathbf{h}) \cdot \mathbf{A}_{T,T} \end{array} \right],$$

can be defined, where diag($\mathbf{h}$) is a matrix having $\mathbf{h}$ on the diagonal and zeros elsewhere, and $\mathbf{h}$ is the expected holding time vector for the timed states, defined by $\mathbf{h}_i = (\sum_{j \neq i} \mathbf{A}_{i,j})^{-1}$. The holding time in any instantaneous state is instead zero, of course.

Then, steady-state instantaneous or cumulative analysis can be performed on the DTMC described by $\tilde{\mathbf{P}}$, resulting in the vector $\tilde{\pi}$, and the corresponding values for the underlying CTMC can be computed as:

$$\pi_i = \frac{\tilde{\pi}_i \cdot \mathbf{h}_i}{\sum_{j \in \mathcal{S}_T} \tilde{\pi}_j \cdot \mathbf{h}_j}$$

This can be justified by considering the model as a semi-Markov process [Çinlar, 1975] where the sojourn times are restricted to be zero or exponentially distributed.

When using iterative methods to find the equilibrium probabilities, ithe preservation approach seems to be inferior to the elimination approach [Ciardo et al., 1991], unless the model contains many instantaneous cycles and switches. One of the reasons is that the convergence rate of the numerical solution, at least when using simple iterative methods such as Gauss-Seidel, is negatively affected by the larger diameter of the graph describing the embedded process. The potential for increasing the number of nonzero entries in $\mathbf{Q}$, however, can be an issue for extremely large models. Selective elimination has been proposed in [Ciardo et al., 1991], but not implemented in any tool, to the best of our knowledge. The preservation method is limited to steady-state analysis, because the approach of weighting the discrete-time process probabilities or sojourn times by the expected holding times cannot be applied in the case of a finite time t.

5 MANAGING MODEL COMPLEXITY

We now discuss in detail algorithmic and storage issues connected with the generation and solution of the large CTMCs that are needed in practical applications.

5.1 The State Space

The storage of the state-space S must be addressed already in procedure $BuildRS$, presented in Fig. 1. Two factors are critical:

- The overall storage required for S should be as small as possible, ideally at most of the same order as the probability vector needed in the subsequent numerical solution.

- Testing whether a newly generated state has already been discovered (statement 8 in $BuildRS$) should require little computation.

Since the size of S is usually not known beforehand, hash tables are not the best choice to store S during the execution of $BuildRS$. Balanced search trees can instead grow dynamically and are often used for this purpose. However, both hash tables and search trees require $O(|S| \cdot \sum_{k=1}^{K} b_k)$ bits of storage when implemented in a straightforward way, where b_k is the number of bits used to store the k-th state component.

A better alternative is to use the multilevel data structure shown in Fig. 10 [Ciardo and Miner, 1997] during the state-space exploration. To find whether state $(i_1, \ldots, i_K)$ has already been found, we simply apply the following steps

1. Search for i_1 in the (only) level-1 tree. If found, we follow the pointer, to a level-2 tree.

2. Search for i_2 in this tree. If found, follow the pointer, to a level-3 tree.

3. Continue to follow the pointers until either a local state i_k is not found or until i_K is found. In the former case, $(i_1, \ldots, i_K)$ is a new state that must be inserted by creating a new tree at each level k through K, each tree containing a single node, labeled i_k through i_K, respectively. In the latter case, $(i_1, \ldots, i_K)$ had already been inserted previously and no action must be taken.

Once $BuildRS$ has completed execution, the dynamic data structure can be compressed into K static arrays that require only little over $|S| \cdot b_K$ bits, assuming the trees at level K are "not too small". In other words, with an appropriate partition of the model, the memory usage is dominated by the array at level K.

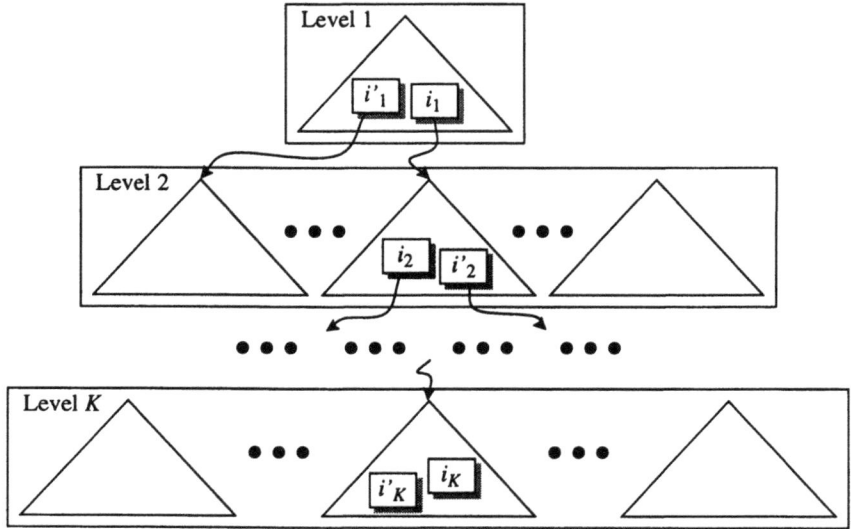

Figure 10 A multilevel data structure to store $\mathcal{S}$.

5.2 The Transition Rate Matrix

$\mathcal{S}$ is potentially very large, but normally only a few transitions are possible from each state. For example, in the CTMC obtained from a SPN with no immediate transitions, each row can have at most $|\mathcal{T}|$ nonzeros. A sparse format should then be used to store the off-diagonal elements of $\mathbf{Q}$ (which we indicate with $\mathbf{R}$, for transition *rate* matrix), while the diagonal of $\mathbf{Q}$ is stored as a full vector (in practice, the inverse of the diagonal, $\mathbf{h}$, is actually stored).

A sparse row-wise or column-wise format is usually employed for $\mathbf{R}$, which requires only to store $|\mathcal{S}| + 1 + \eta(\mathbf{R})$ integers and $\eta(\mathbf{R})$ reals, where $\eta(\mathbf{R})$ is the number of nonzeros in $\mathbf{R}$. Thus matrices with 10^5 or even 10^6 states can be stored on a modern workstation, provided the number of nonzeros is manageable. An additional advantage of using sparse storage is that most iterative solution algorithms essentially require the multiplication of a full vector by $\mathbf{R}$, and this can be performed in $O(\eta(\mathbf{R}))$ operations, instead of the $O(|\mathcal{S}|^2)$ operations that would be required if $\mathbf{R}$ were stored as a full matrix.

Even when $\mathbf{R}$ is stored in sparse format, though, its memory requirements are the main practical limitation to the size of the problems that can be solved.

Thus, several research efforts have focused on the possibility of encoding $\mathbf{R}$ as the sum of tensor products [Amoia et al., 1981, Davio, 1981] of much smaller matrices. This idea has been applied to high-level formalisms such as stochastic automata networks (SANs) [Fernandes et al., 1996, Plateau, 1985, Plateau and Atif, 1991, Plateau et al., 1988, Stewart et al., 1995], GSPNs [Buchholz and Kemper, 1995, Ciardo and Tilgner, 1996, Donatelli, 1991, Donatelli, 1994, Kemper, 1996], and various queuing models [Bause et al., 1995, Buchholz, 1991, Buchholz, 1994].

We refer to Chapter 5 in this book, which describes the idea applied to SANs. The main advantage of this approach is that the storage for **R** becomes negligible, thus the main limitation remains the size of the iteration vectors.

It is worth mentioning two alternative approaches that, even if not compositional in nature, achieve a similar objective and have analogous limitations (restricted type of access to the entries of **R**, additional overhead) to the tensor-based approach.

In [Deavours and Sanders, 1997a], **R** is generated and stored to disk, then appropriate block-iterative methods are employed that allow the retrieval of the blocks from disk to overlap with the numerical computation. This partially masks the impact of the slow disk access times, as compared to memory access times.

In [Deavours and Sanders, 1997b], instead, the high-level model is used to directly generate the entries of **R** at each iteration of the numerical method employed for the solution. In principle, this is of course possible because the high-level model is the information used to generate **R** in the first place. However, the amount of overhead involved in doing so at each iteration, instead of just once, can be quite high. This is especially true for models such as GSPNs, where a rate $\mathbf{R}_{i,j}$ might be due to the firing of one timed transition followed by an arbitrary number of immediate transitions.

If **R** is not stored explicitly, the size of the problems that can be solved is mainly limited by the size of the state space $|\mathcal{S}|$, not (much) by that of $\eta(\mathbf{R})$. In practical applications, this allows an order of magnitude increase in the size of the state-space that can be solved, at the cost of additional execution overhead.

5.3 Parallel Implementations

Another way to increase the size of the models that can be solved numerically is to make use of multiple processors. In addition to increasing the amount of available memory, distributed algorithms have, of course, the potential of speeding up the solution, if they can make good use of the overall processing power.

In [Caselli et al., 1995], both data-parallel and message-passing parallel implementations for the state-space exploration are studied. [Ciardo et al., 1998] focuses on distributed generation of the state-space over N processors using message-passing and a user-defined "partitioning function" $f : \hat{\mathcal{S}} \to \{0, \ldots N-1\}$. Through f, a processor can determine which processor is responsible for storing and exploring a state i without having to consult a centralized directory. At the end of the exploration, each processor n stores the set of states $\mathcal{S}^{(n)} = \{i \in \mathcal{S} : f(i) = n\}$ and the block $\mathbf{R}_{\mathcal{S}, \mathcal{S}^{(n)}}$ of the transition rate matrix, that is, the columns of **R** corresponding to the states that were assigned to it. An alternative approach is to avoid the partitioning function altogether and use instead heuristics that dynamically balance either the memory or the execution load.

The main advantage of these distributed algorithms is that they scale up almost linearly in terms of memory, and almost as well in terms of execution time, provided a reasonable balance is achieved. However, the distributed numerical solution of the stochastic process is nontrivial, since, for true scalability, the iteration vector must also be partitioned over the N processors. The presence of nonzeros in $\mathbf{R}_{\mathcal{S}(m),\mathcal{S}(n)}$, for $n \neq m$, complicates the algorithm, since they require processor n to know the corresponding entries in the portion of the iteration vector assigned to processor m. Block-iterative methods should then be used to reduce communication overhead.

References

[Ajmone Marsan et al., 1989] Ajmone Marsan, M., Balbo, G., Bobbio, A., Chiola, G., Conte, G., and Cumani, A. (1989). The effect of execution policies on the semantics and analyis of Stochastic Petri Nets. *IEEE Trans. Softw. Eng.*, 15(7):832–846.

[Ajmone Marsan et al., 1995] Ajmone Marsan, M., Balbo, G., Conte, G., Donatelli, S., and Franceschinis, G. (1995). *Modelling with generalized stochastic Petri nets*. John Wiley & Sons.

[Amoia et al., 1981] Amoia, V., De Micheli, G., and Santomauro, M. (1981). Computer-oriented formulation of transition-rate matrices via Kronecker algebra. *IEEE Trans. Rel.*, 30:123–132.

[Baskett et al., 1975] Baskett, F., Chandy, K. M., Muntz, R. R., and Palacios-Gomez, F. (1975). Open, Closed, and Mixed networks of queues with different classes of customers. *J. ACM*, 22(2):335–381.

[Bause et al., 1995] Bause, F., Buchholz, P., and Kemper, P. (1995). QPN-Tool for the specification and analysis of hierarchically combined queueing Petri nets. In Beilner, H. and Bause, F., editors, *Quantitative Evaluation of Computing and Communication Systems, 8th Int. Conf. Modelling Techniques and Tools*, LNCS 977, pages 224–238.

[Berson et al., 1991] Berson, S., de Souza e Silva, E., and Muntz, R. R. (1991). A methodology for the specification and generation of Markov models. In Stewart, W. J., editor, *Numerical Solution of Markov Chains*, pages 11–36. Marcel Dekker, Inc., New York, NY.

[Blum et al., 1994] Blum, A. M., Goyal, A., Heidelberger, P., and Lavenberg, S. S. (1994). Modeling and analysis of system dependability using the System Availability Estimator. In *Proc. 24th Int. Symp. on Fault-Tolerant Computing*, pages 137–141, Austin, TX.

[Bobbio and Cumani, 1984] Bobbio, A. and Cumani, A. (1984). Discrete state stochastic systems with phase-type distributed transition times. In *Proc. 1984 AMSE Int. Conf. on Modelling and Simulation*, Athens, Greece.

[Buchholz, 1991] Buchholz, P. (1991). Numerical solution methods based on structured descriptions of Markovian models. In Balbo, G. and Serazzi, G., editors, *Computer performance evaluation*, pages 251–267. Elsevier Science Publishers B.V. (North-Holland).

[Buchholz, 1994] Buchholz, P. (1994). A class of hierarchical queueing networks and their analysis. *Queueing Systems.*, 15:59–80.

[Buchholz and Kemper, 1995] Buchholz, P. and Kemper, P. (1995). Numerical analysis of stochastic marked graphs. In *Proc. 6th Int. Workshop on Petri Nets and Performance Models (PNPM'95)*, pages 32–41, Durham, NC. IEEE Comp. Soc. Press.

[Caselli et al., 1995] Caselli, S., Conte, G., and Marenzoni, P. (1995). Parallel state space exploration for GSPN models. In De Michelis, G. and Diaz, M., editors, *Application and Theory of Petri Nets 1995, Lecture Notes in Computer Science 935 (Proc. 16th Int. Conf. on Applications and Theory of Petri Nets, Turin, Italy)*, pages 181–200. Springer-Verlag.

[Cassandras, 1993] Cassandras, C. G. (1993). *Discrete Event Systems: Modeling and Performance Analysis.* Aksen Associates.

[Çinlar, 1975] Çinlar, E. (1975). *Introduction to Stochastic Processes.* Prentice-Hall.

[Chen et al., 1989] Chen, P.-z., Bruell, S. C., and Balbo, G. (1989). Alternative methods for incorporating non-exponential distributions into stochastic timed Petri nets. In *Proc. 3rd Int. Workshop on Petri Nets and Performance Models (PNPM'89)*, Kyoto, Japan. IEEE Comp. Soc. Press.

[Chiola et al., 1993] Chiola, G., Dutheillet, C., Franceschinis, G., and Haddad, S. (1993). Stochastic well-formed colored nets and symmetric modeling applications. *IEEE Trans. Comp.*, 42(11):1343–1360.

[Choi et al., 1994] Choi, H., Kulkarni, V. G., and Trivedi, K. S. (1994). Markov regenerative stochastic Petri nets. *Perf. Eval.*, 20(1-3):337–357.

[Ciardo, 1989] Ciardo, G. (1989). *Analysis of large stochastic Petri net models.* PhD thesis, Duke University, Durham, NC.

[Ciardo, 1994] Ciardo, G. (1994). Petri nets with marking-dependent arc multiplicity: properties and analysis. In Valette, R., editor, *Application and Theory of Petri Nets 1994, Lecture Notes in Computer Science 815 (Proc. 15th Int. Conf. on Applications and Theory of Petri Nets, Zaragoza, Spain)*, pages 179–198. Springer-Verlag.

[Ciardo, 1995] Ciardo, G. (1995). Discrete-time Markovian stochastic Petri nets. In Stewart, W. J., editor, *Computations with Markov Chains*, pages 339–358. Kluwer, Boston, MA.

[Ciardo et al., 1993a] Ciardo, G., Blakemore, A., Chimento, P. F. J., Muppala, J. K., and Trivedi, K. S. (1993a). Automated generation and analysis of Markov reward models using Stochastic Reward Nets. In Meyer, C. and Plemmons, R. J., editors, *Linear Algebra, Markov Chains, and Queueing Models*, volume 48 of *IMA Volumes in Mathematics and its Applications*, pages 145–191. Springer-Verlag.

[Ciardo et al., 1993b] Ciardo, G., German, R., and Lindemann, C. (1993b). A characterization of the stochastic process underlying a stochastic Petri net. In

Proc. 5th Int. Workshop on Petri Nets and Performance Models (PNPM'93), pages 170–179, Toulouse, France. IEEE Comp. Soc. Press.

[Ciardo et al., 1998] Ciardo, G., Gluckman, J., and Nicol, D. (1998). Distributed state-space generation of discrete-state stochastic models. *INFORMS J. Comp.*, 10(1):82–93.

[Ciardo and Miner, 1997] Ciardo, G. and Miner, A. S. (1997). Storage alternatives for large structured state spaces. In Marie, R., Plateau, B., Calzarossa, M., and Rubino, G., editors, *Proc. 9th Int. Conf. on Modelling Techniques and Tools for Computer Performance Evaluation*, LNCS 1245, pages 44–57, St. Malo, France. Springer-Verlag.

[Ciardo et al., 1991] Ciardo, G., Muppala, J. K., and Trivedi, K. S. (1991). On the solution of GSPN reward models. *Perf. Eval.*, 12(4):237–253.

[Ciardo and Tilgner, 1996] Ciardo, G. and Tilgner, M. (1996). On the use of Kronecker operators for the solution of generalized stochastic Petri nets. ICASE Report 96-35, Institute for Computer Applications in Science and Engineering, Hampton, VA.

[Ciardo and Zijal, 1996] Ciardo, G. and Zijal, R. (1996). Well-defined stochastic Petri nets. In *Proc. 4th Int. Workshop on Modeling, Analysis and Simulation of Computer and Telecommunication Systems (MASCOTS'96)*, pages 278–284, San Jose, CA, USA. IEEE Comp. Soc. Press.

[Clocksin and Mellish, 1984] Clocksin, W. F. and Mellish, C. S. (1984). *Programming in Prolog*. Springer-Verlag.

[Cox, 1955] Cox, D. (1955). A use of complex probabilities in the theory of stochastic processes. *Proc. of the Cambridge Philosophical Society*, 51:313–319.

[Cumani, 1985] Cumani, A. (1985). ESP - A package for the evaluation of stochastic Petri nets with phase-type distributed transitions times. In *Proc. Int. Workshop on Timed Petri Nets*, Torino, Italy.

[Davio, 1981] Davio, M. (1981). Kronecker products and shuffle algebra. *IEEE Trans. Comp.*, C-30:116–125.

[de Souza e Silva and Gail, 1989] de Souza e Silva, E. and Gail, H. R. (1989). Calculating availability and performability measures of repairable computer systems using randomization. *J. ACM.*, 36(1):171–193.

[Deavours and Sanders, 1997a] Deavours, D. D. and Sanders, W. H. (1997a). An efficient disk-based tool for solving very large Markov models. In Marie, R., Plateau, B., Calzarossa, M., and Rubino, G., editors, *Proc. 9th Int. Conf. on Modelling Techniques and Tools for Computer Performance Evaluation*, LNCS 1245, pages 58–71, St. Malo, France. Springer-Verlag.

[Deavours and Sanders, 1997b] Deavours, D. D. and Sanders, W. H. (1997b). "On-the-fly" solution techniques for stochastic Petri nets and extensions. In *Proc. 7th Int. Workshop on Petri Nets and Performance Models (PNPM'97)*, pages 132–141, St. Malo, France. IEEE Comp. Soc. Press.

[Donatelli, 1991] Donatelli, S. (1991). Superposed Stochastic Automata: a class of stochastic Petri nets amenable to parallel solution. In *Proc. 4th Int. Workshop on Petri Nets and Performance Models (PNPM'91)*, pages 54–63, Melbourne, Australia. IEEE Comp. Soc. Press.

[Donatelli, 1994] Donatelli, S. (1994). Superposed generalized stochastic Petri nets: definition and efficient solution. In Valette, R., editor, *Application and Theory of Petri Nets 1994, Lecture Notes in Computer Science 815 (Proc. 15th Int. Conf. on Applications and Theory of Petri Nets)*, pages 258–277, Zaragoza, Spain. Springer-Verlag.

[Feller, 1962] Feller, W. (1962). *An Introduction to Probability Theory and Its Applications*. John Wiley, New York. Second Edition.

[Fernandes et al., 1996] Fernandes, P., Plateau, B., and Stewart, W. J. (1996). Numerical issue for stochastic automata networks. In *Proc. of the 4rd Workshop on Process Algebra and Performance Modelling (PAPM)*, Torino, Italy. CLUT.

[Grassmann, 1991] Grassmann, W. K. (1991). Finding transient solutions in Markovian event systems through randomization. In Stewart, W. J., editor, *Numerical Solution of Markov Chains*, pages 357–371. Marcel Dekker, Inc., New York, NY.

[Grassmann and Wang, 1995] Grassmann, W. K. and Wang, Y. (1995). Immediate events in Markov chains. In Stewart, W. J., editor, *Computations with Markov Chains*, pages 163–176. Kluwer, Boston, MA.

[Gross and Miller, 1984] Gross, D. and Miller, D. (1984). The randomization technique as a modeling tool and solution procedure for transient Markov processes. *Oper. Res.*, 32:343–361.

[Jensen, 1953] Jensen, A. (1953). Markoff Chains as an Aid in the Study of Markoff Processes. *Skand. Aktuarietidskr.*, 36:87–91.

[Jensen, 1987] Jensen, K. (1987). Coloured Petri nets. In *Petri Nets: Central Models and Their Properties, Lecture Notes in Computer Science 254*, pages 248–299. Springer-Verlag.

[Kemper, 1996] Kemper, P. (1996). Numerical analysis of superposed GSPNs. *IEEE Trans. Softw. Eng.*, 22(4):615–628.

[Molloy, 1985] Molloy, M. K. (1985). Discrete time stochastic Petri nets. *IEEE Trans. Softw. Eng.*, 11(4):417–423.

[Murata, 1989] Murata, T. (1989). Petri Nets: properties, analysis and applications. *Proc. of the IEEE*, 77(4):541–579.

[Page et al., 1989] Page, T. W., Berson, S. E., Cheng, W. C., and Muntz, R. R. (1989). An object-oriented modeling environment. In *OOPSLA '89 Proc.*

[Plateau, 1985] Plateau, B. (1985). On the stochastic structure of parallelism and synchronisation models for distributed algorithms. In *Proc. 1985 ACM SIGMETRICS Conf. on Measurement and Modeling of Computer Systems*, pages 147–153, Austin, TX, USA.

[Plateau and Atif, 1991] Plateau, B. and Atif, K. (1991). Stochastic Automata Network for modeling parallel systems. *IEEE Trans. Softw. Eng.*, 17(10):1093–1108.

[Plateau et al., 1988] Plateau, B., Fourneau, J.-M., and Lee, K. H. (1988). PEPS: a package for solving complex Markov models of parallel systems. In Puigjaner, R., editor, *Proc. 4th Int. Conf. Modelling Techniques and Tools*, pages 341–360.

[Qureshi et al., 1995] Qureshi, M. A., Sanders, W. H., van Morsel, A. P. A., and German, R. (1995). Algorithms for the generation of state-level representations of stochastic activity networks with general reward structures. In *Proc. 6th Int. Workshop on Petri Nets and Performance Models (PNPM'95)*, pages 180–190, Durham, NC. IEEE Comp. Soc. Press.

[Sanders, 1988] Sanders, W. H. (1988). *Construction and solution of performability models based on Stochastic Activity Networks*. PhD thesis, Department of Computer Science and Engineering, University of Michigan, Ann Arbor, MI.

[Sanders and Meyer, 1991] Sanders, W. H. and Meyer, J. F. (1991). Reduced base model construction methods for stochastic activity networks. *IEEE J. Sel. Areas in Comm.*, 9(1):25–36.

[Stewart, 1991] Stewart, W. J. (1991). MARCA: Markov Chain Analyser, a software package for Markov modelling. In Stewart, W. J., editor, *Numerical Solution of Markov Chains*, pages 37–61. Marcel Dekker, Inc., New York, NY.

[Stewart et al., 1995] Stewart, W. J., Atif, K., and Plateau, B. (1995). The numerical solution of stochastic automata networks. *Europ. J. of Oper. Res.*, 86:503–525.

[Van Dijk, 1991] Van Dijk, N. M. (1991). Truncation of Markov chains with applications to queueing. *Operations Research*, 39(6):1018–1026.

[Wallace and Rosenberg, 1994] Wallace, V. L. and Rosenberg, R. (1994). RQA-1, The recursive queue analyser. Technical Report 2, Systems Engineering Laboratory, University of Michigan, Ann Arbor.

Gianfranco Ciardo Gianfranco Ciardo is an Associate Professor in the Department of Computer Science at the College of William and Mary, Williamsburg, Virginia. He received a Laurea from Università di Torino, Italy, in 1982 and a Ph.D. from Duke University, in 1989. He is interested in theory and tools for the behavioral, performance, and dependability modeling of complex hardware/software systems, with an emphasis on stochastic Petri nets. He has explored both Markovian and non-Markovian models, and has recently begun working on distributed algorithms and compositional approaches for coping with large state spaces. Dr. Ciardo was Program Co-Chair of the 1995 Petri Nets and Performance Models (PNPM) Workshop and will be Organization Chair of the 1999 International Conference on Application and Theory of Petri Nets (ICATPN). He is a designer and developer of SPNP, the Stochastic Petri Net Package, in use at some 50 sites worldwide, and has consulted for both industry and government.

3 TRANSIENT SOLUTIONS FOR MARKOV CHAINS

Edmundo de Souza e Silva[1] and H. Richard Gail[2]

[1] NCE and CS Department
Federal University of Rio de Janeiro
Cx.P. 2324, CEP 20001-970, Rio de Janeiro
Brazil
edmundo@nce.ufrj.br

[2] IBM
T. J. Watson Research Center
POB 704
Yorktown Heights, NY 10598
rgail@watson.ibm.com

1 DEFINITION OF TRANSIENT MEASURES

Much of the theory developed for solving Markov chain models is devoted to obtaining steady state measures, that is, measures for which the observation interval $(0, t)$ is "sufficiently large" ($t \to \infty$). These measures are indeed approximations of the behavior of the system for a finite, but long, time interval, where long means with respect to the interval of time between occurrences of events in the system. However, an increasing number of applications requires the calculation of measures during a relatively "short" period of time. These are the so-called *transient measures*. In these cases the steady state measures are not good approximations for the transient, and one has to resort to different techniques to obtain the desired quantities.

The object of this chapter is to survey several methods that have been applied to obtain transient measures of Markovian models. After defining the notation used and the transient measures of interest we first devote our attention to the calculation of a simple measure, the transition probabilities at a finite time t, using different approaches. Then we comment on a few of the problems encountered in obtaining this transient measure and present some solutions to them. Algorithms used to calculate complex measures are then presented. We also discuss approximations that have recently been proposed. We conclude the chapter with a few application examples.

Consider a time-homogeneous continuous-time Markov chain (or CTMC) $\mathcal{X} = \{X(t), t \geq 0\}$ with finite state space $\mathcal{S} = \{s_i : i = 1, \ldots, M\}$ and infinitesimal generator

$$\mathbf{Q} = \begin{bmatrix} -q_1 & q_{12} & \cdots & q_{1M} \\ q_{21} & -q_2 & \cdots & q_{2M} \\ \vdots & \vdots & \ddots & \vdots \\ q_{M1} & q_{M2} & \cdots & -q_M \end{bmatrix} \quad (1)$$

Here $q_i = \sum_{j \neq i} q_{ij}$ represents the (exponential) rate out of state s_i, while $q_{i,j}$ represents the (exponential) rate from s_i to s_j. The transition function of the chain is the family of $M \times M$ matrices $\mathbf{\Pi}(t)$, $t \geq 0$, where, because of the time-homogeneity, for any $t' \geq 0$

$$\mathbf{\Pi}(t)_{(i,j)} = P\{X(t'+t) = s_j | X(t') = s_i\}. \quad (2)$$

Since the state space $\mathcal{S}$ is finite, $\mathbf{\Pi}(t)$ satisfies (see [Çinlar, 1975]) both Kolmogorov's backward equations

$$\mathbf{\Pi}'(t) = \mathbf{Q}\mathbf{\Pi}(t), \quad (3)$$

and Kolmogorov's forward equations

$$\mathbf{\Pi}'(t) = \mathbf{\Pi}(t)\mathbf{Q}. \quad (4)$$

The vector

$$\boldsymbol{\pi}(t) = [\pi_1(t), \ldots, \pi_M(t)] \quad (5)$$

gives the state probabilities for the chain $\mathcal{X}$ at time t, i.e., $\pi_i(t) = P\{X(t) = s_i\}$. These vectors may be calculated from the transition function as

$$\boldsymbol{\pi}(t) = \boldsymbol{\pi}(0)\mathbf{\Pi}(t). \quad (6)$$

The entries of $\boldsymbol{\pi}(t)$ can be expressed in terms of an indicator random variable as

$$\pi_i(t) = P\{\mathcal{I}\{X(t) = s_i\} = 1\} = E[\mathcal{I}\{X(t) = s_i\}]. \quad (7)$$

More generally, the probability that the chain is in a given set of states $\mathcal{O} \subset \mathcal{S}$ at time t is

$$\pi_{\mathcal{O}}(t) \triangleq \sum_{i:s_i \in \mathcal{O}} \pi_i(t) =$$
$$P\{X(t) \in \mathcal{O}\} = P\{\mathcal{I}\{X(t) \in \mathcal{O}\} = 1\} = E[\mathcal{I}\{X(t) \in \mathcal{O}\}]. \quad (8)$$

Other random variables of interest include $\mathcal{T}_\mathcal{O}(t)$, the time during a finite interval $(0,t)$ that the Markov chain spends in a set of states $\mathcal{O}$, and the corresponding fraction of time $\mathcal{F}_\mathcal{O}(t) = \mathcal{T}_\mathcal{O}(t)/t$ spent in $\mathcal{O}$ during the period. When $\mathcal{O} = \{s_i\}$, we use the simplified notation $\mathcal{T}_i(t)$ and $\mathcal{F}_i(t)$. Then, we have

$$\mathcal{T}_\mathcal{O}(t) = \int_0^t \mathcal{I}\{X(\tau) \in \mathcal{O}\}d\tau. \tag{9}$$

In particular, the expected amount of time spent in these states during $(0,t)$ is

$$E[\mathcal{T}_\mathcal{O}(t)] = \int_0^t P\{X(\tau) \in \mathcal{O}\}d\tau. \tag{10}$$

Note that, by assigning reward rate 1 for states in $\mathcal{O}$ and reward rate 0 for states outside of $\mathcal{O}$, the time $\mathcal{T}_\mathcal{O}(t)$ can be thought of as the reward accumulated over $(0,t)$.

In fact, Markov chains with a more general reward structure have been studied, and we now describe these chains. We assume there are $K+1$ rate rewards $r_1 > r_2 > \cdots > r_{K+1}$, and state $s \in \mathcal{S}$ has a reward rate $r_{c(s)}$. For instance, if $\mathcal{S} = \{1,2,3,4,5\}$, states 1, 2 and 3 may have reward rate r_1, and states 4 and 5 reward rate r_2. Hence, $c(s) = 1$ for $s = 1,2,3$, and $c(s) = 2$ for $s = 4,5$. A reward of r_j is accumulated per unit time in any state with this rate corresponding to it. Without loss of generality, we can make the normalization $r_1 = 1$, $r_{K+1} = 0$, by replacing r_i with $(r_i - r_{K+1})/(r_1 - r_{K+1})$, for $i = 1, \ldots, K+1$. The instantaneous reward rate at time t is the random variable

$$IR(t) = r_{c(X(t))}, \tag{11}$$

and the cumulative rate based reward during $(0,t)$ is

$$CR(t) = \int_0^t r_{c(X(\tau))}d\tau. \tag{12}$$

When the reward is averaged over the length of the observation period, we obtain

$$ACR(t) = \frac{1}{t}CR(t). \tag{13}$$

The cumulative reward $CR(t)$ has been called *performability* when the model being analyzed attempts to capture the effect of gracefully degradable systems on the system performance. In this case, each state in the model represents a given system structure, and the reward associated with a state is related to some performance measurement at that state [Meyer, 1980]. The name *performability* has also been used in a broader sense to refer to a set of measures that jointly incorporate both dependability and performance.

Let τ_n, $n = 1, 2, \ldots$, be the time of the nth transition (with $\tau_0 = 0$), so that $X(t) = X(\tau_n)$ in the interval $[\tau_n, \tau_{n+1})$. Let $\sigma_n = (X(\tau_{n-1}), X(\tau_n))$, $n = 1, 2, \ldots$, be the pair of states that represent the nth transition. Impulse rewards can be associated with transitions of the Markov chain, i.e., with pairs of states. For example, in a fault tolerant system model, impulse rewards can be used to count events that lead to a system failure. We assume there are $\widehat{K}+1$ impulse rewards $\rho_1 > \rho_2 > \cdots > \rho_{\widehat{K}+1}$ associated with the transitions of the Markov chain. A reward of ρ_j is earned for any transition with this impulse reward corresponding to it. Furthermore, we can assume that $\rho_{\widehat{K}+1} = 0$, similar to the rate reward case. For any transition $s \to s'$, i.e., a pair of states (s, s'), we let $\rho_{\widehat{c}(s,s')}$ be the impulse reward for that transition, where $\widehat{c}(s, s') \in \{1, \ldots, \widehat{K}+1\}$ is the index of the impulse reward corresponding to $s \to s'$. Let $N(t)$ be the number of transitions that occur during $(0,t)$. Then the cumulative impulse based reward during $(0,t)$ is

$$CI(t) = \sum_{n=1}^{N(t)} \rho_{\widehat{c}(\sigma_n)} \qquad (14)$$

(note that $CI(t) = 0$ if $N(t) = 0$). Averaging the impulse based reward over t yields

$$ACI(t) = \frac{1}{t}CI(t). \qquad (15)$$

The total reward accumulated during $(0,t)$ is

$$CIR(t) = CR(t) + CI(t), \qquad (16)$$

while when averaged over the length t we have

$$ACIR(t) = ACR(t) + ACI(t). \qquad (17)$$

The special case of two rate rewards $1 = r_1 > r_2 = 0$ is of particular interest. Suppose that r_1 is assigned to states in $\mathcal{O} \subset \mathcal{S}$, while r_0 is assigned to states not in $\mathcal{O}$. Then, as mentioned above, the cumulative reward $CR(t)$ is simply $T_\mathcal{O}(t)$, the random variable that represents the amount of time during $(0,t)$ that the chain spends in the subset of states $\mathcal{O}$. Averaging the reward over t gives $\mathcal{F}_\mathcal{O}(t)$, the fraction of time spent in $\mathcal{O}$, i.e., $ACR(t)$ is this fraction. The instantaneous reward rate at t indicates whether or not the chain is in the set of states $\mathcal{O}$, and in fact it is the random variable $\mathcal{I}\{X(t) \in \mathcal{O}\}$. In the context of modeling fault tolerant computer systems, suppose that a reward of 1 is assigned to states representing an operational (up) system, while 0 is assigned to states corresponding to a failed (down) system. In this dependability case, the random variable $ACR(t)$ is the availability of the system over the observation period $(0,t)$, while $CR(t)$ is the cumulative operational time.

For models with only rate rewards present, two other random variables of interest are $\Theta(r)$, the time until a given reward level r is achieved, and $\Phi_r(t)$, the total time during $(0,t)$ spent in states with reward level at least r. Since $\Theta(r) < t$ if and only if $CR(t) > r$, the distribution of $\Theta(r)$ is given directly in terms of that of $CR(t)$. To find $\Phi_r(t)$, consider the indicator random variable that is 1 if the instantaneous reward rate is above r at a designated time τ and 0 otherwise. Then

$$\Phi_r(t) = \int_0^t \mathcal{I}\{r_{c(X(\tau))} > r\}d\tau. \tag{18}$$

Note that the cumulative operational time is $\Phi_r(t)$ where $r_1 = 1 > r > r_2 = 0$, and any r between 0 and 1 may be used. On the other hand, specializing to the case of impulse rewards, a random variable representing the number of events of a certain type can be defined. As an example, in a fault tolerant model, one may wish to count the number of occurrences of a particular type of component failure that caused the system to go down. Simply assign impulse rewards $\hat{r}_1 = 1$ to transitions that are of interest, and $\hat{r}_2 = 0$ to all other transitions.

Suppose our interest is in obtaining the first passage time to a designated set of states in the Markov chain. If we change the original Markov chain by making these states absorbing, then by assigning reward rate 0 to the absorbing states and 1 to the non-absorbing states, the first passage time can be considered as the cumulative operational time in this chain with absorbing states. More generally, assume that the Markov chain under consideration has a set $\mathcal{A}$ of absorbing states. For example, in a model of a non-repairable computer system, the absorbing states correspond to a failed system. The probability that the chain is not in an absorbing state at the end of an observation period $(0,t)$ is

$$R(t) = P\{\mathcal{I}\{X(t) \notin \mathcal{A}\} = 1\} = E[\mathcal{I}\{X(t) \notin \mathcal{A}\}]. \tag{19}$$

For fault tolerant computer systems, the quantity $R(t)$ is the reliability of the system, or the probability it does not fail during the period $(0,t)$. A random variable of interest in this absorbing case is the lifetime $L(t)$, which is the time it takes during $(0,t)$ to reach the absorbing states. In the non-repairable system model, $L(t)$ is the time until system failure (hence the name lifetime). Then

$$L(t) = \int_0^t \mathcal{I}\{X(\tau) \notin \mathcal{A}\}d\tau, \tag{20}$$

and thus the expected lifetime is given in terms of the reliability as

$$E[L(t)] = \int_0^t R(\tau)d\tau. \tag{21}$$

Note that, by setting $\mathcal{O} = \mathcal{S} \setminus \mathcal{A} = \{s \in \mathcal{S} : s \notin \mathcal{A}\}$, the reliability is simply $\pi_\mathcal{O}(t)$ and the lifetime is $\mathcal{T}_\mathcal{O}(t)$.

Measures of interest include means, variances and distributions of the random variables above (moments can be calculated as well in some cases). We have already given expressions for certain of the random variables that have been introduced. As another example, the expected reward at time t is

$$E[IR(t)] = E[r_{c(X(t))}] = \mathbf{r}^* \cdot \boldsymbol{\pi}(t),$$

where $\mathbf{r}^* = [r_{c(s_1)}, \ldots, r_{c(s_M)}]$ is the vector of rate rewards associated with the states of the Markov chain. Here $\mathbf{x} \cdot \mathbf{y} = \sum_i x_i y_i$ denotes the inner product of the vectors $\mathbf{x} = [x_1, \ldots, x_M]$ and $\mathbf{y} = [y_1, \ldots, y_M]$. Also, the expected cumulative reward over $(0, t)$ is

$$E[CR(t)] = \int_0^t E[IR(\tau)]d\tau = \mathbf{r}^* \cdot \int_0^t \boldsymbol{\pi}(\tau)d\tau.$$

In the remainder of the paper we will describe methods to calculate these expectations, along with other performance measures.

2 TRANSIENT ANALYSIS OF STATE PROBABILITIES

In this section we describe different methods that can be used to compute the transition function $\boldsymbol{\Pi}(t)$ and thus, from (6), the state probability vector $\boldsymbol{\pi}(t)$. However, most approaches can also be employed to calculate more complex measures. Several of the methods are based on general techniques for ordinary differential equations, and many of the techniques lead to identical equations when used to solve the backward Kolmogorov equations (3).

2.1 Solutions Based on Ordinary Differential Equations

An equation of the type

$$y'(t) = f(t, y) \tag{22}$$

is called an ordinary differential equation (or ODE). Setting $y(t) = \boldsymbol{\Pi}(t)$ and $f(t, y) = f(y) = \mathbf{Q}y$, we see that (3) has this form, and so ODE methods may be applied in solving Kolmogorov's backward equations.

The simplest approximation technique for obtaining a solution to an ODE is Euler's method. First, the observation interval $(0, t)$ is divided into subintervals of length h, and then an approximate value of $y((n+1)h)$ is obtained from $y(nh)$ by approximating the derivative at the point $(y(nh), nh)$ with the difference quotient $[y((n+1)h) - y(nh)]/h$. The result for (3) is

$$\boldsymbol{\Pi}(t) = [\mathbf{I} + \mathbf{Q}h]\boldsymbol{\Pi}(t - h), \tag{23}$$

where $\boldsymbol{\Pi}(0) = \mathbf{I}$. The matrix $\boldsymbol{\Pi}(t)$ can be easily obtained recursively from (23). Given $\boldsymbol{\Pi}(nh)$, multiply on the left by $\mathbf{I} + \mathbf{Q}h$ to obtain the next term $\boldsymbol{\Pi}((n+1)h)$ in the sequence.

It is interesting to observe that there is a probabilistic interpretation for this recursive approach. First note that, for h sufficiently small, the matrix $\mathbf{I} + \mathbf{Q}h$ is positive, and the (i, j) element $q_{i,j}h$ for $i \neq j$ approximates the probability that the system makes a transition to state s_j from state s_i in an interval of length h. In fact, the approach simply involves discretizing time into sufficiently small slots and noting that the probability that an exponential event with rate λ occurs in a small interval h is $\lambda h + o(h)$. A weakness of Euler's method is that the slot size might be quite small in order to achieve an acceptable accuracy.

As a consequence, the computational cost may be very high. Furthermore, no error bounds are obtained. An advantage is the simplicity of the approach.

Euler's method is a *single step* method, since $y((n+1)h)$ is computed from $y(nh)$. In *multi-step* methods, $y((n+1)h)$ is computed from values obtained at several previous steps. For example, consider the formula for numerical differentiation

$$y'(t_n) = \frac{y(t_{n+1}) - y(t_{n-1})}{2h} + o(h^2), \qquad (24)$$

which approximates the derivative at a point using the slope obtained from two neighboring points. This may provide a better approximation to the derivative than the one used in Euler's method. Applying (24) to (3), we obtain

$$\Pi(t) = \Pi(t - 2h) + 2h\mathbf{Q}\Pi(t - h). \qquad (25)$$

This is a two-step method, since two previous values are used to calculate $\Pi(t)$. Note that the initial value $\Pi(h)$ must be obtained before the recursion given by (25) can begin, and for that we may use the formula (23). Unfortunately, this method is subject to oscillation, and modifications to it must be made to dampen the oscillations. One such example, the *modified midpoint method*, uses (23) at each step $m, 2m, \ldots$ of the algorithm. In other words, $\Pi((m-1)h)$ is calculated from (25), but $\Pi(mh)$ is a weighted sum of the values calculated from (23) and (25).

The methods above are called explicit, since $y(nh)$ is computed from values at previous time points. The *trapezoid* method is an example of an implicit method, since $y(nh)$ appears implicitly in the right hand side of the equation

$$\frac{y((n+1)h) - y(nh)}{h} = \frac{f(nh, y(nh)) + f((n+1)h, y((n+1)h))}{2} \qquad (26)$$

that is used. From (3) and (26) we obtain

$$\left[\mathbf{I} - \frac{h}{2}\mathbf{Q}\right]\Pi(t) = \left[\mathbf{I} + \frac{h}{2}\mathbf{Q}\right]\Pi(t - h). \qquad (27)$$

From (27) we see that at each step of the recursion a system of linear equations of the form $\mathbf{Ax} = \mathbf{b}$ must be solved in order to calculate a column of $\Pi(t)$. We can also invert $\mathbf{I} - (h/2)\mathbf{Q}$ before starting the recursion. The better accuracy achieved by the use of the trapezoidal formula (26) in comparison to Euler's method is obtained at the expense of greater computational cost.

The matrix $\Pi(t)$ can also be found through the use of Taylor series expansion. We first recall that an infinitely differentiable function $y(t)$ can be expanded around t as

$$y(t + h) = y(t) + hy'(t) + \frac{h^2}{2!}y'' + \frac{h^3}{3!}y''' + \cdots. \qquad (28)$$

The ODE of interest is $\Pi'(t) = \mathbf{Q}\Pi(t)$, and thus $\Pi^{(n)}(t) = \mathbf{Q}^n\Pi(t)$ for all n. Substituting these derivatives into (28), we obtain

$$\Pi(t + h) = \left[\mathbf{I} + h\mathbf{Q} + \frac{h^2}{2!}\mathbf{Q}^2 + \frac{h^3}{3!}\mathbf{Q}^3 + \cdots\right]\Pi(t). \qquad (29)$$

Euler's method corresponds to taking the first two terms in this Taylor series expansion.

Runge-Kutta methods for solving ODEs are based on computing $f(t, y)$ at selected points in the interval $(t, t+h)$ and then combining these values in order to obtain a good approximation for the increment $y(t + h) - y(t)$. A simple example of an order two method is

$$\begin{aligned} k_1 &= hf(t, y) \\ k_2 &= hf(t + h, y + k_1) \\ y(t + h) - y(t) &= (k_1 + k_2)/2. \end{aligned}$$

Applying these equations to (3) gives

$$\begin{aligned} k_1 &= h\mathbf{Q}\mathbf{\Pi}(t) \\ k_2 &= h\mathbf{Q}\left[\mathbf{\Pi}(t) + h\mathbf{Q}\mathbf{\Pi}(t)\right] \\ \mathbf{\Pi}(t + h) &= \left[\mathbf{I} + h\mathbf{Q} + \frac{h^2}{2}\mathbf{Q}^2\right]\mathbf{\Pi}(t). \end{aligned}$$

The most widely used Runge-Kutta method is

$$\begin{aligned} k_1 &= hf(t, y) \\ k_2 &= hf(t + h/2, y + k_1/2) \\ k_3 &= hf(t + h/2, y + k_2/2) \\ k_4 &= hf(t + h, y + k_3) \\ y(t + h) - y(t) &= (k_1 + 2k_2 + 2k_3 + k_4)/6. \end{aligned}$$

These equations can be used recursively by calculating the values of k_i at each step. From the backward Kolmogorov equations, replacing the values of k_i in the above expression for $y(t + h) - y(t)$, we have (see [Grassmann, 1977a])

$$\mathbf{\Pi}(t + h) = \left[\mathbf{I} + h\mathbf{Q} + \frac{h^2\mathbf{Q}^2}{2!} + \frac{h^3\mathbf{Q}^3}{3!} + \frac{h^4\mathbf{Q}^4}{4!}\right]\mathbf{\Pi}(t). \tag{30}$$

Note that (30) is identical to the expression obtained by taking the first four terms of the Taylor series expansion for $\mathbf{\Pi}(t)$.

2.2 Solutions Based on the Exponential of a Matrix

It is easy to verify that the system of Kolmogorov backward equations (3) has the solution

$$\mathbf{\Pi}(t) = e^{\mathbf{Q}t}, \tag{31}$$

where $e^{\mathbf{Q}t}$ is the matrix exponential defined as

$$e^{\mathbf{Q}t} = \sum_{n=0}^{\infty} \mathbf{Q}^n \frac{t^n}{n!}. \tag{32}$$

There are several methods to calculate the exponential of a matrix, but most of them lead to numerical problems (see [Golub and van Loan, 1989, Moler and van Loan, 1978] for a survey). In fact, the authors of [Moler and van Loan, 1978] mention that none of the existing methods are completely satisfactory. We present a few ways to obtain $\Pi(t)$ using (31). In one such method, the definition (32) of the exponential of a matrix is applied directly by calculating the powers of $\mathbf{Q}$ and truncating the infinite series in (32) at some point N. This can be done recursively by noting that, from (31) and (32),

$$\Pi(t) = \sum_{n=0}^{N} \mathbf{F}_n(t) + \mathbf{E}(N). \tag{33}$$

Here $\mathbf{E}(N)$ is a matrix that represents the error introduced when the infinite series in (32) is truncated at N, and $\mathbf{F}_n(t)$ is defined by the recursion

$$\mathbf{F}_n(t) = \mathbf{F}_{n-1}(t)\mathbf{Q}\frac{t}{n}. \tag{34}$$

Note that the matrix $\mathbf{F}_n(t)$ may not be sparse, even if the original matrix $\mathbf{Q}$ is. We can avoid storing a non-sparse matrix and also replace the matrix multiplication operation in (34) by a vector matrix operation, if our interest is in calculating $\pi(t) = \pi(0)\Pi(t)$ instead of $\Pi(t)$. In this case, (33) and (34) are replaced by $\pi(t) = \sum_{n=0}^{N} \mathbf{f}_n(t) + \mathbf{e}(N)$ and $\mathbf{f}_n(t) = \mathbf{f}_{n-1}(t)\mathbf{Q}t/n$ with $\mathbf{f}_0(t) = \pi(0)$, where $\mathbf{e}(N)$ is the vector that corresponds to the truncation error.

A related approach is to discretize time into small slots of length h such that $t = Nh$ and then apply (31). Since the Markov chain under consideration is time-homogeneous, we have

$$\pi((l+1)h) = \pi(lh)e^{\mathbf{Q}h} = \sum_{n=0}^{N} \mathbf{f}_n(lh) + \mathbf{e}(N), \tag{35}$$

where N is the truncation point of the infinite series in (32) for an interval of length h. Comparing (35) with (30), it is easy to see that this approach is equivalent to the Runge-Kutta method. Unfortunately, severe roundoff errors may be encountered when using these methods due to the negative diagonal elements of $\mathbf{Q}$.

Another conceptually simple approach is to perform a similarity transformation of $\mathbf{Q}$. Assume that all the eigenvalues of $\mathbf{Q}$ are distinct, or more generally that $\mathbf{Q}$ is diagonalizable. Then we may write $\mathbf{Q}$ in Jordan canonical form as [Horn and Johnson, 1985]

$$\mathbf{Q} = \mathbf{V}\mathbf{\Lambda}\mathbf{V}^{-1}, \tag{36}$$

where $\mathbf{\Lambda}$ is a diagonal matrix with the eigenvalues λ_i of $\mathbf{Q}$, $i = 1, \ldots, M$, on its diagonal, and $\mathbf{V}$ is a matrix for which its ith column is a right eigenvector of $\mathbf{Q}$ corresponding to λ_i (the ith row of $\mathbf{V}^{-1}$ is a left eigenvector). Therefore, we may write (see also Chapter 10)

$$\Pi(t) = \sum_{n=0}^{\infty} \mathbf{Q}^n \frac{t^n}{n!} = \mathbf{V}\left[\sum_{n=0}^{\infty} \mathbf{\Lambda}^n \frac{t^n}{n!}\right] \mathbf{V}^{-1} \qquad (37)$$
$$= \mathbf{V} e^{\mathbf{\Lambda} t} \mathbf{V}^{-1} = \mathbf{V}\,\mathrm{diag}\{e^{\lambda_1 t}, \ldots, e^{\lambda_M t}\} \mathbf{V}^{-1}.$$

Then, the problem of finding $\Pi(t)$ reduces to the problem of finding the eigenvalues and eigenvectors of $\mathbf{Q}$. Note that the matrix $\mathbf{Q}$ has an eigenvalue equal to 0, and so the corresponding eigenvector gives the steady state solution ($t \to \infty$). Furthermore, the convergence to the steady state solution is determined by the subdominant eigenvalue of $\mathbf{Q}$.

2.3 Laplace Transform methods

An alternative way of solving differential equations is by using the Laplace transform, and so this method of solution can be applied to the Kolmogorov backward equations. In order to illustrate the approach, transform the equation (3) to obtain

$$(s\mathbf{I} - \mathbf{Q})\Pi^*(s) = \mathbf{I}, \qquad (38)$$

where $\Pi^*(s)$ is the Laplace transform of $\Pi(t)$, and $s\Pi^*(s) - \Pi(0) = s\Pi^*(s) - \mathbf{I}$ is therefore the transform of $\Pi'(t)$. If the modulus of s is large enough, then $\mathbf{Q}/s$ has spectral radius < 1, and so $\mathbf{I} - \mathbf{Q}/s$ is invertible. Furthermore, applying Corollary 5.6.16 of [Horn and Johnson, 1985], we have

$$\Pi^*(s) = \frac{1}{s}(\mathbf{I} - \mathbf{Q}/s)^{-1} = \sum_{n=0}^{\infty} \frac{\mathbf{Q}^n}{s^{n+1}}. \qquad (39)$$

Inverting (39) gives

$$\Pi(t) = \sum_{n=0}^{\infty} \frac{(\mathbf{Q}t)^n}{n!} = e^{\mathbf{Q}t}, \qquad (40)$$

which is the solution obtained before in (31).

Several approaches for finding transform equations satisfied by more complicated performance measures than the transition function $\Pi(t)$ involve conditioning on the time of the first transition and then applying the Laplace transform to the resulting equation. Below we follow these steps in the case of $\Pi(t)$ in order to illustrate the basic ideas. Although the final result is identical to that obtained above, the development serves as an introduction to the material covered in later sections of this chapter.

As before, consider the transition function for time t and condition on τ_1, the time of the first transition. Then we have

$$\pi_{ij}(t) = e^{-q_i t}\mathcal{I}\{i = j\} + \int_{\tau=0}^{t} q_i e^{-q_i \tau} \sum_{k \neq i} p_{ik} \pi_{kj}(t - \tau) d\tau, \qquad (41)$$

where $p_{ik} = q_{ik}/q_i$ ($i \neq k$) is the probability that, given a transition took place, it occurred from state s_i to s_k. Taking the Laplace transform of (41) and noting that the integral on the righthand side is a convolution, we obtain

$$\pi_{ij}^*(s) = \frac{1}{s+q_i}\mathcal{I}\{i=j\} + \frac{q_i}{s+q_i}\sum_{k\neq i}p_{ik}\pi_{kj}^*(s). \quad (42)$$

Equation (42) can be written in matrix form as

$$\Pi^*(s) = \text{diag}\{\mathbf{w}(s)\} + \text{diag}\{\mathbf{w}(s)\}\text{diag}\{\mathbf{q}\}\mathbf{P}\Pi^*(s), \quad (43)$$

where we have defined the vectors $\mathbf{w}(s) = [1/(s+q_i)]$ and $\mathbf{q} = [q_i]$. By definition, $p_{ii} = 0$ for all i, and so $\mathbf{P} = \mathbf{I} + \text{diag}\{\mathbf{q}\}^{-1}\mathbf{Q}$. Thus we have

$$[\mathbf{I} - \text{diag}\{\mathbf{w}(s)\}\text{diag}\{\mathbf{q}\} - \text{diag}\{\mathbf{w}(s)\}\mathbf{Q}]\Pi^*(s) = \text{diag}\{\mathbf{w}(s)\}.$$

Noting that $\mathbf{I} - \text{diag}\{\mathbf{w}(s)\}\text{diag}\{\mathbf{q}\} = \text{diag}\{\mathbf{w}(s)\}s$, we obtain

$$\text{diag}\{\mathbf{w}(s)\}(s\mathbf{I} - \mathbf{Q})\Pi^*(s) = \text{diag}\{\mathbf{w}(s)\}.$$

This yields equation (39), since $\text{diag}\{\mathbf{w}(s)\}$ is invertible when $s \neq -q_i$ for all i.

2.4 Krylov Subspace Method

Let $\mathcal{K}_m$ and $\mathcal{L}_m$ be two m-dimensional subspaces of $\mathcal{R}^M$, and suppose we are interested in solving the linear system $\mathbf{A}\mathbf{x} = \mathbf{b}$, where $\mathbf{A}$ is an $M \times M$ matrix. A general projection method finds an approximate solution $\tilde{\mathbf{x}}_m$ to this system by imposing the conditions that $\tilde{\mathbf{x}}_m \in \mathcal{K}_m$ and that the residual vector $\mathbf{b} - \mathbf{A}\tilde{\mathbf{x}}_m$ is orthogonal to $\mathcal{L}_m$. A Krylov subspace method is a projection method that uses the Krylov subspace $\mathcal{K}_m(\mathbf{A}, \mathbf{v}) \triangleq \text{span}\{\mathbf{v}, \mathbf{A}\mathbf{v}, \mathbf{A}^2\mathbf{v}\ldots, \mathbf{A}^{m-1}\mathbf{v}\}$ for the subspace $\mathcal{K}_m$. Roughly, the method approximates $\mathbf{A}^{-1}\mathbf{b}$ by $p_{m-1}(\mathbf{A})\mathbf{b}$, where p_{m-1} is a polynomial of degree $m-1$.

Our interest is in finding an approximation for $\pi(t) = \pi(0)\Pi(t) = \pi(0)e^{\mathbf{Q}t}$, where we have used the solution (31). To correspond to the above notation, we set $\mathbf{A} = \mathbf{Q}^T$, $\mathbf{v} = \pi(0)^T$, and for convenience we (temporarily) suppress the time variable t. We then seek an approximate solution $\tilde{\mathbf{x}}_m$ to $\mathbf{x} = e^{\mathbf{A}}\mathbf{v}$ of the form $p_{m-1}(\mathbf{A})\mathbf{v}$, so that $\tilde{\mathbf{x}}_m$ is an element of $\mathcal{K}_m(\mathbf{A}, \mathbf{v})$.

We start from Arnoldi's method (see Chapter 4) which is used to find an orthonormal basis $\mathbf{v}_1, \ldots, \mathbf{v}_m$ of the Krylov subspace $\mathcal{K}_m$ [Saad, 1995]. Let $\mathbf{V}_m$ be the $M \times m$ matrix with column vectors $\mathbf{v}_1, \ldots, \mathbf{v}_m$, and let $\mathbf{H}_m$ be the $m \times m$ Hessenberg matrix obtained from Arnoldi's method. The following equalities are an immediate consequence of the method

$$\mathbf{A}\mathbf{V}_m = \mathbf{V}_m\mathbf{H}_m + \mathbf{w}_m\mathbf{e}_m^T \quad (44)$$
$$\mathbf{V}_m^T\mathbf{A}\mathbf{V}_m = \mathbf{H}_m, \quad (45)$$

where $\mathbf{w}_m = \mathbf{A}\mathbf{v}_j - \sum_{i=1}^{j}h_{ij}\mathbf{v}_i$ (from Arnoldi's method) and $\mathbf{e}_i$ is the m dimensional vector with 1 in the i-th entry and 0 in all others.

If **v** (suitably normalized) is chosen as one of the column vectors of $\mathbf{V}_m$, then we can find the solution **x** by calculating $e^\mathbf{A} \mathbf{V}_m$. The key to the approximate solution $\tilde{\mathbf{x}}_m$ is the equation (see [Saad, 1992])

$$\mathbf{V}_m^T e^\mathbf{A} \mathbf{V}_m \approx e^{\mathbf{V}_m^T \mathbf{A} \mathbf{V}_m} = e^{\mathbf{H}_m}. \tag{46}$$

To motivate this approximation, note that if we expand $e^\mathbf{A}$ in an infinite series and then use (44) and (45), we can write $\mathbf{V}_m^T e^\mathbf{A} \mathbf{V}_m$ as $e^{\mathbf{H}_m}$ plus an infinite sum of terms that depend on $\mathbf{A}$, $\mathbf{H}_m$ and $\mathbf{V}_m$.

Now, choosing $\mathbf{v}_1 = \mathbf{v}/\|\mathbf{v}\|_2$ for Arnoldi's method, setting $\beta = \|\mathbf{v}\|_2$, noting that $\mathbf{V}_m^T \mathbf{V}_m = \mathbf{I}$ and using (46) we obtain

$$e^\mathbf{A} \mathbf{v} = e^\mathbf{A} \beta \mathbf{V}_m \mathbf{e}_1 \approx \beta (\mathbf{V}_m^T)^{-1} e^{\mathbf{H}_m} \mathbf{e}_1 = \beta \mathbf{V}_m e^{\mathbf{H}_m} \mathbf{e}_1. \tag{47}$$

(see also [Philippe and Sidge, 1995, Stewart, 1994].) Because of equalities (44) and (45), introducing t in (47) causes no problem.

From (47) we see that an approximation of the solution **x** can be obtained by evaluating the exponential of a (small) $m \times m$ matrix instead of the exponential of a large matrix. It can be shown [Saad, 1992] that the error of the procedure is bounded as

$$\|e^\mathbf{A}\mathbf{v} - \beta \mathbf{V}_m e^{\mathbf{H}_m} \mathbf{e}_1\|_2 \leq 2\beta \frac{\rho^m e^\rho}{m!}, \tag{48}$$

where $\rho = \|\mathbf{A}\|_2$. Consequently, the approximation is reasonable when $\|\mathbf{Q}^T t\|_2$ is small.

2.5 Uniformization

The uniformization technique, also known as the randomization technique and Jensen's method, is widely used to obtain $\pi(t)$. This technique has also been used to calculate a variety of transient measures. The method was proposed by Jensen [Jensen, 1953] and has become very popular in the last ten years (see [de Souza e Silva and Gail, 1996, Grassmann, 1991] and the references therein).

Consider a continuous-time Markov chain $\mathcal{X}$ with finite state space $\mathcal{S}$ of cardinality M and generator **Q**. Let $\Lambda \geq \max_i\{q_i\}$, and define the matrix $\mathbf{P} = \mathbf{I} + \mathbf{Q}/\Lambda$. Thus **P** is stochastic by choice of Λ, and from (31) and (32) we have

$$\pi(t) = \pi(0) \sum_{n=0}^\infty e^{-\Lambda t} \frac{(\Lambda t)^n}{n!} \mathbf{P}^n. \tag{49}$$

This equation can be evaluated recursively by defining

$$\mathbf{v}(n) = \pi(0) \mathbf{P}^n \tag{50}$$

and noting that $\mathbf{v}(n) = \mathbf{v}(n-1)\mathbf{P}$.

Equation (49) has several nice properties. Note that the operations in equation (49) involve only positive elements. This contrasts with the approaches we discussed so far, except for Euler's method given by (23), provided that h is sufficiently small. As a consequence, the numerical robustness of the method is very good. Furthermore, the recursion to calculate $\mathbf{v}(n)$ preserves the structure of the matrix $\mathbf{P}$, which is important in solving large problems when $\mathbf{P}$ is sparse. Another advantage is that the error when we truncate the infinite series of (49) is easily bounded by the tail of the Poisson distribution.

But perhaps, one of the main advantages of (49) is its probabilistic interpretation that has been used as the basis to calculate complex transient measures, such as transient cumulative distributions. Consider process $\mathcal{X}$ and let $\mathcal{X}'$ be constructed from $\mathcal{X}$ such that both processes have the same state space, the time in any state of $\mathcal{X}'$ until a transition occurs (either to the same or to a different state) is exponentially distributed with rate Λ, and the transition probability from a state i to j in $\mathcal{X}'$ is equal to q_{ij}/Λ. It is not difficult to see that: (a) the probability that $\mathcal{X}'$ transitions from state i to j, for $j \neq i$, is identical to the equivalent probability for process $\mathcal{X}$ and; (b) the residence time in state i before transitioning to another state is exponentially distributed with rate q_i, and thus the residence time has distribution identical to the distribution of the equivalent random variable for process $\mathcal{X}$. This last statement can be proved by noting that, given that $\mathcal{X}'$ made n jumps in state i before leaving i, the residence time in i has an Erlangian distribution, and that the number of jumps before leaving i is geometrically distributed. As a consequence, the two processes are equivalent.

Since the times between jumps in $\mathcal{X}'$ are all exponential with rate Λ, and because the one-step transition probability matrix corresponding to these jumps is $\mathbf{P} = \mathbf{I} + \mathbf{Q}/\Lambda$, we can think of $\mathcal{X}'$ as equivalent to a discrete-time Markov chain $\mathcal{Z} = \{Z_n : n = 0, 1, \ldots\}$ (with transition probability matrix $\mathbf{P}$) subordinated to a Poisson process $\mathcal{N} = \{N(t) : t \geq 0\}$ with rate Λ which is independent of $\mathcal{Z}$. In other words, $X(t) = X'(t) = Z_{N(t)}$. From this interpretation, if we condition on the number of transitions by time t, equation (49) is easily obtained. Note that $\mathbf{v}(n)$ is the transient state probability vector for $\mathcal{Z}$.

2.6 Remarks

We surveyed several approaches to obtain the state probabilities by time t, (or equivalently, to solve equation (3)). In general, Laplace transform methods are costly and prone to numerical problems, but are very useful if one can invert explicitly the transform equation (for recent progress in transform inversion, see Chapter 8). ODE solvers are costly since they may require a very small value of h to obtain accurate results. Nevertheless they have been used with success to solve non-homogeneous Markov models. Methods based on the exponential of a matrix may suffer from numerical problems and are also costly to solve. A promising approach is that based on Krylov subspaces.

Uniformization is simple to implement and numerically robust. It is the method of choice in many cases. However, it is costly for large Λt values and to obtain solutions for non-homogeneous Markov chain models. In [Grassmann, 1987, Reibman et al., 1989] a few computational approaches are examined in terms of accuracy and their computational cost (see also [Stewart, 1994] Table

8.5). Several of these methods, namely Laplace transform methods, ODEs and uniformization have been used to obtain other transient measures of interest, and in the following sections we discuss the approaches.

3 MISCELLANEOUS ISSUES

In the previous section we surveyed several methods to obtain the state probabilities at time t, for a finite state, continuous-time, time-homogeneous Markov chain. Even for this basic measure, several issues have to be overcome and are still subject to recent investigation. One of these issues is to solve *stiff* models, i.e., models with transition rates that differ in orders of magnitude. Another difficulty is to solve non-homogeneous Markov chain models. Most solutions proposed for stiff models are based on the uniformization technique. In the case of non-homogeneous Markov chains, ODE, Laplace transform methods and uniformization have been employed. We also consider models with infinite state space.

3.1 Stiff Models

Stiff models may greatly increase the computational cost of several techniques, including uniformization. For instance, in order to use equation (49) we need to truncate the infinite series. It is not difficult to see that the number of terms in (49) that must be computed to obtain the desirable accuracy is proportional to Λt. Stiff models give rise to large Λt values and consequently large number of operations. ODE solvers have similar problems, since in this case a very small value for h should be used.

We first consider the Uniformization Power method proposed by Abdallah and Marie [Abdallah and Marie, 1993]. Suppose we subdivide the observation period $(0, t)$ in sub-intervals of length $t_0 = t/2^n$. We choose n such that Λt_0 is small (say 0.1). Then the number of terms needed in (49) to calculate the state probabilities at t_0 (starting from a given state) is small. Let $t_k = 2^k t_0$ and recall that $\Pi(t_k)$ is the matrix whose (i,j) element is the probability that the Markov chain $\mathcal{X}$ is in state j at time t_k starting initially from state i. From the Chapman-Kolmogorov equations we have $\Pi(t_k) = \Pi(t_{k-1})^2$, and so $\pi(t_n) = \pi(0)\Pi(t_n)$ can be calculated by performing n matrix multiplications starting from $\Pi(t_0)$.

The procedure above is simple, and it requires a much smaller number of steps than uniformization for Λt very large. However, at each step, a matrix by matrix multiplication is performed unlike uniformization. One should also note that the matrix $\Pi(t_k)$ is not sparse even if the original generator $\mathbf{Q}$ is, and so the storage requirements may be excessive for large problems. This contrasts with uniformization which preserves the sparseness of the transition matrix of the model. Preserving sparseness also reduces the time complexity per iteration considerably since only the non-zero entries need to be evaluated.

Concerning the error tolerance, we should note that $\Pi(t_0)$ is obtained within a given error tolerance using (49). But we must evaluate the error after performing the matrix multiplications to calculate the desired result. In [Abdallah and Marie, 1993] an approach is developed to bound the final solution based on bounding the infinite norm of the error matrix that results from a matrix product.

Adaptive Uniformization [van Moorsel and Sanders, 1994] is another method that aims at reducing the computational requirements to calculate $\pi(t)$ for stiff models. Consider a Markov process $\mathcal{X}$. In [Grassmann, 1991] it was noted that the calculations performed at step n of the uniformization technique only consider the set of states that are reachable in n steps from a given initial set. Let $\mathcal{A}(n)$ be the set of reachable states in n steps (which are called *active states*) and $\mathbf{P}^{(n)}$ be the uniformized matrix corresponding to $\mathcal{A}(n)$. Adaptive Uniformization uses the notion of active states and, at step n, the largest output rate $\Lambda(n)$ is calculated from $\mathcal{A}(n)$. Since it is always true that $\Lambda(n) \leq \Lambda$, possible computational savings may be obtained by using $\mathbf{P}^{(n)}$ instead of $\mathbf{P}$. Similar to equation (49), we have

$$\mathbf{p}(t) = \sum_{n=0}^{\infty} W_n(t)\mathbf{v}(n) \qquad (51)$$

where $\mathbf{v}(n) = \mathbf{v}(n-1)\mathbf{P}^{(n)}$ and $W_n(t)$ is the probability of exactly n transitions by time t occurring in a pure birth process with possibly different rates for each state. It is not always easy to calculate $W_n(t)$, and the usefulness of the method to obtain large savings depends on the particular application problem.

A technique based on the theory of aggregation has also been proposed for dealing with stiff models [Bobbio and Trivedi, 1986]. The approach gives approximate results and will be surveyed in Section 5.

Consider equation (49), and recall that $\mathbf{v}(n) = \mathbf{v}(n-1)\mathbf{P}$. If $\mathbf{v}(n)$ approaches its steady state limit for $n = K < N$, where N is the truncation point of the infinite series in (49), then $\mathbf{v}(K+1) \approx \mathbf{v}(K)$ and we can rewrite (49) as

$$\pi(t) \approx \sum_{n=0}^{K-1} e^{-\Lambda t}\frac{(\Lambda t)^n}{n!}\mathbf{v}(n) + \left[1 - \sum_{n=0}^{K-1} e^{-\Lambda t}\frac{(\Lambda t)^n}{n!}\right]\mathbf{v}(K). \qquad (52)$$

This approach was proposed by Ciardo *et al* in [Ciardo et al., 1993] and is called steady state detection. Note that a large Λt value implies a large value of N (since $N \approx \Lambda t$ for Λt large). It was observed in [Ciardo et al., 1993] that, for many stiff problems, $\mathbf{v}(n)$ converges rapidly, and $K \ll N$. In this case, significant computational savings result when (52) is applied.

The main problem with this method is the difficulty in detecting convergence if the $\mathbf{v}(n)$ converge slowly [Ciardo et al., 1993], and error bounds are not computable unlike the standard uniformization method. However, if one is interested not in the state distribution at time t but in the probability that the model is in a subset of states by time t, error bounds can be easily obtained by performing little extra computation as follows [Sericola, 1999].

Let $\mathbf{y}(n) = \mathbf{P}^n\mathbf{1}_\mathcal{U}$ where the i-th entry of $\mathbf{1}_\mathcal{U}$ is equal to 1 if state i is in subset $\mathcal{U} \subset \mathcal{S}$ and 0 otherwise (recall $\mathcal{S}$ is the state space of the chain). Let

$m_n = \min_{i \in S}\{y_i(n)\}$, $M_n = \max_{i \in S}\{y_i(n)\}$. Since $y_i(n+1) = \sum_{j=1}^{M} p_{ij} y_j(n)$, it follows that $m_n \leq y_i(n+1) \leq M_n$ for all i, $m_{n+1} \geq m_n$, $M_{n+1} \leq M_n$, that is, the sequences m_n and M_n are non-decreasing and non-increasing, respectively. Now we note that $P\{X(t) \in \mathcal{U}\}$, the measure of interest, is obtained from $\mathbf{y}(n)$ by

$$P\{X(t) \in \mathcal{U}\} = \sum_{n=0}^{\infty} e^{-\Lambda t} \frac{(\Lambda t)^n}{n!} \pi(0) \mathbf{y}(n). \tag{53}$$

A bound on the final measure easily follows from $m_n \leq \pi(0)\mathbf{y}(n) \leq M_n$.

A method called Regenerative Randomization was developed by Carrasco and Calderón also for handling models with large Λt values [Carrasco and Calderón, 1995]. In order to present the intuition behind the approach, consider a continuous-time Markov chain $\mathcal{X}$ and select a state u as the *regeneration state*. Let $\mathcal{Z}$ be the discrete-time Markov chain obtained from $\mathcal{X}$ after uniformization, $\mathcal{W}$ be a discrete-time chain obtained from $\mathcal{Z}$ such that state i indicates that $\mathcal{Z}$ made i transitions since its last visit to u. Let $\mathcal{W}^*$ be the continuous-time chain for which the transition rates out of any state of $\mathcal{W}^*$ are equal to Λ and the state space and transition probabilities between any two states of $\mathcal{W}^*$ are the same as $\mathcal{W}$. Finally, construct an auxiliary transient chain $\mathcal{V}$ which is identical to $\mathcal{Z}$ except that u is an absorbing state. The transition probabilities of $\mathcal{W}$ can be easily obtained from $\mathcal{V}$. In [Carrasco and Calderón, 1995] it is shown that the probability vector $\pi(t)$ can be obtained from $\mathcal{W}^*$ and $\mathcal{V}$. Intuitively if we know that $\mathcal{W}^*$ is in state n by time t, then the probability that $\mathcal{X}$ is in state l by that time is equal to the probability that $\mathcal{V}$ is in l after n steps, given that it has not reached the absorbing state in n steps. The computational advantage of the approach over standard uniformization can be roughly explained as follows. If u is a frequently visited state, then the probability of making a large number of transitions between visits to u is small and so we need only to evaluate the probability of being in a state of $\mathcal{V}$ for a small number of steps, to obtain $\pi(t)$ within the desired tolerance. An important issue is the choice of the regeneration state u.

3.2 Infinite State Space

Thus far we have considered models with finite state space, but many examples with infinite state space can be found in the literature, for instance, queueing models with infinite buffer capacity. One approach to handle such models is to truncate the infinite state space and aggregate the remaining states into a single absorbing state. It is not difficult to see that the error introduced by this approximation is bounded by the probability that the system is in the absorbing state at the end of the observation period.

The work reported in [van Dijk, 1992a] considers the uniformization technique applied to infinite state space models and unbounded transition rates. The approach is based on both truncation of the state space and the transition rates (similar to the simple approach mentioned above). Then bounds are calculated following techniques as in [van Dijk, 1990].

3.3 Non-Homogeneous Markov Chains

A non-homogeneous continuous-time Markov chain has a time-dependent infinitesimal generator. Equation (3) is then modified to

$$\Pi'(t) = \mathbf{Q}(t)\Pi(t), \tag{54}$$

which is still of the type $y'(t) = f(t,y)$, with $y(t) = \Pi(t)$ and $f(t,y) = \mathbf{Q}(t)y$. Therefore, the solution methods based on ODEs are again applicable. For instance, the equation obtained with Euler's method is simply equation (23), but with $\mathbf{Q}$ replaced by $\mathbf{Q}(t-h)$.

When the changes in the transition rate matrix $\mathbf{Q}$ occur only at a finite number of time points $\tau_1, \tau_2, \ldots, \tau_K$, and all rates are constant during the intervals (τ_{i-1}, τ_i), $i = 1, \ldots, K$, we can immediately apply the uniformization technique. Let $\mathbf{P}(i)$ be the uniformized transition probability matrix corresponding to the interval (τ_{i-1}, τ_i). The state probability vector at time τ_i, $\pi(\tau_i)$, can be computed from $\pi(t_{i-1})$ using equation (49) as

$$\pi(\tau_i) = \sum_{n=0}^{\infty} e^{-\Lambda \delta_i} \frac{(\Lambda \delta_i)^n}{n!} \mathbf{v}(i,n), \tag{55}$$

where $\delta_i = \tau_i - \tau_{i-1}$, $\mathbf{v}(i,n) = \mathbf{v}(i, n-1)\mathbf{P}(i)$ and $\mathbf{v}(i,0) = \pi(\tau_{i-1})$.

The infinite series in (55) has to be truncated for each interval (say, at N_i), and it is not difficult to show that the total error for the interval $(0, t)$ is the sum of the truncation error for each interval ($\epsilon_i(N_i)$). Thus the values of N_i must be chosen such that the given error tolerance is greater than $\sum_i \epsilon_i(N_i)$.

In [van Dijk, 1992b] uniformization was also employed to obtain $\pi(t)$ for non-homogeneous Markov chains with time-dependent infinitesimal generator $\mathbf{Q}(t)$. The approach is interesting from a theoretical point of view, but computationally very costly. Assume that all output rates are bounded by Λ and let $\mathbf{P}(t)$ be the uniformized matrix obtained from $\mathbf{Q}(t)$. Conditioning on n transitions by time t and on the times $\tau_1, \ldots, \tau_n$ of these transitions, then $\Pi(t)|(n, \tau_1, \ldots, \tau_n) = \mathbf{P}(\tau_1)\mathbf{P}(\tau_2) \cdots \mathbf{P}(\tau_n)$. The joint distribution of $\tau_1, \ldots, \tau_n$ is equal to the distribution of the order statistics of a set of n random variables uniform on $(0, t)$, and the number of transitions in the observation interval has a Poisson distribution. Unconditioning we obtain an expression with multiple integrals, which is evaluated by numerical integration.

4 CALCULATION OF OTHER TRANSIENT MEASURES

In previous sections we have seen how to compute the state probability vector $\pi(t)$ using various methods. Of course, $\pi_{\mathcal{O}}(t)$ as defined in (7), where $\mathcal{O}$ is a given set of states, is easily computed once $\pi(t)$ has been calculated. An example for fault tolerant models is the point availability, which is the probability that the system is operational at time t (here $\mathcal{O}$ is the set of operational states). We now consider the calculation of other measures of interest. These include means, variances, moments and distributions of some of the random variables introduced in Section 1.

4.1 Expected Values

From (49) we have seen that uniformization can be used to calculate the state probability vector $\pi(t)$ in a numerically stable manner. This equation also has the following probabilistic interpretation. In (49), the index of summation n represents the number of transitions of the Poisson process $\mathcal{N}$ in the time interval $(0,t)$, and $\mathbf{v}(n)$ represents the (transient) state probability vector at the nth transition of the discrete-time Markov chain $\mathcal{Z}$. Thus, as mentioned at the end of Section 2.5, the equation (49) may be derived directly through probabilistic arguments. Expressions for other transient measures can also be obtained probabilistically and then efficiently calculated using the uniformization technique. For example, the expected fraction of time spent in state s_i during an observation period $(0,t)$ is, from (10) and (49)

$$E[\mathcal{F}_i(t)] = \frac{1}{t}\int_0^t \pi_i(\tau)d\tau = \sum_{n=0}^{\infty} e^{-\Lambda t}\frac{(\Lambda t)^n}{n!}\left[\frac{\sum_{j=0}^n v_i(j)}{n+1}\right]. \qquad (56)$$

To interpret this expression, suppose that n transitions occur during $(0,t)$. These transitions split the interval into $n+1$ subintervals, and since the transition times correspond to events from a Poisson process, the lengths of the subintervals are *exchangeable* random variables (see [Ross, 1983]). In particular, each subinterval has expected length $1/(n+1)$. Finally, the probability of being in state s_i during the jth subinterval, $j = 0, \ldots, n$, is simply $v_i(j)$.

Expectations for quantities in Markov reward models can also be obtained using uniformization. The expected instantaneous reward at time t (the point performability in fault tolerant terms) is

$$E[IR(t)] = \sum_{n=0}^{\infty} e^{-\Lambda t}\frac{(\Lambda t)^n}{n!}[\mathbf{r} \cdot \mathbf{v}(n)], \qquad (57)$$

while the expected cumulative reward averaged over the length t of the time interval $(0,t)$ is

$$E[ACR(t)] = \sum_{n=0}^{\infty} e^{-\Lambda t}\frac{(\Lambda t)^n}{n!}\left[\frac{\sum_{j=0}^n \mathbf{r} \cdot \mathbf{v}(j)}{n+1}\right]. \qquad (58)$$

This latter expression can also be written in terms of the $n+1$-stage Erlangian distribution $E_{n+1,\Lambda}(t) = 1 - \sum_{j=0}^n e^{-\Lambda t}(\Lambda t)^j/j!$ as

$$E[ACR(t)] = \frac{1}{\Lambda t}\sum_{n=0}^{\infty} E_{n+1,\Lambda}(t)[\mathbf{r} \cdot \mathbf{v}(n)]. \qquad (59)$$

According to [Grassmann, 1987], (59) converges faster than (58). Note that, given a specified error tolerance ϵ, an integer $N = N(\epsilon)$ based on the Poisson distribution can be found so that $E[ACR(t)]$ can be calculated to within ϵ by truncating the infinite series in (58) at N. Of course, the same holds true when calculating $E[IR(t)]$ using (57). This ability to guarantee the amount of error that will occur is another advantage of the uniformization technique.

4.2 Moments

Grassmann [Grassmann, 1987] proposed an efficient algorithm to evaluate the second moment (and the variance) of the total accumulated reward in $(0,t)$, averaged over t, of a Markovian process, i.e. $ACR(t)$ defined in (13). First, note that

$$ACR(t)^2 = \left[\frac{1}{t}\int_0^t r_{c(X(\tau))}d\tau\right]^2 = \frac{2}{t^2}\int_{\tau=0}^t \int_{\gamma=0}^\tau r_{c(X(\gamma))}r_{c(X(\tau))}d\gamma d\tau \quad (60)$$

By taking expectations on both sides of (60) and using the following equation, obtained from similar arguments as in (49)

$$E[r_{c(X(\gamma))}r_{c(X(\tau))}] = \sum_{j=0}^\infty \sum_{i=0}^j E[r_{c(Z_i)}r_{c(Z_j)}]e^{-\Lambda\gamma}\frac{(\Lambda\gamma)^i}{i!}e^{-\Lambda(\tau-\gamma)}\frac{(\Lambda(\tau-\gamma))^j}{j!} \quad (61)$$

(where we recall that Z_i is the state of the model in the i-th step after uniformizing the process $X(t)$) we get, after evaluating the integrals in (60),

$$E[ACR(t)^2] = \frac{2}{(\Lambda t)^2}\sum_{j=0}^\infty E_{j+2,\Lambda}(t)D_j \quad (62)$$

where $D_j = \sum_{i=0}^j E[r_{c(Z_i)}r_{c(Z_j)}]$. A recursive expression for calculating D_j is obtained in [Grassmann, 1987] as well as an error bound when the infinite series in (62) is truncated.

Grassmann also considered Markov reward systems in [Grassmann, 1993]. He obtained a recursive expression for $Var[\sum_{n=0}^m r_{c(Z_n)}]$ using similar arguments as above. He also found linear equations to calculate the difference of the expected reward accumulated in an interval from the steady state value.

We now consider higher moments of the reward accumulated during $(0,t)$. Let the random variable $CR_i(t) = CR(t)|X(0) = i$ be the reward accumulated in $(0,t)$ starting in state i. The nth moment of $CR_i(t)$ is $m_i^n(t) = E[(CR_i(t))^n]$ and $\mathbf{m}^n(t) = [m_1^n(t), \ldots, m_M^n(t)]$ is the corresponding vector.

Several methods can be employed to calculate $\mathbf{m}^n(t)$. We start by finding an expression for the Laplace transform of $F_i(y,t) = P\{CR_i(t) \le y|X(0) = i\}$. Let us condition on τ_1, the time of the first transition. Similar to equation (41), we have

$$F_i(y,t) = e^{-q_i t}u(y - r_i t) + \int_{\tau=0}^t q_i e^{-q_i \tau}\sum_{k=1}^M p_{ik}F_j(y - r_i\tau, t - \tau)d\tau, \quad (63)$$

where $u(x)$ is the unit step function. Taking the Laplace-Stieltjes transform of $F_i(y,t)$ in the y variable, and then on the t variable, we have

$$h_i^{**}(s,\delta) = \frac{1}{q_i + \delta + r_i s} + \frac{q_i}{q_i + \delta + r_i s}\sum_{j=1}^M p_{ij}h_j^{**}(s,\delta). \quad (64)$$

If we write (64) in matrix form and recall that $\mathbf{P} = \mathbf{I} + (\text{diag}\{\mathbf{q}\})^{-1}\mathbf{Q}$, we obtain

$$\mathbf{h}^{**}(s, \delta) = [\delta \mathbf{I} + s\,\text{diag}\{\mathbf{r}\} - \mathbf{Q}]^{-1}\mathbf{1}, \tag{65}$$

where $\mathbf{1}$ is the vector with all entries equal to one. Equation (65) was first obtained by Puri [Puri, 1971] and later by Kulkarni et al [Kulkarni et al., 1986]. Since $e^{\mathbf{A}t}$ is the inverse of the transform $[\delta \mathbf{I} - \mathbf{A}]^{-1}$, we can invert (65) in the δ parameter [Iyer et al., 1986]

$$\mathbf{l}^{*}(s) = \left[e^{\mathbf{Q}t - s\,\text{diag}\{\mathbf{r}\}t} \right] \mathbf{1}. \tag{66}$$

We return to the calculation of the moments. We first rewrite (65) as

$$[\delta \mathbf{I} + s\,\text{diag}\{\mathbf{r}\} - \mathbf{Q}]\,\mathbf{h}^{**}(s,\delta) = \mathbf{1},$$

then we differentiate $n+1$ times with respect to s and set $s = 0$

$$\mathbf{m}^{(*)n+1}(\delta) = (n+1)(\delta \mathbf{I} - \mathbf{Q})^{-1}\,\text{diag}\{\mathbf{r}\}\mathbf{m}^{(*)n}(\delta), \tag{67}$$

where $\mathbf{m}^{(*)n}(\delta)$ is the vector $(-1)^n \partial^n/\partial s^n \mathbf{h}^{**}(s,\delta)|_{s=0}$ and so it is equal to the Laplace transform of $\mathbf{m}^n(t)$. Using the same observation that led to equation (66), we can invert (67) and obtain

$$\mathbf{m}^{n+1}(t) = (n+1) \int_0^t e^{\mathbf{Q}(t-\tau)}\,\text{diag}\{\mathbf{r}\}\mathbf{m}^n(\tau)d\tau, \tag{68}$$

where $\mathbf{m}^0(t) = \mathbf{1}$.

Pattipati and Shah [Pattipati and Shah, 1990] also found a recursion for the moments of $CR(t)$, but for non-homogeneous Markov chains. They obtained a differential equation for the moments in the following manner. Taking the derivative of $(CR(t))^n = (\int r_{c(X(s))})^n ds$, taking the expected value of $(CR(t))^n$ and using (3) yields

$$-\frac{d\mathbf{m}^n(t)}{dt} = \mathbf{Q}(t)\mathbf{m}^n(t) + n\,\text{diag}\{\mathbf{r}\}\mathbf{m}^{n-1}(t). \tag{69}$$

Note the similarity between (69) and (68) for the homogeneous case. ODE solvers can be employed for calculating the moments from equation (69).

4.3 Distributions

We consider the calculation of distributions in the reward case for various random variables of interest. Pattipati et al [Pattipati et al., 1993] obtained the distribution of the total accumulated reward in $(0, t)$ (the distribution of performability in dependability modeling terms), for non-homogeneous Markov chains and for time varying reward rates, by solving a partial differential equation they derived. Let $\mathbf{F}(y, t)$ be the joint distribution matrix such that

$F_{ij}(y,t) = P\{CR(t) \leq y, X(t) = j|X(0) = i\}$ and $\mathbf{W}(y,t)$ be the corresponding density matrix. An equation for $\mathbf{W}(y,t) = P\{CR(t) = y, X(t) = j|X(0) = i\}$ was obtained as follows. Using Bayes' rule we obtain

$$P\{CR(t) = y, X(t) = j|X(0) = i\} =$$
$$\int_z \sum_{k=1}^{M} P\{CR(t) = y, X(t) = j|CR(t-h) = z, X(t-h) = k, X(0) = i\}$$
$$P\{CR(t-h) = z, X(t-h) = k|X(0) = i\}dz. \tag{70}$$

From the Markov property and noting that $X(t)$ and $CR(t)$ are independent random variables given $(CR(t-h), X(t-h))$ for $h \to 0$, and also that $X(t)$ is independent of $CR(t-h)$ given $X(t-h)$, we have

$$P\{CR(t) = y, X(t) = j|X(0) = i\} =$$
$$\int_z \sum_{k=1}^{M} P\{CR(t) = y|CR(t-h) = z, X(t-h) = k\}$$
$$P\{X(t) = j|X(t-h) = k\}$$
$$P\{CR(t-h) = z, X(t-h) = k|X(0) = i\}dz. \tag{71}$$

But, for h small, $CR(t) = CR(t-h) + hr_{c(X(t-h))} + o(h)$ and so

$$P\{CR(t) = y|CR(t-h) = z, X(t-h) = k\} = \delta(y - (z + hr_{c(k)}(t-h)) + o(h)) \tag{72}$$

where $\delta()$ is the Dirac delta function and $r_{c(k)}(t-h)$ is the reward rate at time $t-h$ associated with state k. Using (72) in (70)

$$W_{ij}(y,t) =$$
$$\int_z \sum_{k=1}^{M} W_{ik}(z, t-h)[\delta_{kj} + q_{kj}(t-h)h + o(h)]$$
$$\delta(y - (z + hr_{c(k)}(t-h)) + o(h))dz. \tag{73}$$

Now, noting that

$$\int_z W_{ik}(z, t-h)\delta(y - (z + hr_{c(k)}(t-h)) + o(h))dz = W_{ik}(y - hr_{c(k)}(t-h), t-h), \tag{74}$$

and after simple algebraic manipulations and taking the limit as $h \to 0$, we obtain

$$\frac{\partial \mathbf{W}(y,t)}{\partial t} = -\frac{\partial \mathbf{W}(y,t)}{\partial y}\mathbf{R}(t) + \mathbf{W}(y,t)\mathbf{Q}(t), \tag{75}$$

where $\mathbf{R}(t) = \text{diag}\{[r_1(t), \ldots, r_M(t)]\}$. The equation for $\mathbf{F}(y,t)$ is identical to (75) with $\mathbf{W}(y,t)$ replaced by $\mathbf{F}(y,t)$.

We now discuss the methodology introduced in [de Souza e Silva and Gail, 1989], and extended in [de Souza e Silva and Gail, 1998], based on uniformization for the homogeneous case. Consider a rate plus impulse reward model, and suppose that the distribution of $ACIR(t)$, the total reward averaged over t that has been accumulated during $(0,t)$, is to be calculated. To calculate $P\{ACIR(t) \leq r\}$, note that if a total impulse reward of $\widehat{r}$ is accumulated during $(0,t)$, then at most $r-\widehat{r}$ must be "made up" from the rate rewards. That is, we may write $P\{ACIR(t) \leq r|\widehat{r}\} = P\{ACR(t) \leq r-\widehat{r}\}$. To find the distribution of cumulative rate reward, we condition on n transitions occurring during the interval $(0,t)$. As discussed above, the transitions divide the interval into $n+1$ subintervals, the lengths of which are exchangeable random variables. Each subinterval has a reward associated with it corresponding to the state of the discrete-time Markov chain $\mathcal{Z}$. Let $\mathbf{k} = [k_1, \ldots, k_{K+1}]$ record the number of subintervals that have reward r_i, $i = 1, \ldots, K+1$, so that $\|\mathbf{k}\| = n+1$ where $\|\mathbf{k}\| = k_1 + \cdots + k_{K+1}$. The vector $\mathbf{k}$ is called a rate coloring, since subintervals can be thought of as being assigned a color based on its reward. Because the subinterval lengths are exchangeable, the distribution of the cumulative reward does not depend on the order in which the rewards are earned, but only on the number of subintervals associated with each reward (i.e., only on the rate coloring).

For $n = 0, 1, \ldots$, define $\mathcal{K}_n = \{\mathbf{k} : \|\mathbf{k}\| = n+1\}$. Conditioning on n transitions and a rate coloring $\mathbf{k} \in \mathcal{K}_n$, it is shown in [de Souza e Silva and Gail, 1989] that the conditional distribution of $ACR(t)$ is given in terms of a linear combination of uniform order statistics on $(0,1)$. Weisberg [Weisberg, 1971] determined the distribution of such a linear combination, and using his result it is shown in [de Souza e Silva and Gail, 1998] that

$$P\{ACR(t) > r|n, \mathbf{k}\} = \sum_{i:r_i > r} f_i[r_i, n, \mathbf{k}, r, k_i - 1] \qquad (76)$$

$$P\{ACR(t) = r|n, \mathbf{k}\} = \sum_{i:r_i = r} f_i[r_i, n, \mathbf{k}, r, k_i - 1] \qquad (77)$$

$$P\{ACR(t) < r|n, \mathbf{k}\} = \sum_{i:r_i < r} f_i[r_i, n, \mathbf{k}, r, k_i - 1]. \qquad (78)$$

Here, for $n \geq 0$, $\mathbf{k} \in \mathcal{K}_n$, $l \geq 0$, $i = 1, \ldots, K+1$, the functions f_i are defined as

$$f_i[x, n, \mathbf{k}, r, l] = \frac{1}{l!} \cdot \frac{d^l}{dx^l} \left\{ \frac{(x-r)^n}{\prod_{\substack{j=1 \\ j \neq i}}^{K+1} (x - r_j)^{k_j}} \right\}. \qquad (79)$$

Unconditioning on n, $\mathbf{k}$ and the total impulse reward $\widehat{r}$, we have

$$P\{ACIR(t) \leq r\} =$$
$$\sum_{n=0}^{\infty} e^{-\Lambda t} \frac{(\Lambda t)^n}{n!} \sum_{\widehat{r} \leq r} \sum_{\mathbf{k} \in \mathcal{K}_n} \Theta[n, \mathbf{k}, \widehat{r}] \sum_{i:r_i \leq r-\widehat{r}} f_i[r_i, n, \mathbf{k}, r - \widehat{r}, k_i - 1]. \qquad (80)$$

Here $\Theta[n, \mathbf{k}, \widehat{r}]$ is the probability, given n transitions, of a rate coloring $\mathbf{k}$ and an average accumulated impulse reward of $\widehat{r}$. Specializing to the rate reward case yields

$$P\{ACR(t) \leq r\} = \sum_{n=0}^{\infty} e^{-\Lambda t}\frac{(\Lambda t)^n}{n!} \sum_{\mathbf{k} \in \mathcal{K}_n} \Theta[n, \mathbf{k}] \sum_{i: r_i \leq r} f_i[r_i, n, \mathbf{k}, r, k_i - 1], \quad (81)$$

where we have used the notation $\Theta[n, \mathbf{k}] = \Theta[n, \mathbf{k}, 0]$. Once recursions for Θ and f_i are obtained, the distribution of $ACR(t)$ can be calculated using (81).

Let us first consider the case of two rewards [de Souza e Silva and Gail, 1986], $r_1 = 1$, $r_2 = 0$ (for example, the dependability case). Let $\mathcal{O}$ be the subset of states which have reward rate 1 ($\mathcal{O}$ consists of the operational states in fault tolerant system models). In this case, a rate coloring $\mathbf{k}$ can be identified with the number of intervals, say k, with reward 1. The distribution in (81) becomes for $0 < r < 1$ (note that if $k = n + 1$ then $ACR(t) = 1$)

$$P\{ACR(t) \leq r\} = \sum_{n=0}^{\infty} e^{-\Lambda t}\frac{(\Lambda t)^n}{n!} \sum_{k=0}^{n} \Theta[n, k] \sum_{i=k}^{n} \binom{n}{i} r^i (1-r)^{n-i}. \quad (82)$$

Now $\Theta[n, k] = \sum_{s \in \mathcal{S}} \Theta_s[n, k]$, where we define $\Theta_s[n, k]$ to be the probability, given n transitions, of k intervals with reward 1 and the last state visited is s. If the recursive calculations are organized in tabular form, so that n corresponds to a row of the table and k corresponds to a column, then we already know that a row truncation point N can be specified in advance to meet any given error tolerance. Rubino and Sericola [Rubino and Sericola, 1993] found a way to also truncate the columns at a point $k = C$ in advance to a prespecified error tolerance. For models of highly available systems, it is likely that most intervals will correspond to operational states and thus have reward 1. As a result few columns should be calculated before reaching the desired error bound.

Although (81) yielded efficient numerical procedures in the case of two rewards, note that the evaluation of the sums over $\mathbf{k} \in \mathcal{K}_n$ required by a direct application of (81) leads to algorithms that are exponential in the number of rewards [de Souza e Silva and Gail, 1989]. However, recently this approach has been modified in [de Souza e Silva and Gail, 1998] to obtain recursions that greatly reduce the computational complexity and, in fact, are polynomial in the number of rewards. Instead of developing recursive schemes based on the rate colorings $\mathbf{k}$, these vectors are aggregated and recursions are found involving the resulting sets of vectors. Specifically, for $i = 1, \ldots, K + 1$, define the partition $G_g[i, n] = \{\mathbf{k} \in \mathcal{K}_n : k_i = g\}$, $g = 0, \ldots, n + 1$. Then $G_g[i, n]$ represents all possible paths for which there are exactly g intervals with reward rate r_i. Using these partitions, define for $s \in \mathcal{S}$, $i = 1, \ldots, K+1$, $n = 0, 1, \ldots$, $u = 0, \ldots, n$,

$$\Upsilon_s[i, n, u] = \sum_{g=n+1-u}^{n+1} \sum_{\mathbf{k} \in G_g[i,n]} \Theta_s[n, \mathbf{k}] f_i[r_i, n, \mathbf{k}, r, g - (n + 1 - u)]. \quad (83)$$

It is shown in [de Souza e Silva and Gail, 1998] that

$$P\{ACR(t) > r\} = \sum_{n=0}^{\infty} e^{-\Lambda t}\frac{(\Lambda t)^n}{n!} \sum_{i: r_i > r} \|\Upsilon[i, n, n]\| \quad (84)$$

$$P\{ACR(t) = r\} = \sum_{n=0}^{\infty} e^{-\Lambda t} \frac{(\Lambda t)^n}{n!} \sum_{i:r_i=r} \|\Upsilon[i,n,n]\| \qquad (85)$$

$$P\{ACR(t) < r\} = \sum_{n=0}^{\infty} e^{-\Lambda t} \frac{(\Lambda t)^n}{n!} \sum_{i:r_i<r} \|\Upsilon[i,n,n]\|. \qquad (86)$$

Here the vector $\Upsilon[i,n,u] = [\Upsilon_{s_1}[i,n,u], \ldots, \Upsilon_{s_M}[i,n,u]]$. A simple recursion for $\Upsilon_s[i,n,u]$ is derived in [de Souza e Silva and Gail, 1998] (see also [de Souza e Silva et al., 1995a]).

Models that incorporate both rate and impulse based rewards were first introduced by Qureshi and Sanders [Qureshi and Sanders, 1994]. They extended the methodology from [de Souza e Silva and Gail, 1989] by incorporating the notion of an impulse coloring. By conditioning on both rate and impulse colorings, they derived recursions for the rate plus impulse model. However, these algorithms are exponential not only in the number of rate rewards, but also in the number of impulse rewards.

In [de Souza e Silva and Gail, 1989] it was shown that obtaining the distribution of $ACR(t)$ can be reduced to finding the distribution of a linear combination of uniform order statistics on $(0,1)$. As mentioned above, Weisberg developed a recursive method to calculate the distribution of this linear combination. Later, Matsunawa [Matsunawa, 1985] derived a closed form expression for such a distribution that involves mixtures of beta functions. Qureshi and Sanders used the results of Matsunawa to calculate the distribution of $ACR(t)$, but their recursion is based on rate colorings, and so it is exponential in the number of rate rewards. They also introduced a procedure based on the concept of most probable paths, which considerably reduces the computational complexity of their method in some cases. However, this technique seems to be problem specific, and any potential computational savings cannot be determined in advance.

Another approach to the rate reward case (no impulse rewards present in the model) was developed by Donatiello and Grassi in [Donatiello and Grassi, 1991] and combines both the uniformization technique and Laplace transform methods. Instead of directly transforming the original Markov chain $\mathcal{X}$, it is first uniformized, i.e., it is considered to be of the form $X(t) = Z_{N(t)}$. As before, an equation is found for $F_i(y,t) = P\{CR(t) \le y|X(0) = i\}$ by conditioning on the time τ_1 of the first transition. Conditioning on n transitions in $(0,t)$, we have

$$F_i(y,t) = \sum_{n=0}^{\infty} e^{-\Lambda t} \frac{(\Lambda t)^n}{n!} F_i(y,t|n). \qquad (87)$$

Since the transitions of the uniformized chain $\mathcal{X}$ are given by a Poisson process of rate Λ, the random variable τ_1, given n transitions, has the same distribution as the minimum of n independent uniform random variables on $(0,t)$. Thus, using this density for τ_1, we obtain

$$F_i(y,t|n) = \int_0^t \sum_{j=1}^M p_{ij} F_j(y - r_i\tau, t - \tau|n-1) \frac{n}{t}\left[1 - \frac{\tau}{t}\right]^{n-1} d\tau. \qquad (88)$$

Taking the Laplace-Stieltjes transform of (88) on the y and t variables, simplifying the result using partial fractions, and inverting gives an expression for $F_i(y,t|n)$ in terms of quantities that can be calculated in a recursive manner. Using this approach, Donatiello and Grassi were the first to show that the distribution of $CR(t)$ could be computed with an algorithm that is polynomial in the number of rate rewards.

Recently Nabli and Sericola [Nabli and Sericola, 1996] found a new polynomial algorithm in the rate reward case. They also uniformize the chain and condition on the time τ_1 of the first transition, which yields an integral renewal equation. A recursion is then found after integrating by parts the equation obtained from conditioning on τ_1. Specifically, they consider the complementary distribution $\widetilde{F}(y,t) = P\{CR(t) > y\}$, which satisfies an equation similar to (87), and show given n transitions in $(0,t)$ that

$$\widetilde{F}(y,t|n) = \sum_{k=0}^n \sum_{j=1}^m \binom{n}{k} y_j^k (1-y_j)^{n-k} b^{(j)}(n,k) \mathcal{I}\{r_{j-1}t \leq y < r_j t\}. \qquad (89)$$

Here $y_j = (y - r_{j-1}t)/(r_j t - r_{j-1}t)$ and the coefficients $b^{(j)}(n,k)$ are given recursively. This method allows for error tolerances to be specified in advance, as is the case for algorithms based solely on the uniformization technique.

The algorithms of [de Souza e Silva et al., 1995a], [Donatiello and Grassi, 1991] and [Nabli and Sericola, 1996] were developed for the rate reward case. Although polynomial in nature, they have somewhat different costs. For instance, the storage requirements in [Donatiello and Grassi, 1991] and [Nabli and Sericola, 1996] may be as large as $O(KMN)$, where K is the number of distinct reward rates, M is the cardinality of the state space, and N is the truncation point of the infinite series in the algorithms, while the storage cost of the algorithm in [de Souza e Silva et al., 1995a] is $O(MN)$. The algorithm of [de Souza e Silva et al., 1995a] is based on probabilistic arguments, while the recursion of [Nabli and Sericola, 1996] involves only real numbers in the interval $[0,1]$ and, as such, has nice numerical properties.

5 APPROXIMATIONS

Most methods surveyed in previous sections either produce error bounds for the measure of interest or obtain exact Laplace transform expressions for the specific measure, or partial differential equations. In this section we briefly discuss a few techniques based on approximations lacking error bounds.

One example of such an approximation is based on the theory of aggregation to deal with stiff models [Bobbio and Trivedi, 1986]. In this approach, the states of the model are divided into two sets: the set of *fast* states which contains states with at least one transition with a high rate (i.e., a rate that is much greater than $1/t$, where t is the length of the observation period); and the set of *slow* states which are those that do not have any transitions with a high rate. All transitions with slow rates are eliminated from the set of fast states. From the remaining state transition diagram, the states are subdivided into two subsets called fast recurrent and fast transient, according to the standard terminology of recurrent and transient states for Markov chains. The key assumption is that the subset of fast recurrent states reaches steady state by time t, and then this subset is aggregated. Fast transient states are eliminated by standard techniques and the transient solution for the slow states is found from the aggregate model. The solution for the fast states is found from a disaggregation step.

Ross [Ross, 1987] developed an approximate equation for calculating $\pi_{ij}(t)$ assuming the observation interval $(0,t)$ is not deterministic, but instead it is a random variable with an Erlangian distribution. Consider a continuous-time Markov chain $\mathcal{X}$ and let $\mathcal{E}$ be an event which occurs at times $\tau_1, \tau_2, \ldots$, where the intervals $\tau_i - \tau_{i-1}$, for $i > 0$, are independent and identically distributed random variables with exponential rate λ and also independent of process $\mathcal{X}$. Assume $\mathcal{X}$ is initially in state i. Then either a transition of $\mathcal{X}$ occurs before event $\mathcal{E}$, or $\mathcal{E}$ occurs first. Clearly, the first event occurs with probability $q_i/(q_i + \lambda)$ and the second with probability $\lambda/(q_i + \lambda)$. From the memoryless property of the exponential distribution and conditioning on which event occurs first we have

$$\pi_{ij}(\tau_1) = \frac{q_i}{q_i + \lambda} \sum_{k \neq i} \pi_{kj}(\tau_1) \frac{q_{ik}}{q_i} + \frac{\lambda}{q_i + \lambda} \delta_{ij}, \qquad (90)$$

where $\delta_{ii} = 1$ and $\delta_{ij} = 0$ for $i \neq j$. Rewriting (90) in matrix form gives

$$\mathbf{\Pi}(\tau_1) = \left(\mathbf{I} - \frac{\mathbf{Q}}{\lambda}\right)^{-1}. \qquad (91)$$

From the Chapman-Kolmogorov equations

$$\mathbf{\Pi}(\tau_r) = \left[\left(\mathbf{I} - \frac{\mathbf{Q}}{\lambda}\right)^{-1}\right]^r. \qquad (92)$$

The parameter r is chosen to approximate λt. Furthermore, if r is chosen as a power of 2, $\mathbf{\Pi}(\tau_r)$ can be calculated by performing $\log_2 r$ matrix multiplications, similar to the uniformized power method of Section 3. In fact these two approaches are related. The intervals corresponding to the step size in Ross' method are exponentially distributed, while they are constant in the uniformized power method.

Ross provides reasonable evidence that the approach is accurate, even if a small number of steps is chosen for the recursion. Observe also that, if one is interested in computing $\pi_j(t) = [\pi_{1j}(t), \ldots \pi_{Mj}(t)]$, then only r vector by matrix operations are needed. To see this, we simply multiply both sides of (92) by $\mathbf{1}_j$.

Yoon and Shanthikumar [Yoon and Shanthikumar, 1989] compared several approximations for $\Pi(t)$. Most of them are based on discretizing time and using the Chapman-Kolmogorov equations, and so they have the same flavor as the uniformized power method. They also were able to bound the error of the approximations, but the bounds are based on eigenvalue computation and are expensive to obtain for large models. However, they serve to illustrate the error behavior of such approximations.

The work reported in [Carmo et al., 1996] proposed efficient solutions to compute state probabilities at time t based on Ross' approximation. Their approach is very efficient for stiff models, and much cheaper than the uniformization technique in these cases. They also show how to compute expected performability measures using the same methodology.

6 APPLICATION EXAMPLES

Quoting Grassmann in [Grassmann, 1991], "In this world everything is in constant change, and equilibrium solutions are therefore nonexistent". As a consequence, one is always interested in transient solutions, but in general they are more difficult to obtain than equilibrium solutions. This is probably the main reason why many more studies are based on equilibrium analysis. Equilibrium solutions are good approximations of the transient behavior of the system, provided that the convergence to steady state is fast in comparison to the observation interval, but many examples exist for which it is imperative to perform transient analysis. In this section we present several application examples that use transient analysis.

One of the major areas that requires analyzing a system during a relatively "short" observation interval is dependability modeling [Trivedi, 1982]. Among the measures of interest are the reliability, point availability, cumulative operational time, etc. Many tools have been built that include transient solution techniques (e.g., [Goyal et al., 1986, Johnson Jr. and Malek, 1988, Stewart, 1991, Sanders et al., 1995, Carmo et al., 1998, Lindemann, 1992, Ciardo et al., 1989]).

With the advent of gracefully degradable systems, structural changes that can occur in order to maintain operation after a fault may cause the system to operate at different levels of performance. Instead of developing separate models to study the effect of faults on the ability of the system to perform useful work (dependability modeling) and to study the system performance under a given configuration (performance modeling), Meyer [Meyer, 1980] proposed a modeling framework to study the effect of configuration changes on the system performance. A set of new measures called *performability* measures were proposed. Roughly, as proposed by Meyer, performability is the set of measures used to quantify the ability of the system to perform at a particular level, in a given subset of the possible performance levels, during a specified time interval. Many of these measures are defined in Section 1.

A considerable body of work has been published that deals with performability modeling (e.g., see [Meyer, 1992, Meyer, 1995, de Souza e Silva and Gail, 1992] and the references therein). Many interesting application models have appeared in the literature. Early work includes the paper [Meyer et al., 1980], in which a fault-tolerant aircraft computer was modeled. An example of a multiprocessor model was proposed in [Meyer, 1982] and used later by many authors. The model consists of a number of processors that execute tasks that are stored in a buffer after arrival. Both the processors and the buffer are subject to failures which affect the performance of the system. Another multiprocessor model can be found in [Lopez-Benitez and Trivedi, 1993]. Several other application examples exist, such as real-time systems [Islam and Ammar, 1991], database systems [Tai and Meyer, 1996] and communication protocols [Prodromides and Sanders, 1993]. See also [Meyer, 1995] for an extensive list of applications.

Sensitivity analysis of a measure with respect to model parameters plays an important role in studying the effect of uncertainties in estimating parameter values, and it is also useful in determining the influence of the model parameters for the measure being considered. Heidelberger and Goyal [Heidelberger and Goyal, 1988] showed how to calculate the derivative of $\pi(t)$ with respect to a parameter of the transition rate matrix. Their solution is based on uniformization, and examples taken from dependability modeling were presented.

Queueing systems have been extensively used in computer and communication models. A discussion of several methods for computing transient measures of the M/M/1 queue appears in [Abate and Whitt, 1989], and recently in [Leguesdron et al., 1993] an expression for the point probabilities at time t was found based on uniformization and generating functions. The transient queue length and workload distributions for the M/G/1 and BMAP/G/1 (batch Markovian arrival process) queues were considered in [Choudhury et al., 1993] and [Lucantoni et al., 1994], respectively. Markovian queues such as M/M/1, $M/E_r/1$, tandem queues, etc., were solved in [Grassmann, 1977b] using uniformization. Grassmann [Grassmann, 1982] applied transient analysis to find the transition probability matrix for the GI/PH/1 queue. The matrix entries were obtained by constructing an embedded Markov chain at arrival instants. When the time between arrivals is constant, equation (49) can be immediately applied to obtain the transition probability from one state to another from the Markov chain modeling the departure process. Randomly distributed interarrival times can be handled by first conditioning on the time between arrivals and then using (49) (see [Grassmann, 1982] for details).

Transient analysis was also used to study the dynamic characteristics of complex communication systems. In [van As, 1986] van As was concerned with the foreground-background congestion control mechanism in a packet switching network under a two-level global control mechanism. He illustrated the potential of a priority deadlock in a queueing system with common storage. The measures of interest calculated were based on the state probabilities at time t, and the Runge-Kutta ODE technique was used as the basic solution method.

Ren and Kobayashi [Ren and Kobayashi, 1995] and Tanaka *et al* [Tanaka et al., 1995] consider the transient solution of a fluid flow queueing model of an ATM switch. Their fluid flow model consists of a queue fed by sources with variable rate, modulated by a Markov chain. They obtain a partial differential equation for the distribution of the queue size at time t as follows.

Consider a continuous-time Markov chain $\mathcal{X}$ with generator $\mathbf{Q}$ and associate an arrival rate λ_i to state i. This Markov chain with rewards is a model of the arrival rate process. Let $q(t)$ be the number of bits in the queue at time t, and let $F_i(t,r) = P\{X(t) = i, q(t) \leq r\}$. Given that $X(t) = i$, in $t + \Delta t$ either a phase transition occurs and the next state is j (with probability $q_{ij}\Delta t$) or no transitions occur (with probability $1 - q_i\Delta t$). In both cases $q(t)$ changes by an amount $(\lambda_i - c)\Delta t$, where c is the service rate of the queue. Using this observation and taking the limit as $\Delta t \to 0$, for $r > 0$ we have (see [Tanaka et al., 1995] for details)

$$\frac{\partial F_j(t,r)}{\partial t} = -(\lambda_i - c)\frac{\partial F_j(t,r)}{\partial x} + \sum_{i=1}^{M} q_{ij} F_i(t,r). \tag{93}$$

The solution of (93) was obtained by Laplace transform techniques.

Several modeling applications require the evaluation of the time that it takes an application to complete when running in a computer system. In Section 1 we mentioned that the events "time to achieve a given reward level $> t$" and "accumulated reward by $t >$ given reward level" were equivalent under certain conditions. The first event can be thought of as the completion time of a task, and its distribution may be evaluated as the distribution of cumulative reward. In general, for example when work losses are allowed, evaluating the distribution of the task completion time requires different solution techniques. In [Bobbio and Telek, 1993] Bobbio and Telek present a comprehensive treatment of this transient measure and include several application examples.

An interesting class of application examples for which transient analysis has been employed is in calculating steady state measures of certain non-Markovian models. In analyzing these models, embedded points are found such that the behavior of the system between these points is Markovian. The transition probability matrix between the embedded points requires the transient analysis of certain submodels. This is equivalent to obtaining point probabilities at the end of an interval, which can be deterministic or random, depending on the overall model. This was done, for example, in [Grassmann, 1982] for the GI/PH/1 queue as mentioned above. Once the transition probability matrix is obtained, the remaining issue is to calculate the measures of interest. These calculations often require the use of transient measures as we briefly describe in what follows.

Consider a non-Markovian continuous-time process $\mathcal{X} = \{X(t) : t \geq 0\}$ with finite state space $\mathcal{S}$, and assume that there are E types of non-exponential events and at most one of these events can be enabled when $\mathcal{X}$ is in a given state. In order to simplify the exposition, assume that non-exponential events are deterministic. Let Δ_j for $j = 1, \ldots, E$ be an interval which starts when the deterministic event ϕ_j is enabled and ends when the event triggers or is disabled by an exponential event. Let Δ_0 be an interval in which all deterministic events are disabled. The points corresponding to the beginning and end of each interval define an embedded Markov chain $\mathcal{Y}$. During any interval Δ_j, note that the process is Markovian, since all other events in the process are exponential.

Let $P_\mathcal{R}$ be the probability as $t \to \infty$ that the original process $\mathcal{X}$ is in a particular subset $\mathcal{R} \subset \mathcal{S}$ of states, let $U_s^{(j)}$ be the cumulative time that the process $\mathcal{X}$ spends in the set $\mathcal{R}$ during an interval Δ_j which starts in state s, and let $\delta_s^{(j)}$ be the length of this interval. It can be shown that (see [de Souza e Silva et al., 1995b] for a proof and an intuitive interpretation)

$$P_\mathcal{R} = \frac{\sum_{j=0}^{E} \sum_{s \in \mathcal{S}'(j)} E[U_s^{(j)}] \beta_s}{\sum_{j=0}^{E} \sum_{s \in \mathcal{S}'(j)} E[\delta_s^{(j)}] \beta_s}, \qquad (94)$$

where $\mathcal{S}'(j)$ is the set of possible initial states for the interval Δ_j, and β_s is the sth entry of the stationary probability vector β for the embedded chain $\mathcal{Y}$. The quantities $E[U_s^{(j)}]$ and $E[\delta_s^{(j)}]$ are calculated using transient analysis, once the distribution of the time between embedded points is obtained. The steps follow the approach described in Section 4.

Example applications of the basic theory outlined above can be found in [de Souza e Silva and Gail, 1990], where scheduled maintenance policies of computer systems were analyzed. In [de Souza e Silva et al., 1995c, de Souza e Silva et al., 1998] polling models with deterministic timeouts are considered.

Deterministic and stochastic Petri Nets (DSPNs) were introduced by Ajmone Marsan *et al* in [Ajmone Marsan and Chiola, 1987]. This class of Petri nets allows the occurrence of deterministic transitions in addition to transitions with exponential firing distributions. They can be analyzed using the embedded chain approach discussed above (see [de Souza e Silva et al., 1995b]). Other related work can be found in [Lindemann, 1991, Choi et al., 1994].

Acknowledgments

The work of E. de Souza e Silva was partially supported by grants ProTem-CC and Pronex from CNPq(Brazil).

References

[Abate and Whitt, 1989] Abate, J. and Whitt, W. (1989). Calculating time-dependent performance measures for the M/M/1 queue. *IEEE Trans. on Communications*, 37(10):891–904.

[Abdallah and Marie, 1993] Abdallah, H. and Marie, R. (1993). The uniformized power method for transient solutions of Markov processes. *Computers & Operations Research*, 20(5):515–526.

[Ajmone Marsan and Chiola, 1987] Ajmone Marsan, M. and Chiola, G. (1987). On Petri nets with deterministic and exponentially distributed firing times. In Rozenberg, G., editor, *Lecture Notes in Computer Science 266: Advances in Petri Nets 1987*, pages 132–145. Springer-Verlag.

[Bobbio and Telek, 1993] Bobbio, A. and Telek, M. (1993). Task completion time. In *Proceedings of the 2nd Int'l Workshop on Performability Modeling of Computer and Communication Systems*, Le Mont Saint-Michel, France.

[Bobbio and Trivedi, 1986] Bobbio, A. and Trivedi, K. (1986). An aggregation technique for the transient analysis of stiff Markov chains. *IEEE Trans. on Computers*, C-35(9):803–814.

[Carmo et al., 1998] Carmo, R., de Carvalho, L., de Souza e Silva, E., Diniz, M., and Muntz, R. (1998). Performance/availability modeling with the TANGRAM-II modeling environment. *Performance Evaluation*, 33:45–65.

[Carmo et al., 1996] Carmo, R., de Souza e Silva, E., and Marie, R. (1996). Efficient solutions for an approximation technique for the transient analysis of Markovian models. Technical report, IRISA, no. 3055, Campus universitaire de Beaulieu.

[Carrasco and Calderón, 1995] Carrasco, J. and Calderón, A. (1995). Regenerative randomization: Theory and application examples. In *Proc. Performance '95 and 1995 ACM SIGMETRICS Conf.*, pages 241–262.

[Çinlar, 1975] Çinlar, E. (1975). *Introduction to Stochastic Processes*. Prentice-Hall.

[Choi et al., 1994] Choi, H., Kulkarni, V., and Trivedi, K. (1994). Markov regenerative stochastic Petri nets. *Performance Evaluation*, 20:337–357.

[Choudhury et al., 1993] Choudhury, G., Lucantoni, D., and Whitt, W. (1993). Multi-dimensional transform inversion with applications to the transient M/G/1 queue. Technical report, AT&T Bell Labs, AT&T Bell Labs, Holmdel.

[Ciardo et al., 1993] Ciardo, G., Blakemore, A., Chimento, P., Muppala, J., and Trivedi, K. (1993). Automated generation and analysis of Markov reward models using stochastic reward nets. In Meyer, C. and Plemmons, R., editors, *Linear Algebra, Markov Chains, and Queueing Models*, pages 145–191. Springer-Verlag.

[Ciardo et al., 1989] Ciardo, G., Muppala, J., and Trivedi, K. (1989). SPNP: stochastic Petri net package. In *Proceedings of the Third International Workshop on Petri-nets and Performance Models*, pages 142–151.

[de Souza e Silva and Gail, 1986] de Souza e Silva, E. and Gail, H. (1986). Calculating cumulative operational time distributions of repairable computer systems. *IEEE Trans. on Computers*, C-35(4):322–332.

[de Souza e Silva and Gail, 1989] de Souza e Silva, E. and Gail, H. (1989). Calculating availability and performability measures of repairable computer systems using randomization. *Journal of the ACM*, 36(1):171–193.

[de Souza e Silva and Gail, 1990] de Souza e Silva, E. and Gail, H. (1990). Analyzing scheduled maintenance policies for repairable computer systems. *IEEE Trans. on Computers*, 39(11):1309–1324.

[de Souza e Silva and Gail, 1992] de Souza e Silva, E. and Gail, H. (1992). Performability analysis of computer systems: from model specification to solution. *Performance Evaluation*, 14:157–196.

[de Souza e Silva and Gail, 1996] de Souza e Silva, E. and Gail, H. (1996). The uniformization method in performability analysis. Technical report, IBM Research Report RC 20383, Yorktown Heights, N. Y.

[de Souza e Silva and Gail, 1998] de Souza e Silva, E. and Gail, H. (1998). An algorithm to calculate transient distributions of cumulative rate and impulse based reward. *Communications in Statistics - Stochastic Models*, 14(3):509–536.

[de Souza e Silva et al., 1995a] de Souza e Silva, E., Gail, H., and Campos, R. V. (1995a). Calculating transient distributions of cumulative reward. In *Proc. Performance '95 and 1995 ACM SIGMETRICS Conf.*, pages 231–240.

[de Souza e Silva et al., 1995b] de Souza e Silva, E., Gail, H., and Muntz, R. (1995b). Efficient solutions for a class of non-Markovian models. In Stewart, W., editor, *Computations with Markov Chains*, pages 483–506. Kluwer Academic Publishers.

[de Souza e Silva et al., 1995c] de Souza e Silva, E., Gail, H., and Muntz, R. (1995c). Polling systems with server timeouts and their application to token passing networks. *IEEE Trans. on Networking*, 3(5):560–575.

[de Souza e Silva et al., 1998] de Souza e Silva, E., Gail, H., and Muntz, R. (1998). Gated time-limited polling systems. In Hasegawa, T., Takagi, H., and Takahashi, Y., editors, *Performance and Management of Complex Ciommunication Networks*, pages 253–274. Chapman & Hall.

[Donatiello and Grassi, 1991] Donatiello, L. and Grassi, V. (1991). On evaluating the cumulative performance distribution of fault-tolerant computer systems. *IEEE Trans. on Computers*, 40(11):1301–1307.

[Golub and van Loan, 1989] Golub, G. and van Loan, C. (1989). *Matrix Computations*. Johns Hopkins University Press, second edition.

[Goyal et al., 1986] Goyal, A., Carter, W., de Souza e Silva, E., Lavenberg, S., and Trivedi, K. (1986). The system availability estimator. In *Proceedings of FTCS-16*, pages 84–89.

[Grassmann, 1977a] Grassmann, W. (1977a). Transient solutions in Markovian queueing systems. *Computers & Operations Research*, 4:47–53.

[Grassmann, 1977b] Grassmann, W. (1977b). Transient solutions in Markovian queues. *European Journal of Operational Research*, 1:396–402.

[Grassmann, 1982] Grassmann, W. (1982). The GI/PH/1 queue: a method to find the transition matrix. *INFOR*, 20(2):144–156.

[Grassmann, 1987] Grassmann, W. (1987). Means and variances of time averages in Markovian environments. *European Journal of Operational Research*, 31:132–139.

[Grassmann, 1991] Grassmann, W. (1991). Finding transient solutions in Markovian event systems through randomization. In Stewart, W. J., editor, *Numerical Solution of Markov Chains*, pages 357–371. Marcel Dekker, Inc.

[Grassmann, 1993] Grassmann, W. (1993). Means and variances in Markov reward systems. In Meyer, C. and Plemmons, R., editors, *Linear Algebra, Markov Chains, and Queueing Models*, pages 193–204. Springer-Verlag.

[Heidelberger and Goyal, 1988] Heidelberger, P. and Goyal, A. (1988). Sensitivity analysis of continuous time Markov chains using uniformization. In Iazeolla, G., Courtois, P., and Boxma, O., editors, *Computer Performance and Reliability*, pages 93–104. North-Holland.

[Horn and Johnson, 1985] Horn, R. and Johnson, C. (1985). *Matrix Analysis*. Cambridge Univ. Press.

[Islam and Ammar, 1991] Islam, S. and Ammar, H. (1991). Performability analysis of distributed real-time systems. *IEEE Trans. on Computers*, 40(11):1239–1251.

[Iyer et al., 1986] Iyer, B., Donatiello, L., and Heidelberger, P. (1986). Analysis of performability for stochastic models of fault-tolerant systems. *IEEE Trans. on Computers*, C-35(10):902–907.

[Jensen, 1953] Jensen, A. (1953). Markoff chains as an aid in the study of Markoff processes. *Skandinavsk Aktuarietidskrift*, 36:87–91.

[Johnson Jr. and Malek, 1988] Johnson Jr., A. and Malek, M. (1988). Survey of software tools for evaluating reliability, availability and serviceability. *ACM Computing Surveys*, 20:227–271.

[Kulkarni et al., 1986] Kulkarni, V., Nicola, V., Smith, R., and Trivedi, K. (1986). Numerical evaluation of performability and job completion time in repairable fault-tolerant systems. In *Proceedings of FTCS-16*, pages 252–257.

[Leguesdron et al., 1993] Leguesdron, P., Pellaumail, J., Rubino, G., and Sericola, B. (1993). Transient analysis of the M/M/1 queue. Technical report, IRISA, no. 720, Campus universitaire de Beaulieu.

[Lindemann, 1991] Lindemann, C. (1991). An improved numerical algorithm for calculating steady-state solutions of deterministic and stochastic Petri net models. In *Proceedings of the 4th International Workshop on Petri-nets and Performance Models*, pages 176–184.

[Lindemann, 1992] Lindemann, C. (1992). DSPNexpress: A software package for the efficient solution of deterministic and stochastic Petri nets. In *Proceedings of the Sixth International Conference on Modelling Techniques and Tools for Computer Systems Performance Evaluation*, pages 15–29, Edinburgh, Great Britain.

[Lopez-Benitez and Trivedi, 1993] Lopez-Benitez, N. and Trivedi, K. (1993). Multiprocessor performability analysis. *IEEE Trans. on Reliability*, 42(4):579–587.

[Lucantoni et al., 1994] Lucantoni, D., Choudhury, G., and Whitt, W. (1994). The transient BMAP/G/1 queue. *Communications in Statistics - Stochastic Models*, 10(1).

[Matsunawa, 1985] Matsunawa, T. (1985). The exact and approximate distributions of linear combinations of selected order statistics from a uniform distribution. *Ann. Inst. Statist. Math.*, 37:1–16.

[Meyer, 1980] Meyer, J. (1980). On evaluating the performability of degradable computing systems. *IEEE Trans. on Computers*, C-29(8):720–731.

[Meyer, 1982] Meyer, J. (1982). Closed-form solutions of performability. *IEEE Trans. on Computers*, C-31(7):648–657.

[Meyer, 1992] Meyer, J. (1992). Performability: A retrospective and some pointers to the future. *Performance Evaluation*, 14:139–156.

[Meyer, 1995] Meyer, J. (1995). Performability evaluation: Where it is and what lies ahead. In *IPDS'95*, pages 334–343.

[Meyer et al., 1980] Meyer, J., Furchtgott, D., and Wu, L. (1980). Performability evaluation of the SIFT computer. *IEEE Trans. on Computers*, C-29(6):501–509.

[Moler and van Loan, 1978] Moler, C. and van Loan, C. (1978). Nineteen dubious ways to compute the exponential of a matrix. *SIAM Review*, 20(4):801–836.

[Nabli and Sericola, 1996] Nabli, H. and Sericola, B. (1996). Performability analysis: a new algorithm. *IEEE Trans. on Computers*, 45(4):491–494.

[Pattipati et al., 1993] Pattipati, K., Li, Y., and Blom, H. (1993). A unified framework for the performability evaluation of fault-tolerant computer systems. *IEEE Trans. on Computers*, 42(3):312–326.

[Pattipati and Shah, 1990] Pattipati, K. and Shah, S. (1990). On the computational aspects of performability models of fault-tolerant computer systems. *IEEE Trans. on Computers*, 39(6):832–836.

[Philippe and Sidge, 1995] Philippe, B. and Sidge, R. (1995). Transient solutions of Markov processes by Krylov subspaces. In Stewart, W., editor, *Computations with Markov Chains*, pages 95–119. Kluwer Academic Publishers.

[Prodromides and Sanders, 1993] Prodromides, K. and Sanders, W. (1993). Performability evaluation of CSMA/CD and CSMA/CR protocols under transient fault conditions. *IEEE Trans. on Reliability*, 42(1):116–127.

[Puri, 1971] Puri, P. (1971). A method for studying the integral functionals of stochastic processes with applications: I. Markov chain case. *Journal of Applied Probability*, 8(10):331–343.

[Qureshi and Sanders, 1994] Qureshi, M. and Sanders, W. (1994). Reward model solution methods with impulse and rate rewards: An algorithm and numerical results. *Performance Evaluation*, 20(4):413–436.

[Reibman et al., 1989] Reibman, A., Trivedi, K., Kumar, S., and Ciardo, G. (1989). Analysis of stiff Markov chains. *ORSA Journal on Computing*, 1(2):126–133.

[Ren and Kobayashi, 1995] Ren, Q. and Kobayashi, H. (1995). Transient solutions for the buffer behavior in statistical multiplexing. *Performance Evaluation*, 23:65–87.

[Ross, 1983] Ross, S. (1983). *Stochastic Processes*. John Wiley & Sons.

[Ross, 1987] Ross, S. (1987). Approximating transition probabilities and mean occupation times in continuous-time Markov chains. *Probability in the Engineering and Informational Sciences*, 1:251–264.

[Rubino and Sericola, 1993] Rubino, G. and Sericola, B. (1993). Interval availability distribution computation. In *Proceedings of FTCS-23*, pages 48–55.

[Saad, 1992] Saad, Y. (1992). Analysis of some Krylov subspace approximations to the matrix exponential operator. *SIAM Journal on Numerical Analysis*, 29(1):208–227.

[Saad, 1995] Saad, Y. (1995). *Iterative Methods for Sparse Linear Systems*. PWS Publishing Company.

[Sanders et al., 1995] Sanders, W., II, W. O., Qureshi, M., and Widjanarko, F. (1995). The UltraSAN modeling environment. *Performance Evaluation*, 24:89–115.

[Sericola, 1999] Sericola, B. (1999). Availability analysis and stationary regime detection of Markov processes. *IEEE Trans. on Computers, (to appear)*.

[Stewart, 1991] Stewart, W. (1991). MARCA Markov chain analyzer, a software package for Markov chains. In *Numerical Solution of Markov Chains*, pages 37–61. Marcel Dekker, Inc.

[Stewart, 1994] Stewart, W. (1994). *Introduction to the Numerical Solution of Markov Chains*. Princeton University Press.

[Tai and Meyer, 1996] Tai, A. and Meyer, J. (1996). Performability management in distributed database systems: an adaptive concurrency control protocol. In *Proceedings of the 4th Int'l Workshop on Modeling, Analysis, and Simulation of Computer and Telecommunication Systems*, pages 212–216, San Jose, California.

[Tanaka et al., 1995] Tanaka, T., Hashida, O., and Takahashi, Y. (1995). Transient analysis of fluid model for ATM statistical multiplexer. *Performance Evaluation*, 23:145–162.

[Trivedi, 1982] Trivedi, K. (1982). *Probability and Statistics with Reliability, Queueing and Computer Science Applications*. Prentice-Hall.

[van As, 1986] van As, H. (1986). Transient analysis of Markovian queueing systems and its application to congestion control modeling. *IEEE Journal on Selected Areas in Communications*, SAC-4(6):891–904.

[van Dijk, 1990] van Dijk, N. (1990). The importance of bias-terms for error bounds and comparison results. In *The First International Conference on the Numerical Solution of Markov Chains*, pages 640–663.

[van Dijk, 1992a] van Dijk, N. (1992a). Approximate uniformization for continuous-time Markov chains with an application to performability analysis. *Stochastic Processes and Their Applications*, 40(2):339–357.

[van Dijk, 1992b] van Dijk, N. (1992b). Uniformization for nonhomogeneous Markov chains. *Operations Research Letters*, 12:283–291.

[van Moorsel and Sanders, 1994] van Moorsel, A. and Sanders, W. (1994). Adaptive uniformization. *Communications in Statistics - Stochastic Models*, 10(3):619–648.

[Weisberg, 1971] Weisberg, H. (1971). The distribution of linear combinations of order statistics from the uniform distribution. *Annals Math. Stat.*, 42:704–709.

[Yoon and Shanthikumar, 1989] Yoon, B. and Shanthikumar, J. (1989). Bounds and approximations for the transient behavior of continuous time Markov chains. *Probability in the Engineering and Informational Sciences*, 3:175–198.

Edmundo de Souza e Silva Edmundo de Souza e Silva received the B.Sc. and M.Sc. degrees in electrical engineering, both from Pontifical Catholic University of Rio de Janeiro (PUC/RJ), and the Ph.D. degree in computer science from the University of California, Los Angeles in 1984. During 1984-1985 he was a Visiting Scientist at the IBM Thomas J. Watson Research Center, Yorktown Heights, NY, and during the summer of 1987 he was a Visiting Faculty at the same Center, working in the Systems Analysis Department. He was a lecturer with the UCLA Department of Computer Science during the winter of 1987 and 1995, a Visiting Scientist at the IBM Tokyo Research Laboratory in the summer of 1990, and a lecturer at the Electrical Engineering/Computer Science Department of the University of Southern California during the fall of 1993. During 1993-1994 he was on sabbatical leave at the UCLA Computer Science Department. He was also a visiting researcher at the Politecnico di Torino (May-1995), Chinese University of Hong Kong (June-1996) and IRISA/INRIA-Rennes (Oct-1996). Currently he is with the Federal University of Rio de Janeiro, NCE and Computer Science Department. His areas of interest include the modeling and analysis of computer systems and computer communication multimedia networks.

H. Richard Gail received the Ph.D. degree in Engineering from the University of California, Los Angeles. He is a research staff member at the IBM T. J. Watson Research Center in Yorktown Heights, NY. His research interests include queueing theory and computer systems modeling and analysis.

4 NUMERICAL METHODS FOR COMPUTING STATIONARY DISTRIBUTIONS OF FINITE IRREDUCIBLE MARKOV CHAINS

William J. Stewart

Department of Computer Science
North Carolina State University
Raleigh, N.C. 27695-8206
USA

billy@csc.ncsu.edu

1 INTRODUCTION

In this chapter our attention will be devoted to computational methods for computing stationary distributions of finite irreducible Markov chains. We let q_{ij} denote the rate at which an n-state Markov chain moves from state i to state j. The $n \times n$ matrix Q whose off-diagonal elements are q_{ij} and whose i^{th} diagonal element is given by $-\sum_{j=1, j \neq i}^{n} q_{ij}$ is called the *infinitesimal generator* of the Markov chain. It may be shown that the stationary probability vector π, a row vector whose k-th element denotes the stationary probability of being in state k, can be obtained by solving the homogeneous system of equations $\pi Q = 0$. Alternatively, the problem may be formulated as an eigenvalue problem $\pi P = \pi$, where $P = Q\Delta t + I$ is the stochastic matrix of transition probabilities, (Δt must be chosen sufficiently small so that the probability of two or more transitions occurring in time Δt is small, i.e., of order $o(t)$). Mathematically, the problem is therefore quite simple. Unfortunately, problems arise from the computational point of view because of the large number of states which many systems may occupy. As indicated in Chapters 1 and 2, it is not uncommon for thousands of states to be generated even for simple applications.

We begin our discussion with an examination of the relative advantages and disadvantages of iterative and direct solution methods. We shall show that iterative methods are generally preferred, unless the infinitesimal generator has some special structure which makes a direct method more efficient. In Section 3, we discuss direct methods and show how to implement them in a computationally efficient manner. Basic single vector iteration methods are considered in Section 4. In particular, we examine the power method forward and backward Gauss-Seidel and SOR and preconditioned power iterations. Block single vector iterative methods are also considered. Iterative methods that incorporate a subspace of vectors are treated in Section 5. These go under the more generic name of *projection techniques*, and have been shown in comparison testing to be among the most effective for general Markov chain problems. The final methods considered for the computation of stationary solutions are the decompositional methods described in Section 6. These are valuable when the matrix is *nearly-completely-decomposable*, NCD, a situation which arises often in practice.

It is obviously impossible in a single chapter to cover the entire field of numerical procedures for computing stationary distributions of Markov chains, a field that has been in constant evolution from the initial papers of Wallace and his colleagues, ([Wallace and Rosenberg, 1966a], [Wallace and Rosenberg, 1966b], [Wallace, 1973]). To make up for this, we have provided an extensive bibliography at the end of the chapter. In particular, we cite the recent text of the author of this chapter, [Stewart, 1994]. We hope that these references will provide the reader with all the further reading material likely to be needed.

2 ITERATIVE AND DIRECT SOLUTION METHODS

Iterative methods of one type or another are by far the most commonly used methods for obtaining the stationary probability vector from either the stochastic transition probability matrix or from the infinitesimal generator. There are several important reasons for this choice. First, an examination of the iterative methods usually employed shows that the only operation in which the matrices are involved, are multiplications with one or more vectors, or with preconditioners. These operations do not alter the form of the matrix. Thus compact storage schemes, which minimize the amount of memory required to store the matrix and which in addition are well suited to matrix multiplication, may be conveniently implemented. Since the matrices involved are usually large and very sparse, the savings made by such schemes can be considerable. With direct equation solving methods, the elimination of one nonzero element of the matrix during the reduction phase often results in the creation of several nonzero elements in positions which previously contained zero. This is called fill-in and not only does it make the organization of a compact storage scheme more difficult, since provision must be made for the deletion and the insertion of elements, but in addition, the amount of fill-in can often be so extensive that available memory is quickly exhausted. A successful direct method must incorporate a means of overcoming these difficulties.

Iterative methods have other advantages. Use may be made of good initial approximations to the solution vector and this is especially beneficial when a series of related experiments is being conducted. In such circumstances the parameters of one experiment often differ only slightly from those of the previous; many will remain unchanged. Consequently, it is to be expected that the solution to the new experiment will be close to that of the previous and it is advantageous to use the previous result as the new initial approximation. If indeed there is little change, we should expect to compute the new result in relatively few iterations.

An iterative process may be halted once a prespecified tolerance criterion has been satisfied, and this may be relatively lax. For example, it may be wasteful to compute the solution of a mathematical model correct to full machine precision when the model itself contains errors of the order of 5-10%. A direct method is obligated to continue until the final specified operation has been carried out.

And lastly, with iterative methods, the matrix is never altered and hence the build-up of rounding error is, to all intents and purposes, non-existent.

For these reasons, iterative methods have traditionally been preferred to direct methods. However, iterative methods have a major disadvantage in that often they require a very long time to converge to the desired solution. More advanced iterative techniques such as the method of Arnoldi, have helped to alleviate this problem but much research still remains to be done, particularly in estimating a priori, the number of iterations, and hence the time, required for convergence. Direct methods have the advantage that an upper bound on the time required to obtain the solution may be determined before the calculation is initiated. More important, for certain classes of problems, direct methods often result in a much more accurate answer being obtained in *less time*. Since iterative method will in general require less memory than direct methods, these latter can only be recommended if they obtain the solution in less time. Unfortunately, it is often difficult to predict when a direct solver will be more efficient than an iterative solver.

3 DIRECT SOLUTION METHODS

We are concerned with obtaining the stationary probability vector π from the equations

$$\pi Q = 0, \quad \pi \geq 0, \quad \pi e = 1, \tag{1}$$

i.e., a homogeneous system of n linear equations in n unknowns. The vector e is a column vector whose elements are all equal to 1. Also, we shall need to introduce the vector e_i, a column vector whose elements are all equal to zero except the ith which is equal to 1.

3.1 LU Decompositions

The system of equations (1) has a solution other than the trivial solution ($\pi_i = 0$, for all i) if and only if the determinant of the coefficient matrix is zero, i.e., if and only if the coefficient matrix is singular. Since the determinant of a matrix is equal to the product of its eigenvalues and since Q possesses a zero eigenvalue, the singularity of Q and hence the existence of a non-trivial solution, follows. It is known that if the matrix Q is irreducible, there exists lower and upper triangular matrices L and U such that

$$Q^T = LU.$$

Once an LU decomposition has been determined, a forward substitution step followed by a backward substitution is usually sufficient to determine the solution of the system of equations.

For example, suppose we are required to solve $Ax = b$ with $det(A) \neq 0$ and $b \neq 0$ and suppose further that the decomposition $A = LU$ is available so that $LUx = b$. By setting $Ux = z$, the vector z may be obtained by forward substitution on $Lz = b$, since both L and b are known quantities. The solution x may subsequently be obtained from $Ux = z$ by backward substitution since by this time both U and z are known quantities. However, in the case of the numerical solution of Markov chains, the system of equations, $\pi Q = 0$, is homogeneous, i.e., $b = 0$, and the coefficient matrix is singular. In this case, the final row of U (assuming that the decomposition is such that the diagonal elements of L are set to unity) is equal to zero. Proceeding as indicated above for the non-homogeneous case, we have

$$Q^T x = (LU)x = 0.$$

If we now set $Ux = z$ and attempt to solve $Lz = 0$ we find that, since L is nonsingular, we must have $z = 0$. Let us now proceed to the back substitution on $Ux = z = 0$ when $u_{nn} = 0$. It is evident that we may assign any nonzero value to x_n, say $x_n = \eta$, and then determine, by simple back-substitution, the remaining elements of the vector x in terms of η. We have $x_i = c_i \eta$ for some constants $c_i, i = 1, 2, ..., n$, and $c_n = 1$. Thus the solution obtained depends on the value of η. There still remains one equation that the elements of a probability vector must satisfy, namely that the sum of the probabilities must be one. Consequently, normalizing the solution obtained from solving $Ux = 0$ so that the conservation of probability condition holds, yields the desired unique stationary probability vector π corresponding to the infinitesimal generator Q.

An alternative approach to this use of the normalization equation is to replace the last equation of the original system with $\pi e = 1$. If the Markov chain is irreducible, this will ensure that the coefficient matrix is nonsingular. Furthermore, the system of equations will no longer be homogeneous (since the right hand side is now e_n), and so the solution may be computed without problem.

Of course, it is not necessary to replace the last equation of the system by the normalization equation. Indeed, any equation could be replaced. However, this is generally undesirable, for it will entail more numerical computation. For example, if the first equation is replaced, the first row of the coefficient matrix will contain all ones and the right hand side will be e_1. The first consequence of this is that during the forward substitution stage, the entire sequence of operations must be performed to obtain the vector z; whereas if the last equation is replaced, it is simply possible to read off the solution immediately, i.e., $z_1 = z_2 = \cdots = z_{n-1} = 0$ and $z_n = 1$. The second and more damaging consequence is that substantial fill-in will occur since a multiple of the first row, which contains all ones, must be added to *all* remaining rows and a cascading effect will undoubtedly occur in all subsequent reduction steps.

For more information on these and other direct methods, including sensitivity analyses, the reader should consult the following references: [Funderlic and Meyer, 1986], [Funderlic and Plemmons, 1986], [Golub and Meyer, 1986], [Grassmann et al., 1985], [Harrod and Plemmons, 1984], [Heyman, 1987], [Stewart, 1978] and [Stewart, 1994].

3.2 Compact Storage Schemes for Direct Methods

Frequently the matrices generated from Markov models are too large to permit regular two-dimensional arrays to be used to store them in computer memory. Since these matrices are usually very sparse, it is economical, and indeed necessary, to use some sort of packing scheme whereby only the nonzero elements and their positions in the matrix are stored.

Assume, as is usually the case, that the coefficient matrix can be derived row by row. Then, immediately after the second row has been obtained, it is possible to eliminate the element in position (2,1) by adding a multiple of the first row to it. This process may be continued so that when the i-th row of the coefficient matrix is generated, rows 1 through $(i-1)$ have been derived and are already reduced to upper triangular form. The first $(i-1)$ rows may therefore be used to eliminate all nonzero elements in row i from column positions $(i, 1)$ through $(i, i-1)$, thus putting it into the desired triangular form. Note that since this reduction is performed on Q^T, it is the columns of the infinitesimal generator that are required to be generated one at a time and not its rows.

This method has a distinct advantage in that once a row has been generated in this fashion, no more fill-in will occur into this row. It is suggested that a separate storage area be reserved to hold temporarily a single unreduced row. The reduction is performed in this storage area. Once completed, the reduced row may be compacted into any convenient form and appended to the rows which have already been reduced. In this way no storage space is wasted holding subdiagonal elements which, due to elimination, have become zero, nor in reserving space for the inclusion of additional elements. The storage scheme should be chosen bearing in mind the fact that these rows will be used in the reduction of further rows and also later in the algorithm during the back-substitution phase.

Since the form of the matrix will no longer be altered, the efficient storage schemes which are used with many iterative methods can be adopted. Note that this approach can not be used for solving general systems of linear equations because it inhibits a pivoting strategy from being implemented. It is valid when solving irreducible Markov chains since pivoting is not required in order that the LU decomposition be performed in a stable manner.

3.3 The GTH Advantage

It is appropriate at this point to mention a variant of Gaussian elimination that has attributes that make it even more stable than the usual version. This procedure is commonly referred to as the GTH (Grassmann–Taksar–Heyman) algorithm [Grassmann et al., 1985, Sheskin, 1985]. In GTH the diagonal elements are obtained by summing the off-diagonal elements rather than performing a subtraction; it is known that subtractions can sometimes lead to loss of significance in the representation of real numbers. The concept evolved from probabilistic arguments, and the originally suggested implementation is a backward (rather than forward) elimination procedure. However, other implementations are possible.

The key to the GTH algorithm is that the properties that characterize the infinitesimal generator of an irreducible Markov chain,

$$q_{ii} < 0, \quad q_{ij} \geq 0, \quad \text{and} \quad \sum_{j=1}^{n} q_{ij} = 0 \quad \text{for all } i,$$

are invariant under the elementary row operations carried out when Gaussian elimination is applied to Q. (The reader should carefully note that it is the matrix Q and *not* Q^T that is the subject of discussion in this paragraph.) This may be seen by considering the first two rows of Q and the elimination of q_{21}, which for obvious reasons we assume to be nonzero.

$$\text{Row 1}: \quad q_{11} \quad q_{12} \quad q_{13} \quad \cdots \quad q_{1n} \quad \text{with} \quad q_{11} = \sum_{k=2}^{n} q_{1k}$$

$$\text{Row 2}: \quad q_{21} \quad q_{22} \quad q_{23} \quad \cdots \quad q_{2n} \quad \text{with} \quad q_{22} = \sum_{k \neq 2}^{n} q_{2k}$$

To eliminate $q_{21} \neq 0$, we add a multiple of row 1 into row 2. This multiple is given by $-q_{21}/q_{11} = q_{21}/\sum_{k=2}^{n} q_{1k}$, which is a positive quantity. Since $q_{12}, q_{13}, \ldots, q_{1n}$ are positive, this elementary row operation causes a positive quantity to be added into each element q_{22} through q_{2n} (q_{21} is *set* to zero). The elements in row 2 become

$$q_{2j} \leftarrow q_{2j} + q_{21} \frac{q_{1j}}{\sum_{k=2}^{n} q_{1k}} = q_{2j} + q_{21} w_{1j}, \quad \text{for } j = 2, \ldots, n, \qquad (2)$$

which shows that the quantity q_{21} is distributed over the other elements of the second row according to their weight w_{1j} in row 1. Consequently it follows that the sum of the elements in row 2 remains zero. Also, the off-diagonal elements in row 2 do not decrease (since they were greater than or equal to zero and nonnegative quantities are added), and the diagonal element, which had been strictly less than zero, can only move closer to zero — it cannot exceed zero, for otherwise the sum across the row could not be zero. The only operation in (2) that involves a negative number occurs with $j = 2$. We have

$$q_{22} \leftarrow q_{22} + q_{21} \frac{q_{12}}{\sum_{k=2}^{n} q_{1k}} \qquad (3)$$

and q_{22} is negative. However, instead of using (3), q_{22} may be found by summing the new off-diagonal elements and then negating this sum. We get

$$q_{22} = -\sum_{j=3}^{n}\left(q_{2j} + q_{21}\frac{q_{1j}}{\sum_{k=2}^{n}q_{1k}}\right). \qquad (4)$$

This procedure may be extended in the obvious manner to cover the entire Gaussian elimination algorithm. Thus, as the Gaussian elimination algorithm is unfolding, a diagonal element may be computed either in the usual fashion, by subtracting a multiple of the element immediately above it from its current value, or by summing the off-diagonal elements in the row once it has been reduced, and negating this sum. The latter approach involves no subtractions and yields a more stable algorithm. Notice that in an implementation, it would be redundant to actually insert a minus sign before the computed diagonal element.

The GTH implementation requires more numerical operations than the standard implementation (compare equations 3, and 4), but this may be offset by a gain in precision when the matrix Q is ill-conditioned. The extra additions are not very costly when compared with the overall cost of the elimination procedure, which leads to the conclusion that the GTH advantage should be exploited where possible. It is also possible to apply the GTH approach to Q^T. If the transition rate matrix is stored in a two-dimensional or band storage structure, access is easily available to both the rows and columns of Q, and there is no difficulty in applying GTH to obtain an LU decomposition of either Q or Q^T.

Unfortunately, difficulties arise in implementing GTH when computer memory is at a premium and sparse compact storage schemes, such as those described in Section 3.2, must be used [Dayar and Stewart, 1996]. Suppose first that Q is stored by rows and an LU decomposition of Q is sought. Both L and U need to be kept during the GTH reduction stage: the rows of U are needed to eliminate nonzero elements in the unreduced part of the coefficient matrix, while L is needed to compute the solution from $\pi L = e_n$ by forward substitution. This is no different from an implementation of regular Gaussian elimination applied to Q. Nor does it change if we store Q by columns. An LU reduction of Q always requires storage for L. Let us now consider the possibilities when an LU decomposition of Q^T is sought, since in this case only U need be kept and the multipliers may be discarded immediately after they have been used. If Q^T is stored by rows, then in order to obtain the diagonal elements by adding off-diagonal terms, access is also needed to the columns of Q^T. If the storage scheme does not provide convenient access to both the rows and columns, then this approach cannot be easily used. Incorporating link pointers to provide such access can have a serious detrimental effect on computation time. That leaves us only with the case in which Q^T is stored by columns. The following algorithm, requiring access only to the rows of Q, has been proposed for implementing GTH to obtain an LU decomposition of Q^T. To accomplish this goal, it keeps a running sum of the elements that contribute to the pivot.

During the ith step of the elimination, the sum for the $(i+1)^{st}$ column is accumulated and taken as the pivot element for the $(i+1)^{st}$ elimination step. Thus, diagonal elements are computed just before they are needed.

Algorithm: GTH Applied to Q^T

1. Compute $q_{11} = q_{12} + q_{13} + \ldots + q_{1n}$.
2. For $i = 1, 2, \ldots, n-1$ do
    - Set $\sigma = 0$.
    - For $j = i+1, i+2, \ldots, n$ do
        - Compute $\mu = q(j,i)/q(i,i)$.
        - For $k = i+1, i+2, \ldots, n$ do
            * $q(j,k) = q(j,k) + \mu \times q(i,k)$.
        - If $j > i+1$, then compute $\sigma = \sigma + q(j, i+1)$.
    - Set $q(i+1, i+1) = \sigma$.

A careful examination of this algorithm will reveal that although it implements the GTH approach on Q^T, it requires access only to the *rows* of Q! However, it suffers from the drawback that *all* rows below the ith are modified during the ith step, so that a process of expansion and recompaction of the unreduced portion must be performed continuously throughout the algorithm.

We may well ask ourselves where this leaves us with respect to GTH. We may conclude that if two-dimensional or band storage is implemented, then GTH should be used, because the additional time required to obtain the multipliers by adding off-diagonal elements is likely to be an insignificant part of the total computation. On the other hand, if compact storage schemes are used, then compared with the best possible implementation of Gaussian elimination (an LU decomposition of Q^T with Q^T stored by rows), GTH is likely to require either significantly more memory (if an LU decomposition of Q is computed), significantly more time (if an LU decomposition of Q^T is formed), or both. Since Gaussian elimination is known to be stable, the real need for GTH occurs only when the problem is known to be ill-conditioned.

4 SINGLE VECTOR ITERATIONS

4.1 The Power Method

The simplest iterative method for computing the dominant eigenvector of a matrix A is the single vector iteration

$$x^{(k+1)} = \frac{1}{\xi^{(k)}} A x^{(k)},$$

where $\xi^{(k)}$ is a normalizing factor, typically the component of the vector $Ax^{(k)}$ that has the largest modulus. In our case the matrix of interest A is P^T. Since we know that the matrix P has row sums equal to 1 and has 1 as the dominant eigenvalue, we can safely skip the normalizing factor and the above iteration takes the form

$$x^{(k+1)} = P^T x^{(k)}$$

One problem with this simple scheme is that its rate of convergence can be very slow. The convergence factor for the eigenvector corresponding to the dominant eigenvalue λ_1 is given by $|\lambda_2|/|\lambda_1|$, where λ_2 is the subdominant eigenvalue. In situations where the eigenvalues cluster around λ_1, as is the case for nearly decomposable systems, for example, the convergence can be unacceptably slow.

4.2 Gauss-Seidel Iteration and Successive Overrelaxation

Relaxation schemes are based on the decomposition

$$Q^T = D - E - F,$$

where D is the diagonal of Q^T, $-E$ is the strictly lower triangular part of Q^T and $-F$ its strictly upper triangular part. Given that $Q^T x = 0$, the Gauss-Seidel iteration then takes the form

$$(D - E)x^{(k+1)} = Fx^{(k)}. \tag{5}$$

This corresponds to correcting the j-th component of the current approximate solution, for $j = 1, 2, \ldots, n$, i.e., from top to bottom. To denote specifically the direction of solution this is sometimes referred to as *forward* Gauss-Seidel. A *backward* Gauss-Seidel iteration takes the form

$$(D - F)x^{(k+1)} = Ex^{(k)},$$

and corresponds to correcting the components from bottom to top.

Note that convergence of the above (forward) iteration is governed by the spectral radius of $(D - E)^{-1}F$. Convergence may sometimes be improved by using the alternative splitting

$$\omega Q^T = (D - \omega E) - (\omega F + (1 - \omega)D),$$

which leads to the iteration, called successive overrelaxation, (SOR)

$$(D - \omega E)x^{(k+1)} = (\omega F + (1 - \omega D))\, x^{(k)}.$$

A backward SOR relaxation may also be written.

For many problems there exists a value of ω which provides the best possible convergence rate. The resulting optimal convergence rate can be a considerably improvement over Gauss-Seidel. The choice of an optimal, or even a reasonable, value for ω has been the subject of much study, especially for problems arising in the numerical solution of partial differential equations [Young, 1971]. Some results have been obtained for certain classes of matrices. Unfortunately, very little is known at present for arbitrary non-symmetric linear systems.

As a general rule, it is best to use a forward iterative method when the preponderance of the elemental mass is to be found below the diagonal, for in this case the iterative method essentially works with the inverse of the lower triangular portion of the matrix and intuitively, the closer this is to the inverse of the entire matrix, the faster the convergence. On the other hand, the backward iterative schemes work with the inverse of the upper triangular portion and these methods work best when the nonzero mass lies above and on the diagonal. We point out that some specialized counter examples exist which makes the above recommendations only rules of thumb, [Kaufman, 1983].

Little information is available on the effect of the ordering of the state space on the convergence of these iterative methods. Examples are available in which Gauss-Seidel works well for one ordering but not at all for an opposing ordering, [Mitra and Tsoucas, 1988]. In these examples the magnitude of the nonzero elements appears to have little effect on the speed of convergence. It appears that an ordering that in some sense preserves the direction of probability flow works best. For further information on convergence and convergence properties, see [Barker and Plemmons, 1986], [Berman and Plemmons, 1979], [Courtois and Semal, 1986], [Lubachevsky and Mitra, 1986] and [Stewart, 1994].

4.3 SSOR Iteration

The Symmetric Successive Overrelaxation method (SSOR) consists of following a relaxation sweep from top down by a relaxation sweep from bottom up. Thus, the case $\omega = 1$ corresponding to a SGS (Symmetric Gauss Seidel) scheme would be as follows:

$$\begin{aligned}(D - E)x^{(k+1/2)} &= Fx^{(k)} \\ (D - F)x^{(k+1)} &= Ex^{(k+1/2)},\end{aligned}$$

while for arbitrary ω, it is:

$$\begin{aligned}(D - \omega E)x^{(k+1/2)} &= (\omega F + (1 - \omega D))\, x^{(k)} \\ (D - \omega F)x^{(k+1)} &= (\omega E + (1 - \omega D))\, x^{(k+1/2)}.\end{aligned}$$

The main attraction of SSOR is that the iteration matrix is similar to a symmetric matrix when the original matrix Q^T is symmetric. This situation rarely occurs in Markov chain models. SSOR does however, help to reduce poor convergence behavior that results from a badly ordered state space.

4.4 Preconditioned Power Iterations

As was already mentioned the power method can be slow to converge when the subdominant eigenvalue is very close to one. The relaxation schemes described above typically have a better convergence rate. This means that the iteration matrices corresponding to these schemes have an eigenvalue λ_2 farther away from 1 than the original matrix.

Preconditioning is a technique whereby the original system of equations is modified in such a way that the solution is unchanged but the distribution of the eigenvalues is better suited to iterative methods. In a general context, a preconditioning technique consists of replacing a system $Ax = b$ by a modified system such as $M^{-1}Ax = M^{-1}b$. Here M is a preconditioning matrix for which the solution of $Mx = y$ is inexpensive, [Alexsson, 1985]. When the coefficient matrix is singular and the right hand side is zero, the method turns out to be equivalent to the power method applied to the matrix $(I - M^{-1}A)$.

For the numerical solution of Markov chain problems, the power method may be written as

$$x^{(k+1)} = P^T x^{(k)} = (I - (I - P^T))x^{(k)} \qquad (6)$$

Here preconditioning involves premultiplying the matrix $I - P^T$ with a matrix M^{-1}, generally chosen so that M approximates $I - P^T$ but is such that its LU decomposition can be efficiently determined. In this case, the iteration matrix, $I - M^{-1}(I - P^T)$ has one unit eigenvalue and the remaining eigenvalues are (hopefully) all close to zero, leading to a rapidly converging iterative procedure. We refer to such methods as *preconditioned power iterations*, or *fixed point iterations*.

4.5 Gauss-Seidel, SOR and SSOR Preconditionings

A look at (5) reveals an interesting connection with the power method. We can rewrite (5) as

$$\begin{aligned} x^{(k+1)} &= (D - E)^{-1} F x^{(k)} \\ &= (D - E)^{-1} \left((D - E) - Q^T\right) x^{(k)} \\ &= x^{(k)} - (D - E)^{-1} Q^T x^{(k)}. \end{aligned}$$

Comparing this with equation (6), we observe that the above iteration is simply the power method applied to the matrix

$$I - (D - E)^{-1} Q^T. \qquad (7)$$

Thus $(D - E)$ performs the role of a preconditioning matrix. As a result we may view the Gauss-Seidel method as a preconditioned power iteration. It is an attempt to reduce λ_2, without changing the eigenvector.

The solution to the above system is identical with that of the original one. Its rate of convergence, on the other hand, may be substantially faster than that of the original problem. For this reason we will refer to the system (7) as the Gauss-Seidel preconditioned version of $Q^T x = 0$. Similarly one may define an SOR preconditioning and an SSOR preconditioning.

4.6 ILU Preconditioning

By far the most popular preconditioning techniques are the incomplete LU factorization techniques. These are sometimes also referred to as *combined direct-iterative* methods. Such methods are composed of two phases. First we start by initiating an LU decomposition of Q^T. At various points in the computation, nonzero elements may be omitted according to various rules. Some possibilities are discussed in the following paragraphs. In all cases, instead of arriving at an exact LU decomposition, what we obtain is of the form

$$Q^T = LU - E, \tag{8}$$

where E, called the remainder, is expected to be small in some sense. When this has been achieved, the *direct* phase of the computation is completed. In the second phase, this (incomplete) factorization is incorporated into an iterative procedure by writing

$$Q^T x = (LU - E)x = 0$$

and then using

$$LUx^{(k+1)} = Ex^{(k)},$$

or, by equation (8),

$$x^{(k+1)} = x^{(k)} - (LU)^{-1} Q^T x^{(k)},$$

as the iteration scheme.

Here we describe three different incomplete factorizations. The first, which is called ILU(0), has been widely discussed and found to be successful especially when applied to systems of equations that arise in the solution of elliptic partial differential equations. Given the matrix Q^T, this ILU factorization consists of performing the usual Gaussian Elimination factorization and dropping any fill-in that occurs during the process. In other words, we compute a factorization

$$Q^T = LU - E$$

in which L is unit lower triangular, U is upper triangular, and $L + U$ has the same zero structure as the matrix Q^T.

The second incomplete factorization is a threshold based scheme. Here the decomposition proceeds in a manner similar to that described for the implementation of GE in Section 3.2. However, after a row of the matrix has been reduced and before that row is recompacted and stored, each nonzero element is examined. If the absolute value of any element in the row is less than a prespecified threshold, then it is replaced by zero. Similarly, if any of the multipliers formed during the reduction are less than the threshold, they are also dropped from further consideration. The only exception to this drop threshold are the diagonal elements which are kept no matter how small they become. We refer to this incomplete factorization technique as ILUTH.

The final type of incomplete factorization we mention is based on the realization that only a fixed amount of memory may be available to store the incomplete factors, L and U, so only a fixed number of nonzero elements are kept in each row. These are usually chosen to be the largest in magnitude. The algorithm proceeds in the same way as ILUTH. When a row has been reduced, a search is conducted to find the K largest elements in absolute value. This search is conducted over both the multipliers and the nonzero elements to the right of the diagonal element. As before, the diagonal elements are kept regardless of their magnitude. This incomplete factorization is referred to as ILUK.

Although the above three ILU factorizations are the only ones we cite, there are other possibilities (see [Funderlic and Plemmons, 1984], [Saad, 1995b] for example). However, we believe that, for the moment, ILU0, ILUTH and ILUK are the most effective for Markov chain problems.

4.7 Block Iterative Methods

In Markov chain problems it is frequently the case that the state space can be meaningfully partitioned into subsets. Perhaps the states of a subset interact only infrequently with the states of other subsets, or perhaps the states possess some property that merits special consideration. In these cases it is possible to partition the transition rate matrix accordingly and to develop iterative methods that are based on this partition. In general such block iterative methods require more computation per iteration, but this is offset by a faster rate of convergence.

Let us partition the defining homogeneous system of equations $\pi Q = 0$ as

$$(\pi_1, \pi_2, \ldots, \pi_N) \begin{pmatrix} Q_{11} & Q_{12} & \cdots & Q_{1N} \\ Q_{21} & Q_{22} & \cdots & Q_{2N} \\ \vdots & \vdots & \ddots & \vdots \\ Q_{N1} & Q_{N2} & \cdots & Q_{NN} \end{pmatrix} = 0$$

We now introduce the *block* splitting:

$$Q^T = D_N - (L_N + U_N),$$

where D_N is a *block* diagonal matrix and L_N and U_N are respectively strictly lower and upper *block* triangular matrices. We have

$$D_N = \begin{pmatrix} D_{11} & 0 & \cdots & 0 \\ 0 & D_{22} & \cdots & 0 \\ \vdots & \vdots & \ddots & \vdots \\ 0 & 0 & \cdots & D_{NN} \end{pmatrix}$$

$$L_N = \begin{pmatrix} 0 & 0 & \cdots & 0 \\ L_{21} & 0 & \cdots & 0 \\ \vdots & \vdots & \ddots & \vdots \\ L_{N1} & L_{N2} & \cdots & 0 \end{pmatrix}, \quad U_N = \begin{pmatrix} 0 & U_{12} & \cdots & U_{1N} \\ 0 & 0 & \cdots & U_{2N} \\ \vdots & \vdots & \ddots & \vdots \\ 0 & 0 & \cdots & 0 \end{pmatrix}$$

In analogy with equation (5), the block Gauss–Seidel method is given by

$$(D_N - L_N)x^{(k+1)} = U_N x^{(k)}.$$

If we write this out in full, we get

$$D_{ii}x_i^{(k+1)} = \left(\sum_{j=1}^{i-1} L_{ij}x_j^{(k+1)} + \sum_{j=i+1}^{N} U_{ij}x_j^{(k)}\right), \quad i = 1,2,\ldots,N$$

where the subvectors x_i are partitioned conformally with D_{ii}, $i = 1,2,\ldots,N$. This implies that at each iteration we must now solve N systems of linear equations

$$D_{ii}x_i^{(k+1)} = z_i, \quad i = 1,2,\ldots,N, \qquad (9)$$

where

$$z_i = \left(\sum_{j=1}^{i-1} L_{ij}x_j^{(k+1)} + \sum_{j=i+1}^{N} U_{ij}x_j^{(k)}\right), \quad i = 1,2,\ldots,N.$$

The right-hand side, z_i, may always be computed before the ith system has to be solved.

In a similar vein to block Gauss–Seidel, we may also define a *block* Jacobi method

$$D_{ii}x_i^{(k+1)} = \left(\sum_{j=1}^{i-1} L_{ij}x_j^{(k)} + \sum_{j=1+1}^{N} U_{ij}x_j^{(k)}\right), \quad i = 1,2,\ldots,N,$$

and a *block* SOR method

$$x_i^{(k+1)} = (1-\omega)x_i^{(k)} + \omega\left\{D_{ii}^{-1}\left(\sum_{j=1}^{i-1} L_{ij}x_j^{(k+1)} + \sum_{j=i+1}^{N} U_{ij}x_j^{(k)}\right)\right\}, \quad i = 1,2,\ldots,N.$$

If the matrix Q is irreducible, the N systems of equations (9) are nonhomogeneous and have nonsingular coefficient matrices. We may use either direct or iterative methods to solve them. Naturally, there is no requirement to use the same method to solve all the diagonal blocks. Instead, it is possible to tailor methods to the particular block structures.

If a direct method is to be used, then an LU decomposition of each block D_{ii} may be formed once and for all before beginning the iteration, so that solving $D_{ii}x_i^{(k+1)} = z_i$, $i = 1,\ldots,N$ in each iteration simplifies to a forward and backward substitution. The nonzero structure of the blocks may be such that this is a particularly attractive approach. For example, if the diagonal blocks are themselves diagonal matrices, or if they are upper or lower triangular matrices or even tridiagonal matrices, then it is very easy to obtain their LU decomposition, and a block iterative method becomes very attractive. The LU factors may overwrite the storage arrays used to hold the diagonal blocks.

If the diagonal blocks do not possess such a structure, and when they are of large dimension, it may be appropriate to use an iterative method to solve each of the block systems. In this case, we have many inner iterative methods (one per block) within an outer (or global) iteration. A number of tricks may be used to speed up this process. First, the solution computed for any block D_{ii} at iteration k should be used as the initial approximation to the solution of this same block at iteration $k+1$. Second, it is hardly worthwhile computing a highly accurate solution in early (outer) iterations. We should require only a small number of digits of accuracy until the global process begins to converge. One convenient way to achieve this is to carry out only a fixed, small number of iterations for each inner solution. Initially, this will not give much accuracy, but when combined with the first suggestion, the accuracy achieved will increase from one outer iteration to the next.

Intuitively, it is expected that for a given transition rate matrix Q, the larger the block sizes (and thus the smaller the number of blocks), the fewer the (outer) iterations needed to achieve convergence. This has been shown to be true under fairly general assumptions on the coefficient matrix for general systems of equations (see [Varga, 1962]). In the special case of only one block, the method degenerates to a standard direct method and we compute the solution in a single "iteration."

The reduction in the number of iterations that usually accompanies larger blocks is offset to a certain degree by an increase in the number of operations that must be performed at each iteration. However, in some important cases it may be shown that there is no increase. For example, when the matrix is block tridiagonal (as arises in quasi-birth-death processes) and the diagonal blocks are also tridiagonal, it may be shown that the computational effort per iteration is the same for both point and block iterative methods. In this case the reduction in the number of iterations makes the block methods very attractive indeed.

5 PROJECTION TECHNIQUES

5.1 General Projection Processes

An idea that is basic to sparse linear systems and eigenvalue problems is that of projection processes [Saad, 1982]. Such processes have begun to be applied successfully to Markov chain problems [Philippe et al., 1992]. Whereas iterative methods begin with an approximate solution vector that is modified at each iteration and which (supposedly) converges to a solution, projection methods create vector subspaces and search for the best possible approximation to the solution that can be obtained from that subspace. With a given subspace, for example, it is possible to extract a vector z that is a linear combination of a set of basis vectors for that space and which minimizes $\|b - Az\|$ in some vector norm. This vector z may then be taken as an approximation to the solution of $Ax = b$.

More precisely, given a subspace $\mathcal{K}$ spanned by a system of m vectors $V \equiv [v_1, \ldots, v_m]$, a projection process onto $\mathcal{K} \equiv \text{span}\{V\}$ finds an approximation to the original problem in the subspace $\mathcal{K}$. For a linear system $Ax = b$, this is done by writing $x = Vy$ and requiring that the residual vector $b - AVy$ be orthogonal to some subspace $\mathcal{L}$, not necessarily equal to $\mathcal{K}$. If a basis for $\mathcal{L}$ is $W = \text{span}\{w_1, w_2, \ldots, w_m\}$ then this gives

$$W^T(b - AVy) = 0$$

We solve for y and multiply the result by V to obtain

$$x = Vy = V[W^T AV]^{-1} W^T b.$$

For an eigenvalue problem $Ax = \lambda x$, we seek an approximate eigenvalue $\tilde{\lambda} \in \mathcal{C}$ and an approximate eigenvector $\tilde{x} \in \mathcal{K}$ such that the residual vector $A\tilde{x} - \tilde{\lambda}\tilde{x}$ is orthogonal to the subspace $\mathcal{L}$. Again writing $x = Vy$ this yields,

$$W^T(AVy - \tilde{\lambda}Vy) = 0,$$

or

$$W^T AVy = \tilde{\lambda} W^T Vy,$$

which is a generalized eigenvalue problem of dimension m. The minimum assumptions that must be made in order for these projection processes to be feasible are that $W^T AV$ be nonsingular for linear systems and that $W^T V$ be nonsingular for eigenvalue problems. Clearly this will provide m approximate eigenpairs λ_i, x_i. In most algorithms, the matrix $W^T V$ is the identity matrix, in which case the approximate eigenvalues of A are the eigenvalues of the $m \times m$ matrix $C = W^T AV$. The corresponding approximate eigenvectors are the vectors Vy_i where y_i are the eigenvectors of C. Similarly the approximate Schur vectors are the vector columns of VU, where $U = [u_1, u_2, \ldots, u_m]$ are the Schur vectors of C, i.e., $U^H CU$ is quasi-upper triangular, (see, for example, [Saad, 1995a], pp. 17–18). A common particular case is when $\mathcal{K} = \mathcal{L}$ and $V = W$ is an orthogonal basis of $\mathcal{K}$. This is then referred to as an orthogonal projection process.

5.2 Subspace Iteration

One method for computing invariant subspaces is called subspace iteration. In its simplest form, subspace iteration can be described as follows; see [Jennings and Stewart, 1975], [Jennings and Stewart, 1981], [Stewart, 1976], [Stewart and Jennings, 1981] for details.

Algorithm: Subspace Iteration

1. Choose an initial orthonormal system $V_0 \equiv [v_1, v_2, \ldots, v_m]$ and an integer k;

2. Compute $X = A^k V_0$ and orthonormalize X to get V.

3. Perform an orthogonal projection process onto span$\{V\}$.

4. Test for convergence. If satisfied then exit, else continue.

5. Take $V_0 = VU$, the set of approximate Schur vectors, (alternatively take $V_0 = VY$, the set of approximate eigenvectors), choose a new k and go to 2.

The above algorithm utilizes the matrix A only to compute successive matrix by vector products $w = Av$, so sparsity can be exploited. Often the parameter k is set equal to one, although in some cases it may be advantageous to choose it to be somewhat larger. This results in less iterations but at a greater cost in terms of memory and time per iteration. However, subspace iteration can be a rather slow method, often much slower than some of the alternatives to be described next. In fact a more satisfactory alternative is to use a Chebyshev-Subspace iteration: step 2 is replaced by $X = t_k(A)V_0$, where t_k is obtained from the Chebyshev polynomial of the first kind of degree k, by a linear change of variables. The three-term recurrence of Chebyshev polynomials allows us to compute a vector $w = t_k(A)v$ at almost the same cost as $A^k v$. Performance can be dramatically improved. Details on implementation and some experiments are described in [Saad, 1984].

5.3 Arnoldi's Method

A second technique discussed in the literature is Arnoldi's method [Arnoldi, 1951, Saad, 1980] which is an orthogonal projection process onto $\mathcal{K}_m = \text{span}\{v_1, Av_1, \ldots, A^{m-1}v_1\}$. The algorithm starts with some nonzero vector v_1 and generates the sequence of vectors v_i from the following algorithm,

Algorithm: Arnoldi

1. *Initialize:*

   Choose an initial vector v_1 of norm unity.

2. *Iterate:* Do $j = 1, 2, \ldots, m$

   (a) Compute $w := Av_j$

   (b) Compute a set of j coefficients h_{ij} so that $w := w - \sum_{i=1}^{j} h_{ij} v_i$ is orthogonal to all previous v_i's.

   (c) Compute $h_{j+1,j} = \|w\|_2$ and $v_{j+1} = w/h_{j+1,j}$.

By construction, the above algorithm produces an orthonormal basis of the so-called Krylov subspace $\mathcal{K}_m = \text{span}\{v_1, Av_1, \ldots, A^{m-1}v_1\}$. Notice that this space is constructed from successive iterates of the power method which are then orthogonalized with respect to one another. The $m \times m$ upper Hessenberg matrix H_m consisting of the coefficients h_{ij} computed by the algorithm represents the restriction of the linear transformation A to the subspace $\mathcal{K}_m$, with respect to this basis, i.e., we have $H_m = V_m^T A V_m$, where $V_m = [v_1, v_2, \ldots, v_m]$. Approximations to some of the eigenvalues of A can be obtained from the eigenvalues of H_m. This is Arnoldi's method in its simplest form.

As m increases, the eigenvalues of H_m that are located in the outmost part of the spectrum start converging towards corresponding eigenvalues of A. In practice, however, one difficulty with the above algorithm is that as m increases cost and storage increase rapidly. One solution is to use the method iteratively: m is fixed and the initial vector v_1 is taken at each new iteration as a linear combination of some of the approximate eigenvectors. Moreover, there are several ways of accelerating convergence by preprocessing v_1 by a Chebyshev iteration before restarting, i.e., by taking $v_1 = t_k(A)z$ where z is again a linear combination of eigenvectors.

5.4 Preconditioned GMRES for Singular Systems

In this section we adopt the viewpoint that we are trying to solve the homogeneous system
$$Ax = 0 \tag{10}$$
The case of interest to us is when there is a non-trivial solution to (10), i.e., when A is singular. Then the solution is clearly non-unique and one may wonder whether or not this can cause the corresponding iterative schemes to fail. The answer is usually no and we will illustrate in this section how standard Krylov subspace methods can be used to solve (10), [Saad, 1981]. We start by describing the GMRES (generalized minimum residual) algorithm, [Saad and Schultz, 1986], for solving the more common linear system
$$Ax = b. \tag{11}$$
in which A is nonsingular. GMRES is a least squares procedure for solving (11) on the Krylov subspace $\mathcal{K}_m$. More precisely, assume that we are given an initial guess x_0 to (11) with residual $r_0 = b - Ax_0$. Let us take $v_1 = r_0/\|r_0\|_2$ and perform m steps of Arnoldi's method as described earlier. We seek an approximation to (11) of the form $x_m = x_0 + \delta_m$ where δ_m belongs to $\mathcal{K}_m$. Moreover, we need this approximation to minimize the residual norm, $\|b - Ax_k\|_2$, over $\mathcal{K}_m$. Writing $\delta_m = V_m y_m$ we see that y_m must minimize the following function of y,
$$\begin{aligned} J(y) &= \|b - A(x_0 + V_m y)\|_2 \\ &= \|r_0 - AV_m y\|_2 \\ &= \|\|r_0\|e_1 - AV_m y\|_2. \end{aligned}$$

Setting $\beta \equiv \|r_0\|_2$ and using the fact that $AV_m = V_{m+1}\overline{H}_m$ this becomes

$$J(y) = \|V_{m+1}[\beta e_1 - \overline{H}_m y]\|_2 = \|\beta e_1 - \overline{H}_m y\|_2$$

from the orthogonality of V_{m+1}. As a result the vector y_m can be obtained inexpensively by solving a $(m+1) \times m$ least squares problem. We should point out that this procedure is also a projection process. More precisely, the minimization of $J(y)$ is equivalent to imposing the Gram condition that

$$r_0 - AV_m y \perp v \quad \forall\, v \in \text{span}\{AV_m\},$$

which means that we are solving $A\delta = r_0$ with a projection process with

$$\mathcal{K} = \text{span}\{r_0, Ar_0, \cdots, A^{m-1} r_0\}$$

and $\mathcal{L} = A\mathcal{K}$. Details of this algorithm can be found in [Saad and Schultz, 1986].

5.5 Preconditioned Arnoldi and GMRES Algorithms

Preconditioning techniques can also be used to improve the convergence rates of Arnoldi's method and GMRES. This typically amounts to replacing the original system (10) by, for example, the system,

$$M^{-1} Ax = 0,$$

where M is a matrix such that $M^{-1}w$ is inexpensive to compute for any vector w.

Thus in both the Arnoldi and the GMRES case, we only need to replace the original matrix A in the corresponding algorithm by the preconditioned matrix $M^{-1}A$. We may also precondition from the right, i.e., we may replace A by AM^{-1}. In this situation if $AM^{-1}z = 0$ we should note that the solution to the original problem is $M^{-1}z$, which requires one additional solve with M. If $M = LU$ then the preconditioning can also be split between left and right, by replacing the original matrix by the preconditioned matrix $L^{-1}AU^{-1}$.

6 DECOMPOSITIONAL METHODS

A decompositional approach to solving Markov chains is intuitively very attractive since it appeals to the principle of divide and conquer: if the model is too large or complex to analyze in toto, it is divided into subsystems each of which is analyzed separately and a global solution then constructed from the partial solutions. Some of these methods have been applied to large economic models and most recently to the analysis of computer systems. Currently, there is a very large research effort being devoted to these methods, by many different research groups. (See, for example, [Balsamo and Pandolfi, 1988], [Cao and Stewart, 1985], [Chatelin, 1984b], [Chatelin, 1984a], [Chatelin and Miranker, 1982], [Courtois and Semal, 1984], [Dayar and Stewart, 1997], [Feinberg and Chiu, 1987], [Haviv, 1987], [Haviv, 1989], [Kim and Smith, 1991],

[Koury et al., 1984], [Krieger, 1995], [Leutenegger and Horton, 1995], [McAllister et al., 1984], [Meyer, 1989], [Rieders, 1995], [Schweitzer, 1983], [Schweitzer, 1986], [Schweitzer and Kindle, 1986], [Stewart, 1983], [Stewart et al., 1993], [Stewart, 1991], [Sumita and Rieders, 1991], [Takahashi, 1975], [Vantilborgh, 1985] and [Zarling, 1976]. With the advent of parallel and distributed computing systems, their advantages are immediately obvious.

Ideally the problem is broken into subproblems that can be solved independently and the global solution obtained by "pasting" together the subproblem solutions. Although it is rare to find Markov chains that can be divided into independent subchains, it is not unusual to have Markov chains in which this condition almost holds. An important class of problems that frequently arise in Markov modeling are those in which the state space may be partitioned into disjoint subsets with strong interactions among the states of a subset but with weak interactions among the subsets themselves. Such problems are sometimes referred to as *nearly-completely-decomposable (NCD), nearly uncoupled,* or *nearly separable.* It is apparent that the assumption that the subsystems are independent and can therefore be solved separately does not hold. Consequently an error arises. This error will be small if the assumption is approximately true.

The pioneering work on NCD systems was performed by Simon and Ando, [Simon and Ando, 1961], in investigating the dynamic behavior of linear systems as they apply to economic models. The concept was later extended to Markov chains and the performance analysis of computer systems by Courtois, [Courtois, 1977]. The technique is founded on the idea that it is easy to analyze large systems in which all the states can be partitioned into groups in which

- interactions among the states of a group may be studied as if the interactions among groups do not exist, and

- interactions among groups may be studied without reference to the interactions which take place within groups.

Simon and Ando showed that in NCD systems (in which the above conditions are approximated), the dynamic behavior of the system may be divided into a *short-run* dynamics period, and a *long-run* dynamics period. Specifically, they proved the following results:

- In the short-run dynamics, the strong interactions within each subsystem are dominant and quickly force each subsystem to a local equilibrium almost independently of what is happening in the other subsystems.

- In the long-run dynamics, the strong interactions within each subsystem maintain approximately the relative equilibrium attained during the short-run dynamics but now the weak interactions among groups begin to become apparent and the whole system moves towards a global equilibrium in which the relative equilibrium values attained by the states at the end of the short run dynamics period are maintained.

Strong interactions among the states of a group and weak interactions among the groups themselves imply that the states of a nearly-completely-decomposable Markov chain can be ordered so that the stochastic matrix of transition probabilities has a block structure in which the nonzero elements of the off-diagonal blocks are small compared to those of the diagonal blocks. The irreducible, stochastic matrix P of order n, may be written as

$$P = \begin{pmatrix} P_{11} & P_{12} & \cdots & P_{1N} \\ P_{21} & P_{22} & \cdots & P_{2N} \\ \vdots & \vdots & \vdots & \vdots \\ P_{N1} & P_{N2} & \cdots & P_{NN} \end{pmatrix},$$

in which the subblocks P_{ii} are square, of order $n_i, i = 1, 2, \ldots, N$ with $n = \sum_{i=1}^{N} n_i$. We shall assume that

$$\|P_{ii}\| = O(1), \quad i = 1, 2, \ldots, N, \tag{12}$$

$$\|P_{ij}\| = O(\epsilon), \quad i \neq j, \tag{13}$$

where $\|.\|$ denotes the spectral norm of a matrix and ϵ is a sufficiently small positive number. We partition π conformally with P, i.e., $\pi = (\pi_1, \pi_2, \ldots, \pi_N)$ and π_i is a (row) vector of length n_i.

Following the reasoning of Simon and Ando, an initial approach to determining the solution of $\pi P = \pi$ in the more general case when $P_{ij} \neq 0$ is to assume that the system is completely decomposable and to compute the stationary probability distribution for each component. A first problem that arises with this approach is that the P_{ii} are not stochastic but rather strictly substochastic. A possible solution to this problem is to make them stochastic by adding the probability mass which is to be found on the off-diagonal blocks, $P_{ij}, j = 1, \ldots, N$ and $j \neq i$, into the diagonal block P_{ii} on a row by row basis. This off-diagonal probability mass can be accumulated into the diagonal block in a number of ways. For example, it can be simply added into the diagonal elements of the diagonal block (but this can lead to reducible blocks and numerical difficulties); it can be added into the reverse diagonal elements of the diagonal block to reduce the chance that the diagonal block becomes reducible; it can be distributed along the elements of a row of the block in a random fashion, etc. The way in which it is added to the diagonal block will have an effect on the accuracy of the results obtained. In particular, it may be noted that there exists a distribution of this probability mass that results in the exact answer π_i being obtained up to a multiplicative constant, [Meyer, 1989]. Unfortunately it is not known how to determine this distribution without either a knowledge of the stationary probability vector π itself, or performing extensive calculations possibly in excess of that required to compute the exact solution. A simple way around the problem of distributing the probability mass is to simply ignore it; i.e., to work directly with the substochastic matrices P_{ii} themselves. In other words, we may use the normalized eigenvector corresponding to the Perron root (the eigenvalue closest to 1) of block P_{ii} as the probability vector whose elements denote the probabilities of being in the states of this block, conditioned on the system occupying one of the states of the block.

A second problem is that once we have computed the stationary probability vector for each block, simply concatenating them together will not give a probability vector. The elements of each subvector sum to one. We still need to weigh each of the probability subvectors by a quantity that is the probability of being in that subblock of states. In other words, the probability distribution computed from the P_{ii} are conditional probabilities in the sense that they express the probability of being in a given state of the subset conditioned on the fact that the system is in one of the states of that subset. We need to remove that condition.

To determine the probability of being in a given block of states we need to construct a matrix whose element ij gives the probability of a transition from block i to block j. This must be an $(N \times N)$ stochastic matrix, and, in accordance with the Simon and Ando theory, should characterize the interactions among blocks. To construct this matrix we need to shrink each block P_{ij} of P down to a single element. This is accomplished by first replacing each row of each block by the sum of the elements in that block row. The sum of the elements of row k of block ij gives the probability of making a transition from state k of block i into one of the states of block j. It no longer matters to us which particular state of block j is this destination state. Mathematically, the operation performed for each block is $P_{ij}e$.

To complete the operation, we need to reduce each column vector, $P_{ij}e$, to a scalar. As we have just noted, the kth element of the vector in position ij is the probability of leaving state k of block i and entering into block j. To determine the total probability of leaving (any state of) block i to enter into (any state of) block j we need to sum the elements of this vector after each element has been weighed by the probability of being in that state, (given that the system is in one of the states of that block). These weighing factors may be obtained from the elements of the stationary probability vector. They are the components of $\pi_i/||\pi_i||_1$. The ij^{th} element of the reduced $(N \times N)$ matrix is therefore given by

$$(C)_{ij} = \frac{\pi_i}{||\pi_i||_1} P_{ij} e = \phi_i P_{ij} e,$$

where $\phi_i = \pi_i/||\pi_i||_1$. The matrix C is often referred to as the *aggregation matrix* or *coupling matrix*.

If P is an irreducible stochastic matrix, then C also is irreducible and stochastic. Let ξ denote its left eigenvector, i.e., $\xi C = \xi$ and $\xi e = 1$. The ith component of ξ is the stationary probability of being in (one of the states of) block i. It is easy to show that

$$\xi = (||\pi_1||_1, ||\pi_2||_1, ..., ||\pi_N||_1).$$

Of course, the vector π is not yet known, so that it is not possible to compute the weights $||\pi_i||_1$. However they may be approximated by using the probability vector computed from each of the individual P_{ii}. Consequently, the weights ξ_i can be estimated and an approximate solution to the stationary probability vector, π, can be obtained.

After this sequence of operations is performed, the result is an approximation

$$(\xi_1 u_1, \xi_2 u_2, ..., \xi_N u_N)$$

to π, in which the u_i are approximations to $\pi_i / \|\pi_i\|_1$. The question now arises as to whether we can incorporate this approximation back into the decomposition algorithm to get an even better approximation. Note, however, that the u_i are used to compute the aggregation matrix and that using $\xi_i u_i$ in their place will have no effect on the probability vector which we compute from the aggregation matrix. It was found, however, that applying a power step to the approximation before plugging it back into the decomposition method had a very salutary effect. Later this power step was replaced by a block Gauss-Seidel step and became known as a disaggregation step; forming and solving the matrix C being the aggregation step. The complete procedure follows. The iteration number is indicated by a superscript in parenthesis on the appropriate variable names.

Algorithm: Iterative Aggregation/Disaggregation

1. Let $\pi^{(0)} = (\pi_1^{(0)}, \pi_2^{(0)}, ..., \pi_N^{(0)})$ be a given initial approximation to the solution π, and set $m = 1$.

2. Compute $\phi^{(m-1)} = (\phi_1^{(m-1)}, \phi_2^{(m-1)}, ..., \phi_N^{(m-1)})$, where

$$\phi_i^{(m-1)} = \frac{\pi_i^{(m-1)}}{\|\pi_i^{(m-1)}\|_1}; \quad i = 1, 2, ..., N.$$

3. Construct the aggregation matrix $C^{(m-1)}$ whose elements are given by

$$(C^{(m-1)})_{ij} = \phi_i^{(m-1)} P_{ij} e.$$

4. Solve the eigenvector problem

$$\xi^{(m-1)} C^{(m-1)} = \xi^{(m-1)}, \quad \|\xi^{(m-1)}\|_1 = \sum_{i=1}^{N} \xi_i^{(m-1)} = 1 \quad (14)$$

5. (a) Compute the row vector

$$z^{(m)} = (\xi_1^{(m-1)} \phi_1^{(m-1)}, \xi_2^{(m-1)} \phi_2^{(m-1)}, ..., \xi_N^{(m-1)} \phi_N^{(m-1)}).$$

   (b) Solve the following N systems of equations to find $\pi^{(m)}$:

$$\pi_k^{(m)} = \pi_k^{(m)} P_{kk} + \sum_{j>k} z_j^{(m)} P_{jk} + \sum_{j<k} \pi_j^{(m)} P_{jk}, \quad k = 1, 2, ..., N. \quad (15)$$

6. Conduct a test for convergence. If the estimated accuracy is sufficient, then stop and take $\pi^{(m)}$ to be the required solution vector. Otherwise set $m = m + 1$ and go to step 2.

Let the matrix $(I - P)$ have the decomposition
$$(I - P) = D - L - U,$$
where D, L, and U are respectively block-diagonal, strictly block-lower-triangular and strictly block-upper-triangular matrices. In other words:
$$D = Diag\{I - P_{11}, I - P_{22}, ..., I - P_{NN}\},$$

$$\begin{aligned} L_{ij} &= P_{ij} \text{ if } i > j; \\ &= 0 \text{ otherwise}; \end{aligned}$$

$$\begin{aligned} U_{ij} &= P_{ij} \text{ if } i < j; \\ &= 0 \text{ otherwise}. \end{aligned}$$

Let
$$I^{(m-1)} = Diag\left\{\frac{\xi_1^{(m-1)}}{||\pi_1^{(m-1)}||_1}I, \ldots, \frac{\xi_N^{(m-1)}}{||\pi_N^{(m-1)}||_1}I\right\}.$$
Then
$$z^{(m)} = \pi^{(m-1)} I^{(m-1)}.$$
Furthermore, it is easy to show that the block Gauss-Seidel iteration method applied to step 5b can be written as
$$\pi^{(m)} = z^{(m)} L(D - U)^{-1}.$$
Therefore, this aggregation/disaggregation method is equivalent to the iterative formula
$$\pi^{(m)} = \pi^{(m-1)} I^{(m-1)} L(D - U)^{-1}. \tag{16}$$
It may be shown that π, the exact stationary probability vector, is a fixed point of equation (16).

We now turn our attention to some implementation details. The critical points are steps 3 through 5. In step 3, it is more efficient to compute $P_{ij}e$ only once for each block and to store it somewhere for use in all future iterations. This is only possible if sufficient memory is available; otherwise it is necessary to compute it each time it is needed.

To obtain the vector ξ from equation (14), the coupling system in step 4, any of the methods discussed in the previous section may be used, for the vector ξ

is simply the stationary probability vector of an irreducible stochastic matrix C.

In step 5, each of the N systems of equations in (15) can be written as $Bx = r$ where $B = (I - P_{kk})^T$ and

$$r = \sum_{j>k} z_j P_{jk} + \sum_{j<k} \pi_j P_{jk}, \quad k = 1, 2, \ldots, N.$$

In all cases, P_{kk} is a strictly substochastic matrix so that B is nonsingular. The vector r will have small norm if the system is NCD. If a direct method is used, the LU decomposition of $(I - P_{kk})$, $k = 1, 2, \ldots, N$ need only be performed once, since this remains unchanged from one iteration to the next. If an iterative method is used we have an iteration algorithm within an iteration algorithm. In this case it is advantageous to perform only a small number of iterations, (e.g., 30–40 of Gauss-Seidel) each time a solution of $(I - P_{kk})x = r$ is needed but to use the final approximation at one step as the initial approximation the next time the solution of that same subsystem is needed.

In these methods, it is important to order the states so that the matrix has the block structure of equations (12) and (13). Only after reordering the states, can we guarantee that the resulting transition matrix will have the property that directly reflects the structural characteristics of the NCD system. This may be accomplished by treating the Markov chain as a directed graph and utilizing some of the graph algorithms of Tarjan [Tarjan, 1972]. We have been successful in using a *depth-first search* (DFS) algorithm which searches in the forward (deeper) direction as long as possible. Details of the non-recursive algorithm of DFS are given in [Aho et al., 1974]. Coding details for DFS are given in [Hopcroft and Tarjan, 1973]. The complexity of this algorithm is $O(|V|+|E|)$, where $|V|$ is the number of vertices and $|E|$ is the number of edges in the graph.

For more information on Iteration/Aggregation algorithms, including implementation details and the results of many test problems, the interested reader is referred to [Stewart and Wu, 1992].

References

[Aho et al., 1974] Aho, A., Hopcroft, J., and Ullman, J. (1974). *The Design and Analysis of Computer Algorithms*. Addison-Wesley, Reading, Mass.

[Alexsson, 1985] Alexsson, O. (1985). A survey of preconditioned iterative methods for linear systems of algebraic equations. *BIT*, 25:166–187.

[Arnoldi, 1951] Arnoldi, W. (1951). The principle of minimized iteration in the solution of the matrix eigenvalue problem. *Quart. Appl. Math*, 9:17–29.

[Balsamo and Pandolfi, 1988] Balsamo, S. and Pandolfi, B. (1988). Bounded aggregation in Markovian networks. In *Computer Performance and Reliability*, pages 73–92. North Holland, Amsterdam.

[Barker and Plemmons, 1986] Barker, G. and Plemmons, R. (1986). Convergent iterations for computing stationary distributions of Markov chains. *SIAM J. Alg. Disc. Meth*, 7:390–398.

[Berman and Plemmons, 1979] Berman, A. and Plemmons, R. (1979). *Nonnegative Matrices in the Mathematical Sciences*. Academic Press, New York.

[Cao and Stewart, 1985] Cao, W. and Stewart, W. (1985). Iterative aggregation/disaggregation techniques for nearly uncoupled Markov chains. *J. Assoc. Comp. Mach*, 32 3:702–719.

[Chatelin, 1984a] Chatelin, F. (1984a). Iterative aggregation/disaggregation methods. In Iazeolla, G., Courtois, P., and Hordijk, A., editors, *Mathematical Computer Performance and Reliability*, pages 199–207. North Holland, Amsterdam.

[Chatelin, 1984b] Chatelin, F. (1984b). *Spectral Approximation of Linear Operators*. Academic Press, New York.

[Chatelin and Miranker, 1982] Chatelin, F. and Miranker, W. (1982). Acceleration by aggregation of successive approximation methods. *Linear Algebra Appl*, 43:17–47.

[Courtois, 1977] Courtois, P. (1977). *Decomposability; Queueing and Computer System Applications*. Academic Press, Orlando, Florida.

[Courtois and Semal, 1984] Courtois, P. and Semal, P. (1984). Bounds for the positive eigenvectors of nonnegative matrices and their approximation by decomposition. *J. Assoc. Comp. Mach.*, 31:804–825.

[Courtois and Semal, 1986] Courtois, P. and Semal, P. (1986). Block iterative algorithms for stochastic matrices. *Linear Algebra Appl*, 76:59–70.

[Dayar and Stewart, 1996] Dayar, T. and Stewart, W. (1996). On the effects of using the Grassmann-Taksar-Heyman method in iterative aggregation-disaggregation. *SIAM Journal on Scientific Computing*, 17:1–17.

[Dayar and Stewart, 1997] Dayar, T. and Stewart, W. (1997). Quasi-lumpability, lower bounding coupling matrices and nearly completely decomposable Markov chains. *SIAM Journal on Matrix Analysis and Applications*, 18:482–498.

[Feinberg and Chiu, 1987] Feinberg, B. and Chiu, S. (1987). A method to calculate steady state distributions of large Markov chains by aggregating states. *Operations Research*, 35:282–290.

[Funderlic and Meyer, 1986] Funderlic, R. and Meyer, C. (1986). Sensitivity of the stationary distribution vector for an ergodic Markov chain. *Linear Algebra Appl*, 76:1–17.

[Funderlic and Plemmons, 1984] Funderlic, R. and Plemmons, R. (1984). A combined direct-iterative method for certain M-matrix linear systems. *SIAM J. Alg. Disc. Meth*, 5(1):32–42.

[Funderlic and Plemmons, 1986] Funderlic, R. and Plemmons, R. (1986). Updating LU factorizations for computing stationary distributions. *SIAM J. Alg. Disc. Meth*, 7:30–42.

[Golub and Meyer, 1986] Golub, G. and Meyer, C. (1986). Using the QR factorization and group inversion to compute, differentiate and estimate the sensitivity of stationary distributions for Markov chains. *SIAM J. Alg. Disc. Meth*, 7:273–281.

[Grassmann et al., 1985] Grassmann, W., Taksar, M., and Heyman, D. (1985). Regenerative analysis and steady state distributions for Markov chains. *Operations Research*, 33:1107–1116.

[Harrod and Plemmons, 1984] Harrod, W. and Plemmons, R. (1984). Comparisons of some direct methods for computing stationary distributions of Markov chains. *SIAM J. Sci. Comput*, 5:453–469.

[Haviv, 1987] Haviv, M. (1987). Aggregation/disaggregation methods for computing the stationary distribution of a Markov chain. *SIAM J. Numer. Anal*, 24:952–966.

[Haviv, 1989] Haviv, M. (1989). More on a Rayleigh-Ritz refinement technique for nearly uncoupled stochastic matrices. *SIAM J. Matrix Anal. Appl*, 10:287–293.

[Heyman, 1987] Heyman, D. (1987). Further comparisons of direct methods for computing stationary distributions of Markov chains. *SIAM J. Alg. Disc. Meth*, 8:226–232.

[Hopcroft and Tarjan, 1973] Hopcroft, J. and Tarjan, R. (1973). Efficient algorithms for graph manipulation. *CACM*, 16(6):372–378.

[Jennings and Stewart, 1975] Jennings, A. and Stewart, W. (1975). Simultaneous iteration for partial eigensolution of real matrices. *J. IMA*, 15:351–361.

[Jennings and Stewart, 1981] Jennings, A. and Stewart, W. (1981). A simultaneous iteration algorithm for real matrices. *ACM Trans. of Math. Software*, 7:184–198.

[Kaufman, 1983] Kaufman, L. (1983). Matrix methods for queueing problems. *SIAM J. Sci. Comput*, 4:525–552.

[Kim and Smith, 1991] Kim, D. and Smith, R. (1991). An exact aggregation algorithm for mandatory set decomposable Markov chains. In Stewart, W. J., editor, *Numerical Solution of Markov Chains*. Marcel Dekker, New York, NY.

[Koury et al., 1984] Koury, R., McAllister, D., and Stewart, W. (1984). Iterative methods for computing stationary distributions of nearly completely decomposable Markov chains. *SIAM J. Alg. Disc. Math*, 5(2):164–186.

[Krieger, 1995] Krieger, U. (1995). Numerical solution of large finite Markov chains by algebraic multigrid techniques. In Stewart, W. J., editor, *Computations with Markov Chains*. Kluwer Academic Publishers, Boston.

[Leutenegger and Horton, 1995] Leutenegger, S. and Horton, G. (1995). On the utility of the multi-level algorithm for the solution of nearly completely decomposable Markov chains. In Stewart, W. J., editor, *Computations with Markov Chains*. Kluwer Academic Publishers, Boston.

[Lubachevsky and Mitra, 1986] Lubachevsky, B. and Mitra, D. (1986). A chaotic asynchronous algorithm for computing the fixed point of a nonnegative matrix of unit spectral radius. *J. Asoc. Comput. Mach*, 33:130–150.

[McAllister et al., 1984] McAllister, D., Stewart, G., and Stewart, W. (1984). On a Raleigh-Ritz refinement technique for nearly uncoupled stochastic matrices. *Linear Alg. Applications*, 60:1–25.

[Meyer, 1989] Meyer, C. (1989). Stochastic complementation, uncoupling Markov chains and the theory of nearly reducible systems. *SIAM Rev*, 31:240–272.

[Mitra and Tsoucas, 1988] Mitra, D. and Tsoucas, P. (1988). Relaxations for the numerical solutions of some stochastic problems. *Stochastic Models*, 4(3):387–419.

[Philippe et al., 1992] Philippe, B., Saad, Y., and Stewart, W. (1992). Numerical methods in Markov chain modelling. *Operations Research*, 40(6):1156–1179.

[Rieders, 1995] Rieders, M. (1995). State space decomposition for large Markov chains. In Stewart, W. J., editor, *Computations with Markov Chains*. Kluwer Academic Publishers, Boston.

[Saad, 1980] Saad, Y. (1980). Variations on Arnoldi's method for computing eigenelements of large unsymmetric matrices. *Lin. Alg. Appl*, 34:269–295.

[Saad, 1981] Saad, Y. (1981). Krylov subspace methods for solving large unsymmetric linear systems. *Math. Comp*, 37:105–126.

[Saad, 1982] Saad, Y. (1982). Projection methods for solving large sparse eigenvalue problems. In Kagstrom, B. and Ruhe, A., editors, *Matrix Pencils, Proceedings, Pitea Havsbad*, pages 121–144. University of Umea, Sweden, Springer Verlag, Berlin.

[Saad, 1984] Saad, Y. (1984). Chebyshev acceleration techniques for solving non-symmetric eigenvalue problems. *Mathematics of Computation*, 42:567–588.

[Saad, 1995a] Saad, Y. (1995a). *Iterative Methods for Sparse Linear Systems*. PWS Publishing Company, Boston, MA.

[Saad, 1995b] Saad, Y. (1995b). Preconditioned Krylov subspace methods for the numerical solution of Markov chains. In Stewart, W. J., editor, *Computations with Markov Chains*. Kluwer Academic Publishers, Boston.

[Saad and Schultz, 1986] Saad, Y. and Schultz, M. (1986). GMRES: A generalized minimal residual algorithm for solving non-symmetric linear systems. *SIAM J. Sci. Stat. Comput*, 7:856–869.

[Schweitzer, 1983] Schweitzer, P. (1983). Aggregation methods for large Markov chains. In *Models of Computer Communication Systems*, pages 225–234. University of Pisa, Italy. International Workshop on Applied Mathematics and Performance Reliability.

[Schweitzer, 1986] Schweitzer, P. (1986). Perturbation series expansions for nearly completely decomposable Markov chains. In Boxma, O., Cohen, J., and Tijms, H., editors, *Teletraffic Analysis and Computer Performance Evaluation*, pages 319–328. Elsevier North-Holland, Amsterdam.

[Schweitzer and Kindle, 1986] Schweitzer, P. and Kindle, K. (1986). An iterative aggregation-disaggregation algorithm for solving linear systems. *Applied Math. and Comp*, 18:313–353.

[Sheskin, 1985] Sheskin, T. (1985). A Markov chain partitioning algorithm for computing steady state probabilities. *Operations Research*, 33:228–235.

[Simon and Ando, 1961] Simon, H. and Ando, A. (1961). Aggregation of variables in dynamic systems. *Econometrica*, 29:111–138.

[Stewart, 1976] Stewart, G. (1976). Simultaneous iteration for computing invariant subspaces of non-Hermitian matrices. *Numer. Mat*, 25:123–136.

[Stewart, 1983] Stewart, G. (1983). Computable error bounds for aggregated Markov chains. *J. Assoc. Comp. Mach.*, 30:271–285.

[Stewart, 1991] Stewart, G. (1991). On the sensitivity of nearly uncoupled Markov chains. In Stewart, W. J., editor, *Numerical Solution of Markov Chains*. Marcel Dekker, New York, NY.

[Stewart et al., 1993] Stewart, G., Stewart, W., and McAllister, D. (1993). A two stage iteration for solving nearly uncoupled Markov chains. In *Recent Advances in Iterative Methods*, volume 60 of *IMA Volumes in Mathematics and its Applications*, pages 201–216. Springer Verlag, New York.

[Stewart, 1978] Stewart, W. (1978). A comparison of numerical techniques in Markov modelling. *Comm. ACM*, 21:144–151.

[Stewart, 1994] Stewart, W. (1994). *An Introduction to the Numerical Solution of Markov Chains*. Princeton University Press,, New Jersey.

[Stewart and Jennings, 1981] Stewart, W. and Jennings, A. (1981). A simultaneous iteration algorithm for real matrices. *ACM Trans. Math. Software*, 7:184–198.

[Stewart and Wu, 1992] Stewart, W. and Wu, W. (1992). Numerical experiments with iteration and aggregation for Markov chains. *ORSA Journal on Computing*, 4:336–350.

[Sumita and Rieders, 1991] Sumita, U. and Rieders, M. (1991). A comparison of the replacement process with aggregation-disaggregation. In Stewart, W. J., editor, *Numerical Solution of Markov Chains*. Marcel Dekker, New York, NY.

[Takahashi, 1975] Takahashi, Y. (1975). A lumping method for numerical calculation of stationary distributions of Markov chains. Technical report, Department of Information Sciences, Tokyo Institute of Technology, Tokyo, Japan.

[Tarjan, 1972] Tarjan, R. (1972). Depth first search and linear graph algorithms. *SIAM J. Comput.*, 1(2):146–160.

[Vantilborgh, 1985] Vantilborgh, H. (1985). Agreggation with an error of $o(\epsilon^2)$. *JACM*, 32:161–190.

[Varga, 1962] Varga, R. (1962). *Matrix Iterative Analysis*. Prentice Hall, Englewood Cliffs, NJ.

[Wallace, 1973] Wallace, V. (1973). Towards an algebraic theory of Markovian networks. In *Proc. Symp. Computer-Communication Networks and Teletraffic*. Polytechnic Press, New York.

[Wallace and Rosenberg, 1966a] Wallace, V. and Rosenberg, R. (1966a). Markovian models and numerical analysis of computer system behavior. In *Proc. AFIPS Spring Joint Computer Conference*. AFIPS Press, New Jersey.

[Wallace and Rosenberg, 1966b] Wallace, V. and Rosenberg, R. (1966b). RQA-1: The recursive queue analyzer. Technical Report 2, Systems Engineering Laboratory, University of Michigan, Ann Arbor, Michigan.

[Young, 1971] Young, D. (1971). *Iterative solution of large linear systems*. Academic Press, New York.

[Zarling, 1976] Zarling, R. (1976). *Numerical solutions of nearly completely decomposable queueing networks*. PhD thesis, Department of Computer Science, University of North Carolina.

William J. Stewart William J. Stewart is director of the Operations Research Program and Professor of Computer Science at North Carolina State University. His research interests are in the numerical solution of Markov chains and he is the author of the text *An Introduction to the Numerical Solution of Markov Chains*, published by Princeton University Press. Stewart organized the first two international meetings in this area (in 1990 and 1995 respectively) and edited the proceedings of both. He is currently involved in organizing the third which will be held in September 1999, in Zaragoza, Spain. Also, Stewart has developed a software package, *MARCA* — *Markov Chain Analyzer* for the generation, characterization and numerical analysis of Markov chains. This package has been successfully employed in a number of areas and especially in the development and performance analysis of telecommunication models.

5 STOCHASTIC AUTOMATA NETWORKS

Brigitte Plateau[1]
and William J. Stewart[2]

[1]LMC - IMAG - INPG
Campus Universitaire, BP 53
38041 - Grenoble - Cedex 9
FRANCE
Brigitte.Plateau@imag.fr

[2]Department of Computer Science
Box 8206
North Carolina State University
Raleigh, NC 27695-802
billy@csc.ncsu.edu

1 INTRODUCTION

A *Stochastic Automata Network* (SAN) consists of a number of individual stochastic automata that operate more or less independently of each other. Each individual automaton, $\mathcal{A}$, is represented by a number of states and rules that govern the manner in which it moves from one state to the next. The state of an automaton at any time t is just the state it occupies at time t and the state of the SAN at time t is given by the state of each of its constituent automata.

The use of stochastic automata networks is becoming increasingly important in performance modeling issues related to parallel and distributed computer systems. As such models become increasingly complex, so also does the complexity of the modeling process. Although systems analysts have a number of other modeling strategems at their disposal, it is not unusual to discover that these are inadequate. The use of queueing network modeling is limited by the constraints imposed by assumptions needed to keep the model tractable.

The results obtained from the myriad of available approximate solutions are frequently too gross to be meaningful. Simulations can be excessively expensive. This leaves models that are based on Markov chains, but here also, the difficulties are well documented. As shown in Chapters 1 and 2, the size of the state space generated is so large that it effectively prohibits the computation of a solution. This is true whether the Markov chain results from a stochastic Petri net formalism, or from a straightforward Markov chain analyzer.

In many instances, the SAN formalism is an appropriate choice. Parallel and distributed systems are often viewed as collections of components that operate more or less independently, requiring only infrequent interaction such as synchronizing their actions, or operating at different rates depending on the state of parts of the overall system. This is exactly the viewpoint adopted by SANs. The components are modeled as individual stochastic automata that interact with each other. Furthermore, the state-space explosion problem associated with Markov chain models is mitigated by the fact that the state transition matrix is not stored, nor even generated. Instead, it is represented by a number of much smaller matrices, one for each of the stochastic automata that constitute the system, and from these all relevant information may be determined without explicitly forming the global matrix. The implication is that a considerable saving in memory is realized by storing the matrix in this fashion. We do not wish to give the impression that we regard SANs as a panacea for all modeling problems, just that there is a niche that it fills among the tools that modelers may use. It is fairly obvious that their memory requirements are minimal; it remains to show that this does not come at the cost of a prohibitive amount of computation time.

Stochastic Automata Networks and the related concept of *Stochastic Process Algebras* have become a hot topic of research in recent years. This research has focused on areas such as the development of languages for specifying SANs and their ilk, [Hermanns and Rettelbach, 1994, Hillston, 1995], and on the development of suitable solution methods that can operate on a transition matrix given as a compact SAN descriptor. The development of languages for specifying stochastic process algebras is mainly concerned with structural properties of the nets (compositionality, equivalence, etc.) and with the mapping of these specifications onto Markov chains for the computation of performance measures [Hillston, 1995, Baccelli et al., 1995, Buchholz, 1995]. Although a SAN may be viewed as a stochastic process algebra, its original purpose was to provide an efficient and convenient methodology for computing performance measures rather than a means of deriving algebraic properties of complex systems, [Plateau and Atif, 1991]. Nevertheless, computational results such as those discussed in this chapter can also be applied in the context of stochastic process algebras.

There are two overriding concerns in the application of any Markovian modeling methodology, viz., memory requirements and computation time. Since these are frequently functions of the number of states, a first approach is to develop techniques that minimize the number of states in the model. In SANs, it is possible to make use of symmetries as well as lumping and various superpositioning of the automata to reduce the computational burden, [Atif, 1992, Buchholz, 1992, Siegle, 1992]. Furthermore, in [Fourneau and Quessette, 1995], structural properties of the Markov chain graph (specificially the occurrence of cycles) are used to compute steady state solutions. We point out that similar, and even more extensive results have previously been developed in the context of Petri nets and stochastic activity networks. For example, in [Buchholz, 1993, Buchholz, 1992, Chiola et al., 1993, Franceschinis and Muntz, 1993, Sanders and Meyer, 1991, Simone and Marsan, 1991], equivalence relations and symmetries are used to decrease the computational burden of obtaining performance indices. In [Henderson and Lucic, 1993], reduction techniques for Petri nets are used in conjunction with insensitivity results to enable all computations to be performed on a reduced set of markings. In [Ciardo and Trivedi, 1991], nearly independent subnets are exploited in an iterative procedure in which a global solution is obtained from partial solutions. In [Donatelli, 1993] it is shown that the tensor structure of the transition matrix may be extracted from a stochastic Petri net, and in [Kemper, 1995] that this can be used efficiently to work with the reachable state space in an iterative procedure.

Once the number of states has effectively been fixed, the problem of memory and computation time still must be addressed, for the number of states left may still be large. With SANs, the use of a compact descriptor goes a long way to satisfying the first of these, although with the need to keep a minimum of two vectors of length equal to the global number of states, and considerably more than two for more sophisticated procedures such as the GMRES method, (see Chapter 4), we cannot afford to become complacent about memory requirements. As far as computation time is concerned, since the numerical methods used are iterative, it is important to keep both the number of iterations and the amount of computation per iteration to a minimum.

This chapter is laid out as follows. Sections 2 through 4 present an informal introduction to SANs and tensor algebra. This is followed in Section 5 with the presentation of a number of sufficient conditions for the existence of product forms in SANs, for in some restricted cases, product forms can indeed be found. We note in passing that in [Balbo et al., 1994, Boucherie, 1994, Donatelli and Serano, 1992, Frosh and Natarajan, 1991, Henderson and Taylor, 1991, Lazar and Robertazzi, 1991] product forms have been found in Petri net models, using either the structure of the state space or flow properties. The numerical issues of computation time and memory requirements in computing stationary distributions by means of iterative methods are discussed in Sections 6 through 10.

2 BASIC PROPERTIES OF TENSOR ALGEBRA

Define two matrices A and B as follows:

$$A = \begin{pmatrix} a_{11} & a_{12} \\ a_{21} & a_{22} \end{pmatrix} \quad \text{and} \quad B = \begin{pmatrix} b_{11} & b_{12} & b_{13} & b_{14} \\ b_{21} & b_{22} & b_{23} & b_{24} \\ b_{31} & b_{32} & b_{33} & b_{34} \end{pmatrix}.$$

The *tensor product* $C = A \otimes B$ is given by

$$C = \begin{pmatrix} a_{11}B & a_{12}B \\ a_{21}B & a_{22}B \end{pmatrix} \qquad (1)$$

In general, to define the tensor product of two matrices, A of dimensions $(\rho_1 \times \gamma_1)$ and B of dimensions $(\rho_2 \times \gamma_2)$, it is convenient to observe that the tensor product matrix has dimensions $(\rho_1 \rho_2 \times \gamma_1 \gamma_2)$ and may be considered as consisting of $\rho_1 \gamma_1$ blocks each having dimensions $(\rho_2 \times \gamma_2)$, i.e., the dimensions of B. To specify a particular element, it suffices to specify the block in which the element occurs and the position within that block of the element under consideration. Thus, in the above example, the element c_{47} $(= a_{22}b_{13})$ is in the $(2,2)$ block and at position $(1,3)$ of that block.

Note that if there are two independent discrete time Markov chains with transition matrices A and B, then the joint process has $A \otimes B$ as its transition matrix. The states are ordered lexicographically, with the states of A used as the primary sort criterion. The matrix $B \otimes A$ describes the same process, except that the states of B are used as the primary sort criterion. Due to the different order of states, $A \otimes B$ tends to be different from $B \otimes A$.

In the case of continuous time Markov chains, the *tensor sum* takes the place of the tensor product. The *tensor sum* of two *square* matrices A and B is defined in terms of tensor products as

$$A \oplus B = A \otimes I_{n_2} + I_{n_1} \otimes B \qquad (2)$$

where n_1 is the order of A, n_2 the order of B, I_{n_i} the identity matrix of order n_i and "+" represents the usual operation of matrix addition. Since both sides of this operation (matrix addition) must have identical dimensions, it follows that tensor addition, is defined for square matrices only. For example, with

$$A = \begin{pmatrix} a_{11} & a_{12} \\ a_{21} & a_{22} \end{pmatrix} \quad \text{and} \quad B = \begin{pmatrix} b_{11} & b_{12} & b_{13} \\ b_{21} & b_{22} & b_{23} \\ b_{31} & b_{32} & b_{33} \end{pmatrix}, \qquad (3)$$

the tensor sum $C = A \oplus B$ is given by

$$C = \left(\begin{array}{ccc|ccc} a_{11}+b_{11} & b_{12} & b_{13} & a_{12} & 0 & 0 \\ b_{21} & a_{11}+b_{22} & b_{23} & 0 & a_{12} & 0 \\ b_{31} & b_{32} & a_{11}+b_{33} & 0 & 0 & a_{12} \\ \hline a_{21} & 0 & 0 & a_{22}+b_{11} & b_{12} & b_{13} \\ 0 & a_{21} & 0 & b_{21} & a_{22}+b_{22} & b_{23} \\ 0 & 0 & a_{21} & b_{31} & b_{32} & a_{22}+b_{33} \end{array} \right).$$

Some important properties of tensor products and sums are

1. Associativity:
   $A \otimes (B \otimes C) = (A \otimes B) \otimes C \quad \text{and} \quad A \oplus (B \oplus C) = (A \oplus B) \oplus C.$

2. Distributivity over (ordinary matrix) addition:
   $(A + B) \otimes (C + D) = A \otimes C + B \otimes C + A \otimes D + B \otimes D.$

3. Compatibility with (ordinary matrix) multiplication:
   $(A \times B) \otimes (C \times D) = (A \otimes C) \times (B \otimes D).$

4. Compatibility with (ordinary matrix) inversion:
   $(A \otimes B)^{-1} = A^{-1} \otimes B^{-1}.$

The associativity property implies that the operations $\bigotimes_{k=1}^{N} A^{(k)}$ and $\bigoplus_{k=1}^{N} A^{(k)}$ are well defined. In particular, observe that the tensor sum of N terms may be written as the (usual) matrix sum of N terms, each term consisting of an N-fold tensor product. We have

$$\bigoplus_{k=1}^{N} A^{(k)} = \sum_{k=1}^{N} I_{n_1} \otimes \cdots \otimes I_{n_{k-1}} \otimes A^{(k)} \otimes I_{n_{k+1}} \otimes \cdots \otimes I_{n_N},$$

where n_k is the order of the matrix $A^{(k)}$ and I_{n_k} is the identity matrix of order n_k. These laws (specifically, number 3) may be used to prove the following useful result for square matrices

$$(A^n \otimes B^n) = (A \otimes B)^n.$$

This implies that the transient probabilities of two independent discrete-time Markov chains can be obtained by taking the tensor products of the individual chains. Further information concerning the properties of tensor algebra may be found in Davio [Davio, 1981].

3 STOCHASTIC AUTOMATA NETWORKS

3.1 Non-Interacting Stochastic Automata

Consider the case of a system that may be modeled by two completely independent stochastic automata, each of which may be represented by a discrete-time Markov chain. Let us assume that the first automaton, denoted $\mathcal{A}^{(1)}$, has n_1 states and that its stochastic transition probability matrix is given by $P^{(1)} \in \overline{S}^{n_1 \times n_1}$. Similarly, let $\mathcal{A}^{(2)}$ denote the second automaton; n_2, the number of states in its representation and $P^{(2)} \in \overline{S}^{n_2 \times n_2}$, its stochastic transition probability matrix. The state of the overall (two-dimensional) system may be represented by the pair (i, j) where $i \in \{1, 2, \ldots, n_1\}$ and $j \in \{1, 2, \ldots, n_2\}$, and the stochastic transition probability matrix of the two-dimensional system is given by $P^{(1)} \otimes P^{(2)}$. Similarly, if the stochastic automata are continuous in time, characterized by the infinitesimal generators, $Q^{(1)}$ and $Q^{(2)}$ respectively, the infinitesimal generator of the two-dimensional system is given by $Q^{(1)} \oplus Q^{(2)}$. Throughout this chapter, we present results on the basis of continuous-time SANs, although the results are equally valid in the context of discrete-time SANs.

Given N *independent* stochastic automata, $\mathcal{A}^{(1)}, \mathcal{A}^{(2)}, \ldots, \mathcal{A}^{(N)}$, with associated infinitesimal generators, $Q^{(1)}, Q^{(2)}, \ldots, Q^{(N)}$, and probability distributions $\pi^{(1)}(t), \pi^{(2)}(t), \ldots, \pi^{(N)}(t)$ at time t, the infinitesimal generator of the N-dimensional system, which we shall refer to as the *global generator*, is given by

$$Q = \bigoplus_{k=1}^{N} Q^{(k)} = \sum_{k=1}^{N} I_{n_1} \otimes \cdots \otimes I_{n_{k-1}} \otimes Q^{(k)} \otimes I_{n_{k+1}} \otimes \cdots \otimes I_{n_N}. \quad (4)$$

The probability that the system is in state $(i_1, i_2, \ldots, i_N)$ at time t, where i_k is the state of the kth automaton at time t with $1 \leq i_k \leq n_k$ and n_k is the number of states in the kth automaton, is given by $\prod_{k=1}^{N} \pi_{i_k}^{(k)}(t)$ where $\pi_{i_k}^{(k)}(t)$ is the probability that the kth automaton is in state i_k at time t. Furthermore, the probability distribution of the N-dimensional system, $\pi(t)$, is given by the tensor product of the probability vectors of the individual automaton at time t, i.e.,

$$\pi(t) = \bigotimes_{k=1}^{N} \pi^{(k)}(t). \quad (5)$$

To solve N-dimensional systems that are formed from independent stochastic automata is therefore very simple. It suffices to solve for the probability distributions of the individual stochastic automata and to form the tensor product of these distributions. This resolves the case of independent stochastic automata, and we now turn our attention to automata that interact with each other.

3.2 Interacting Stochastic Automata

There are two ways in which stochastic automata interact:

1. The rate at which a transition occurs may be a *function* of the state of a set of automata. Such transitions are called *functional* transitions. Transitions that are not functional are said to be *constant*.

2. A transition in one automaton may *force* a transition to occur in one or more other automata. We allow for both the possibility of a *master/slave* relationship, in which an action in one automaton (the master) actually occasions a transition in one or more other automata (the slaves), and for the case of a *rendez-vous* in which the presence (or absence) of two or more automata in designated states causes (or prevents) transitions to occur. We refer to such transitions collectively under the name of *synchronized* transitions. Synchronized transitions are triggered by a synchronizing event; indeed, a single synchronizing event will generally cause multiple synchronized transitions. Transitions that are not synchronized are said to be *local*.

The elements in the matrix representation of any individual stochastic automaton are either constants, i.e., nonnegative real numbers, or functions from the global state space to the nonnegative reals. Transition rates that depend only on the state of the automaton itself, and not on the state of any other automaton, are to all intents and purposes, constant transition rates. A synchronized transition may be either functional or constant. The same is true for local transitions.

Consider, for example, a simple queueing network consisting of two service centers in tandem and an arrival process that is Poisson at rate λ. Each service center consists of an infinite queue and a single server. The service time distribution of the first server is assumed to be exponential at fixed rate μ, while the service time distribution at the second is taken to be exponential with a rate ν that varies with the number and distribution of customers in the network. Since a state of the network is completely described by the pair (n_1, n_2) where n_1 denotes the number of customers at station 1 and n_2 the number at station 2, the service rate at station 2 is more properly written as $\nu(n_1, n_2)$.

We may define two stochastic automata $\mathcal{A}^{(1)}$ and $\mathcal{A}^{(2)}$ corresponding to the two different service centers. The state space of each is given by the set of nonnegative integers $\{0, 1, 2, \ldots, \}$ since any nonnegative number of customers may be in either station. Transitions in $\mathcal{A}^{(2)}$ depend on the first automaton in two ways. Firstly the rate at which customers are served in the second station depends on the number of customers in the network and hence, in particular, on the number at the first station. Thus $\mathcal{A}^{(2)}$ contains functional transition rates, $(\nu(n_1, n_2))$. Secondly, when a departure occurs from the first station, a customer enters the second and therefore instantaneously forces a transition to occur within the second automaton. The state of the second automaton is instantaneously changed from n_2 to $n_2 + 1$! This entails transitions of the second type, namely synchronized transitions. The event, "departure from station 1", is a synchronizing event.

3.3 Building Generators using Synchronizing Events

To build generators using synchronizing events, we need the following observations:

1. Consider a matrix C, which is equal to a tensor product $A \otimes B$, except that block ij is 0. To obtain C, form A^-, which is equal to A, except that the element of A in row i, column j is set to zero 0. C is now given as $A^- \otimes B$. Setting several elements of A to 0 will cause the blocks corresponding to any of these elements to become 0.

2. To create a matrix C that is 0, except that the block ij is identical to a given matrix A, one forms a matrix D in which $d_{ij} = 1$ and all other elements are zero. C can now be obtained by forming $D \otimes A$.

3. To change a single block in the tensor product $A \otimes B$ from $a_{ij}B$ to $a_{ij}E$, one first replaces $A = [a_{mn}]$ by a matrix A^-, which is equal to A, except that for the given values i and j, $a_{ij} = 0$. One then forms the matrix D,

which is zero, except that in row i, column j, it contains a_{ij}. The the matrix $A^- \otimes B + D \otimes B$ yields the desired result.

4. To find a matrix C which is equal to $A \otimes B$, except that several blocks must be changed from $a_{ij}B$ to $a_{ij}E$, one separates A into two parts, $A^{(l)}$ and $A^{(e)}$, where $A^{(l)}$ contains all entries which must be $a_{ij}B$ in C, and $A^{(e)}$ contains all entries which must be $a_{ij}E$ in C. One has

$$C = A^{(l)} \otimes B + A^{(e)} \otimes E.$$

We begin with a small example of two interacting stochastic automata, $\mathcal{A}^{(1)}$ and $\mathcal{A}^{(2)}$, whose infinitesimal generator matrices are given by

$$Q^{(1)} = \begin{pmatrix} -\lambda_1 & \lambda_1 \\ \lambda_2 & -\lambda_2 \end{pmatrix} \quad \text{and} \quad Q^{(2)} = \begin{pmatrix} -\mu_1 & \mu_1 & 0 \\ 0 & -\mu_2 & \mu_2 \\ \mu_3 & 0 & -\mu_3 \end{pmatrix}$$

respectively. At the moment, neither contains synchronizing events nor functional transition rates. The infinitesimal generator of the global, two-dimensional system is therefore given as

$$Q^{(1)} \oplus Q^{(2)} =$$

$$\left(\begin{array}{ccc|ccc} -(\lambda_1 + \mu_1) & \mu_1 & 0 & \lambda_1 & 0 & 0 \\ 0 & -(\lambda_1 + \mu_2) & \mu_2 & 0 & \lambda_1 & 0 \\ \mu_3 & 0 & -(\lambda_1 + \mu_3) & 0 & 0 & \lambda_1 \\ \hline \lambda_2 & 0 & 0 & -(\lambda_2 + \mu_1) & \mu_1 & 0 \\ 0 & \lambda_2 & 0 & 0 & -(\lambda_2 + \mu_2) & \mu_2 \\ 0 & 0 & \lambda_2 & \mu_3 & 0 & -(\lambda_2 + \mu_3) \end{array} \right).$$

(6)

Let us now observe the effect of introducing synchronizing events. Suppose that each time automaton $\mathcal{A}^{(1)}$ generates a transition from state 2 to state 1 (at rate λ_2), it forces the second automaton into state 1. The transition from 2 to 1 affects the $(2,1)$ block, and in this block, a transition is made to state 1 at rate λ_2. Consequently, one obtains the generator

$$\left(\begin{array}{ccc|ccc} -(\lambda_1 + \mu_1) & \mu_1 & 0 & \lambda_1 & 0 & 0 \\ 0 & -(\lambda_1 + \mu_2) & \mu_2 & 0 & \lambda_1 & 0 \\ \mu_3 & 0 & -(\lambda_1 + \mu_3) & 0 & 0 & \lambda_1 \\ \hline \lambda_2 & 0 & 0 & -(\lambda_2 + \mu_1) & \mu_1 & 0 \\ \lambda_2 & 0 & 0 & 0 & -(\lambda_2 + \mu_2) & \mu_2 \\ \lambda_2 & 0 & 0 & \mu_3 & 0 & -(\lambda_2 + \mu_3) \end{array} \right).$$

If, in addition, the second automaton $\mathcal{A}^{(2)}$ initiates a synchronizing event each time it moves from state 3 to state 1 (at rate μ_3), by for example forcing the first automaton into state 1, we obtain the following global generator.

$$\left(\begin{array}{ccc|ccc} -(\lambda_1+\mu_1) & \mu_1 & 0 & \lambda_1 & 0 & 0 \\ 0 & -(\lambda_1+\mu_2) & \mu_2 & 0 & \lambda_1 & 0 \\ \mu_3 & 0 & -(\lambda_1+\mu_3) & 0 & 0 & \lambda_1 \\ \hline \lambda_2 & 0 & 0 & -(\lambda_2+\mu_1) & \mu_1 & 0 \\ \lambda_2 & 0 & 0 & 0 & -(\lambda_2+\mu_2) & \mu_2 \\ \lambda_2+\mu_3 & 0 & 0 & 0 & 0 & -(\lambda_2+\mu_3) \end{array}\right).$$

Note that the move from 3 to 1 in $\mathcal{A}^{(2)}$ only affect the elements $3,1$ of each block, and since $\mathcal{A}^{(2)}$ moves to state 1, only the blocks $(1,1)$ and $(2,1)$ are affected.

Our immediate reaction in observing these altered matrices may be to assume that a major disadvantage of incorporating synchronized transitions is to remove the possibility of representing the global transition rate matrix as a (sum of) tensor products. However, Plateau [Plateau, 1985] has shown that, by separating local transitions from synchronized transitions, this is not necessarily so; that the global transition rate matrix can still be written as a (sum of) tensor products. To observe this we proceed as follows, using observations 1 through 4 specified above.

The transitions at rates λ_1, μ_1 and μ_2 are not synchronized transitions, but rather *local* transitions. The part of the global generator that consists uniquely of local transitions may be obtained by forming the tensor sum of infinitesimal generators $Q_l^{(1)}$ and $Q_l^{(2)}$ that represent only local transitions:

$$Q_l^{(1)} = \begin{pmatrix} -\lambda_1 & \lambda_1 \\ 0 & 0 \end{pmatrix} \quad \text{and} \quad Q_l^{(2)} = \begin{pmatrix} -\mu_1 & \mu_1 & 0 \\ 0 & -\mu_2 & \mu_2 \\ 0 & 0 & 0 \end{pmatrix},$$

with tensor sum

$$Q_l = Q_l^{(1)} \oplus Q_l^{(2)} = \left(\begin{array}{ccc|ccc} -(\lambda_1+\mu_1) & \mu_1 & 0 & \lambda_1 & 0 & 0 \\ 0 & -(\lambda_1+\mu_2) & \mu_2 & 0 & \lambda_1 & 0 \\ 0 & 0 & -\lambda_1 & 0 & 0 & \lambda_1 \\ \hline 0 & 0 & 0 & -\mu_1 & \mu_1 & 0 \\ 0 & 0 & 0 & 0 & -\mu_2 & \mu_2 \\ 0 & 0 & 0 & 0 & 0 & 0 \end{array}\right).$$

The rates λ_2 and μ_3 are associated with two synchronizing events that we call e_1 and e_2 respectively. The part of the global generator that is due to the first synchronizing event is given by

$$Q_{e_1} = \left(\begin{array}{ccc|ccc} 0 & 0 & 0 & 0 & 0 & 0 \\ 0 & 0 & 0 & 0 & 0 & 0 \\ 0 & 0 & 0 & 0 & 0 & 0 \\ \hline \lambda_2 & 0 & 0 & -\lambda_2 & 0 & 0 \\ \lambda_2 & 0 & 0 & 0 & -\lambda_2 & 0 \\ \lambda_2 & 0 & 0 & 0 & 0 & -\lambda_2 \end{array}\right)$$

which is the (ordinary) matrix sum of two tensor products:

$$Q_{e_1} = \begin{pmatrix} 0 & 0 \\ \lambda_2 & 0 \end{pmatrix} \otimes \begin{pmatrix} 1 & 0 & 0 \\ 1 & 0 & 0 \\ 1 & 0 & 0 \end{pmatrix} + \begin{pmatrix} 0 & 0 \\ 0 & -\lambda_2 \end{pmatrix} \otimes \begin{pmatrix} 1 & 0 & 0 \\ 0 & 1 & 0 \\ 0 & 0 & 1 \end{pmatrix}.$$

The first factor of the first tensor product indicates that a change in $\mathcal{A}^{(1)}$ from 2 to 1 causes a change of state in $\mathcal{A}^{(2)}$, namely, the change described by the second factor. The second product corrects the diagonal elements.

The part of the global generator due to synchronizing event e_2 is

$$Q_{e_2} = \left(\begin{array}{ccc|ccc} 0 & 0 & 0 & 0 & 0 & 0 \\ 0 & 0 & 0 & 0 & 0 & 0 \\ \mu_3 & 0 & -\mu_3 & 0 & 0 & 0 \\ \hline 0 & 0 & 0 & 0 & 0 & 0 \\ 0 & 0 & 0 & 0 & 0 & 0 \\ \mu_3 & 0 & 0 & 0 & 0 & -\mu_3 \end{array} \right)$$

which may be obtained from a sum of tensor products as

$$Q_{e_2} = \begin{pmatrix} 1 & 0 \\ 1 & 0 \end{pmatrix} \otimes \begin{pmatrix} 0 & 0 & 0 \\ 0 & 0 & 0 \\ \mu_3 & 0 & 0 \end{pmatrix} + \begin{pmatrix} 1 & 0 \\ 0 & 1 \end{pmatrix} \otimes \begin{pmatrix} 0 & 0 & 0 \\ 0 & 0 & 0 \\ 0 & 0 & -\mu_3 \end{pmatrix}.$$

This can be explained in a similar way as the expression leading to Q_{e_1}, except that it is now $\mathcal{A}^{(2)}$, which corresponds to the second factor in each term, that initiates the transition in the other automaton. The global infinitesimal generator Q is now given by

$$Q = Q_l + Q_{e_1} + Q_{e_2}.$$

The general principle as it applies to N automata should now be clear. First, one identifies the part of each automaton which is not affected by synchronizing events. If the corresponding matrices are $Q_l^{(k)}$, then one forms the tensor sum $\bigoplus_{k=1}^{N} Q_l^{(k)}$. Next, one forms, for each synchronizing event, two tensor products, each containing a factor for every automaton. One factor of each product is associated with the automaton causing the event, the *owner*. The other factors relate to the other automata.

In the tensor product making up the first term, the matrix of the owner contains for each row and column the rate at which the synchronizing event occurs. The other factors describe the transitions which are forced upon them by the synchronizing event. Specifically, if the synchronizing event forces an automaton from i to j, there is a 1 in position (i,j) of the corresponding matrix. If, in some state i, the event has no effect, then there is a 1 on the diagonal. Consequently, if an event has no effect at all on a certain automaton, the corresponding matrix is the identity matrix. A row without any entry means that the event is prevented from occurring in this state.

The product corresponding to the diagonal correction is found as follows: Let A be the factor of the owner in the first product. The diagonal elements of the matrix of the owner in the diagonal correction are then row sums of A, multiplied by -1. All other matrices are identity matrices, except if the row of the corresponding matrix in the first product is empty, in which case the diagonal element is zero.

In conclusion, one has a standard sum which contains, in addition to the tensor sum of the matrices $Q_l^{(i)}$, $2E$ terms, where E is the number of synchronizing events. Since the tensor sum of N terms can be expressed as a (standard) matrix sum with N terms, each term being a tensor product, the infinitesimal generator of a system containing N stochastic automata with E synchronizing events (and no functional transition rates) may be written as

$$\sum_{j=1}^{2E+N} \bigotimes_{i=1}^{N} Q_j^{(i)}. \qquad (7)$$

This quantity is referred to as the *descriptor* of the stochastic automata network.

The computational burden imposed by synchronizing events is now apparent and is two-fold. Firstly, the number of terms in the *descriptor* is increased, — two for each synchronizing event. We may therefore conclude that the SAN approach is not well suited to models in which there are many synchronizing events. On the other hand, it may still be useful for systems that may be modeled with several stochastic automata that operate mostly independently and only infrequently need to synchronize their operations, such as those found in many models of highly parallel machines.

A second and even greater burden is that the simple form of the solution, equation (5), no longer holds. Although we have been successful in writing the descriptor in a compact form as the sum of tensor products, the solution is not simply the sum of the vectors computed as the tensor product of the solutions of the individual $Q_j^{(i)}$. Other methods for computing solutions must be found. The usefulness of the SAN approach will be determined solely by our ability to solve this problem. We now turn our attention to functional transition rates, for these may appear not only in local transitions, but also in synchronized transitions.

3.4 Building Generators using Functional Transitions

We return to the two original automata given in equation (6) and consider what happens when one of the transition rates of the second automaton becomes a functional transition rate. Suppose, for example, that the rate of transition from state 2 to state 3 in the second automaton is $\hat{\mu}_2$ when the first automaton is in state 1 and $\tilde{\mu}_2$ when the first automaton is in state 2. The global infinitesimal generator is now

$$\left(\begin{array}{ccc|ccc} -(\lambda_1+\mu_1) & \mu_1 & 0 & \lambda_1 & 0 & 0 \\ 0 & -(\lambda_1+\hat{\mu}_2) & \hat{\mu}_2 & 0 & \lambda_1 & 0 \\ \mu_3 & 0 & -(\lambda_1+\mu_3) & 0 & 0 & \lambda_1 \\ \hline \lambda_2 & 0 & 0 & -(\lambda_2+\mu_1) & \mu_1 & 0 \\ 0 & \lambda_2 & 0 & 0 & -(\lambda_2+\hat{\mu}_2) & \hat{\mu}_2 \\ 0 & 0 & \lambda_2 & \mu_3 & 0 & -(\lambda_2+\mu_3) \end{array}\right).$$

If, in addition, the rate at which the first automaton produces transitions from state 1 to state 2 is $\bar{\lambda}_1$, $\hat{\lambda}_1$ and $\tilde{\lambda}_1$ depending on whether the second automaton is in state 1, 2 or 3, the two-dimensional infinitesimal generator is given by

$$\left(\begin{array}{ccc|ccc} -(\bar{\lambda}_1+\mu_1) & \mu_1 & 0 & \bar{\lambda}_1 & 0 & 0 \\ 0 & -(\hat{\lambda}_1+\hat{\mu}_2) & \hat{\mu}_2 & 0 & \hat{\lambda}_1 & 0 \\ \mu_3 & 0 & -(\tilde{\lambda}_1+\mu_3) & 0 & 0 & \tilde{\lambda}_1 \\ \hline \lambda_2 & 0 & 0 & -(\lambda_2+\mu_1) & \mu_1 & 0 \\ 0 & \lambda_2 & 0 & 0 & -(\lambda_2+\hat{\mu}_2) & \hat{\mu}_2 \\ 0 & 0 & \lambda_2 & \mu_3 & 0 & -(\lambda_2+\mu_3) \end{array}\right).$$

A moment's reflection should convince the reader that the introduction of functional transition rates has no effect on the *structure* of the global transition rate matrix other than when functions evaluate to zero in which case a degenerate form of the original structure is obtained. The associativity and distributivity axioms of tensor products described in Section 2 remain valid and carry over to what we will later call generalized tensor products and vector sums. However, even if the structure is preserved, the actual values of the nonzero elements prevents us from writing the solution in the simple form of equation (5). Nevertheless it is still possible to profit from this unaltered nonzero structure. This is the concept behind the extended (generalized) tensor algebraic approach, [Plateau and Fourneau, 1991]. The descriptor is still written as in equation (7), but now the elements of $Q_j^{(i)}$ may be functions. This means that it is necessary to track elements that are functions and to substitute (or recompute) the appropriate numerical value each time the functional rate is needed.

4 EXAMPLES

We now introduce two fairly large models that we will use for purposes of illustration. The first is a model of resource sharing that includes functional transitions. The second is a finite queueing network model with both functional transitions and synchronizing events.

4.1 A Model of Resource Sharing

In this model, N distinguishable processes share a certain resource. Each of these processes alternates between a *sleeping* state and a resource *using* state.

However, the number of processes that may concurrently use the resource is limited to P where $1 \leq P \leq N$ so that when a process wishing to move from the sleeping state to the resource using state finds P processes already using the resource, that process fails to access the resource and returns to the sleeping state. Notice that when $P = 1$ this model reduces to the usual mutual exclusion problem. When $P = N$, all of the the processes are independent. Let $\lambda^{(i)}$ be the rate at which process i awakes from the sleeping state wishing to access the resource, and let $\mu^{(i)}$ be the rate at which this same process releases the resource when it has possession of it.

In our SAN representation, each process is modeled by a two state automaton $\mathcal{A}^{(i)}$, the two states being *sleeping* and *using*. We shall let $s\mathcal{A}^{(i)}$ denote the current state of automaton $\mathcal{A}^{(i)}$. Also, we introduce the function

$$f = \mathcal{I}\left\{\sum_{i=1}^{N} \mathcal{I}\{s\mathcal{A}^{(i)} = using\} < P\right\},$$

where $\mathcal{I}\{b\}$ is indicator function that has the value 1 if the boolean b is true, and the value 0 otherwise. Thus the function f has the value 1 when access is permitted to the resource and has the value 0 otherwise. Figure 1 provides a graphical illustration of this model.

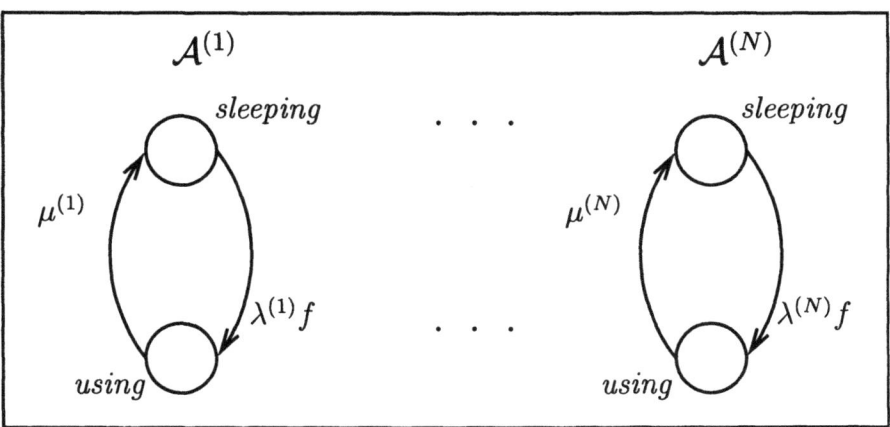

Figure 1 Resource Sharing Model

The local transition matrix for automaton $\mathcal{A}^{(i)}$ is

$$Q_l^{(i)} = \begin{pmatrix} -\lambda^{(i)} f & \lambda^{(i)} f \\ \mu^{(i)} & -\mu^{(i)} \end{pmatrix}.$$

Note that f depends on the state of other automata, which introduces a functional relation into $Q_l^{(i)}$. The overall descriptor for the model is

$$Q = \bigoplus_{g}{}_{i=1}^{N} Q_l^{(i)} = \sum_{i=1}^{N} I_2 \otimes_g \cdots \otimes_g I_2 \otimes_g Q_l^{(i)} \otimes_g I_2 \otimes_g \cdots \otimes_g I_2,$$

where $\otimes_g$ denotes the generalized tensor operator, a precise definition of which is given in Section 7.

The SAN product state space for this model is of size 2^N. Notice that when $P = 1$, the reachable state space is of size $N + 1$, which is considerably smaller than the product state space, while when $P = N$, the reachable state space is the entire product state space. Other values of P give rise to intermediate cases.

4.2 A Queueing Network with Blocking and Priority Service

The second model we shall use is an open queueing network of three finite capacity queues and two customer classes. Class 1 customers arrive from the exterior to queue 1 according to a Poisson process with rate λ_1. Arriving customers are lost if they arrive and find the buffer full. Similarly, class 2 customers arrive from outside the network to queue 2, also according to a Poisson process, but this time at rate λ_2 and they also are lost if the buffer at queue 2 is full. The servers at queues 1 and 2 provide exponential service at rates μ_1 and μ_2 respectively. Customers that have been served at either of these queues try to join queue 3. If queue 3 is full, class 1 customers are blocked (blocking after service) and the server at queue 1 must halt. This server cannot begin to serve another customer until a slot becomes available in the buffer of queue 3 and the blocked customer is transferred. On the other hand, when a (class 2) customer has been served at queue 2 and finds the buffer at queue 3 full, that customer is lost. Queue 3 provides exponential service at rate μ_{3_1} to class 1 customers and at rate μ_{3_2} to class 2 customers. In queue 3, class 1 customers have preemptive priority over class 2 customers. Customers departing after service at queue 3 leave the network. We shall let $C_k - 1$, $k = 1, 2, 3$ denote the finite buffer capacity at queue k.

Queues 1 and 2 can each be represented by a single automaton ($\mathcal{A}^{(1)}$ and $\mathcal{A}^{(2)}$ respectively) with a one-to-one correspondence between the number of customers in the queue and the state of the associated automaton. Queue 3 requires two automata for its representation; the first, $\mathcal{A}^{(3_1)}$, provides the number of class 1 customers and the second, $\mathcal{A}^{(3_2)}$, the number of class 2 customers present in queue 3. Figure 2 illustrates this model.

This SAN has two synchronizing events: the first corresponds to the transfer of a class 1 customer from queue 1 to queue 3 and the second, the transfer of a class 2 customer from queue 2 to queue 3. These are synchronizing events since a change of state in automaton $\mathcal{A}^{(1)}$ or $\mathcal{A}^{(2)}$ occasioned by the departure of a customer, must be synchronized with a corresponding change in automaton $\mathcal{A}^{(3_1)}$ or $\mathcal{A}^{(3_2)}$, representing the arrival of that customer to queue 3. We shall denote these synchronizing events as s_1 and s_2 respectively. In addition to these synchronizing events, this SAN required two functions. They are:

$$f = \mathcal{I}\{s\mathcal{A}^{(3_1)} + s\mathcal{A}^{(3_2)} < C_3 - 1)\}$$

$$g = \mathcal{I}\{s\mathcal{A}^{(3_1)} = 0\}$$

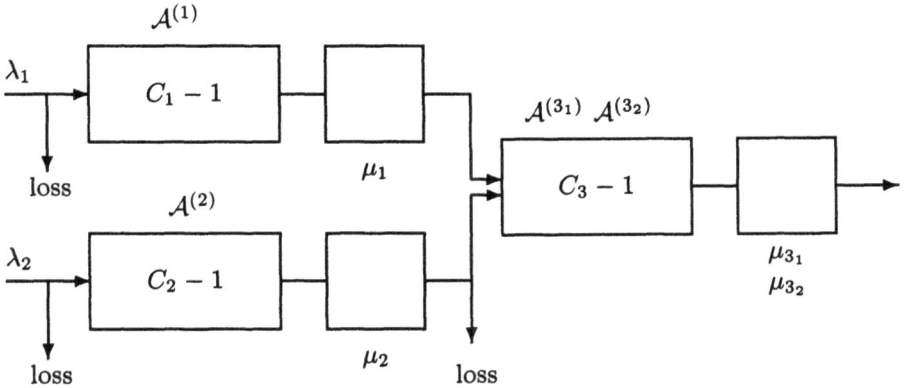

Figure 2 Network of Queues Model

The function f has the value 0 when queue 3 is full and the value 1 otherwise, while the function g has the value 0 when a class 1 customer is present in queue 3, thereby preventing a class 2 customer in this queue from receiving service. It has the value 1 otherwise.

Since there are two synchronizing events, each automaton will give rise to *five* separate matrices in our representation. For each automaton k we will have a matrix of local transitions, denoted by $Q_l^{(k)}$; a matrix corresponding to each of the two synchronizing events, $Q_{s_1}^{(k)}$ and $Q_{s_2}^{(k)}$, and a diagonal corrector matrix for each synchronizing event, $\overline{Q}_{s_1}^{(k)}$ and $\overline{Q}_{s_2}^{(k)}$. In these last two matrices, nonzero elements can appear only along the diagonal; they are defined in such a way as to make $\left(\bigotimes_k Q_{s_j}^{(k)}\right) + \left(\bigotimes_k \overline{Q}_{s_j}^{(k)}\right)$, $j = 1, 2$, generator matrices (row sums equal to zero). The five matrices for each of the four automata in this SAN are as follows (where we use I_m to denote the identity matrix of order m). For $\mathcal{A}^{(1)}$:

$$Q_l^{(1)} = \begin{pmatrix} -\lambda_1 & \lambda_1 & 0 & \cdots & 0 \\ 0 & -\lambda_1 & \lambda_1 & \cdots & 0 \\ \vdots & \ddots & \ddots & \ddots & \vdots \\ 0 & \cdots & 0 & -\lambda_1 & \lambda_1 \\ 0 & \cdots & 0 & 0 & 0 \end{pmatrix}$$

$$Q_{s_1}^{(1)} = \begin{pmatrix} 0 & 0 & 0 & \cdots & 0 \\ \mu_1 & 0 & 0 & \cdots & 0 \\ \vdots & \ddots & \ddots & \ddots & \vdots \\ 0 & \cdots & \mu_1 & 0 & 0 \\ 0 & \cdots & 0 & \mu_1 & 0 \end{pmatrix}$$

128   COMPUTATIONAL PROBABILITY

$$\overline{Q}^{(1)}_{s_1} = \begin{pmatrix} 0 & 0 & 0 & \cdots & 0 \\ 0 & -\mu_1 & 0 & \cdots & 0 \\ \vdots & \ddots & \ddots & \ddots & \vdots \\ 0 & \cdots & 0 & -\mu_1 & 0 \\ 0 & \cdots & 0 & 0 & -\mu_1 \end{pmatrix}$$

$$Q^{(1)}_{s_2} = I_{C_1} = \overline{Q}^{(1)}_{s_2}.$$

For $\mathcal{A}^{(2)}$:

$$Q^{(2)}_l = \begin{pmatrix} -\lambda_2 & \lambda_2 & 0 & \cdots & 0 \\ 0 & -\lambda_2 & \lambda_2 & \cdots & 0 \\ \vdots & \ddots & \ddots & \ddots & \vdots \\ 0 & \cdots & 0 & -\lambda_2 & \lambda_2 \\ 0 & \cdots & 0 & 0 & 0 \end{pmatrix}$$

$$Q^{(2)}_{s_2} = \begin{pmatrix} 0 & 0 & 0 & \cdots & 0 \\ \mu_2 & 0 & 0 & \cdots & 0 \\ \vdots & \ddots & \ddots & \ddots & \vdots \\ 0 & \cdots & \mu_2 & 0 & 0 \\ 0 & \cdots & 0 & \mu_2 & 0 \end{pmatrix},$$

$$\overline{Q}^{(2)}_{s_2} = \begin{pmatrix} 0 & 0 & 0 & \cdots & 0 \\ 0 & -\mu_2 & 0 & \cdots & 0 \\ \vdots & \ddots & \ddots & \ddots & \vdots \\ 0 & \cdots & 0 & -\mu_2 & 0 \\ 0 & \cdots & 0 & 0 & -\mu_2 \end{pmatrix}$$

$$Q^{(2)}_{s_1} = I_{C_2} = \overline{Q}^{(2)}_{s_1}.$$

For $\mathcal{A}^{(3_1)}$:

$$Q^{(3_1)}_l = \begin{pmatrix} 0 & 0 & 0 & \cdots & 0 \\ \mu_{3_1} & -\mu_{3_1} & 0 & \cdots & 0 \\ \vdots & \ddots & \ddots & \ddots & \vdots \\ 0 & \cdots & \mu_{3_1} & -\mu_{3_1} & 0 \\ 0 & \cdots & 0 & \mu_{3_1} & -\mu_{3_1} \end{pmatrix}$$

$$Q^{(3_1)}_{s_1} = \begin{pmatrix} 0 & f & 0 & \cdots & 0 \\ 0 & 0 & f & \cdots & 0 \\ \vdots & \ddots & \ddots & \ddots & \vdots \\ 0 & \cdots & 0 & 0 & f \\ 0 & \cdots & 0 & 0 & 0 \end{pmatrix}$$

$$\overline{Q}_{s_1}^{(3_1)} = \begin{pmatrix} f & 0 & 0 & \cdots & 0 \\ 0 & f & 0 & \cdots & 0 \\ \vdots & \ddots & \ddots & \ddots & \vdots \\ 0 & \cdots & 0 & f & 0 \\ 0 & \cdots & 0 & 0 & 0 \end{pmatrix}$$

$$Q_{s_2}^{(3_1)} = I_{C_3} = \overline{Q}_{s_2}^{(3_1)}.$$

For $\mathcal{A}^{(3_2)}$:

$$Q_l^{(3_2)} = \begin{pmatrix} 0 & 0 & 0 & \cdots & 0 \\ \mu_{3_2}g & -\mu_{3_2}g & 0 & \cdots & 0 \\ \vdots & \ddots & \ddots & \ddots & \vdots \\ 0 & \cdots & \mu_{3_2}g & -\mu_{3_2}g & 0 \\ 0 & \cdots & 0 & \mu_{3_2}g & -\mu_{3_2}g \end{pmatrix}$$

$$Q_{s_2}^{(3_2)} = \begin{pmatrix} 1-f & f & 0 & \cdots & 0 \\ 0 & 1-f & f & \cdots & 0 \\ \vdots & \ddots & \ddots & \ddots & \vdots \\ 0 & \cdots & 0 & 1-f & f \\ 0 & \cdots & 0 & 0 & 1 \end{pmatrix}$$

$$\overline{Q}_{s_2}^{(3_2)} = I_{C_3} = Q_{s_1}^{(3_2)} = \overline{Q}_{s_1}^{(3_2)}.$$

The overall descriptor for this model is given by

$$Q = \bigoplus_g Q_l^{(i)} + \bigotimes_g Q_{s_1}^{(i)} + \bigotimes_g \overline{Q}_{s_1}^{(i)} + \bigotimes_g Q_{s_2}^{(i)} + \bigotimes_g \overline{Q}_{s_2}^{(i)},$$

where the generalized tensor sum and the four generalized tensor products are taken over the index set $\{1, 2, 3_1 \text{ and } 3_2\}$. For instance,

$$\bigotimes_g \overline{Q}_{s_1}^{(i)} = Q_{s_1}^{(1)} \otimes I_{C_2} \otimes Q_{s_1}^{(3_1)} \otimes I_{C_3},$$

indicating that departures in $\mathcal{A}^{(1)}$, which happen according to the rates given by $Q_{s_1}^{(1)}$, affect only $\mathcal{A}^{(3_1)}$, and this effect is given by $Q_{s_1}^{(3_1)}$. The two identity matrices indicate that $\mathcal{A}^{(2)}$ and $\mathcal{A}^{(3_2)}$ are not affected.

The reachable state space of the SAN is of size $C_1 \times C_2 \times C_3(C_3 + 1)/2$ whereas the complete SAN product state space has size $C_1 \times C_2 \times C_3^2$. Finally, we would like to draw our readers attention to the sparsity of the matrices presented above.

5 PRODUCT FORMS

Stochastic automata networks constitute a general modeling technique and as such they can sometimes inherit results from those already obtained by other modeling approaches. Jackson networks, for example, may be represented by a SAN; the reversibility results of Kelly, [Kelly, 1979], and the competition conditions of Boucherie, [Boucherie, 1994], can be applied to SANs leading to product forms. In this section we shall present sufficient conditions for a SAN to have a product form solution. These conditions extend those given by Boucherie and in addition may be shown to be applicable to truncated state spaces. They apply only to SANs with no synchronizing events, which means that the transitions of the SAN can only be transitions of one automaton at a time. Thus Jackson networks lie outside their scope of applicability. The way we proceed is to work on global balance equations and search for sufficient conditions on the functional transition rates to obtain a product form solution.

Let us state the problem more formally. Consider a SAN with N automata and local transition matrices $Q_l^{(k)}$, $k = 1, 2, \ldots, N$. The states of $\mathcal{A}^{(k)}$ are denoted $i_k \in \mathcal{S}^{(k)}$, and a state of the SAN is denoted $i = (i_1, \ldots, i_N)$. A state of the SAN without automaton $\mathcal{A}^{(k)}$ is denoted $\overline{i_k} = (i_1, \ldots, i_{k-1}, i_{k+1}, \ldots, i_N)$. A state i in which the kth component is replaced by i'_k is denoted by $\overline{i_k}|i'_k$. For $i_k \neq i'_k$, the elements of $Q_l^{(k)}$ are assumed to be of the following form

$$Q_l^{(k)}(i_k, i'_k) = q^{(k)}(i_k, i'_k) f^{(k)}(i, i'),$$

where $q^{(k)}(i_k, i'_k)$ is a constant transition rate and $f^{(k)}(i, i')$ is any positive function. Notice that when the transition of the SAN is occasioned by a transition of its kth automaton, the function $f^{(k)}(i, i')$ actually depends only on i and i'_k.

Assume now that the $q^{(k)}(i_k, i'_k)$ satisfy balance equations in the sense that there exist a set of positive numbers $\pi^{(k)}(i_k)$ which sum to 1 and which satisfy either local balance equations within each automaton

$$\forall i_k, i'_k \in \mathcal{S}^{(k)}, \quad \pi^{(k)}(i_k) q^{(k)}(i_k, i'_k) = \pi^{(k)}(i'_k) q^{(k)}(i'_k, i_k), \tag{8}$$

or a global balance equation for each automaton

$$\forall i_k \in \mathcal{S}^{(k)}, \quad \sum_{i'_k \in \mathcal{S}^{(k)}} \left(\pi^{(k)}(i_k) q^{(k)}(i_k, i'_k) - \pi^{(k)}(i'_k) q^{(k)}(i'_k, i_k) \right) = 0. \tag{9}$$

Note that equation (8) is stronger, and implies, equation (9). The SAN generator is $Q = \bigoplus_{k=1}^{N} Q_l^{(k)}$. Its transition rates, for $i \neq i'$, are given by

$$q(i, i') = \begin{cases} q^{(k)}(i_k, i'_k) f^{(k)}(i, \overline{i_k}|i'_k) & \text{if } i' = \overline{i_k}|i'_k \\ 0 & \text{otherwise} \end{cases}$$

The reachable state space of the SAN is denoted by $\overline{\mathcal{S}}$ and, because of the effect of functional transitional rates, can be strictly smaller that $\mathcal{S} = \prod_1^N \mathcal{S}^{(k)}$. The global balance equations for the SAN are, for $i \in \overline{\mathcal{S}}$,

$$\sum_{i' \in \mathcal{S}} (\pi(i) q(i, i') - \pi(i') q(i', i)) = 0. \tag{10}$$

Substituting for $q(i, i')$ and $q(i', i)$ in (10) yields

$$\sum_{k=1}^{N} \sum_{i'_k \in S^{(k)}} \left(\pi(i) q^{(k)}(i_k, i'_k) f^{(k)}(i, \overline{i_k}|i'_k) - \pi(\overline{i_k}|i'_k) q^{(k)}(i'_k, i_k) f^{(k)}(\overline{i_k}|i'_k, i) \right) = 0.$$

(11)

This SAN has a product form solution if, for some normalizing constant C, $\pi(i) = C \prod_{1}^{N} \pi^{(k)}(i_k)$ is a solution of these balance equations. Substituting this into (11) gives

$$\sum_{k=1}^{N} \prod_{j=1, j \neq k}^{N} \pi^{(j)}(i_j) \sum_{i'_k \in S^{(k)}} \left(\pi^{(k)}(i_k) q^{(k)}(i_k, i'_k) f^{(k)}(i, \overline{i_k}|i'_k) \right.$$

$$\left. - \pi^{(k)}(i'_k) q^{(k)}(i'_k, i_k) f^{(k)}(\overline{i_k}|i'_k, i) \right) = 0.$$

Now it only remains to find sufficient conditions on the functions $f^{(k)}$ for which

$$\sum_{i'_k \in S^{(k)}} \left(\pi^{(k)}(i_k) q^{(k)}(i_k, i'_k) f^{(k)}(i, \overline{i_k}|i'_k) - \pi^{(k)}(i'_k) q^{(k)}(i'_k, i_k) f^{(k)}(\overline{i_k}|i'_k, i) \right)$$

(12)

is equal to zero, knowing the balance equations, (8) or (9).

First case: The set of balance equations (8) is satisfied and the functions $f^{(k)}$ express a truncation of the state space of the SAN (similar to that described by Kelly). That is to say, the functions are equal to the indicator function of the reachable state space $\overline{S}$: $f^{(k)}(i, i') = \mathcal{I}\{i, i' \in \overline{S}\}$. Thus the expression (12) is trivially equal to zero: either $\overline{i_k}|i'_k \in \overline{S}$ (the functions are equal to 1) and we have the local balance equations, or $\overline{i_k}|i'_k \notin \overline{S}$ and the functions themselves are zero. The normalizing constant C is the inverse of $\sum_{i \in \overline{S}} \prod_{k=1}^{N} \pi(i)$ and might be difficult to compute if $\overline{S}$ is large.

Second case: The set of balance equations (9) is satisfied and the functions $f^{(k)}$ depend only on $\overline{i_k}$ and not on the current state of automaton $\mathcal{A}^{(k)}$. This means that the decomposition $Q_l^{(k)}(i_k, i'_k) = q^{(k)}(i_k, i'_k) f^{(k)}(i, i')$ is a real product form. The variable $q^{(k)}(i_k, i'_k)$ is the local transition rate of automaton k and $f^{(k)}(i, i') = f^{(k)}(\overline{i_k})$ expresses the interaction of the rest of the SAN when it is in state $\overline{i_k}$. In essence, the functions $f^{(k)}(\overline{i_k})$ either force the system to halt, if they evaluate to zero (this can also be interpreted as a truncation of the state space), or else permit the automaton $\mathcal{A}^{(k)}$ to execute independently, albeit with modified rates: the function uniformly slows down or speeds up the automaton for a given $\overline{i_k}$. When $\overline{i_k}$ changes, the slowing/speeding factor changes. The balance equations are given by

$$f^{(k)}(\overline{i_k}) \prod_{j=1, j \neq k}^{N} \pi^{(j)}(i_j) \sum_{i'_k \in S^{(k)}} \left(\pi^{(k)}(i_k) q^{(k)}(i_k, i'_k) - \pi^{(k)}(i'_k) q^{(k)}(i'_k, i_k) \right) = 0$$

which must hold because the local balance equations themselves, (9), hold. This second case is a generalization of the Boucherie competition conditions. The constant C is equal to 1 when the reachability space is the product space; otherwise it must be chosen so that the individual probabilities sum to one.

Third case: Notice that the two previous cases are not overlapping, and this for two reasons:

- In case 1, $f^{(k)}(i,i') = \mathcal{I}\{i,i' \in \overline{S}\}$ which reduces to $\mathcal{I}\{\overline{i_k}|i'_k \in \overline{S}\}$ for each balance equation (8). In general, the function $\mathcal{I}\{\overline{i_k}|i'_k \in \overline{S}\}$ depends not only on $\overline{i_k}$, but on i'_k as well.

- In case 2, we have a "uniform" modification of the rates of $\mathcal{A}^{(k)}$ while in case 1, they are either 0 or unchanged, for a given $\overline{i_k}$.

This presents the possibility of combining cases 1 and 2 to yield a third case:

$$f^{(k)}(i,i') = \mathcal{I}\{i,i' \in \overline{S}\}f^{(k)}(\overline{i_k}).$$

Using the notation described above, we may summarize these results in the following theorem:

Theorem 1 *Given a SAN having no synchronizing events in which the elements of $Q_l^{(k)}$ are of product form type; i.e., for $i_k \neq i'_k$,*

$$Q_l^{(k)}(i_k, i'_k) = q^{(k)}(i_k, i'_k) f^{(k)}(i, i'),$$

each of the following sufficient conditions leads to a product form solution, $\pi(i) = C \prod_1^N \pi^{(k)}(i_k)$:

- *Case 1:* $f^{(k)}(i,i') = \mathcal{I}\{i,i' \in \overline{S}\}$
- *Case 2:* $f^{(k)}(i,i') = f^{(k)}(\overline{i_k})$
- *Case 3:* $f^{(k)}(i,i') = \mathcal{I}\{i,i' \in \overline{S}\} f^{(k)}(\overline{i_k})$

The $\pi(i)$ satisfy the global balance equations for the SAN, C is a normalizing constant and the $\pi^{(k)}(i_k)$ are solutions of the local balance equations (8) for cases 1 and 3 and a solution of (9) for case 2.

Examples:

1. The resource sharing example of Section 4.1 falls under case 2 with

$$f^{(k)}(i,i') = \mathcal{I}\left\{\sum_{j=1, j\neq k}^{N} \mathcal{I}\{i_j = using\} < P\right\} \quad (13)$$

$$= \mathcal{I}\left\{\sum_{j=1, j\neq k}^{N} \mathcal{I}\{i'_j = using\} < P\right\}. \quad (14)$$

2. Consider a number, N, of identical processes each represented by a three state automaton. State 1 represents the state in which the process computes independently; state 2 represents an interacting state (int) in which the automaton computes and sends messages to the other automata and state 3 is a state in which the process has exclusive access in order to write (w) to a critical resource. Each process may move from any of the three states to any other according to well defined rates of transition. To provide for mutually exclusive access to the critical resource, the rates of transition to state 3 must be multiplied by a function g defined as

$$g^{(k)}(i,i') = \mathcal{I}\left\{\sum_{j=1, j\neq k}^{N} \mathcal{I}\{i_j = w\} = 0\right\}.$$

To provide for the effect of communication overhead, all transition rates within the automaton $\mathcal{A}^{(k)}, k = 1, 2, \ldots, N$ must be multiplied by a function $f^{(k)}$ defined as

$$f^{(k)}(i,i') = \frac{C}{\sum_{j=1, k\neq j}^{N} \mathcal{I}\{i_j = int\}}.$$

Such a SAN therefore also falls under case 2.

3. The examples provided in the paper of Boucherie, [Boucherie, 1994]; viz: the dining philosophers problem, locking in a database system, and so on, all fall under case 2. In these examples, the functions $f^{(k)}(\overline{i_k})$ express a reachable state space *and* yield a uniform multiplicative factor. Other examples may be found in [Donatelli and Serano, 1992, Frosh and Natarajan, 1991, Henderson and Taylor, 1991, Lazar and Robertazzi, 1991].

4. Our final example explicitly displays the dependence of the function on i' and falls under case 1. Consider a system consisting of P units of resource and N identical processes, each represented by a three-state Markov chain. While in state 0, a process may be considered to be sleeping; in state 1, it uses a single unit of resource; while in state 2, the process uses 2 units of the resource. The transitions among its three states are such that while in the sleeping state (0), it can move directly to either state 1 or 2, (i.e., it may request, and receive, a single unit of resource, or two units of resource). From state 2, the process can move to state 1 or to state 0, (i.e., it may release one or both units of the resource). Finally, from state 1, it can move to state 0 or to state 2 (i.e., with a single unit of resource, it has the choice of releasing this unit or acquiring a second). To cater to the case in which sufficient resources are not available, the rates of transition towards states 1 and 2 must each be multiplied by the function

$$f^{(k)}(i, i'_k) = \mathcal{I}\left\{(i'_k + \sum_{j=1}^{N} i_j) \leq P\right\}.$$

For certain values of the transition rates, (e.g., $q(0,1) = q(1,2)$; $q(2,1) = q(1,0)$ and $q(0,2) = q(2,0)$), there exists a solution to the balance equations (8) and the SAN has product form.

6 VECTOR-DESCRIPTOR MULTIPLICATIONS

Typically, product form solutions are not available and the analyst must turn to other solutions procedures. When the global infinitesimal generator of a SAN is available only in the form of a SAN descriptor, the most general and suitable methods for obtaining probability distributions are numerical iterative methods, [Stewart et al., 1995]. Thus, the underlying operation, whether we wish to compute the stationary distribution, or the transient solution at any time t, is the product of a vector, with a matrix. If x is a vector, one has

$$xQ = x \sum_{j=1}^{2E+N} \bigotimes_{i=1}^{N} Q_j^{(i)} = \sum_{j=1}^{2E+N} x \bigotimes_{i=1}^{N} Q_j^{(i)}.$$

In this expression, the basic operation is

$$x \bigotimes_{i=1}^{N} Q^{(i)},$$

where, for notational convenience, we have removed the subscripts of the $Q_j^{(i)}$. It is essential that this operation be implemented as efficiently as possible. The following theorem is proven in [Plateau and Fourneau, 1991].

Theorem 2 *The product*

$$x \bigotimes_{i=1}^{N} Q^{(i)},$$

where $Q^{(i)}$, of order n_i, contains only constant terms and x is a real vector of length $\prod_{i=1}^{N} n_i$, may be computed in ρ_N multiplications, where

$$\rho_0 = 0$$

$$\rho_N = n_N \times (\rho_{N-1} + \prod_{i=1}^{N} n_i) = \prod_{i=1}^{N} n_i \times \sum_{i=1}^{N} n_i, \ N > 0$$

Proof. The proof of this theorem is done by first counting the number of operations needed to evaluate $x(A \otimes B)$, where A and B have dimensions n and m, respectively, and where x is a vector of size nm of the form $x = [x_1, x_2, \ldots x_n]$, where x_i, $i = 1, 2, \ldots, n$ are vectors of size m. One has, as is easily verified

$$x(A \otimes B) = \left[\sum_{i=1}^{n} a_{ij}(x_i B); \ j = 1, 2, \ldots, n \right].$$

Here, the a_{ij} are the entries of the matrix A. Suppose now that for any x_i, x_iB can be evaluated in $\rho(B)$ multiplications. Since there are n different vectors x_i, one needs $n\rho(B)$ multiplications to evaluate them all. The x_iB are vectors of dimension m, and since there are n^2 values of a_{ij} with which these vectors have to be multiplied, one needs another n^2m multiplications. Hence,

$$\rho(A \otimes B) = n^2m + n\rho(B) \tag{15}$$

This relation can be used to prove the theorem by complete induction. First, note that the theorem is correct for $N = 1$, in which case ρ_N is merely $n_1 n_1$, and this is exactly the number of multiplications needed to multiply a vector of dimension n_1 with a matrix of the same dimension. Now, let $q_{ij}^{(i)}$ denote the elements of $Q^{(i)}$. It follows

$$x \bigotimes_{i=1}^{N+1} Q^{(i)} = \left[\sum_{k=2}^{n_1} q_{kj}^{(1)} x_k \bigotimes_{i=2}^{N+1} Q^{(i)}; \; j = 1, 2, \ldots, n_1 \right]$$

We identify $Q^{(1)}$ with A and $\bigotimes_{i=2}^{N+1} Q^{(i)}$ with B and apply (15). According to the inductive hypothesis, $\rho(\bigotimes_{i=2}^{N+1} Q^{(i)})$ is $\prod_{i=1}^{N} n_i \times \sum_{i=1}^{N} n_i$, and the result follows easily from (15) with $n = n_1$ and $m = n_2 n_3 \ldots n_{N+1}$.

To compare this with the number of multiplications needed when advantage is not taken of the special structure resulting from the tensor product, observe that to naively expand $\bigotimes_{i=1}^{N} Q^{(i)}$ requires $\left(\prod_{i=1}^{N} n_i \right)^2$ multiplications. In the case of calculating $x(A \otimes B)$, these savings were obtained by calculating the vector products x_iB only once for each i. If B is again a vector product, the same method can be used to evaluate x_iB for each i. Note that in a program, one starts evaluating vector-matrix products in reverse order.

Let us now examine the effect of introducing functional rates. The savings made in the computation of $x \bigotimes_{i=1}^{N} Q^{(i)}$ are due to the fact that once a product is formed, it may be used in several places without having to re-do the multiplication. Even with functional rates, which imply that the elements in the matrices change according to their context, this same savings is sometimes possible [Stewart et al., 1995]. This leads to an extension of some of the properties of tensor products and to the concept of *Generalized Tensor Products (GTPs)* as opposed to *Ordinary Tensor Products (OTP)*.

7 GENERALIZED TENSOR PRODUCTS

We assume throughout that all matrices are square. We shall use $B[\mathcal{A}]$ to indicate that the matrix B may contain transitions that are a function of the state of the automaton $\mathcal{A}$. More generally, $A^{(m)}[\mathcal{A}^{(1)}, \mathcal{A}^{(2)}, \ldots, \mathcal{A}^{(m-1)}]$ indicates that the matrix $A^{(m)}$ may contain elements that are a function of one or more of the states of the automata $\mathcal{A}^{(1)}, \mathcal{A}^{(2)}, \ldots, \mathcal{A}^{(m-1)}$. We shall use the notation $\otimes_g$ to denote a generalized tensor product. Thus $A \otimes_g B[\mathcal{A}]$ denotes the generalized tensor product of the matrix A with the functional matrix $B[\mathcal{A}]$ and we have

$$A \otimes_g B[\mathcal{A}] = \begin{pmatrix} a_{11}B(a_1) & a_{12}B(a_1) & \cdots & a_{1n_a}B(a_1) \\ a_{21}B(a_2) & a_{22}B(a_2) & \cdots & a_{2n_a}B(a_2) \\ \vdots & \vdots & \ddots & \vdots \\ a_{n_a1}B(a_{n_a}) & a_{n_a2}B(a_{n_a}) & \cdots & a_{n_an_a}B(a_{n_a}) \end{pmatrix}, \quad (16)$$

where $B(a_k)$ represents the matrix B when its functional entries are evaluated with the argument a_k, $k = 1, 2, \ldots, n_a$, the a_i being the states of automaton $\mathcal{A}$. Also,

$$A[\mathcal{B}] \otimes_g B = \begin{pmatrix} a_{11}[\mathcal{B}]I_{n_b} \times B & a_{12}[\mathcal{B}]I_{n_b} \times B & \cdots & a_{1n_a}[\mathcal{B}]I_{n_b} \times B \\ a_{21}[\mathcal{B}]I_{n_b} \times B & a_{22}[\mathcal{B}]I_{n_b} \times B & \cdots & a_{2n_a}[\mathcal{B}]I_{n_b} \times B \\ \vdots & \vdots & \ddots & \vdots \\ a_{n_a1}[\mathcal{B}]I_{n_b} \times B & a_{n_a2}[\mathcal{B}]I_{n_b} \times B & \cdots & a_{n_an_a}[\mathcal{B}]I_{n_b} \times B \end{pmatrix},$$

where

$$a_{ij}[\mathcal{B}]I_{n_b} = diag\{a_{ij}(b_1), a_{ij}(b_2), \ldots, a_{ij}(b_{n_b})\}$$

and $a_{ij}(b_k)$ is the value of the ij^{th} element of the matrix A when its functional entries are evaluated with the argument b_k, $k = 1, 2, \ldots, n_b$. Finally, when both automata are functional we have

$$A[\mathcal{B}] \otimes_g B[\mathcal{A}] =$$

$$\begin{pmatrix} a_{11}[\mathcal{B}]I_{n_b} \times B(a_1) & a_{12}[\mathcal{B}]I_{n_b} \times B(a_1) & \cdots & a_{1n_a}[\mathcal{B}]I_{n_b} \times B(a_1) \\ a_{21}[\mathcal{B}]I_{n_b} \times B(a_2) & a_{22}[\mathcal{B}]I_{n_b} \times B(a_2) & \cdots & a_{2n_a}[\mathcal{B}]I_{n_b} \times B(a_2) \\ \vdots & \vdots & \ddots & \vdots \\ a_{n_a1}[\mathcal{B}]I_{n_b} \times B(a_{n_a}) & a_{n_a2}[\mathcal{B}]I_{n_b} \times B(a_{n_a}) & \cdots & a_{n_an_a}[\mathcal{B}]I_{n_b} \times B(a_{n_a}) \end{pmatrix}.$$

For $A[\mathcal{B}] \otimes_g B[\mathcal{A}]$, the generic entry (l,k) within block (i,j) is $a_{ij}(b_k) \times b_{kl}(a_i)$.

We now present a number of lemmas concerning generalized tensor products. Their proofs may be found in [Fernandes et al., 1998]. These lemmas are useful for deriving many important properties of generalized tensor products.

Lemma 1 (GTP: Associativity)

$$(A[\mathcal{B},\mathcal{C}] \otimes_g B[\mathcal{A},\mathcal{C}]) \otimes_g C[\mathcal{A},\mathcal{B}] = A[\mathcal{B},\mathcal{C}] \otimes_g (B[\mathcal{A},\mathcal{C}] \otimes_g C[\mathcal{A},\mathcal{B}])$$

Lemma 2 (GTP: Distributivity over Addition)

$$(A_1[\mathcal{B}] + A_2[\mathcal{B}]) \otimes_g (B_1[\mathcal{A}] + B_2[\mathcal{A}]) =$$

$$(A_1[\mathcal{B}] \otimes_g B_1[\mathcal{A}] + A_1[\mathcal{B}] \otimes_g B_2[\mathcal{A}] + A_2[\mathcal{B}] \otimes_g B_1[\mathcal{A}] + A_1[\mathcal{B}] \otimes_g B_2[\mathcal{A}])$$

As for ordinary tensor products, compatibility with multiplication usually does not hold for generalized tensor products either. However, there exists three degenerate compatibility forms when some of the factors are identity matrices and not all of the factors have functional entries. They are

Lemma 3 (GTP: Compatibility over Multiplication: I Two Factors)

$$(A[\mathcal{C}] \times B[\mathcal{C}]) \otimes_g I_{n_c} = (A[\mathcal{C}] \otimes_g I_{n_c}) \times (B[\mathcal{C}] \otimes_g I_{n_c})$$

Similarly,

$$I_{n_c} \otimes_g (A[\mathcal{C}] \times B[\mathcal{C}]) = (I_{n_c} \otimes_g A[\mathcal{C}]) \times (I_{n_c} \otimes_g B[\mathcal{C}]).$$

Lemma 4 (GTP: Compatibility over Multiplication: II Two Factors)

$$A \otimes_g B[\mathcal{A}] = [A \times I_{n_a}] \otimes_g [I_{n_b} \times B[\mathcal{A}]] = (I_{n_a} \otimes_g B[\mathcal{A}]) \times (A \otimes I_{n_b}).$$

Lemma 5 (GTP: Compatibility over Multiplication: III Two Factors)

$$A[\mathcal{B}] \otimes_g B = [A[\mathcal{B}] \times I_{n_a}] \otimes_g [I_{n_b} \times B] = (A[\mathcal{B}] \otimes_g I_{n_b}) \times (I_{n_a} \otimes B).$$

This lemma still holds if I_{n_a} is replaced by any constant matrix.

In the following lemmas, if n_i is the size of matrix A_i, then $I_{l:k}$ denotes the identity matrix of size $\prod_{i=l}^{k} n_i$.

Lemma 6 (GTP: Compatibility over Multiplication Many Factors)

$$\begin{aligned}
A^{(1)} \otimes_g A^{(2)}[\mathcal{A}^{(1)}] &\otimes_g A^{(3)}[\mathcal{A}^{(1)}, \mathcal{A}^{(2)}] \otimes_g \cdots \otimes_g A^{(m)}[\mathcal{A}^{(1)}, \ldots, \mathcal{A}^{(m-1)}] \\
&= I_{1:m-1} \otimes_g A^{(m)}[\mathcal{A}^{(1)}, \ldots, \mathcal{A}^{(m-1)}] \\
&\times I_{1:m-2} \otimes_g A^{(m-1)}[\mathcal{A}^{(1)}, \ldots, \mathcal{A}^{(m-2)}] \otimes_g I_{m:m} \\
&\times \cdots \\
&\times I_{1:1} \otimes_g A^{(2)}[\mathcal{A}^{(1)}] \otimes_g I_{3:m} \\
&\times A^{(1)} \otimes_g I_{2:m}
\end{aligned} \qquad (17)$$

Similarly, one may use Lemma 5 to find

$$\begin{aligned}
A^{(m)}[\mathcal{A}^{(1)}, \ldots, \mathcal{A}^{(m-1)}] &\otimes_g A^{(m-1)}[\mathcal{A}^{(1)}, \ldots, \mathcal{A}^{(m-2)}] \otimes_g \cdots \otimes_g A^{(2)}[\mathcal{A}^{(1)}] \otimes_g A^{(1)} \\
&= A^{(m)}[\mathcal{A}^{(1)}, \ldots, \mathcal{A}^{(m-1)}] \otimes_g I_{m-1:1} \\
&\times I_{m:m} \otimes_g A^{(m-1)}[\mathcal{A}^{(1)}, \ldots, \mathcal{A}^{(m-2)}] \otimes_g I_{m-2:1} \\
&\times \cdots \\
&\times I_{m:3} \otimes_g A^{(2)}[\mathcal{A}^{(1)}] \otimes_g I_{1:1} \\
&\times I_{m:2} \otimes_g A^{(1)}
\end{aligned} \qquad (18)$$

In Lemma 6, only one automaton can depend on all the $(m-1)$ other automata, only one can depend on at most $(m-2)$ other automata and so on. One automaton must be independent of all the others. This provides a means by which the individual factors on the left-hand side of equation (17) may be ranked; i.e., according to the number of automata on which they *may* depend. An automaton may actually depend on a subset of the automata in its parameter list.

Lemma 7 (GTP: Pseudo-Commutativity) *Let σ be a permutation of the integers $[1, 2, \ldots, N]$, then there exists a permutation matrix, P_σ of order $\prod_{i=1}^{N} n_i$, such that*

$$\bigotimes_{k=1}^{N}{}_g A^{(k)}[\mathcal{A}^{(1)}, \ldots, \mathcal{A}^{(N)}] = P_\sigma \bigotimes_{k=1}^{N}{}_g A^{(\sigma(k))}[\mathcal{A}^{(1)}, \ldots, \mathcal{A}^{(N)}] P_\sigma^T.$$

These lemmas allow the following theorem to be proven. (The proof itself may be found in [Fernandes et al., 1998].)

Theorem 3 (GTP: Algorithm) *The multiplication*

$$x \times \left(A^{(1)} \otimes_g A^{(2)}[\mathcal{A}^{(1)}] \otimes_g A^{(3)}[\mathcal{A}^{(1)}, \mathcal{A}^{(2)}] \otimes_g \cdots \otimes_g A^{(N)}[\mathcal{A}^{(1)}, \ldots, \mathcal{A}^{(N-1)}] \right)$$

where x is a real vector of length $\prod_{i=1}^{N} n_i$ may be computed in $O(\rho_N)$ multiplications, where

$$\rho_N = n_N \times (\rho_{N-1} + \prod_{i=1}^{N} n_i) = \prod_{i=1}^{N} n_i \times \sum_{i=1}^{N} n_i; \qquad \rho_0 = 0.$$

with the algorithm described in Figure 3.

The following algorithm is based directly on Lemma 6 and implements an efficient product of a vector x with a generalized tensor product in which the automata satisfy the functional dependencies described in Lemma 6. In this algorithm, the notation $A^{(i)}[a_{k_1}^{(1)}, \ldots, a_{k_{i-1}}^{(i-1)}]$ implies that the matrix is evaluated under the assumption that automaton $\mathcal{A}^{(j)}$ is in state $a_{k_j}^{(j)}$, for $j = 1, 2, \ldots, i-1$. The cost of the function evaluations is included in the definition of the big Oh formula. We would like to point out that although Lemma 7 allows us to reorganize the terms in the generalized tensor product in any way we wish, the advantage of leaving them in the form given above is precisely that the computation of the state indices, k_j, can be moved outside the innermost summation of the algorithm.

1. Initialize: $nleft = n_1 n_2 \cdots n_{N-1}$; $nright = 1$.
2. For $i = N, \ldots, 2, 1$ do
   - $base = 0$; $jump = n_i \times nright$
   - For $k = 1, 2, \ldots, nleft$ do
     ○ For $j = 1, 2, \ldots, i - 1$ do
       * $k_j = \left(\left[(k-1)/\prod_{l=j+1}^{i-1} n_l\right] \bmod \left(\prod_{l=j}^{i-1} n_l\right)\right) + 1$
     ○ For $j = 1, 2, \ldots, nright$ do
       * $index = base + j$
       * For $l = 1, 2, \ldots, n_i$ do
         · $z_l = x_{index}$; $index = index + nright$
       * Multiply: $z' = z \times A^{(i)}[a_{k_1}^{(1)}, \ldots, a_{k_{i-1}}^{(i-1)}]$
       * $index = base + j$
       * For $l = 1, 2, \ldots, n_i$ do
         · $x_{index} = z'_l$; $index = index + nright$
     ○ $base = base + jump$
   - If $i > 1$ then $nleft = nleft/n_{i-1}$
   - $nright = nright \times n_i$

Figure 3 Algorithm for Vector Multiplication with a Generalized Tensor Product $x\left(A^{(1)} \otimes_g A^{(2)}[\mathcal{A}^{(1)}] \otimes_g A^{(3)}[\mathcal{A}^{(1)}, \mathcal{A}^{(2)}] \otimes_g \cdots \otimes_g A^{(N)}[\mathcal{A}^{(1)}, \ldots, \mathcal{A}^{(N-1)}]\right)$

Proof. Observe that the inner-most loop, marked by the label *multiply* involves computing the product of a vector z, of length n_i, with a matrix of size $n_i \times n_i$. This requires $(n_i)^2$ multiplications, assuming that the matrices are full. This operation lies within consecutively nested i, k and j loops, Since the k and the second j loops together involve $\prod_{l=1}^{N} n_l/n_i$ iterations, and the outermost i loop is executed N times, the total operation count is given by

$$\sum_{i=1}^{N} \left(\frac{\prod_{l=1}^{N} n_l}{n_i}\right) n_i^2 = \prod_{i=1}^{N} n_i \times \sum_{i=1}^{N} n_i.$$

The complexity result of Theorem 3 was computed under the assumption that the matrices are full. However, the number of multiplications may be reduced by taking advantage of the fact that the block matrices are generally sparse. It is immediately apparent that when the matrices are not full, but possess a special structure such as tridiagonal, or contain only one nonzero row or column, etc., or are sparse, then this may be taken into account and this number reduced in consequence. To see how sparsity in the matrices $Q^{(i)}$ may be taken into account let us denote by η_i the number of nonzero entries in

$Q^{(i)}$. Then the multiply operation of the previous algorithm, i.e., the operation inherent in the statement

$$* \text{ Multiply: } z' = z \times Q^{(i)},$$

has complexity of order η_i, so, taking the nested loops into account, the overall operation has a complexity of the order of

$$\sum_{i=1}^{N} \eta_i \prod_{i=1, i \neq j}^{N} n_i = \prod_{i=1}^{N} n_i \sum_{j=1}^{N} \frac{\eta_j}{n_j}.$$

To compare this complexity with a global sparse format, the number of nonzero entries of $\bigotimes_{i=1}^{N} Q^{(i)}$ is $\prod_{i=1}^{N} \eta_i$. It is hard in general to compare the two numbers $\prod_{i=1}^{N} \eta_i$ and $\prod_{i=1}^{N} n_i \sum_{j=1}^{N} \frac{\eta_j}{n_j}$. Note however that if all $\eta_i = N^{1/(N-1)} n_i$, both orders of complexity are equal. If all matrices are sparse (e.g., below this bound), the sparse method is probably better in terms of computation time. This remark is valid for a single tensor product. For a descriptor which is a sum of tensor products and where functions in the matrices $Q^{(i)}$ may evaluate to zero, it is hard to compute, a priori, the order of complexity of each operation. In this case, insight can only be obtained from numerical experiments.

We now introduce one final lemma that allows us to prove a theorem (Theorem 4) concerning the reduction in the cost of a vector-descriptor multiplication in the case when the functional dependencies among the automata do not satisfy the constraints given above.

Lemma 8 (GTP: Decomposability into OTP) *Let $\ell_k(A)$ denote the matrix obtained by setting all elements of A to zero except those that lie on the kth row which are left unchanged. Then*

$$A \otimes_g B[\mathcal{A}] = \sum_{k=1}^{n_a} \ell_k(A) \otimes B[a_k].$$

Thus we may write a generalized tensor product as a sum of ordinary tensor products.

A term $\bigotimes_{i=1g}^{N} A^{(i)}[\mathcal{A}^{(1)}, \ldots, \mathcal{A}^{(N)}]$ involved in the descriptor of a SAN is said to contain a *functional dependency cycle* if it contains a subset of automata $\mathcal{A}^{(p)}, \mathcal{A}^{(p+1)}, \ldots, \mathcal{A}^{(p+c)}$, $c \geq 1$, with the property that the matrix representation of $\mathcal{A}^{(p+i)}$ contains transitions that are a function of $\mathcal{A}^{(p+(i+1) mod_{c+1})}$, for $0 \leq i \leq c$. For example, a SAN with two automata $\mathcal{A}$ and $\mathcal{B}$ contains a term with a functional dependency cycle if and only if the matrix representations are such that A is a function of $\mathcal{B}$, $(A[\mathcal{B}])$, B is a function of $\mathcal{A}$, $(B[\mathcal{A}])$, and $A[\mathcal{B}] \otimes B[\mathcal{A}]$ occurs in the descriptor. Let $\mathcal{G}$ denote a graph whose nodes are the individual automata of a SAN and whose arcs represent dependencies among the automata within a term of the descriptor. Let $\mathcal{T}$ be a *cutset* of the cycles of $\mathcal{G}$, [Berge, 1970]. Then $\mathcal{T}$ is a set of nodes of $\mathcal{G}$ with the property that $\mathcal{G} - \mathcal{T}$ does not contain a cycle where $\mathcal{G} - \mathcal{T}$ is the graph of $\mathcal{G}$ with all arcs that lead into the nodes of $\mathcal{T}$ removed.

Theorem 4 (GTP with Cycles: Complexity of Vector-Descriptor Product) *Given a SAN descriptor containing a term with a functional dependency graph $\mathcal{G}$ and cutset $\mathcal{T}$ of size t, the cost of performing the vector-descriptor product*

$$x \times \bigotimes_{i=1}^{N}{}_g A^{(i)}[\mathcal{A}^{(1)}, \ldots, \mathcal{A}^{(N)}]$$

is

$$\left(\prod_{i=1, i \in \mathcal{T}}^{N} n_i\right)\left(\prod_{i=1}^{N} n_i\right)\left(\sum_{i=1, i \notin \mathcal{T}}^{N} n_i\right).$$

8 APPLICABILITY OF THE MULTIPLICATION THEOREMS

We now return to the context of SANs proper. We have seen that the descriptor of a SAN is a sum of tensor products and we now wish to examine each of the terms of these tensor products in detail to see whether they fall into the categories of Theorem 3 or 4. In the case of SANs in continuous-time, and with no synchronized transitions, the descriptor is given by

$$Q = \bigoplus_{k=1}^{N}{}_g Q^{(k)} = \sum_{k=1}^{N} I_{n_1} \otimes_g \cdots \otimes_g I_{n_{k-1}} \otimes_g Q^{(k)} \otimes_g I_{n_{k+1}} \otimes_g \cdots \otimes_g I_{n_N},$$

and we can apply Theorem 3 directly to each term of the summation. Notice that all but one of the terms in each tensor product is an identity matrix, and as we pointed out in the proof of Theorem 3, advantage can be taken of this to reduce the number of multiplications involved.

Consider now what happens when we add synchronizing events. The part of the SAN descriptor that corresponds to *local* transitions has the same minimal cost as above which means that we need only consider that part which is specifically involved with the synchronizing events. Recall from Section 3.3, that each synchronizing event results in two additional terms in the SAN descriptor. The first of these may be thought of as representing the actual transitions and their rates; the second corresponds to an updating of the diagonal elements in the infinitesimal generator to reflect these transitions. Since the second is more efficiently handled separately and independently, we need only be concerned with the first. It can be written as

$$\bigotimes_{i=1}^{N} Q_j^{(i)}.$$

We must now analyze the matrices that represent the different automata in this tensor product. There are three possibilities depending upon whether they are unaffected by the transition, are the designated automaton with which the transition rate of the synchronizing event is associated, or are affected in some other manner by the event. These matrices have the following structure.

- Matrices $Q_j^{(i)}$ corresponding to automata that do not participate in the synchronizing event are identity matrices of appropriate dimension.

- Associated with each synchronizing event is a particular automaton, called the *owner* of the synchronizing event. We shall let E denote the matrix of transition rates associated with this automaton. For example, if a synchronizing event e arises as a result of the automaton changing from state i to state j, then the matrix E consists entirely of zeros with a single nonzero element in position ij. This nonzero element gives the rate at which the transition occurs. If several of the elements are nonzero, this indicates that the automaton may cause this particular synchronizing event by transitions from and to several states.

- The remaining matrices correspond to automata that are otherwise affected by the synchronizing event. We shall refer to these as Λ matrices. Each of these Λ matrices consists of rows whose nonzero elements are positive and sum to 1 (essentially, they correspond to routing probabilities, not transition rates), or rows whose elements are all equal to zero. The first case corresponds to the automaton being synchronized into different states according to certain fixed probabilities and the state currently occupied by the automaton. In the second case, a row of zeros indicates that this automaton disables the synchronized transition while it is in the state corresponding to the row of zeros. In many cases, these matrices will have a very distinctive structure, such as a column of ones; this case arises when the automaton is forced into a particular state, independent of its current state.

When the only functional rates are those of the synchronizing event, only the elements of the matrix E are functional; the other matrices consist of identity matrices or constant Λ matrices. Thus, once again, this case falls into the scope of Theorem 3.

This leaves the case in which the Λ matrices contain transition probabilities that are functional. If there is no cycle in the functional dependency graph then we can apply Theorem 3; otherwise we must resort to Theorem 4. Notice that if the functional dependency graph is fully connected, Theorem 4 offers no savings compared with ordinary multiplication. This shows that Theorem 4 is only needed when the routing probabilities associated with a synchronizing event are functional and result in cycles within the functional dependency graph, (which we suspect to be rather rare). For SANs in discrete-time, [Plateau and Atif, 1991], it seems that we may not be so fortunate since cycles in the functional dependency graph of the tensor product tend to occur rather more often.

Following up on these results, extensive experiments conducted on a set of small examples, and reported in [Fernandes et al., 1996], provided a rule of thumb for ordering automata in a network to achieve better performance. More precisely, it is not the automata in the SAN that must be ordered, but rather, within each term of the descriptor, a best ordering should be computed independently.

9 THE MEMORY VERSUS CPU-TIME TRADE-OFF

An important advantage that the SAN approach has over others that generate and manipulate the entire state space of the underlying Markov chain is that of minimal memory requirements. The infamous *state-space explosion* problem associated with other approaches is avoided. The price to be paid, of course, is that of increased CPU time. The obvious question that should be asked is whether some sort of compromise can be reached. Such a compromise would take the form of reducing a SAN with a certain (possibly large) number of "natural" automata, each with only a small number of states, to an "equivalent" SAN with less automata in which many (possibly all) have a larger number of states. A natural way to produce such an equivalent SAN is to collect subsets of the original automata into a small number of groups. The limit of this process is a single automaton containing all the states of the Markov chain. However, we do not wish to go to this extreme. Observe that just four automata each of order 100 brings us to the limit of what is currently possible to solve using regular sparse matrix techniques, yet memory requirements for four automata of size 100 remain modest. Furthermore, in the process of grouping automata, a number of simplifications may result. For example, automata may be grouped in such a way that some (or all) of the synchronizing events disappear, or grouped so that functional transition rates become constant, or both. Furthermore, it is frequently the case that the reachable state space of the grouped automata is smaller than the product state space of the automata that constitute the group. To pursue this line of thought, we shall need to define our notion of equivalence among SANs and the grouping process.

Consider a SAN containing N stochastic automata $\mathcal{A}_1, \ldots, \mathcal{A}_N$ of size n_i respectively, E synchronizing events $s_1, \ldots, s_E$, and functional transition rates. Its descriptor may be written as

$$Q = \sum_{j=1}^{N+2E} \bigotimes_{i=1}^{N}{}_g Q_j^{(i)}.$$

Let $1, \ldots, N$ be partitioned in B groups called $b_1, \ldots, b_B$. Without loss of generality, we assume that b_1 contains the indices from 1 to c_2, b_2 the indices from $c_2 + 1$ to c_3, and so on. We also set $c_1 = 0$, $c_{B+1} = N$, and assume $c_{i+1} > c_i$, $i = 1, 2, \ldots, B$. The descriptor can be rewritten, using the associativity of the generalized tensor product, as

$$Q = \sum_{j=1}^{2E+N} \bigotimes_{k=1}^{B}{}_g \left(\bigotimes_{j=c_k+1}^{c_{k+1}}{}_g Q_j^{(i)} \right).$$

The matrices $R_j^{(k)} = \bigotimes_{g\,j=c_k+1}^{c_{k+1}} Q_j^{(i)}$, for $j \in 1, \ldots, 2E + N$, are, by definition, the transition matrices of a compound automaton, called $\mathcal{G}_k$ of size $h_k = \prod_{i=c_k+1}^{c_{k+1}} n_i$. The descriptor may be rewritten as

$$Q = \sum_{j=1}^{2E+N} \bigotimes_{g\,k=1}^{B} R_j^{(k)}.$$

Separating out the terms resulting from local transitions from those resulting from synchronizing events, we obtain

$$Q = \sum_{j=1}^{N+2E} \bigotimes_{g\,i=1}^{N} Q_j^{(i)} = \bigoplus_{g\,i=1}^{N} Q_l^{(i)} + \sum_{j=1}^{E} \left(\bigotimes_{g\,i=1}^{N} Q_{s_j}^{(i)} + \bigotimes_{g\,i=1}^{N} \overline{Q}_{s_j}^{(i)} \right).$$

Grouping by associativity gives

$$Q = \bigoplus_{g\,k=1}^{B} R_l^{(k)} + \sum_{j=1}^{E} \left(\bigotimes_{g\,k=1}^{B} R_{s_j}^{(k)} + \bigotimes_{g\,k=1}^{B} \overline{R}_{s_j}^{(k)} \right),$$

with

$$R_l^{(k)} = \bigoplus_{g\,i=c_k+1}^{c_{k+1}} Q_l^{(i)}; \quad R_{s_j}^{(k)} = \bigotimes_{g\,i=c_k+1}^{c_{k+1}} Q_{s_j}^{(i)}; \quad \overline{R}_{s_j}^{(k)} = \bigotimes_{g\,i=c_k+1}^{c_{k+1}} \overline{Q}_{s_j}^{(i)}.$$

First simplification: Removal of synchronizing events.

Assume that one of the synchronizing events, say s_1, is such that it synchronizes automata within a group, say b_1. As a result, this synchronizing event becomes internal to group b_1 and may be treated as a transition that is local to $\mathcal{G}_1$. In this case, the value of $R_l^{(1)}$ may be changed in order to simplify the formula for the descriptor. Using

$$R_l^{(1)} \Longleftarrow R_l^{(1)} + R_{s_1}^{(1)} + \overline{R}_{s_1}^{(1)},$$

the descriptor may be rewritten as

$$\bigoplus_{g\,k=1}^{B} R_l^{(k)} + \sum_{j=2}^{E} \left(\bigotimes_{g\,i=1}^{B} R_{s_j}^{(i)} + \bigotimes_{g\,i=1}^{B} \overline{R}_{s_j}^{(i)} \right).$$

The descriptor is thus reduced (two terms having disappeared). This procedure can be applied to all identical situations.

Second simplification: Removal of functional terms.

Assume now that the local transition matrix of $\mathcal{G}_1$ is a tensor sum of matrices whose elements are functions only of the states of the automata that are in the subset b_1. Then the functions in $Q_l^{(i)}$ of

$$R_l^{(1)} = \bigoplus_{g\,i=c_1+1}^{c_2} Q_l^{(i)}$$

may be evaluated when performing the generalized tensor operator and $R_l^{(1)}$ becomes a constant matrix. As for the removal of synchronizing events, this process of replacing functions with constants may be applied in all similar situations.

However, if $R_{s_j}^{(1)}$ is the tensor product of matrices that are functions of the states of automata, some of which are in b_1 and some of which are not in b_1, then performing the generalized tensor product $R_{s_j}^{(1)} = \bigotimes_{g\,i=c_1+1}^{c_2} Q_{s_j}^{(i)}$ allows us to only partially evaluate the functions for the arguments in b_1. Others arguments cannot be evaluated. These must be evaluated later when performing the computation $\bigotimes_{g\,i=1}^{B} R_{s_j}^{(i)}$ and may in fact, result in an increased number of function evaluations.

Third simplification: Reduction of the reachable state space.

In the process of grouping, the situation might arise that a compound automaton $\mathcal{G}_i$ has a reachable state space smaller than the product state space, which has a size of $\prod_{i=c_j+1}^{c_{j+1}} n_i$. This happens after simplifications of type 1 or 2 have been performed. For example, functions may evaluate to zero, or synchronizing events may disable certain transitions. In this case, a reachability analysis may be performed in order to compute the reachable state space. In the SAN methodology, the global reachable state space is known in advance and the reachable state space of a group may be computed by means of a simple projection.

A series of numerical experiments conducted on the two examples presented in Section 4 was reported in [Fernandes et al., 1996] that quantify the effect of these simplifications. The goal was to observe the effect on the time required to perform 10 premultiplications of the descriptor by a vector and on the amount of array storage needed.

In the first model, that of resource sharing, the parameters P and N were varied and the automata (recall that all are identical) were grouped in a variety of ways. The results showed a substantial reduction in CPU time as the number of blocks of automata was reduced, — a combined effect of a reduction in the reachable state space, algorithm overhead, and the number of functions that needed to be evaluated, and this with relatively little impact on memory requirements. In regard to memory requirements, two contrasting effects were observed. On the one hand, the reduction in the reachable state space caused a subsequent reduction in the size of the probability vectors and hence an overall reduction in the amount of memory needed. On the other hand, the size of the matrices representing the grouped automata increased thereby increasing the amount of memory needed. This latter effect was observed to become more important as the number of resources (P) approached the number of processes (N), when it became the dominant effect.

The queueing network model was also analyzed under a variety of different parameter values and with two different kinds of grouping:

- A grouping of the automata according to customer class ($\mathcal{A}_1$ and $\mathcal{A}_{3_1}$) and ($\mathcal{A}_2$ and $\mathcal{A}_{3_2}$).

- A grouping of the automata according to queue ($\mathcal{A}_1$ and $\mathcal{A}_2$) and ($\mathcal{A}_{3_1}$ and $\mathcal{A}_{3_2}$);

The results showed that the CPU times obtained with the first grouping was *worse* than in the non-grouped case. This is a result of the fact that this model incorporates functions that cannot be removed using simplification 2. Although the first grouping eliminates the synchronizing events, it results in an increase in the number of functions that must be evaluated and increases the overall time needed. The second grouping allowed for the possibility of a reduction in the state space of the joint automata, ($\mathcal{A}_{3_1}$ and $\mathcal{A}_{3_2}$), since the priority queue is now represented by a single automaton. This, along with the elimination of functional elements from the grouped descriptors, lead to a reduction in CPU-time. Additionally, the elimination of non-reachable states reduced the amount of array storage needed so that this grouping lead to a reduction in CPU-time *and* memory needs.

This series of experiments showed that the benefits that accrue from grouping are non-negligible, so long as the number of function evaluations do not rise drastically as a result. In fact, it seems that function evaluations should be the main concern in choosing which automata to group together. Indirectly, functions also play an important role in identifying non-reachable states, the elimination of which permit important reductions in CPU time and memory. The number of groups should be kept small. Four or less appeared to be optimal in this particular set of experiments.

10 NUMERICAL SOLUTION METHODS

Consider a stochastic automata network consisting of N automata. Let Q be its descriptor, i.e.,

$$Q = \sum_{j=1}^{2E+N} \bigotimes_{i=1}^{N} Q_j^{(i)}.$$

Our goal is to find the stationary probability vector, i.e., a vector π such that $\pi Q = 0$ and $\pi e = 1$. (See also Chapter 4.) Of all the numerical solution methods discussed in [Stewart, 1994], only those involving a product of a vector with a matrix derived from the infinitesimal generator are suitable when using descriptors. Thus, the power method and the various projection methods are easy to implement. It is not easy to see how methods such as Gaussian elimination can be adopted to efficiently solve SANs. Furthermore, as we have already seen in this chapter, current research has lead to efficient descriptor-vector multiplication algorithms.

The simplest numerical solution method in our context is the power method. If P is the stochastic transition probability matrix of an irreducible Markov chain, the power method is described by the iterative procedure

$$\pi^{(k+1)} = \pi^{(k)} P, \tag{19}$$

where $\pi^{(0)}$ is an arbitrary initial approximation to the solution. When an infinitesimal generator Q is available, P may be obtained from

$$P = I + \Delta t Q \tag{20}$$

where $\Delta t \leq 1/\max_i |q_{ii}|$. P may be written as a sum of tensor products, since

$$I + \Delta t Q = \bigotimes_{i+1}^{N} I_{n_i} + \sum_{j=1}^{2E+N} \Delta t \bigotimes_{i=1}^{N} Q_j^{(i)}.$$

Thus the power method becomes

$$\pi^{(k+1)} = \pi^{(k)}(I + \Delta t Q) = \pi^{(k)} + \Delta t \pi^{(k)} \left(\sum_{j=1}^{2E+N} \bigotimes_{i=1}^{N} Q_j^{(i)} \right). \tag{21}$$

This is the form that the power method takes when it is applied to a SAN descriptor.

Projection Methods include the class of methods known as *simultaneous iteration* or *subspace iteration*, [Jennings and Stewart, 1975, Stewart, 1976, Stewart and Jennings, 1981], which iterate continuously with a fixed number of vectors, as well as methods that begin with a single vector and construct a subspace one vector at a time, [Saad, 1981]. The subspace most often used in Markov chain problems is the *Krylov subspace* given by

$$\mathcal{K}_m(P, v) = \mathrm{span}\{v, vP, vP^2, \ldots, vP^{m-1}\}.$$

This subspace is spanned by consecutive iterates of the power method and these vectors may be evaluated using (21). This is the only part of the projection methods that interacts directly with the coefficient matrix.

It is well known that the projection methods, (and the power method) perform best when accompanied with an effective preconditioner. The objective of preconditioning is to modify the eigenvalue distribution of the iteration matrix so that convergence to the solution vector may be attained more quickly. This is an area of current interest and importance. Much research remains to be done.

References

[Atif, 1992] Atif, K. (1992). *Modélisation du Parallélisme et de la Synchronisation*. PhD thesis, Institut National Polytechnique de Grenoble, Grenoble, France.

[Baccelli et al., 1995] Baccelli, F., Jean-Marie, A., and Mitrani, I., editors (1995). *Quantitative Methods in Parallel Systems, Part I : Stochastic Process Algebras*. Basic Research Series. Springer Verlag.

[Balbo et al., 1994] Balbo, G., S.Bruell, and Sereno, M. (1994). Arrival theorems for product-form stochastic Petri nets. In *Proc. of ACM Sigmetrics Conference*, pages 87–97, Nashville.

[Berge, 1970] Berge, C. (1970). *Graphes et Hypergraphes*. Dunod, Paris.

[Boucherie, 1994] Boucherie, R. (1994). A characterization of independence for competing Markov chains with applications to stochastic Petri nets. *IEEE Transactions on Soft. Eng*, 20:536–544.

[Buchholz, 1992] Buchholz, P. (1992). Hierarchical Markovian models – symmetries and aggregation;. In Pooley, R. and J.Hillston, editors, *Modelling Techniques and Tools for Computer Performance Evaluation*, pages 234–246. Edinburgh, Scotland.

[Buchholz, 1993] Buchholz, P. (1993). Aggregation and reduction techniques for hierarchical GCSPNs. In *Proceedings of the 5th International Workshop on Petri Nets and Performance Models, Toulouse, France*, pages 216–225. IEEE Press.

[Buchholz, 1995] Buchholz, P. (1995). Equivalence relations for stochastic automata networks. In Stewart, W., editor, *Computations with Markov Chains; Proceedings of the 2nd International Meeting on the Numerical Solution of Markov Chains*. Kluwer International Publishers, Boston.

[Chiola et al., 1993] Chiola, G., Dutheillet, C., Franceschinis, G., and Haddad, S. (1993). Stochastic well-formed colored nets and symmetric modeling applications. *IEEE Transactions on Computers*, 42(11):1343–1360.

[Ciardo and Trivedi, 1991] Ciardo, G. and Trivedi, K. (1991). Solution of large GSPN models. In Stewart, W. J., editor, *Numerical Solution of Markov Chains*, pages 565–595. Marcel Dekker, New York.

[Davio, 1981] Davio, M. (1981). Kronecker products and shuffle algebra. *IEEE Trans. Comput*, C-30(2):1099–1109.

[Donatelli, 1993] Donatelli, S. (1993). Superposed stochastic automata: A class of stochastic Petri nets with parallel solution and distributed state space. *Performance Evaluation*, 18:21–36.

[Donatelli and Serano, 1992] Donatelli, S. and Serano, M. (1992). On the product form solution for stochastic Petri nets. In *Proc. of the 13th International Conference on Applications and Theory of Petri Nets, Sheffield, UK*, pages 154–172.

[Fernandes et al., 1998] Fernandes, P., Plateau, B., and Stewart, W. Efficient descriptor–vector multiplications in stochastic automata networks. *Journal of the ACM*, 45(3). Technical Report 2935, INRIA. Obtainable by Anonymous ftp *ftp ftp.inria.fr/INRIA/Publication/RR*.

[Fernandes et al., 1996] Fernandes, P., Plateau, B., and Stewart, W. (1996). Numerical issues for stochastic automata networks. In *PAPM 96, Fourth Process Algebras and Performance Modelling Workshop, Torino, Italy*.

[Fourneau and Quessette, 1995] Fourneau, J.-M. and Quessette, F. (1995). Graphs and stochastic automata networks. In Stewart, W., editor, *Computations with Markov Chains; Proceedings of the 2nd International Meeting on the Numerical Solution of Markov Chains*. Kluwer International Publishers, Boston.

[Franceschinis and Muntz, 1993] Franceschinis, G. and Muntz, R. (1993). Computing bounds for the performance indices of quasi-lumpable stochastic well-formed nets. In *Proceedings of the 5th International Workshop on Petri Nets and Performance Models, Toulouse, France*, pages 148–157. IEEE Press.

[Frosh and Natarajan, 1991] Frosh, D. and Natarajan, K. (1991). Product form solutions for closed synchronized systems of stochastic sequential processes. In *Proc. of 1992 International Computer Symposium, Taiwan*, pages 392–402.

[Greenbaum et al., 1979] Greenbaum, A., Dubois, P., and Rodrique, G. (1979). Approximating the inverse of a matrix for use in iterative algorithms on vector processors. *Computing*, 22:257–268.

[Henderson and Lucic, 1993] Henderson, W. and Lucic, D. (1993). Aggregation and disaggregation through insensitivity in stochastic Petri nets. *Performance Evaluation*, 17:91–114.

[Henderson and Taylor, 1991] Henderson, W. and Taylor, P. (1991). Embedded processes in stochastic Petri nets. *IEEE Trans. in Software Engineering*, 17:108–116.

[Hermanns and Rettelbach, 1994] Hermanns, H. and Rettelbach, M. (1994). Syntax, semantics, equivalences, and axioms for MTIPP. In Herzog, U. and Rettelbach, M., editors, *Proc. of the 2nd Workshop on Process Algebras and Performance Modelling*, Arbeitsberichte, volume 27(4), Erlangen.

[Hillston, 1995] Hillston, J. (1995). Computational Markovian modelling using a process algebra. In Stewart, W., editor, *Computations with Markov Chains; Proceedings of the 2nd International Meeting on the Numerical, Solution of Markov Chains*. Kluwer International Publishers, Boston.

[Jennings and Stewart, 1975] Jennings, A. and Stewart, W. (1975). Simultaneous iteration for partial eigensolution of real matrices. *J. Inst. Math. Applics.*, 15:351–361.

[Kelly, 1979] Kelly, F. (1979). *Reversibility and Stochastic Networks*. Wiley, New York.

[Kemper, 1995] Kemper, P. (1995). Closing the gap between classical and tensor based iteration techniques. In Stewart, W., editor, *Computations with Markov Chains; Proceedings of the 2nd International Meeting on the Numerical, Solution of Markov Chains*, Boston. Kluwer International Publishers.

[Lazar and Robertazzi, 1991] Lazar, A. and Robertazzi, T. (1991). Markovian Petri net protocols with product form solutions. *Performance Evaluation*, 12:67–77.

[Plateau, 1985] Plateau, B. (1985). On the stochastic structure of parallelism and synchronization models for distributed algorithms. In *Proc. ACM Sigmetrics Conference on Measurement and Modelling of Computer Systems*, Austin, Texas.

[Plateau and Atif, 1991] Plateau, B. and Atif, K. (1991). Stochastic automata network for modelling parallel systems. *IEEE Trans. on Software Engineering*, 17(10):1093–1108.

[Plateau and Fourneau, 1991] Plateau, B. and Fourneau, J. (1991). A methodology for solving Markov models of parallel systems. *Journal of Parallel and Distributed Computing*, 12:370–387.

[Saad, 1981] Saad, Y. (1981). Krylov subspace methods for solving unsymmetric linear systems. *Mathematics of Computation*, 37:105–126.

[Sanders and Meyer, 1991] Sanders, W. and Meyer, J. (1991). Reduced base model construction methods for stochastic activity networks. *IEEE Jour. on Selected Areas in Communication*, 9(1):25–36.

[Siegle, 1992] Siegle, M. (1992). On efficient Markov modelling. In *Proc. QMIPS Workshop on Stochastic Petri Nets*, pages 213–225, Sophia-Antipolis, France.

[Simone and Marsan, 1991] Simone, C. and Marsan, M. (1991). The application of the EB-equivalence rules to the structural reduction of GSPN models. *Journal of Parallel and Distributed Computing*, 15(3):296–302.

[Stewart, 1976] Stewart, G. (1976). Simultaneous iteration for computing invariant subspaces of non-hermitian matrices. *Numer. Mat.*, 25:123–136.

[Stewart, 1994] Stewart, W. (1994). *An Introduction to the Numerical Solution of Markov Chains*. Princeton University Press, New Jersey.

[Stewart et al., 1995] Stewart, W., Atif, K., and Plateau, B. (1995). The numerical solution of stochastic automata networks. *European Journal of Operations Research*, 86(3):503–525.

[Stewart and Jennings, 1981] Stewart, W. and Jennings, A. (1981). A simultaneous iteration algorithm for real matrices. *ACM Transactions on Mathematical Software*, 7(2):184–198.

Brigitte Plateau received a Master in Applied Mathematics from the University of Paris 6 in 1976, a *Thèse de Troisième Cycle* in Computer Science from the University of Paris 11 in 1980, and a *Thèse d'Etat* in Computer Science from the University of Paris 11 in 1984. She was *Chargé de Recherche* at CNRS (France) from 1981 to 1985, Assistant Professor at the Computer Science Department of the University of Maryland from 1985 to 1987. She currently holds a position of Professor at the engineering school ENSIMAG in Grenoble (France) and is heading a research group whose main interest is parallel programming. Her research interests include modeling and performance evaluation of parallel and distributed computer systems, numerical solution and simulation of large Markov models.

William J. Stewart William J. Stewart is director of the Operations Research Program and Professor of Computer Science at North Carolina State University. His research interests are in the numerical solution of Markov chains and he is the author of the text *An Introduction to the Numerical Solution of Markov Chains*, published by Princeton University Press. Stewart organized the first two international meetings in this area (in 1990 and 1995 respectively) and edited the proceedings of both. He is currently involved in organizing the third which will be held in September 1999, in Zaragoza, Spain. Also, Stewart has developed a software package, *MARCA — Markov Chain Analyzer* for the generation, characterization and numerical analysis of Markov chains. This package has been successfully employed in a number of areas and especially in the development and performance analysis of telecommunication models.

6 MATRIX ANALYTIC METHODS

Winfried K. Grassmann[1] and David A. Stanford[2]

[1] Winfried K. Grassmann
Department of Computer Science
The University of Saskatchewan
Saskatoon, Saskatchewan, Canada
grassman@cs.usask.ca

[2] Department of Statistical & Actuarial Sciences
The University of Western Ontario
London, Ontario, Canada
stanford@fisher.stats.uwo.ca

1 INTRODUCTION

This chapter shows how to find the equilibrium probabilities in processes of GI/M/1 type, and M/G/1 type, and GI/G/1 type by matrix analytic methods. GI/M/1-type processes are Markov chains with transition matrices having the same structure as the imbedded Markov chain of a GI/M/1 queue, except that the entries are matrices rather than scalars. Similarly, M/G/1 type processes have transition matrices of the same form as the imbedded Markov chain of the M/G/1 queue, except that the entries are matrices. In the imbedded Markov chain of the GI/M/1 queue, all columns but the first have the same entries, except that they are displaced so that the diagonal block entry is common to all. Similarly, in the M/G/1 queue, all rows except the first one are equal after proper centering.

The M/G/1 and the GI/M/1 type processes are special cases of a more general process, which we call the GI/G/1 type process because its transition matrix is similar to the transition matrix of a Markov chain describing the waiting time of a discrete-time GI/G/1 queue. In this transition matrix, the probability of going from state i to j is equal to the probability of going from $i + k$ to $j + k$, provided i and j are large enough, and $k \geq 0$. Hence, except for the boundary, the transition matrix is invariant to shifts of k in the South-East direction.

Markov chains of M/G/1 type, GI/M/1 type and GI/G/1 type can be analyzed by two different methodologies. In this chapter, we use *matrix analytic methods*. In other words, we set up equations involving matrices, and express our solutions in terms of matrices. The next chapter investigates how these type of problems can be solved by means of eigenvalues and eigenvectors.

Matrix analytic methods were introduced by Wallace in 1969 in his thesis [Wallace, 1969], where he introduced the notion of a matrix geometric solution, but the foundations of the theory are due to Neuts, whose work culminated into two monographs, one about the GI/M/1-type process [Neuts, 1981], the other one about M/G/1 type processes. Numerous authors have used these methods to analyze a variety of problems, and two conferences have been devoted to these methods [Chakravarthy and Alfa, 1997, Alfa and Chakravarthy, 1998].

This chapter will discuss the methods introduced by Neuts in some detail. However, we feel that much insight can be gained by exploiting the fact that the transition probabilities are invariant to block shifting in a South-East direction, and the notion of censoring best captures this invariance. We have successfully used this approach earlier to solve a variety of problems [Grassmann, 1986, Grassmann and Heyman, 1990, Grassmann and Heyman, 1993, Stanford and Grassmann, 1993, Grassmann, 1985, Jain and Grassmann, 1988, Grassmann, 1998, Grassmann and Jain, 1989], and so did others [Lal and Bhat, 1988, Kao and Lin, 1989, Kao and Lin, 1991, Zhao et al., 1998].

The outline of this chapter is as follows. After describing and motivating the problem in Section 2, we discuss the theoretical aspects of the GI/M/1 and the M/G/1 paradigm in Section 3, and its numerical solutions in Section 4. Section 5 introduces the idea of censoring and state reduction, and it shows how they can be used to generalize the methods discussed in Sections 3 and 4. Section 6 gives an application of interest.

2 PROBLEM FORMULATION AND MOTIVATING EXAMPLES

2.1 Problem Formulation

This section describes various processes under consideration in this chapter. We consider both continuous and discrete time Markov chains. In the case of a discrete-time Markov chain with transition probability matrix $P = [p_{ij}]$, we aim to determine an (infinite dimensional) vector π satisfying

$$\pi P = \pi \tag{1}$$

$$\pi e = 1. \tag{2}$$

Here, e is a column vector with all its entries equal to 1. In the case of a continuous time Markov chain (CTMC), there is a rate matrix, called the infinitesimal generator Q, and the problem is to find π satisfying (2) and

$$\pi Q = 0. \tag{3}$$

Mathematically, there is no significant difference between (1) and (3). In fact, if we set $Q = P - I$, then (1) implies (3). Moreover, if Q is divided by a number q large enough that all entries are less than 1 in absolute value, then $P = Q/q + I$ satisfies (1). Of course, this construct assumes that all the elements of the matrix Q are finite, which is certainly true for finite-state Markov chains, but may not be true if the number of states is infinite.

In the context of matrices with repeating matrix-based forms, we initially group the methods for finding π into three camps, and follow Neuts' [Neuts, 1981] nomenclature, with discrete-time Markov chains being our standard focus. Again following standard terminology, all states of the Markov chain associated with the same block row are said to be at the same *level* of the process. We use alternately the terms *phase* and *sub-level* to distinguish the states at the same level, so that each state of the Markov chain is uniquely determined by a pairing of a particular level and sub-level.

The simplest case we will address in the chapter is the *quasi birth and death* (QBD) process consisting of three repeating square blocks A_2, A_1, A_0 on the sub-diagonal, diagonal, and super-diagonal respectively, with a common finite size m:

$$Q = \begin{bmatrix} B_0 & C_1 & 0 & 0 & \cdots \\ B_1 & A_1 & A_0 & 0 & \cdots \\ 0 & A_2 & A_1 & A_0 & \cdots \\ 0 & 0 & A_2 & A_1 & \cdots \\ \vdots & \vdots & \vdots & \vdots & \ddots \end{bmatrix}. \qquad (4)$$

We note in particular that the process at level 0 (i.e. the boundary states) may have a different number of sub-levels than all other levels, so that different sized matrices may occupy the boundary row and column.

The second case we will consider generalizes the above as the *Markov chain of GI/M/1 type*, so named because of the block form of Q which resembles the one-step transition matrix of the imbedded Markov chain in that queue:

$$Q = \begin{bmatrix} B_0 & C_1 & 0 & 0 & \cdots \\ B_1 & A_1 & A_0 & 0 & \cdots \\ B_2 & A_2 & A_1 & A_0 & \cdots \\ B_3 & A_3 & A_2 & A_1 & \cdots \\ \vdots & \vdots & \vdots & \vdots & \ddots \end{bmatrix}. \qquad (5)$$

Q in this case takes the form of a *lower Hessenberg* matrix, meaning that all blocks above the superdiagonal are are zero.

The last of the classical processes is the dual of the foregoing, and it will be referred to as a M/G/1-type process. Its transition matrix has the following structure:

$$Q = \begin{bmatrix} B_0 & C_1 & C_2 & C_3 & \cdots \\ B_1 & D_1 & D_2 & D_3 & \cdots \\ 0 & D_0 & D_1 & D_2 & \cdots \\ 0 & 0 & D_0 & D_1 & \cdots \\ \vdots & \vdots & \vdots & \vdots & \ddots \end{bmatrix}. \qquad (6)$$

The M/G/1 queue thus leads to a transition matrix Q of upper Hessenberg form.

In addition to these three paradigms popularized by Neuts, we also consider what has been called the GI/G/1 paradigm ([Grassmann and Heyman, 1990]), so called because it generalizes the transition matrix describing the waiting-time process in a discrete GI/G/1 queue:

$$\begin{bmatrix} B_0 & C_1 & C_2 & C_3 & \cdots \\ B_1 & Q_0 & Q_1 & Q_2 & \cdots \\ B_2 & Q_{-1} & Q_0 & Q_1 & \cdots \\ B_3 & Q_{-2} & Q_{-1} & Q_0 & \cdots \\ \vdots & \vdots & \vdots & \vdots & \ddots \end{bmatrix}. \qquad (7)$$

Obviously, these various paradigms are linked: the QBD process is a special case of both the GI/M/1 and the M/G/1 type process, and the GI/M/1 and the M/G/1 type process are in turn special cases of the GI/G/1 type process. Furthermore, it is often possible to move in the opposite direction, at least theoretically if not in practical terms. Specifically, in many practical cases all Q_i in (7) are zero once $i < -g$ or $i > h$, and this allows one to re-block the system to convert a GI/G/1 type process into a QBD process. This can be achieved as follows: let $b = \max(g, h)$, and set

$$A_0 = \begin{bmatrix} Q_b & 0 & \cdots & \cdots & 0 \\ Q_{b-1} & Q_b & 0 \cdots & & 0 \\ \vdots & \vdots & \vdots & \vdots & \vdots \\ Q_1 & Q_2 & \cdots & \cdots & Q_b \end{bmatrix}$$

$$A_1 = \begin{bmatrix} Q_0 & Q_1 & \cdots & \cdots & Q_{b-1} \\ Q_{-1} & Q_0 & \cdots & \cdots & Q_{b-2} \\ \vdots & \vdots & \vdots & \vdots & \vdots \\ Q_{-b+1} & Q_{-b+2} & \cdots & \cdots & Q_0 \end{bmatrix}$$

$$A_2 = \begin{bmatrix} Q_{-b} & Q_{-b+1} & \cdots & Q_{-2} & Q_{-1} \\ 0 & Q_{-b} & \cdots & Q_{-3} & Q_{-2} \\ \vdots & \vdots & \vdots & \vdots & \\ 0 & 0 & \cdots & 0 & Q_{-b} \end{bmatrix}.$$

Though this construct is convenient for theoretical purposes, it increases the number of operations needed to obtain numerical results considerably as shown in [Grassmann, 1998]. For another method to convert GI/M/1 and M/G/1 processes into QBD processes, see [Ramaswami, 1998].

2.2 Recurrence

We next state several standard definitions regarding recurrence as well as two theorems that indicate the conditions required to obtain steady-state distributions. A state i in a Markov chain is said to be *recurrent* if, when starting from i, the probability of returning to i is equal to 1. Non-recurrent states are called *transient*. A recurrent state is called *positive recurrent* if the expected time between the visits of that state is finite, and *null recurrent* otherwise. Only positive recurrent states have non-null steady-state probabilities, that is, meaningful steady-state probabilities exist only if at least some states are recurrent.

If the Markov chain is non-decomposable, then states of the Markov chain are either all positive recurrent, all null-recurrent, or all transient. In this case, we can talk about *positive recurrent Markov chains, null-recurrent Markov chains* and *transient Markov chains*. We will typically deal with positive recurrent Markov chains. There are a number of criteria for deciding if a Markov chain is recurrent. In particular, the following theorem holds for Markov chains of GI/M/1 type which are not decomposable [Neuts, 1981, pages 16-19].

Theorem 1 *Let $\bar{\pi}$ be the equilibrium vector of the Markov chain with transition matrix $\sum_{i=0}^{\infty} A_i$ and define the vector $\beta = \sum_{i=0}^{\infty} i A_i e$, where e is a column vector with all its elements equal to one. Then the Markov chain is positive recurrent if $\bar{\pi}\beta < 0$.*

A similar theorem exists for the M/G/1 type process which is proven in [Neuts, 1989, page 89 and page 137].

Theorem 2 *Let $\bar{\pi}$ be the equilibrium vector of the Markov chain with transition matrix $\sum_{i=0}^{\infty} D_i$ and define the vector $\beta = \sum_{i=0}^{\infty} i D_i e$, where e is a column vector with all its elements equal to one. Then the Markov chain is positive recurrent if $\bar{\pi}\beta < 0$.*

Both theorems are also applicable to the QBD process. Through reblocking, these theorems can be extended to GI/G/1 type processes with banded transition matrices. For further results, see [Zhao et al., 1998]. In what follows, unless otherwise stated, we will assume throughout that the Markov chains we consider are irreducible and positive recurrent.

2.3 Examples

We now give illustrative examples for all four processes. To facilitate the formulation, let $\text{diag}(\gamma_1, \gamma_2, \ldots, \gamma_m)$ denote a diagonal matrix whose ith diagonal entry is $\gamma_i, i = 1, \ldots, m$. Also, let S_r be the matrix with ones on the rth superdiagonal, and with all other elements equal 0. Premultiplying a matrix by S_r shifts all rows up by r, and postmultiplying by S_r shifts all columns r to the right. S_{-r} can be described in a similar fashion.

Example 1 : QBD Process *Consider the M/M/1 queue in a randomly changing environment, where the arrival and service rates are allowed to depend upon the state of an underlying finite-state continuous time Markov chain. Specifically, the environment is given by a CTMC $\hat{Q}$, with states $1, 2, \ldots m$, and when in state i, arrivals to the queue occur at rate λ_i, while departures occur at rate μ_i. In this case, $A_0 = \text{diag}(\lambda_1, \lambda_2, \ldots, \lambda_m)$, and $A_2 = \text{diag}(\mu_1, \mu_2, \ldots, \mu_m)$. A_1 is given by $\hat{Q} - A_0 - A_2$. Also note that $B_1 = A_2$ and $C_1 = A_0$. Finally, $B_0 = \hat{Q} - A_0$.*

Example 2 : QBD Process *There are two queues in sequence with rates μ_1 and μ_2, respectively. Customers arrive to the first queue at a rate of λ, and after finishing service with the first server, they proceed to the second one as long as the line of the second server is less than $m - 1$. If the number in the second queue is $m - 1$, they block the first server, preventing any further work until space becomes available at the second queue. Let the level be the system size of the first queue, and the phase be the system size of the second queue. To formulate this problem, it is useful to note that γS_r represents the effect of an event at rate γ which increases the phase (i.e. sub-level) by r.*

Clearly, arrivals increase the level by 1, leaving the sub-level unchanged so that $A_0 = \lambda I$. A departure from the first queue decreases the level by 1, and increases the sub-level by 1. Hence, $A_2 = \mu_1 S_1$. Finally, $A_1 = \mu_2 S_{-1} - \text{diag}(\gamma_1, \gamma_2, \ldots, \gamma_m)$, where $\gamma_1 = \lambda + \mu_1$, $\gamma_i = \lambda + \mu_1 + \mu_2$, $i = 2, 3, \ldots, m - 1$, and $\gamma_m = \lambda + \mu_2$.

Example 3 : GI/M/1 Process *Examples of this process include the $M/M^Y/1$ queue in a randomly changing environment, where customers are allowed to be served in bulks of size i according to some discrete distribution $\{b_i\}$. Another possibility would be the GI/PH/1 queue length process imbedded at arrival instants, where PH denotes phase type service.*

Example 4 : M/G/1 Process *Examples of this process are the $M^X/M/1$ queue in a randomly changing environment, where customers arrive in bulks of size i according to some discrete distribution $\{a_i\}$. One can similarly define a queue length process imbedded at departure instants, with phase type arrivals and general service — the so-called PH/G/1 queue.*

Example 5 : GI/G/1 Process *If one allows arbitrary increases and decreases in the level, one obtains the GI/G/1 type process [Grassmann and Heyman, 1990, Zhao et al., 1998]. Hence, an $M^X/M^Y/1$ queue in a randomly changing environment is a GI/G/1 type process. However, the prime example of the GI/G/1 type process is the waiting time process in the discrete-time GI/G/1 queue (see also Chapter 10). To formulate this process, let W_n be the waiting time of the nth customer in a GI/G/1 queue, let S_n be the service time of the nth customer, and A_n the time between the arrival of the nth and the $(n+1)$st customer. Then*

$$W_{n+1} = \max(0, W_n + S_n - A_n)$$

Hence, if d_i is the probability that $S_n - A_n$ is equal to i, then W_n increases by i time units with probability d_i. The sequence $\{W_n,\ n = 0,1,\ldots\}$ obviously describes a Markov chain with transition matrix

$$\begin{bmatrix} \Sigma_0 & d_1 & d_2 & d_3 & \cdots \\ \Sigma_1 & d_0 & d_1 & d_2 & \cdots \\ \Sigma_2 & d_{-1} & d_0 & d_1 & \cdots \\ \Sigma_3 & d_{-2} & d_{-1} & d_0 & \cdots \\ \vdots & \vdots & \vdots & \vdots & \ddots \end{bmatrix}.$$

The first column is given again by the fact that the sum across each row must be 1, that is, $\Sigma_i = \sum_{j=i}^{\infty} d_{-j}$. As before, one can add additional factors, such as different customer types with different service times, multiple servers and so on, and the effect of these additions is that the Σ_i and d_i become matrices.

3 THEORETICAL SOLUTION OF THE PARADIGMS OF NEUTS

Let us now focus on solutions to processes of GI/M/1 type (including QBD processes) and M/G/1 type. In this section, we will limit the generality of the boundary matrices given by (4) to (6) slightly to simplify our discussion, and postpone the discussion of the non-specialized matrices to Section 5.

3.1 The GI/M/1 Paradigm

In the GI/M/1 type process, the level can increase at most by one in any step, but it can decrease by more than one. Historically, the GI/M/1 type process arose as a generalization of the GI/M/1 queue, described below.

Consider the epochs immediately before an arrival in this queue. If there are i customers in line before the νth arrival, and if $n < i + 1$ customers are served during the time between the νth and the $(\nu + 1)$st arrival, then the number in the system before the $(\nu + 1)$st arrival is $i + 1 - n$. Let $A(t)$ be the distribution of the interarrival time, and let μ be the service rate. Since the service time is exponential, one finds for the probability of serving n customers during an interarrival time

$$a_n = \int_0^\infty \frac{(\mu t)^n}{n!} e^{-\mu t} dA(t).$$

The a_n determine the transition probability matrix of this imbedded Markov chain, except for the first column. The ith entry of the first column is denoted by Σ_i, where $\Sigma_i = \sum_{n=i+1}^{\infty} a_n$. This value of Σ_i assures that the sums across the rows are all 1. One therefore finds (see [Gross and Harris, 1998])

$$P = \begin{bmatrix} \Sigma_0 & a_0 & 0 & 0 & \cdots \\ \Sigma_1 & a_1 & a_0 & 0 & \cdots \\ \Sigma_2 & a_2 & a_1 & a_0 & \cdots \\ \Sigma_3 & a_3 & a_2 & a_1 & \cdots \\ \vdots & \vdots & \vdots & \vdots & \ddots \end{bmatrix}.$$

This matrix has the same structure as (5), except that all entries are scalar probabilities rather than matrices. However, as soon as the process is allowed to depend on other factors, such as a randomly changing environment, phase-type service times, and so on, the dimension increases. In this case, the a_n and Σ_n are replaced by matrices A_n and B_n, with the position within the matrix given by the second variable, the sublevel. Neuts [Neuts, 1981] gives a number of examples of this nature, including queues having service times of phase type, queues with bulk service and tandem queues.

In the scalar case considered, (1) instantiates to

$$\pi_0 = \sum_{i=0}^{\infty} \pi_i \Sigma_i \tag{8}$$

$$\pi_j = \sum_{i=0}^{\infty} \pi_{i+j-1} a_i, \quad j > 0. \tag{9}$$

In the scalar case, (8) is redundant. As (9) has the form of a difference equation, we use a geometric form for the trial solution,

$$\pi_i = K\xi^i, \quad i \geq 0 \tag{10}$$

where K is a normalizing constant. After substituting, (9) becomes

$$\xi^j = \sum_{i=0}^{\infty} \xi^{i+j-1} a_i$$

or equivalently

$$\xi = \sum_{i=0}^{\infty} \xi^i a_i. \tag{11}$$

Hence, we have an equation for ξ. According to [Gross and Harris, 1998, page 251]), there is a unique root satisfying $\xi < 1$, and this root is positive, provided $\sum_{i=0}^{\infty} i a_i > 1$. Consequently from (10) it follows that

$$\pi_i = \xi^i(1-\xi).$$

The matrix case. We now turn our attention to solving Markov chains whose transition matrix is given by (5), with the further restriction that $C_1 = A_0$, which in turn implies that level 0 has m phases, just like the other levels. The more general case is is postponed until Section 5 because the exceptional treatment for the boundary states detracts from the essential points we wish to highlight. If the A_i are matrices, we partition π conformal with the matrix P as follows:

$$\pi = [\pi_0, \pi_1, \ldots]. \tag{12}$$

This allows us to write the equilibrium equations in matrix form as follows:

$$\pi_0 = \sum_{i=0}^{\infty} \pi_j B_j \tag{13}$$

$$\pi_j = \sum_{i=0}^{\infty} \pi_{i+j-1} A_i, \quad j > 0. \tag{14}$$

We will also need the transient equivalent to equation (14), which becomes

$$\pi_j^{(n)} = \sum_{i=0}^{\infty} \pi_{i+j-1}^{(n-1)} A_i, \quad j > 0. \tag{15}$$

As a trial solution, we take the so called *matrix geometric solution* given by

$$\pi_i = \pi_0 R^i. \tag{16}$$

The matrix R appearing in (16) is called *rate matrix*. Substituting this solution into (14) yields

$$\pi_0 R = \sum_{i=0}^{\infty} \pi_0 R^i A_i. \tag{17}$$

In this matrix-geometric solution, R must have a matrix norm of less than 1, because otherwise, the sum of all probabilities would diverge. We will show that such a matrix R exists, and that it has a probabilistic interpretation. To understand this interpretation, the following definition is useful:

Definition 1 *Divide the state space into two subspaces, the first subspace consisting of the levels n and below, and the second subspace consisting of the levels above n. If the process is, at any time, in a level above n, it is said to be on an above-n sojourn. The above-n sojourn begins immediately after the last epoch in which the level is at or below n, and it ends immediately before it returns to level n or below.*

In Markov chains with transition matrices of the form (4) to (7), the stochastic properties of sojourns above n are essentially independent of n, which makes them useful for the analysis of the problems we are dealing with here. They will be used extensively throughout this chapter. In particular, they allow one to formulate the following theorem on R satisfying (17).

Theorem 3 *Let $(R)_{ij}$ be the expected number of visits to level $n+1$, phase j of an above-n sojourn, given the last state before the start of the sojourn is level n, phase i. Then, R satisfies*

$$R = \sum_{j=0}^{\infty} R^j A_j. \tag{18}$$

Note that (18) implies (17), which means in turn that $\pi_i = \pi_0 R^i$ satisfies (14).

162   COMPUTATIONAL PROBABILITY

Proof. The above-n sojourn describes a Markov chain. At the first epoch of this sojourn, one can only be at level $n + 1$, and the probabilities of of being in the different phases of this level are given by A_0. At the subsequent epochs, the state probabilities can be obtained by using the following transition matrix

	$\leq n$	$n+1$	$n+2$	$n+3$	$\cdots$
$\leq n$	I	0	0	$\cdots$	
$n+1$	$\sum_{i=2}^{\infty} A_i$	A_1	A_0	0	$\cdots$
$n+2$	$\sum_{i=3}^{\infty} A_i$	A_2	A_1	A_0	0
$n+3$	$\sum_{i=4}^{\infty} A_i$	A_3	A_2	A_1	A_0
$\vdots$	$\vdots$	$\vdots$	$\vdots$	$\vdots$	$\vdots$	$\ddots$

The transition matrix is obviously independent of n. Hence, let $\hat{P}_j^{(\nu)}$ give the probabilities to be in the different phases of level $n+j$ at the νth epoch of the sojourn. Clearly

$$\hat{P}_1^{(1)} = A_0 \tag{19}$$

$$\hat{P}_k^{(1)} = 0, \quad k > 1 \tag{20}$$

$$\hat{P}_1^{(\nu)} = \sum_{j=1}^{\infty} \hat{P}_j^{(\nu-1)} A_j, \quad \nu > 1 \tag{21}$$

$$\hat{P}_k^{(\nu)} = \sum_{j=0}^{\infty} \hat{P}_{k+j-1}^{(\nu-1)} A_j, \quad \nu > 1, \; k > 1. \tag{22}$$

Define

$$R^{(k)} = \sum_{\nu=1}^{\infty} \hat{P}_k^{(\nu)}. \tag{23}$$

$(R^{(k)})_{ij}$ is the expected number of visits to state $(n+k, j)$ during the above n sojourn starting from (n, i). In particular, $R^{(1)} = R$. One finds

$$R^{(k)} = \sum_{j=0}^{\infty} R^{(k+j-1)} A_j, \quad k > 0. \tag{24}$$

To prove (24), use (19) and (21) as follows:

$$R^{(1)} = \sum_{\nu=1}^{\infty} P_1^{(\nu)} = A_0 + \sum_{\nu=2}^{\infty} \hat{P}_1^{(\nu)} = A_0 + \sum_{j=1}^{\infty}\sum_{\nu=2}^{\infty} \hat{P}_j^{(\nu-1)} A_j = A_0 + \sum_{j=1}^{\infty} R^{(j)} A_j$$

and (24) is proven for $k = 1$. For $k > 1$, one applies (20) and (22) as follows

$$R^{(k)} = \sum_{\nu=1}^{\infty} P_1^{(\nu)} = \sum_{\nu=2}^{\infty} P_1^{(\nu)} = \sum_{j=0}^{\infty}\sum_{\nu=2}^{\infty} P_{k+j-1}^{(\nu-1)} A_j = \sum_{j=0}^{\infty} R^{(k+j-1)} A_j.$$

and (24) follows.

Equation (24) implies (18) if

$$R^{(k)} = R^k, \quad k \geq 1. \tag{25}$$

Since $R^{(1)} = R$, this holds for $k = 1$, and we now use complete induction to prove (25) for any $k \geq 1$. Note that all visits to level $n+k+1$ during an above n-sojourn are in turn during an above $n+k$-sojourn. Since $\hat{P}_k^{(\tau)}$ provides the probabilities to visit level $n+k$ at time τ of an above n-sojourn, and since $\hat{P}_1^{(\nu-\tau)}$ provides the probabilities of visiting level $n+k+1$ during an $n+k$ sojourn, one finds (after conditioning on the last visit to level $n+k$ prior to visiting level $n+k+1$)

$$\hat{P}_{k+1}^{(\nu)} = \sum_{\tau=1}^{\nu-1} \hat{P}_k^{(\tau)} \hat{P}_1^{(\nu-\tau)}. \tag{26}$$

We now take the sum of (26) over ν to obtain

$$\begin{aligned} R^{(k+1)} &= \sum_{\nu=1}^{\infty} \hat{P}_{k+1}^{(\nu)} \\ &= \sum_{\nu=1}^{\infty} \sum_{\tau=1}^{\nu-1} \hat{P}_k^{(\tau)} \hat{P}_1^{(\nu-\tau)} \\ &= \sum_{\tau=1}^{\infty} \hat{P}_k^{(\tau)} \sum_{\nu=\tau+1}^{\infty} \hat{P}_1^{(\nu-\tau)} \\ &= \sum_{\tau=1}^{\infty} \hat{P}_k^{(\tau)} \sum_{\nu=1}^{\infty} \hat{P}_1^{(\nu)} = R^{(k)} R. \end{aligned}$$

By iterating for $k = 1, 2, \ldots$, the theorem follows.

Methods to compute R from (18) will be discussed in Section 4. As it turns out, there are several solutions to (18), but it has been proven in [Neuts, 1981] that there is only one solution which has all its eigenvalues inside the unit circle, and which therefore has a norm less than 1.

We still need the vector π_0. For finding it, the condition that the sum of all probabilities must be equal to one is no longer sufficient: one must also use (13) explicitly. One obtains, after replacing all π_j by $\pi_0 R^j$

$$\pi_0 = \pi_0 \sum_{j=0}^{\infty} R^j B_j.$$

This equation provides π_0 only up to a factor. To find this factor, one uses

$$1 = \sum_{i=0}^{\infty} \pi_i e = \sum_{i=0}^{\infty} \pi_0 R^i e = \pi_0 (I-R)^{-1} e.$$

In summary, the solution for the models of GI/M/1 type consists of the following three formulas:

$$R = \sum_{j=0}^{\infty} R^j A_j \qquad (27)$$

$$\pi_0 = \pi_0 \sum_{j=0}^{\infty} R^j B_j \qquad (28)$$

$$\pi_0 (I - R)^{-1} e = 1. \qquad (29)$$

Equation (27) uniquely determines R. Given R, (28) and (29) uniquely determine π_0, The vectors π_i, $i > 0$ can then be obtained as $\pi_0 R^i$, or sequentially as $\pi_i = \pi_{i-1} R$.

3.2 The M/G/1 Paradigm

In processes of M/G/1 type, the level cannot decrease by more then one in a single step, but it can increase by more than one. As suggested by the name, the M/G/1 type process historically arose as a generalization of the M/G/1 queue.

Consider the M/G/1 queue immediately after a departure, and let k_n be the probability of n arrivals during a service time. The transition matrix imbedded at the times immediately after a departure is then given as

$$\begin{bmatrix} k_0 & k_1 & k_2 & k_3 & \cdots \\ k_0 & k_1 & k_2 & k_3 & \cdots \\ 0 & k_0 & k_1 & k_2 & \cdots \\ 0 & 0 & k_0 & k_1 & \cdots \\ \vdots & \vdots & \vdots & \vdots & \ddots \end{bmatrix}.$$

If $S(t)$ is the service time distribution, and if λ is the arrival rate, the probability that there are n arrivals during a service time is given as [Gross and Harris, 1998, page 214]

$$k_n = \int_0^{\infty} \frac{(\lambda t)^n}{n!} e^{-\lambda t} dS(t).$$

By adding additional factors, such as randomly changing environments, arrivals of phase-type [Ramaswami and Lucantoni, 1985, Grassmann, 1982], and so on, the k_n are replaced by matrices. (Further models can be found in [Neuts, 1989].) We consider below a matrix model with greater flexibility than that of the matrix above:

$$Q = \begin{bmatrix} C_0 & C_1 & C_2 & C_3 & \cdots \\ D_0 & D_1 & D_2 & D_3 & \cdots \\ 0 & D_0 & D_1 & D_2 & \cdots \\ 0 & 0 & D_0 & D_1 & \cdots \\ \vdots & \vdots & \vdots & \vdots & \ddots \end{bmatrix}. \qquad (30)$$

This matrix is not as general as (6) because we are assuming that all levels have the same number of states, and that $B_1 = D_0$. For notational purposes we have replaced B_0 by C_0. Its first row remains distinct from the other rows, however.

Like Neuts ([Neuts, 1989]), we first consider the scalar case: discrete-time Markov chains with a transition matrix given by (30). In other words, we first assume that the D_i are scalars. The equilibrium equations for $j > 0$ are in this case

$$\pi_j = \sum_{n=0}^{j} \pi_{j-n+1} D_n + \pi_0 C_j. \tag{31}$$

Solving this expression for π_{j+1} yields:

$$\pi_{j+1} = \left(\pi_j - \sum_{n=1}^{j} \pi_{j-n+1} D_n - \pi_0 C_j \right) / D_0. \tag{32}$$

This equations allows one to find π_1 once π_0 is given, π_2 once π_1 is given, and so on. To find π_0, one can use the probability generating function $P(z) = \sum_{j=0}^{\infty} \pi_i z^i$. To obtain $P(z)$, multiply (31) by z^j and sum over all j to obtain

$$\sum_{j=0}^{\infty} \pi_j z^j = \frac{1}{z} \sum_{j=0}^{\infty} \sum_{n=0}^{j} \pi_{j-n+1} z^{j-n+1} D_n z^n + \pi_0 \sum_{j=0}^{\infty} C_j z^j$$

$$= \frac{1}{z} \sum_{n=0}^{\infty} D_n z^n \sum_{j=n}^{\infty} \pi_{j-n+1} z^{j-n+1} + \pi_0 \sum_{j=0}^{\infty} C_j z^j$$

$$= \frac{1}{z} \sum_{n=0}^{\infty} D_n z^n \left(\sum_{j=0}^{\infty} \pi_j z^j - \pi_0 \right) + \pi_0 \sum_{j=0}^{\infty} C_j z^j.$$

Set $C(z) = \sum_{j=0}^{\infty} C_j z^j$ and $D(z) = \sum_{j=0}^{\infty} D_j z^j$ to obtain the following after multiplying by z:

$$zP(z) = (P(z) - \pi_0)D(z) + z\pi_0 C(z). \tag{33}$$

Since $P(1) = 1$, and $D(1) = C(1) = 1$, this should provide an equation for π_0. However, when setting $z = 1$, π_0 drops out. We therefore take the derivatives with respect to z. After setting $z = 1$ in the resulting equation, and solving for π_0, one finds after a brief calculation

$$\pi_0 = (1 - D'(1))/(1 + C'(1) - D'(1)).$$

In this way, we obtain π_0. Once π_0 is known, (32) allows one to find all π_i, $i > 0$ recursively.

The matrix case. The procedure of successively computing π_1, π_2, ..., assuming π_0 is known is still possible in principle in the matrix case via a slight modification of (32). However, in order to determine π_0, one has to overcome one difficulty: to find all elements of π_0, one needs m equations. As the normalizing condition yields only one equation for this purpose, we need to find additional equations. The following theorem provides these.

Theorem 4 *Let $(G)_{ij}$ be the matrix containing the probabilities that starting from a given level above 0 in phase i, the process will be in phase j when entering the next lower level for the first time. In this case, one has*

$$\pi_0 = \pi_0 \sum_{n=0}^{\infty} C_n G^n. \tag{34}$$

Moreover, G is determined by the following matrix equation

$$G = \sum_{k=0}^{\infty} D_k G^k. \tag{35}$$

Methods to find G numerically will be discussed in Section 4. Note that G in (35) plays a role similar to R in (18). In this sense, equations (35) and (18) form a dual pair. Basically, this means that any algorithm for finding R based on (18) can also be used for finding G from (35) and vice-versa.

Proof. Let $G_\nu^{(k)}$ be the probability matrix containing the probabilities $(G_\nu^{(k)})_{ij}$ that after ν time units a process starting in level $n + k$, phase i enters level n for the first time, and it does so by entering phase j of level n. We can see from (6) that this probability is independent of the level n since we have restricted $B_1 = D_0$. Define

$$G^{(k)} = \sum_{\nu=0}^{\infty} G_\nu^{(k)}.$$

We also define $G_\nu^{(1)} = G_\nu$. Obviously, $G = G^{(1)}$. Furthermore, by conditioning on the level reached after the first transition,

$$G_\nu = \sum_{k=0}^{\infty} D_k G_{\nu-1}^{(k)}. \tag{36}$$

This holds since D_k contains the probabilities of moving to level $n + k - 1$, and in this case, the probabilities to reach level $n - 1$ for the first time in remaining $\nu - 1$ epochs are given by the matrix $G_{\nu-1}^{(k)}$. Summing over k yields (36).

If we take the sum of (36) over ν, and notice that for $\nu = 0$ and $k > 0$, $G_0^{(k)} = 0$, we obtain

$$G = \sum_{k=0}^{\infty} D_k G^{(k)}. \tag{37}$$

This becomes (35) if $G^{(k)} = G^k$. This is proved next.

We note that $(G_r^{(k-1)})_{is}$ is the probability of reaching level $n+1$ at step r for the first time, and entering phase s at this moment, given the process started in level $n+k$, phase i. Given this event takes place, then the conditional probability of first hitting level n, phase j, $\nu - r$ time units later is $(G_{\nu-r})_{sj}$. Consequently

$$G_\nu^{(k)} = \sum_{r=1}^{\nu-1} G_r^{(k-1)} G_{\nu-r}.$$

Summing over all ν yields

$$\begin{aligned}
\sum_{\nu=1}^{\infty} G_\nu^{(k)} &= \sum_{\nu=1}^{\infty} \sum_{r=1}^{\nu-1} G_r^{(k-1)} G_{\nu-r} \\
&= \sum_{r=1}^{\infty} \sum_{\nu=r+1}^{\infty} G_r^{(k-1)} G_{\nu-r} \\
&= \sum_{r=1}^{\infty} G_r^{(k-1)} \sum_{\nu=1}^{\infty} G_\nu.
\end{aligned}$$

Since the left side is $G^{(k)}$, and the right side is $G^{(k-1)}G$, we conclude

$$G^{(k)} = G^{(k-1)} G. \tag{38}$$

Solving (38) recursively for $k = 2, 3, \ldots$, one finds that $G^{(k)} = G^k$, and (35) follows.

To prove (34), one can show that the probabilities of starting in level 0 and returning to level 0 are contained in the matrix

$$\sum_{k=0}^{\infty} C_k G^{(k)}. \tag{39}$$

Clearly, C_k contains the probabilities that one moves from level 0 to level i on a given transition, and $G^{(k)}$ contains the probabilities of entering the various phases of level 0 for the first time. Therefore, the sum given by (39) represents the transition matrix of the Markov chain imbedded at the points of level 0, and, as we will show later in greater detail, its equilibrium vector must be proportional to π_0. Hence, (34) holds, and the theorem is proven.

Theorem 4 only allows us to find π_0 up to a factor. To find this factor, define

$$P(z) = \sum_{i=0}^{\infty} \pi_i z^i, \quad C(z) = \sum_{i=0}^{\infty} C_i z^i, \quad D(z) = \sum_{i=0}^{\infty} D_i z^i.$$

We observe that $D(1)$ and $C(1)$ are Markov chains in their own right. Paralleling the derivation of (33) in the scalar case, one obtains:

$$zP(z) = (P(z) - \pi_0)D(z) + z\pi_0 C(z).$$

Taking the derivative of this relation, and setting $z = 1$ yields

$$P(1) + P'(1) = P'(1)D(1) + (P(1) - \pi_0)D'(1) + \pi_0 C(1) + \pi_0 C'(1).$$

If we postmultiply both sides by e, and use $P(1)e = 1$, $C(1)e = D(1)e = e$, we find after some minor simplifications

$$1 - P(1)D'(1)e = \pi_0(I + C'(1) - D'(1))e.$$

This equation yields the desired condition to assure that the sum of all probabilities is equal to 1, and the problem is solved. Moreover, by taking higher derivatives, higher moments regarding the levels can be obtained.

Once the vector π_0 is obtained, one can, in principle, calculate all π_i recursively by using the following matrix-based modification of (32):

$$\pi_{j+1} = \left(\pi_j - \sum_{n=1}^{j} \pi_{j-n+1} D_n - \pi_0 C_j \right) D_0^{-1}.$$

Unfortunately, this recursion is numerically unstable, and except for low values of i, it is not advisable to calculate π_i in this fashion. A better method will be introduced in Section 5.8. At this point, we will also discuss how to deal with more general initial conditions.

4 NUMERICAL METHODS FOR THE PARADIGMS OF NEUTS

4.1 Introduction

This section discusses the most important methods for finding the basic matrices R and G. However, the reader is alerted to the fact that there are additional methods which, though interesting, would require too much background material to be included, In particular, we cannot discuss the methods based on Toeplitz matrices and fast Fourier transforms as they are introduced in [Bini and Meini, 1996, Meini, 1998]. We should also mention a method introduced in [Alfa et al., 1998], which is based on non-linear programming. We omit the treatment of these methods as so far they have not attained the wide use of the methods discussed here.

4.2 Classical methods

We start our discussion of solution methods with the matrix-geometric solution that applies to GI/M/1 and QBD models. One must first determine the matrix R, which is the minimal solution of (18). Once R is found, all other measures of interest can be found readily. In special situations where an explicit solution for R is known (see for instance [Neuts, 1981, pages 84,259], as well as [Miller, 1981, Liu and Zhao, 1996, Ramaswami and Latouche, 1986]), or if R possesses a special structure (see [Chakravarthy and Alfa, 1993]), it is usually best to exploit this solution. However, such cases are relatively rare, and typically one has to determine R iteratively. One starts with a value $R_{(0)}$, and improves this value somehow to find a value $R_{(1)}$. This same method can then be used for improving $R_{(1)}$ to obtain $R_{(2)}$, and so on. This iteration repeats until convergence is achieved. The preferred method for doing this is *successive substitution*. This is achieved by modifying (18) to become

$$R_{(i+1)} = \sum_{j=0}^{\infty} R_{(i)}^j A_j. \tag{40}$$

Typically, $R_{(0)}$ is equal to 0, or, equivalently, $R_{(1)} = A_0$. Latouche [Latouche, 1992] calls the algorithm obtained in this way the *natural algorithm*, as it arises in proofs pertaining to R as found in [Neuts, 1981] and elsewhere. If one uses $R_{(0)} = 0$ as the initial approximation, it is not difficult to see that the $\{R_{(i)}, i \geq 0\}$ form an increasing sequence, and they will therefore converge to some non-negative solution, say X. Neuts [Neuts, 1981] has shown that X is the minimal non-negative solution of (18), that is, $X = R$. Now let $R_{(i)}$ be the ith approximation corresponding to $R_{(0)} = 0$. Take any other matrix between $R_{(0)} = 0$ and $R_{(n)}$ as a starting point, where n is a large number, and call the sequence of approximations obtained in this way $R'_{(i)}$. By complete induction, one readily sees that $R_{(i)} < R'_{(i)} < R_{(n+i)}$. Since $R_{(i)}$ and $R_{(n+i)}$ obviously have the same limit, any initial solution $R'_{(0)}$ between 0 and X will converge to the same limit X, that is, there is no solution below X. This proves that the sequence $\{R_{(n)}\}$ converges to the minimal non-negative solution. The proof can obviously generalized to other monotonously increasing sequences of matrices, and it is know in literature under the name of the *monotone convergence theorem*.

To evaluate the left side of (40), one uses *Horner's method*, which works as follows: Truncate the sum on the left-hand side of (40) to some finite value, say ω, and form

$$S_\omega = A_\omega$$
$$S_i = R(A_i + RS_{i+1}) \quad i < \omega.$$

This continues until $i = 0$.

Since $I - A_1$ is never singular, the above iterative scheme can be improved. Solving

$$R = A_0 + RA_1 + \sum_{j=2}^{\infty} R^i A_j$$

for R yields

$$R = (A_0 + \sum_{i=2}^{\infty} R^i A_i)(I - A_1)^{-1}.$$

This algorithm, called the *traditional algorithm* by Latouche, reduces the number of iterations as shown in [Latouche, 1992]. In contrast to the natural algorithm, it can also be used for continuous-time Markov chains. In this case, one has to replace $I - A_1$ by the corresponding block of the infinitesimal generator.

Latouche also suggested a different algorithm. One has

$$R = \sum_{j=0}^{\infty} R^j A_j = A_0 + R \sum_{j=1}^{\infty} R^j A_j = A_0 + RY$$

where

$$Y = \sum_{j=0}^{\infty} R^{j+1} A_{j+1}. \tag{41}$$

Latouche uses U instead of Y, but U will be used for a different matrix later. Since $R = A_0 + RY$,

$$R = A_0(I - Y)^{-1}. \tag{42}$$

The following iterative scheme can now be used: first, an approximation for R is created, and this leads to an approximation for Y by (41), which in turn leads to a new approximation for R by (42). This yields the following scheme

$$\begin{aligned} Y_{(i)} &= \sum_{j=0}^{\infty} R_{(i)}^j A_{j+1} \\ R_{(i+1)} &= A_0(I - Y_{(i)})^{-1}. \end{aligned}$$

Again, the sum involved can be evaluated by Horner's method. As an initial value, one can use $R_{(0)} = 0$, which leads to $Y_{(0)} = A_1$. Latouche [Latouche, 1992] showed, by both theoretical analysis and numerical experimentation that the algorithm based on Y is better than both the natural and the traditional algorithm.

All the algorithms discussed can also be used, and have been used to find G in the M/G/1 paradigm. The necessary changes to do this are minor, and they will be left to the reader to work out. In the case of the QBD-process, equations (41) and (42) yield

$$Y = A_1 + RA_2 = A_1(I - Y)^{-1} A_2. \tag{43}$$

In this formula, one can find Y by successive substitution. Given Y, R can be found from (42).

4.3 Logarithmic convergence

The number of iterations to achieve acceptable convergence in the algorithms discussed in Section 4.2 has often been in the hundreds and beyond [Daigle and Lucantoni, 1991, Mitrani and Chakka, 1995]. Thus, methods for reducing the number of iterations have been sought. In this respect, the method given in [Latouche and Ramaswami, 1993] is particularly noteworthy. Unfortunately, this method is limited to QBD processes. We describe this method now.

Consider a Markov chain with the structure of (4), and assume this process is observed only when its level changes. Clearly, the probability that the level changes after ν iterations, and that the level decreases at this point by 1 is equal to $A_1^\nu A_2$. When taking the sum of these probabilities, one finds that B_2, the probability of going one level down is equal to $\sum_{\nu=0}^{\infty} A_1^\nu A_2 = (I - A_1)^{-1} A_2$. In a similar fashion, one finds the probability of going up one level after the change to be $B_0 = (I - A_1)^{-1} A_0$. In this way, a new Markov chain is obtained, with the transition matrix

$$\tilde{P} = \begin{bmatrix} 0 & * & 0 & 0 & \cdots \\ * & 0 & B_0 & 0 & \cdots \\ 0 & B_2 & 0 & B_0 & \cdots \\ 0 & 0 & B_2 & 0 & \cdots \end{bmatrix}.$$

Here, the stars denote matrices of no further interest. One has

$$G = B_2 + B_0 G^2. \tag{44}$$

To verify this, just replace B_2 and B_0 by their definitions, and one obtains, after some minor simplifications $G = A_2 + A_1 G + A_0 G^2$. Hence, the G obtained from (44) is also the G corresponding to $\tilde{P}$.

Suppose now the process given by $\tilde{P}$ starts in an even level. Since in the next step, it goes either up or down, it must be in an odd level in the next epoch, and in an even level two epochs from the start. In other words, if we observe the process only at even times, it will be in even levels. Now, consider the process at even epochs. If it is to go down, it must go down two steps, that is, instead of G, we have to consider $G^{(2)}$, and according to (38), $G^{(2)} = G^2$. Hence, this new process provides us with a value for G^2. This value may be inserted in (44) for G^2. Moreover, the new process is again a QBD, and this allows us to express G^2 in (44) above in terms of G^4, and so on. Let us now investigate this iterative procedure.

First, we obviously need the transition matrix of the process observed at times 0, 2, 4, ..., and at even levels. The matrix giving the probabilities of going down two levels is given by B_0^2, and going up two levels is given by B_2^2. The process does not change levels in the two epochs if it goes first up, then down, or first down, then up, and this is expressed by the matrices $B_0 B_2$ and $B_2 B_0$, respectively. Hence, if the levels are renumbered, such that the new level n is equal to the old level $2n$, and if a similar renumbering takes place for the times, one finds a new transition matrix $P^{(1)}$ as

$$P^{(1)} = \begin{bmatrix} * & * & 0 & 0 & \cdots \\ * & A_1^{(1)} & A_0^{(1)} & 0 & \cdots \\ 0 & A_2^{(1)} & A_1^{(1)} & A_0^{(1)} & \cdots \\ 0 & 0 & A_2^{(1)} & A_1^{(1)} & \cdots \\ \vdots & \vdots & \vdots & \vdots & \ddots \end{bmatrix}$$

with $A_0^{(1)} = B_0^2$, $A_1^{(1)} = B_0 B_2 + B_2 B_0$ and $A_2^{(1)} = B_2^2$. As mentioned before, the G of this process corresponds to $G^{(2)}$ of the old process. Hence, (44) applies, with G replaced by $G^{(2)}$, and B_0 and B_2 replaced by $B_0^{(1)}$ and $B_2^{(1)}$, which are

evaluated as before. In other words
$$B_i^{(1)} = (I - A_1^{(1)})^{-1} A_i^{(1)}, \quad i = 0, 2.$$
Equation (44) now becomes
$$G^{(2)} = B_2^{(1)} + B_0^{(1)}(G^{(2)})^2.$$
Since $G^{(2)} = G^2$, we can use this last relation to replace G^2 in (44) to obtain
$$G = B_2 + B_0 B_2^{(1)} + B_0 B_0^{(1)}(G^{(2)})^2.$$
Obviously, this process can be repeated ad infinitum, which yields the following expression for G:
$$G = \sum_{k=0}^{\infty} \left(\prod_{i=0}^{k-1} B_0^{(i)} \right) B_2^{(k)}. \tag{45}$$
where
$$B_i^{(0)} = (I - A_1)^{-1} A_i, \quad i = 0, 2$$
and
$$B_i^{(k+1)} = (I - A_1^{(k+1)})^{-1} A_i^{(k+1)} = (I - B_0^{(k)} B_2^{(k)} - B_2^{(k)} B_0^{(k)})^{-1} (B_i^{(k)})^2, \quad i = 0, 2.$$

The values for $B_0^{(k)}$ and $B_2^{(k)}$ can now be evaluated recursively, starting with $B_i^{(0)} = (I - A_1) A_i$, $i = 0, 2$, and with it, G can be obtained. To find $B_i^{(k+1)}$, $i = 0, 2$, given $B_0^{(k)}$ and $B_2^{(k)}$, one must evaluate $I - B_0^{(k)} B_2^{(k)} - B_2^{(k)} B_0^{(k)}$ first, which requires two matrix multiplications. The inverse of this result must then be multiplied by $(B_i^{(k)})^2$, which requires one matrix inversion and four matrix multiplications, two to find $B_0^{(k+1)}$ and two to find $B_2^{(k+1)}$. Hence, six matrix multiplications and one matrix inversion must be performed to find $B_i^{(k+1)}$, $i = 0, 2$. Each time, a new term of the sum in (45) is evaluated, one must multiply the product in parentheses by $B_0^{(k-1)}$, and multiply this new product by $B_2^{(k)}$, and this requires two more matrix multiplications, leading to a grand total of eight matrix multiplications and one matrix inversion for each k. However, this high number of multiplications is typically offset by a fast convergence.

Though the derivation was given for G, one can use the duality between G and R to find a similar derivation for R. When doing this, the probabilistic interpretations becomes more difficult, however (see [Latouche and Ramaswami, 1993]).

5 STATE ELIMINATION METHODS

So far, we have presented a number of methods to solve the different paradigms of Neuts. However, there are a number of issues which have not been addressed yet. To begin with, we assumed that the boundary matrices B_0, B_1, and C_1 in (4), (5) and (6) have a special structure. Next, we did not derive a stable recursion for the M/G/1 paradigm. Finally, no method was developed for the GI/G/1 queue. To address all of these issues, we make use of *censoring*.

5.1 Censoring

Equations (28) and (34) yield the following results for finding π_0

$$\pi_0 = \pi_0 \sum_{j=0}^{\infty} R^j B_j \quad \text{(GI/M/1 type processes)}$$

$$\pi_0 = \pi_0 \sum_{j=0}^{\infty} C_j G^j \quad \text{(M/G/1 type processes)}$$

These equations can be viewed as the equilibrium equations of certain Markov chains, specifically the Markov chains imbedded in level 0 of their respective processes. In words, if X_ν is the original Markov chain, and if t_i is the time of the ith visit to level 0, we can define the process $\{Y_i, i > 0\}$, where $Y_i = X_{t_i}$. According to the strong Markov property, Y_i is again a Markov chain, and, as it turns out, its transition matrix is $\sum_{j=0}^{\infty} R^j B_j$ for GI/M/1 type processes, and $\sum_{j=0}^{\infty} C_j G^j$ for M/G/1 type processes.

This is an example of a far more general approach to solving Markov chains known as *censoring*. In general, one starts with a Markov chain with a state space that is partitioned into two set E and E'. In the preceding two examples, E consisted of all states at level zero, while E' contained all other states – those that were censored. We denote the transition matrix of the original chain by P, and the matrix of the chain imbedded in E by P^E. To formulate this mathematically, let X_ν be the state of the process at time ν, and suppose the successive visits to E take place at time $t_1, t_2, \ldots$. The process $\{Y_i, i > 0\}$, with $Y_i = X_{t_i}$ is then called the *process imbedded in E*, and the states in E' are said to be *censored*. Whenever the process is in E', we say that we are on a *sojourn* in E'. The censored process consists of the original process interrupted by sojourns in E'. If one thinks in terms of sample paths, any realization of the censored process is equal to the corresponding sample path of the original process with the portions in E' removed. This construct has been studied in [Kemeny et al., 1966], who provide the following important theorem, which is valid for Markov chains in which all states are communicating.

Theorem 5 *Let P be the transition matrix of a Markov chain in which all states communicate. The states of the Markov chain are partitioned into two sets E and E', and this partitioning induces the following partition of P:*

$$P = \begin{bmatrix} T & H \\ L & Q \end{bmatrix}. \qquad (46)$$

Here, T contains the transition probabilities among the states of E, H the transition probabilities from E to E', L the transition probabilities for E' to E, and Q the transition probabilities among the states of E'. The transition matrix of the Markov chain imbedded in E is now given by

$$P^E = T + HNL. \qquad (47)$$

with

$$N = \sum_{\nu=0}^{\infty} Q^{\nu}. \qquad (48)$$

Theorem 5 follows directly from the following lemma, which is interesting in its own right. The combined proof for both follows the lemma.

Lemma 1 *Let* $N = \sum_{\nu=0}^{\infty} Q^{\nu}$. *In this case, we have*

1. $(N)_{ij}$ *gives the expected number of visits to state* $j \in E'$ *prior to the moment the process first enters a state in* E, *given the process started from state* $i \in E'$.

2. $(HN)_{ij}$ *is the expected number of visits to* $j \in E'$ *prior to returning to* E, *given the process started from state* $i \in E$.

3. $(HNL)_{ij}$ *is the probability of having a sojourn in* E' *which ends by entering* $j \in E$, *given the sojourn started from state* $i \in E$.

Proof. $(Q^{\nu})_{ij}$ is the probability that given the process is in $i \in E'$ at time τ, it is in $j \in E'$ at time $\tau + \nu$. Summing over ν gives the expected number of visits before returning to E, and the first item of Lemma 1 is proven. Item 2 of the lemma can be established in a similar fashion. To prove item 3, note that $(HQ^{\nu}L)_{ij}$ gives the probabilities, conditioned on the starting point, of a sojourn in E' lasting for ν time units with the process returning to $j \in E$ at $\nu + 1$. Summing over all ν yield item 3 of Lemma 1. To prove Theorem 5, note that in order to go from $i \in E$ to $j \in E$, one can either go directly, which corresponds to the matrix T, or via a sojourn in E' before going to j, which correspond to the matrix HNL. ∎

Note that the proof does not presuppose ergodicity as discussed in [Kemeny et al., 1966, page 134]. P could even be a substochastic matrix. The validity of the theorem can even be extended to deal with decomposable Markov chains: If Q contains ergodic subchains, then $\sum_{\nu=0}^{\infty} Q^{\nu}$ does not converge. In this case one replaces HNL in (47) by $\sum_{\nu=0}^{\infty} HQ^{\nu}L$. Even if $\sum_{\nu=0}^{\infty} Q^{\nu}$ diverges, $\sum_{\nu=0}^{\infty} HQ^{\nu}L$ may still converge. This is true, for instance, if $L = 0$ or $H = 0$. Kemeny et al. accommodate this case in their theorem by requiring that 0 times ∞ is zero.

As pointed out in [Kemeny et al., 1966, page 5], matrices of infinite dimension in general require great care, because in this domain, matrix multiplication is no longer associative, and the inverse is no longer unique (see also [Gail et al., 1998, Theorems 1 and 2]). Fortunately, the associative law holds for matrices with non-negative elements, even when they are of infinite dimension.

It is always the case that N is an inverse of $I - Q$. We have:

$$N = I + \sum_{\nu=1}^{\infty} Q^{\nu} = I + Q \sum_{\nu=0}^{\infty} Q^{\nu} = I + QN$$

$$N = I + \sum_{\nu=1}^{\infty} Q^\nu = I + \sum_{\nu=0}^{\infty} Q^\nu Q = I + NQ.$$

The first of these two equations implies $(I - Q)N = I$, that is, N is a right inverse of $I - Q$. The second equation similarly implies $N(I - Q) = I$, that is, N is a left inverse of $I - Q$.

In the case of finite-dimensional matrices, the inverse of $I - Q$ is unique. In the case of infinite-dimensional matrices, one can construct other inverses, as shown in [Kemeny et al., 1966, page 111]. This construct, which will not be described here, also shows that N is the minimal non-negative matrix of $I - Q$.

5.2 Censoring as Block Elimination

If the submatrices T, H, L and Q of (46) are all finite, then an alternative method to find P^E is by block elimination. Let $\hat{\pi}$ be the vector of the equilibrium probabilities belonging to E, and $\overline{\pi}$ the ones belonging to E', then one can write the equilibrium equations corresponding to (46) as follows

$$\hat{\pi} = \hat{\pi}T + \overline{\pi}L$$
$$\overline{\pi} = \hat{\pi}H + \overline{\pi}Q.$$

Hence

$$\overline{\pi} = \hat{\pi}H(I - Q)^{-1} \qquad (49)$$
$$\hat{\pi} = \hat{\pi}T + \hat{\pi}H(I - Q)^{-1}L$$
$$= \hat{\pi}(T + H(I - Q)^{-1}L). \qquad (50)$$

According to (47), $T + H(I - Q)^{-1}L = P^E$, which means that (50) implies

$$\hat{\pi} = \hat{\pi}P^E.$$

The vector $\hat{\pi}$ is therefore the equilibrium vector of P^E. Hence, we obtained P^E by a purely mathematical operation, namely a block elimination. The fact that this block elimination converts larger Markov chains into smaller ones will be crucial for our further discussion. Block elimination is restricted to finite blocks due to questions relating the uniqueness of $(I - Q)^{-1}$.

We now show that (49) can be generalized to cover the case that E' is infinite. It is important to note at the outset that the Markov chain may or may not be recurrent. Consequently, we will work with quantities that will be shown to be equivalent to steady-state probabilities in the positive recurrent case, but which are meaningful in their own right in both cases. Following [Zhao et al., 1999], we select a state, say state 0, and we define α_i the expected number of visits to the states i, $i \geq 0$ before the process returns to state 0, which implies that $\alpha_0 = 1$. As it turns out, the α_i, $i \geq 0$ satisfy the equilibrium equations. As an additional benefit, note that the α_i, $i \in E$ are independent of our choice of E.

Using a method given by [Zhao et al., 1999], we first show that the α_i satisfy the equilibrium equations of the Markov chain under consideration. The starting point is Lemma 1, item 2, with $E = \{0\}$ and $E' = \{1, 2, \ldots\}$. In this case, P becomes

$$P = \begin{bmatrix} \begin{bmatrix} p_{00} \\ p_{10} \\ p_{20} \\ p_{30} \\ \vdots \end{bmatrix} & \begin{bmatrix} p_{01} & p_{02} & p_{03} & \cdots \end{bmatrix} \\ & Q \end{bmatrix}.$$

Let
$$\overline{\alpha} = [\alpha_1 \; \alpha_2 \; \alpha_3 \ldots] \quad \text{and} \quad H = [p_{01} \; p_{02} \; p_{03} \; \ldots].$$

By Lemma 1, item 2, one has

$$\overline{\alpha} = HN = H \sum_{\nu=0}^{\infty} Q^{\nu} = H + H \sum_{\nu=1}^{\infty} Q^{\nu-1} Q = H + \overline{\alpha} Q.$$

This expands to

$$\alpha_j = p_{0j} + \sum_{i=1}^{j-1} \alpha_i p_{ij}, \quad j \neq 0.$$

Hence, the α_i satisfy all equilibrium equations, except the first one. However, in the case of P being stochastic, the row sums of the transition matrix are all one, so that the equilibrium equation connected with $j = 0$ is redundant.

The derivation holds even for non-ergodic Markov chains. Indeed, the sum $\sum_{i \neq 0} \alpha_i$ yields the expected number of epochs between two successive visits to state 0, and this number must be finite if the Markov chains is positive recurrent. In this case

$$\pi_i = \alpha_i / \sum_{j=0}^{\infty} \alpha_j, \quad i \geq 0.$$

Let us now generalize (49) to the case where E' infinite. In place of $\hat{\pi}$ and $\overline{\pi}$, we use $\hat{\alpha}$ and $\overline{\alpha}$, where $\hat{\alpha} = [\alpha_i, i \in E]$ and $\overline{\alpha} = [\alpha_i, i \in E']$. We now prove that

$$\overline{\alpha} = \hat{\alpha} H N. \tag{51}$$

We define $\pi_i^{(0,\tau)}$ as the probability that given we start in state 0 at time 0, the system is in state i at time τ, and has not returned to state 0 yet. Hence

$$\alpha_i = \sum_{\tau=0}^{\infty} \pi_i^{(0,\tau)}.$$

If $j \in E'$, one can find $\pi_j^{(0,\tau)}$ according the following formula

$$\pi_j^{(0,\tau)} = \sum_{\nu=0}^{\tau-1} \sum_{i \in E} \pi_i^{(0,\nu)} (HQ^{\tau-\nu-1})_{ij}.$$

Hence

$$\sum_{\tau=0}^{\infty} \pi_j^{(0,\tau)} = \sum_{i\in E}\sum_{\nu=0}^{\infty}\sum_{\tau=\nu+1}^{\infty} \pi_i^{(0,\nu)}(HQ^{\tau-\nu-1})_{ij}$$
$$= \sum_{i\in E}\sum_{\nu=0}^{\infty} \pi_i^{(0,\nu)}\left(H\sum_{\tau=0}^{\infty} Q^\tau\right)_{ij}.$$

This implies

$$\alpha_j = \sum_{i\in E} \alpha_i (HN)_{ij}$$

and (51) follows. Related results can be found in [Miller, 1981] and [Grassmann et al., 1985].

5.3 Application of Censoring to the QBD Process

We now apply Theorem 5 with E consisting of the states of level 0 to level n and E' of the states of all levels above n in a quasi birth and death process. In this case, H contains only one non-zero square matrix, A_0, in the lower left corner of H, corresponding to the transitions from level n to level $n+1$. Similarly, L has only one non-zero block, namely A_2, which can be found in the upper right corner, and which contains the transitions from level $n+1$ to level n. A little thought shows that this implies that HNL has only one non-zero block, which is in the lower right corner, namely $A_0 N_{11} A_2$, where N_{11} is the upper left block of N, indicating the expected number of visits to level $n+1$ when starting in level $n+1$. In short, we have

$$HNL = \begin{bmatrix} 0 & 0 \\ 0 & A_0 N_{11} A_2 \end{bmatrix}.$$

Since the only non-zero block of HNL is in the lower right corner, the only block of P^E which differs from the corresponding block of P is the lower right block. We denote this block by $P_{nn}^{(n)}$, which can be written as

$$P_{nn}^{(n)} = A_1 + A_0 N_{11} A_2 \qquad (52)$$

Observe that $P_{nn}^{(n)}$ is actually independent of the level n, and is therefore equal to $P_{n+1\,n+1}^{(n+1)}$. Denoting the common value of these matrices by Y, one obtains

$$Y = P_{nn}^{(n)} = P_{n+1\,n+1}^{(n+1)} = A_1 + A_0 N_{11} A_2 \qquad (53)$$

As an alternative to censoring at level n, one could proceed in two steps: first, to censor at level $n+1$, and subsequently to remove level $n+1$ itself. When this is done, one finds that the Markov chains censored at levels n and $n+1$ respectively are linked by the equation

$$Y = P_{nn}^{(n)} = A_1 + A_0(I - P_{n+1\,n+1}^{(n+1)})^{-1} A_2 \qquad (54)$$

By (53), this yields
$$Y = A_1 + A_0(I - Y)^{-1}A_2. \tag{55}$$
Note that this is exactly the method suggested by Latouche as applied to the QBD process. In fact, (55) is identical to (43). This provides a new derivation of a formula given earlier.

The fact that Y in (43) coincides with the one used by Latouche can also be seen by referring to its probabilistic interpretation. According to Theorem 5 $(Y)_{ij}$ is the probability that starting from a given level, phase i, the process will return eventually to that level, phase j, avoiding all levels below the given level on this sojourn. This definition is equivalent with the one given by Latouche on several occasions [Latouche, 1987, Latouche, 1992, Latouche and Ramaswami, 1993]. Also note that by Lemma 1, item 2, $A_0(I - Y)^{-1} = A_0 N_{11}$ gives the expected number of visits to the states of level $n+1$ starting from level n before returning to level n, that is, $A_0(I - Y)^{-1} = R$.

The interpretation of Y in the context of censoring suggest a new algorithm: suppose we truncate the transition matrix of the QBD process at some finite level ω, except that we place an arbitrary matrix Y_0 in the lower right corner of this matrix. Hence, our transition matrix looks as follows

$$\begin{bmatrix} B_0 & C_0 & 0 & 0 & \cdots & 0 & 0 \\ B_1 & A_1 & A_0 & 0 & \cdots & \vdots & \vdots \\ 0 & A_2 & A_1 & A_0 & \ddots & \vdots & \vdots \\ 0 & \ddots & A_2 & A_1 & \ddots & \vdots & \vdots \\ \vdots & \ddots & \ddots & \ddots & \ddots & \vdots & \vdots \\ 0 & \cdots & \cdots & \cdots & \cdots & A_1 & A_0 \\ 0 & \cdots & \cdots & \cdots & \cdots & A_2 & Y_0 \end{bmatrix}. \tag{56}$$

We now censor the last level to obtain a matrix in which the last level is $\omega - 1$, and which has Y_1 on the bottom right corner, where

$$Y_1 = A_1 + A_0(I - Y_0)^{-1}A_2.$$

We can continue this procedure, reducing the transition matrix level by level. In each step, we find a matrix just like (56), except that the right corner contains

$$Y_n = A_1 + A_0(I - Y_{n-1})^{-1}A_2. \tag{57}$$

Hence, if the lower right-hand corner looks as follows

$$\begin{bmatrix} A_1 & A_0 \\ A_2 & Y_n \end{bmatrix}$$

Y_{n+1} will show up in the place of A_1 within this matrix after the elimination.

As will be shown in Section 5.4, the Y_n converge toward Y. Once $Y_n = Y$, Y_n will not change any more until one reaches level 1. At this point, one obtains the following transition matrix:

$$\begin{bmatrix} B_0 & C_1 \\ B_1 & Y \end{bmatrix}.$$

This Markov chain can now be either solved directly, or it can be reduced once more by applying (47) with $H = C_1$, $L = B_1$ and $N = (I - Y)^{-1}$.

$$B_0^* = B_0 + C_1(I - Y)^{-1} B_1.$$

The vector α_0 can now be found by solving:

$$\alpha_0 = \alpha_0 B_0^*$$

with $\alpha_{01} = 1$. For α_1, we use (49) with E consisting of all states of level 0, and E' consisting of all states of level 1, and we obtain

$$\alpha_1 = \alpha_0 C_1 (I - Y)^{-1}.$$

Since the number of visits to the non-censored states are not affected by the censoring level, the α_n are also independent of the level at we are censoring. This allows us to use (49) at all levels. For $n > 1$, this means

$$\alpha_{n+1} = \alpha_n A_0 (I - Y)^{-1}.$$

Here, $(I-Y)^{-1}$ must be interpreted as N, and by referring to item 2, Lemma 1, one finds that the matrix $R = A_0(I - Y)^{-1}$ is the rate matrix. In conclusion, we have been able to derive the matrix geometric solution through block elimination, and this has helped us to deal with more complex initial conditions.

5.4 Approximating Infinite Markov Chains

In the preceding section, we used a large, but finite matrix to obtain Y. This approach raises the question about the convergence of the procedure, and, related to this, about the uniqueness of Y. We should mention from the start that Y is not unique in general. However, it will be proven that in the context of recurrent Markov chains, Y is unique in a certain sense. Even in this restricted sense, Y is not unique for non-recurrent Markov chains. The reader may want to check this by considering a simple non-recurrent random walk.

The basic theorem is as follow:

Theorem 6 *Let $P = [p_{ij}, i, j = 0, 1, 2, \ldots]$ be the transition matrix of a recurrent Markov chain on the non-negative integers. Choose an integer ω, and approximate P by $P(\omega) = [p_{ij}(\omega)]$ such that $P(\omega)$ is stochastic or substochastic and satisfies*

$$p_{ij}(\omega) = p_{ij} \quad \text{for} \quad i, j \leq \omega$$

Let $E = \{0, 1, 2, \ldots, n\}$, and let P^E be the Markov chain imbedded in E. Then

$$\lim_{\omega \to \infty} P^E(\omega) = P^E$$

The Markov chain $P(\omega)$ is not necessarily finite, even though in our applications, it typically is. $P(\omega)$ may be substochastic. In particular, we can truncate P by setting to zero all transition probabilities involving states above ω. Another possibility is to select in some way the transition probabilities from or to all states within some range beyond ω, say in the range from $\omega + 1$ to $\omega + k$, and set the transition probabilities involving states beyond this range equal to zero. We refer to the process of selecting the elements $p_{ij}(\omega)$ with i and/or j greater than ω so the the rows sums add to 1 as *complementation*. In other words, complementation ensures that $P(\omega)$ is stochastic. Different schemes for complementation have been investigated by [Zhao and Liu, 1996].

The theorem indicates that in the case of recurrent Markov chains, any method of changing the transition probabilities in the rows and columns above ω, including truncation or complementation, leads to the same censored Markov chain P^E as $\omega \to \infty$, as long as the matrix $P(\omega)$ remains stochastic or substochastic.

We now present the proof of the theorem.

Proof. Note that

$$P^E = T + HNL$$

expands to

$$p^E_{ij} = p_{ij} + (HNL)_{ij} = p_{ij} + d_{ij}, \ i,j \in E,$$

where d_{ij} is the probability that an E' sojourn starting in $i \in E$ ends in $j \in E$. Define d^s_{ij} to be the probability that such a sojourn has a maximum of s. The symbols $d_{ij}(\omega)$ and $d^s_{ij}(\omega)$ are defined similarly for $P(\omega)$. One has

$$\sum_{s \leq \omega} d^s_{ij} \leq d_{ij}(\omega) \leq \sum_{s \leq \omega} d^s_{ij} + \sum_{s > \omega} \sum_{k \in E} d^s_{ik} + d_{i\infty}.$$

Here, $d_{i\infty}$ is the probability that a sojourn starting from i never ends. The left inequality follows because $d^s_{ij} \geq 0$ and $d^s_{ij} = d^s_{ij}(\omega)$ as long as $s \leq \omega$. The inequality on the right gives the probability that all sojourns reaching a state above ω will be redirected to j, and this obviously includes all sojourns that possibly could end in j when using the matrix $P(\omega)$. If the Markov chain is recurrent, then $d_{i\infty} = 0$. Moreover, the double sum on the right also converges to zero as $\omega \to \infty$. Hence, in the limit, $d_{ij}(\omega) = d_{ij}$, which implies the theorem.

The upshot of the theorem is that the procedure suggested in Section 5.3 is sound. First of all, Y is unique, because if there would be two distinct values of Y, say $Y^{(1)}$ and $Y^{(2)}$, both of them satisfying the conditions of the theorem, then starting with $Y_0 = Y^{(1)}$ would give a different result than starting with $Y_0 = Y^{(2)}$, and this contradicts the theorem. In addition to this, the theorem proves the convergence of the method in question.

5.5 State Reduction

In Section 5.2, we have shown that censoring in finite Markov chains is equivalent to block-elimination. Instead of eliminating the states of a level all at once, one could eliminate all its states one by one, which we will refer to as *scalar elimination*. The question naturally arises as to whether the same results are obtained after, on the one hand, performing one iteration of block elimination, while on the other, performing scalar elimination repeatedly until an entire level has been eliminated. Furthermore, does one method have any computational benefits over the other? As will be established below, identical results are obtained from both approaches, and in general they have the same computational complexity. In cases of sparse matrices, it is our observation that scalar elimination is better able to exploit sparsity.

We will perform scalar elimination by a technique called *state reduction*, State reduction is widely used for finding equilibrium probabilities in finite state Markov chains [Sheskin, 1985, Grassmann et al., 1985, Heyman and Reeves, 1989, Sonin, 1999], as well as for obtaining the fundamental matrix [Kohlas, 1986, Sonin and Thornton, 1998, Heyman, 1987, Heyman, 1995, Heyman and O'Leary, 1998]. We will also demonstrate how state reduction can be used to to find Y_{n+1} in (55).

Consider a finite Markov chain with the states $0, 1, 2, \ldots \omega$, and with the transition matrix $P = [p_{ij}, i, j = 0, 1, \ldots, \omega]$. We use the α_i instead of the equilibrium probabilities π_i, where α_i is the expected number of visits to state i between two visits to state 0, and write the equilibrium equations as follows:

$$\alpha_j = \sum_{i=0}^{\omega} \alpha_i p_{ij} \quad j > 0. \tag{58}$$

The ωth equation can be solved for α_ω, yielding

$$\alpha_\omega = \sum_{i=0}^{\omega-1} \alpha_i p_{i\omega} (1 - p_{\omega\omega})^{-1}. \tag{59}$$

The resulting expression can the be used to eliminate α_ω from all earlier equations, leaving a system of ω equations in ω unknowns:

$$\alpha_j = \sum_{i=0}^{\omega-1} \alpha_i \left(p_{ij} + p_{i\omega}(1 - p_{\omega\omega})^{-1} p_{\omega j} \right) \tag{60}$$

for $j = 0, \ldots, \omega - 1$. If we define

$$p_{ij}^{(\omega-1)} = p_{ij} + p_{i\omega}(1 - p_{\omega\omega})^{-1} p_{\omega j} \tag{61}$$

then (60) can be written as

$$\alpha_j = \sum_{i=0}^{\omega-1} \alpha_i p_{ij}^{(\omega-1)}.$$

Equation (61) can be interpreted in terms of censoring, with $E' = \{\omega\}$, $E = \{0, 1, \ldots, \omega-1\}$, $H = [p_{i\omega}, i \in E]$, $N = (1-p_{\omega\omega})^{-1}$, and $L = [p_{\omega j}, j \in E]$. The new Markov chain $P^{(\omega-1)} = [p_{ij}^{(\omega-1)}]$ then represents the Markov chain with the state ω censored. Once $P^{(\omega-1)}$ is know, the Markov chain with transition matrix $P^{(\omega-2)}$ can be determined by censoring $\omega - 1$, and in this way, one can continue until only state 0 is left. In this way, one state after the other is eliminated, hence the name *state reduction*. In general, one has

$$p_{ij}^{(n-1)} = p_{ij}^{(n)} + p_{in}^{(n)}(1 - p_{nn}^{(n)})^{-1} p_{nj}^{(n)}, \quad i, j = 0, 1, 2, \ldots, n - 1, \qquad (62)$$

where the $p_{ij}^{(n)}$ can either be interpreted as the transition probabilities of the Markov chains with the states $n+1$, $n+2$, ... ω censored, or as the coefficients of the system of equations obtained by eliminating α_ω, $\alpha_{\omega-1}$, ..., α_{n+1} form the equilibrium equations. In each step, an equation similar to (60) is obtained:

$$\alpha_n = \sum_{i=0}^{n-1} \alpha_i p_{in}^{(n)} (1 - p_{nn}^{(n)})^{-1} \qquad (63)$$

Equations (62) and (63) lead to the following algorithm. Note that there is only one array p_{ij}, and when calculating a new set of $p_{ij}^{(n-1)}$, the old values $p_{ij}^{(n)}$ are overwritten, except for the $p_{in}^{(n)}$ and the $p_{nj}^{(n)}$, which remain unchanged. Since one only needs the p_{in}^n to calculate the α_n, this causes no problems.

Algorithm SR1

"Elimination Step"
for $n = \omega, \omega - 1, \ldots, 1$
$\quad p_{ij} \leftarrow p_{ij} + p_{in}(1 - p_{nn})^{-1} p_{nj}, \quad i, j < n$
"Back-Substitution Step"
$\alpha_0 \leftarrow 1$
for $n = 1, 2, \ldots, \omega$
$\quad \alpha_n = \sum_{i=0}^{n-1} \alpha_i p_{in}(1 - p_{nn})^{-1}$

After this algorithm is done, the α_i must still be divided by a norming constant to find the π_i.

There are two modifications to this algorithm. First of all, in each iteration, $P^{(n)}$ is a Markov chain, and one therefore has

$$1 - p_{nn}^{(n)} = \sum_{j=0}^{n-1} p_{ij}^{(n)}. \qquad (64)$$

By using the summation, all subtractions can be avoided and this increases the numerical stability as indicated in Chapter 4. The use of (64) is referred to as *GTH advantage*, because it was first published by Grassmann, Taksar and Heyman [Grassmann et al., 1985]. For a detailed analysis of the rounding errors associated with this method, see also [O'Cinneide, 1993, Grassmann, 1993]. A second improvement originates form the observation that the quotient $p_{ij}/(1 - p_{jj})$ is formed twice, first in the elimination step, then in the back-substitution step. This duplication can be avoided, as shown in the algorithm below, which also uses (64) to insure numerical stability.

Algorithm SR2

"Elimination Step"
for $n = \omega, \omega - 1, \ldots, 1$
$\quad s \leftarrow \sum_{j=0}^{n-1} p_{nj}$
$\quad p_{in} \leftarrow p_{in}/s, \quad i = 0, 1, \ldots, n-1$
$\quad p_{ij} \leftarrow p_{ij} + p_{in}p_{nj}, \quad i, j < n$
"Back-Substitution Step"
$\alpha_0 \leftarrow 1$
for $n = 1, 2, \ldots, \omega$
$\quad \alpha_n \leftarrow \sum_{i=0}^{n-1} \alpha_i p_{in}$

Algorithm SR2, which is the standard implementation of state reduction, can be summarized as follows: to eliminate the current highest state n, one sums to the left of the diagonal, then divides the nth column by this sum (obtaining a set of multipliers p_{in}), and then uses the modified column with the row to update all remaining elements (that is those with indices less than n). At the end, the $\alpha_n, n = 1, 2, \ldots, \omega$, are obtained using the multipliers p_{in}.

It is worth noting that the algorithm SR1 can be used with minor changes for block-elimination. Essentially, one has to replace the scalars p_{ij} by matrices, as indicated by the following equations:

$$P_{ij}^{(n-1)} = P_{ij}^{(n)} + P_{in}^{(n)}(I - P_{nn}^{(n)})^{-1}P_{nj}^{(n)}. \tag{65}$$

and

$$\boldsymbol{\alpha}_j = \sum_{i=0}^{j-1} \boldsymbol{\alpha}_i P_{ij}^{(j)}(I - P_{jj}^{(j)})^{-1}. \tag{66}$$

At the end, $\boldsymbol{\alpha}_0 = \boldsymbol{\alpha}_0 P^{(0)}$ must be solved explicitly. Similar changes can be done in SR2, except that s must be replaced by the matrix $I - P_{nn}^{(n)}$, which is no longer obtainable as the summation of $\sum_{j=0}^{n-1} P_{nj}^{(n)}$.

In using state reduction to remove successively the $m^{\text{th}}, (m-1)^{\text{th}}, \ldots, 1^{\text{st}}$ phases of a given level, we have at each step expressed the impact on the non-censored state of the removal of the state being censored. Once this has been repeated m times, we have the collective impact on the remaining uncensored levels of the removed censored level. Equation (65) shows how this can be achieved in one step using matrices.

If one thinks in terms of sample paths, it should be clear that the same censored sample path is obtained regardless of whether one eliminates successively phases m, $m-1$, ..., 1, or collectively all m phases at once. This establishes that the same results are obtained.

We now illustrate how the state reduction algorithm can be applied to QBD processes on an infinite state space. For related references, see [Grassmann, 1986, Grassmann and Heyman, 1990, Grassmann and Heyman, 1993, Grassmann, 1985, Stanford and Grassmann, 1993]. Consider again the lower right corner of the transition matrix of the QBD-process, with Y replaced by Y_n as follows:

$$\begin{bmatrix} A_1 & A_0 \\ A_2 & Y_n \end{bmatrix}.$$

This matrix arises when eliminating the levels of a Markov chain complemented at level ω as indicated in (56). Y_0 must be chosen so that $P(\omega)$ is either stochastic or substochastic. Y_0 can be chosen to be 0, for instance, in which case $Y_1 = A_1$. After n iterations, one finds Y_n at the lower right corner. At this point, the last row refers to level $\omega - n$, and the one above to level $\omega - n - 1$. The problem is to find Y_{n+1}. We now want to censor level $\ell = \omega - n$. We can do this by applying successively (62) for $i = (\ell, j)$, $j = m, m-1, \ldots, 2, 1$, where (ℓ, j) represents the state of level ℓ, phase j. The cumulative effect of these operations is that A_1 is replaced by Y_{n-1}. This follows immediately from the fact that the new chain is now imbedded in levels 1 to $\ell - 1$, and $P^{(\ell-1)}$ has $Y_{n+1} = A_1 + A_0(I - Y_n)^{-1}A_2$ in its bottom right corner.

Of course, the elimination could have been done until level 0 had been reached, but this would have been a waste of effort. The reason is that once two successive matrices Y_n and Y_{n+1} converge, later Y_n's will stay identical, and nothing changes.

We now introduce a different version of state reduction, which will be needed in Section 5.6 when deriving efficient methods for dealing with inverses, and in Section 5.8 when analyzing the GI/G/1 paradigm. The method is closely related to the Crout method of solving numerical equations (see e.g. [Stewart, 1973]). The basic idea is as follows: as an alternative to SR2 in which we update every lesser element of P each time a state is eliminated, we can postpone the changes to p_{ij} right up to the point that $n = \max\{i, j\}$, that is, until its row or column *requires* modification. However, we are free to update p_{ij} at any earlier point in the process of elimination. The following theorem provides the basis of this "Crout-version" of state reduction:

Theorem 7 *Let $P = [p_{ij}]$ be the transition matrix of a recurrent Markov chain, and let $p_{ij}^* = p_{ij}^{(\max\{i,j\})}$. Then, the following relation holds*

$$p_{ij}^{(n)} = p_{ij} + \sum_{k>n}^{\infty} p_{ik}^*(1 - p_{kk}^*)^{-1}p_{kj}^*. \qquad (67)$$

If P has a block structure, that is, if $P = [P_{rs}]$, then one similarly has

$$P_{rs}^{(n)} = P_{rs} + \sum_{k>n} P_{rk}^*(I - P_{kk}^*)^{-1}P_{ks}, \qquad (68)$$

where $P_{rs}^* = P_{rs}^{(\max\{r,s\})}$.

Of particular interest is the case where $n = \max\{i,j\}$, in which case $p_{ij}^{(n)} = p_{ij}^*$ and $P_{ij}^{(n)} = P_{ij}^*$.

Proof. We will prove the theorem only for the scalar case, because the matrix case is essentially identical. In view of Theorem 6, we can restrict ourselves to finite-state Markov chains. Hence, consider a Markov chain with the states 0, 1, ..., ω. Clearly, $p_{ij}^{(\omega)} = p_{ij}$. Equation (62) now yields

$$\begin{aligned}
p_{ij}^{(n)} &= p_{in+1}^{(n+1)}(1 - p_{n+1n+1}^{(n+1)})^{-1}p_{n+1j}^{(n+1)} + p_{ij}^{(n+1)} \\
&= p_{in+1}^*(1 - p_{n+1n+1}^*)^{-1}p_{n+1j}^* + p_{in+2}^{(n+2)}(1 - p_{n+2n+2}^{(n+2)})^{-1}p_{n+2j}^{(n+2)} + p_{ij}^{(n+2)} \\
&= \ldots \\
&= \sum_{k=n+1}^{\omega} p_{ik}^*(1 - p_{kk}^*)^{-1}p_{kj}^* + p_{ij}
\end{aligned}$$

and (67) follows.

Theorem 7 gives rise to several algorithms for finding equilibrium probabilities for Markov chains. If ω is the last state, then one can first use (67) with $i = \omega$ and $n = \max\{i,j\}$ to find all $p_{\omega j}^*$, then one similarly uses (67) except that $i = \omega - 1$ to find all $p_{\omega-1\,j}^*$, and so on, until all p_{ij} are found. In other words, one reduces the matrix row by row, leaving the columns unchanged. Instead of applying the theorem by rows, one can also apply it by columns, that is, one can first find all $p_{i\,\omega}^*$, then all $p_{i\,\omega-1}^*$, and so on. Hence, the method given by Theorem 7 provides some flexibility.

5.6 Dealing with Inverses

Crucial to the methods discussed here is the evaluation of $A_0(I-Y)^{-1}$, or in the context of Theorem 5, the evaluation of $H(I-Q)^{-1}$, and the question arises how to do this efficiently. Numerical analysts advise not to invert $I - Q$ and multiply H by this inverse, and the reason is as follows. The standard method to evaluate an inverse is by using elimination methods. These methods provide a factorization of $I - Q$ into two triangular matrices, and these two matrices can be used to find $H(I-Q)^{-1}$ with the same effort as is required to multiply H by $(I-Q)^{-1}$. Hence, the calculation of the inverse, and the effort to do so, can be avoided.

In our context here, finding the triangular factors of $I - Q$ have is closely related to scalar elimination. To show this, we partition the state space into two sets E and E' and write

$$P = \begin{bmatrix} T & H \\ B & Q \end{bmatrix}. \tag{69}$$

The matrices T, H and Q are defined as in Theorem 5, but the transitions from E' to E are now given by the matrix B. The states of E will be numbered from 1 to n, and the ones of E' from $n+1$ to $n+m$. We eliminate all states of E' by applying (67) with $n = \max\{i,j\}$, that is, we use

$$p^*_{ij} = p_{ij} + \sum_{k>\max\{i,j\}}^{n+m} p^*_{ik}(1-p^*_{kk})^{-1}p^*_{kj}. \tag{70}$$

To obtain the desired factorization, we first write this equation in matrix form. To this end, we define

$$\left.\begin{array}{l} U = [p^*_{ij},\ i,j \in E',\ i < j] \\ L = [p^*_{ij},\ i,j \in E',\ i > j] \\ D = \mathrm{diag}((1-p^*_{ii})^{-1}, i \in E') \end{array}\right\} \tag{71}$$

From this definition, it is clear that U is a strictly upper triangular matrix, L is a strictly lower triangular matrix, and D a diagonal matrix. If $Q^* = [p^*_{ij},\ i,j \in E']$, we have

$$Q^* = U + L + \mathrm{diag}(p^*_{ii}) = U + L - D^{-1} + I. \tag{72}$$

It is easy to verify that (70) for $i,j \in E'$ can be written as

$$Q^* = Q + UDL. \tag{73}$$

From (72) and (73) we obtain

$$U + L - D^{-1} + I = Q + UDL.$$

Consequently

$$I - Q = UDL - U - L + D^{-1} = (I - UD)D^{-1}(I - DL). \tag{74}$$

This "UDL" factorization allows us to write

$$H(I - Q)^{-1} = H(I - DL)^{-1}D(I - UD)^{-1}$$

Hence, to find $H(I - Q)^{-1}$, one first forms $H(I - DL)^{-1}$, one then multiplies the result by D, and one then forms $H(I - DL)^{-1}D(I - UD)^{-1}$.

$H(I - DL)^{-1}$ can be found from (70). To see this, define

$$H^* = [p^*_{ij},\ i \in E,\ j \in E'].$$

We now show that

$$H^* = H(I - DL)^{-1}. \tag{75}$$

To prove (75), we write (70) in matrix form as

$$H^* = H + H^*DL.$$

Solving this for H^* directly yields (75).

To form H^*D, we merely multiply each column of H^* by the appropriate $(1-p_{ii}^*)^{-1}$. Let

$$c_{ij} = (H^*D)_{ij} = p_{ij}^*(1-p_{jj}^*)^{-1}, \quad i \in E, \ j \in E'$$
$$\rho_{ij} = (UD)_{ij} = p_{ij}^*(1-p_{jj}^*)^{-1}, \quad i \in E', \ j \in E', \ i < j.$$

At this point, one can apply the following algorithm to find $H(I-Q)^{-1}$:

Algorithm Mult

for all $i \in E$ having non-zero elements
   for $\nu = n+1, n+2, \ldots, n+m$
      $c_{ij} = c_{ij} + \rho_{i\nu}c_{\nu j}, \quad \nu < j \le n+m$

This method is similar to the standard method to multiply a matrix by a triangular matrix (see e.g. [Stewart, 1973, page 146]). The idea is to convert $I - UD$ into the matrix I. Applying the same operations to H^*D has the effect of multiplying H^*D by $(I - UD)^{-1}$.

We now summarize: in order to find $H(I-Q)^{-1}$, one first applies state reduction to $Q - I$ to obtain the factorization given by (74), with U, L and D given in (71). One then forms $H^* = H(I - DL)^{-1}$ and multiplies H^* by D. Algorithm Mult can then be used to convert H^*D into $H(I-Q)^{-1}$. The procedure requires, when ignoring all terms of lower order, $m^3/3$ operations to find the factorization, and $n_1 m^2$ operations to multiply the result by the inverse of these factors, where n_1 is the number of non-zero rows of H.

Algorithm Mult also provides a method to find R in the GI/M/1 paradigm. Indeed, one has

$$R = H^*D(I-UL)^{-1}.$$

One concludes that in both block elimination and scalar elimination one first forms $H^* = H(I-DL)^{-1}$. So far, block-elimination and scalar elimination are identical. From here on, they differ slightly. In the case of block elimination, one forms $H^*D(I - UD)^{-1}$ and then $H^*D(I - UD)^{-1}B$, whereas in scalar elimination, one forms $(I-UD)^{-1}B$, and premultiplies this by $H^*D(I-UD)^{-1}$, as the reader may verify. The end result is $H(I-Q)^{-1}B$ in both cases.

5.7 Block Elimination and the Paradigms of Neuts

So far, we have applied block-elimination to the QBD process. We now show how the same technique can be applied to analyze GI/M/1 and M/G/1 type processes. The advantages of doing so are similar as before: we can derive new algorithms, and we can change from the matrix methods of Neuts to state reduction to deal, say, with irregular initial conditions. In the case of the M/G/1 type process, we will also develop a stable recursion to find α_n, $n > 0$.

As before, we rely on Theorem 6, and convert the infinite matrix to a finite one. In the finite matrix, one can use (65) to reduce P to $P^{(0)}$ by censoring all states not belonging to level 0, and solve the resulting imbedded Markov chain.

At this point, equation (66) is used for back-substitution. For convenience, we repeat the essential formulas here:

$$P_{ij}^{(n-1)} = P_{ij}^{(n)} + P_{in}^{(n)}(I - P_{nn}^{(n)})^{-1} P_{nj}^{(n)}. \qquad (76)$$

$$\alpha_0 = \alpha_0 P_{00}^{(0)}.$$
$$\alpha_j = \sum_{i=0}^{j-1} \alpha_i P_{ij}^{(j)} (I - P_{jj}^{(j)})^{-1}. \qquad (77)$$

Consider first the GI/M/1 paradigm. Suppose we have approximated the infinite transition matrix by some some finite matrix $P(\omega)$ such that the conditions of Theorem 6 are satisfied. After eliminating n levels, one finds the following matrix:

$$\begin{bmatrix} \ddots & \ddots & \ddots & \ddots & \ddots & \ddots & \vdots \\ \cdots & A_{i-1} & A_{i-2} & A_{i-3} & \cdots & A_0 & 0 \\ \cdots & A_i & A_{i-1} & A_{i-2} & \cdots & A_1 & A_0 \\ \cdots & A_{i+1}^{(n)} & A_i^{(n)} & A_{i-1}^{(n)} & \cdots & A_2^{(n)} & A_1^{(n)} \end{bmatrix}. \qquad (78)$$

It is easily verified that $A_0^{(n)} = A_0$. Using equation (76), one now finds for $i > 0$:

$$A_i^{(n+1)} = A_i + A_0(I - A_1^{(n)})^{-1} A_{i+1}^{(n)}. \qquad (79)$$

As n gets large, $A_i^{(n)}$, $i > 0$ approaches a limiting value A_i^*, and equation (79) yields

$$A_i^* = A_i + A_0(I - A_1^*)^{-1} A_{i+1}^*, \quad i > 0. \qquad (80)$$

Note that by the definition of R, and by Lemma 1, item 2

$$A_0(I - A_1^*)^{-1} = R. \qquad (81)$$

Using (80) and (81), one finds

$$\begin{aligned} A_i^* &= A_i + RA_{i+1}^* = A_i + R(A_{i+1} + RA_{i+2}^*) \\ &= A_i + RA_{i+1} + R^2 A_{i+2}^* = A_i + RA_{i+1} + R^2(A_{i+2} + RA_{i+3}^*) \\ &= A_i + RA_{i+1} + R^2 A_{i+2} + R^3 A_{i+3}^* \\ &\vdots \\ &= \sum_{j=0}^{\infty} R^j A_{i+j} \end{aligned}$$

or

$$A_i^* = \sum_{j=0}^{\infty} R^j A_{i+j}, \quad i > 0. \qquad (82)$$

The equation corresponding to (80) for the B_i's in the first column is

$$B_i^* = B_i + A_0(I - A_1^*)^{-1} B_{i+1}^*. \tag{83}$$

A similar argument applied to (83) yields

$$B_i^* = \sum_{j=0}^{\infty} R^j B_{i+j}. \tag{84}$$

Hence, the matrix censored at n can be obtained explicitly once R is known, and this allows us to change freely from the GI/M/1 paradigms to block-elimination. In particular, equation (28), which was derived earlier to find π_0 in the GI/M/1 paradigm with simplified initial conditions can now easily be obtained because $\alpha_0 = \alpha_0 B_0^*$, where B_0^* is given by (83). The advantage of our approach is that a switch to block-elimination can be done at any level. As a consequence, even GI/M/1 processes with several exceptional rows prior to the repeating rows can be accommodated readily. First, note that the GI/M/1 queue in its form given earlier in equation (7) leads to the following chain once all states outside levels zero and one have been censored:

$$\begin{bmatrix} B_0 & C_1 \\ B_1^* & A_1^* \end{bmatrix}.$$

This transition matrix has the equilibrium vector $[\alpha_0, \alpha_1]$, which can be obtained easily. The α_n, $n > 1$ can then be found as

$$\alpha_n = \alpha_{n-1} A_0 (I - A_1^*)^{-1} = \alpha_{n-1} R.$$

Note that (82) allows us to re-derive (18), the equation determining R. From (81), one finds

$$R = A_0 + R A_1^* \tag{85}$$

and replacing A_1^* by (82) yields (18).

A numerical algorithm to compute (82) and (84) requires a starting point, and this requires us to truncate the infinite sum. To do this consistently, we assume that $A_i = 0$ for $i > g$, band $B_i = 0$ if $i > g - 1$. In this way, the matrix-based subdiagonal $g - 1$ is the last non-zero subdiagonal.

To find the A_i^*, one can use (79). If $A_i = 0$ for $i > g$, there is, however, a faster method. The basic idea is to replace the $A_i^{(n)}$ by their most recently calculated values, that is, one uses $A_i^{(n+1)}$ in place of $A_i^{(n)}$ where possible. To exploit this idea, one works from left to right, starting with $A_g^{(n)}$, which is A_g for all n. Hence, one evaluates for $i = g - 1, g - 2, \ldots 1$

$$A_i^{(n+1)} = A_i + R_{(n)} A_{i+1}^{(n+1)}.$$

It is easy to see that this modification leads to the algorithm given by equations (41) and (42). In the case of the QBD process, the two methods yield identical results as is easily verified.

In the case of the M/G/1 paradigm, one has similar methods to find the Markov chain censored at level n. In fact, the derivation is essentially the dual of the one for the GI/M/1 paradigm. The matrix reduced to n levels is given as

$$\begin{bmatrix} \cdots & C_{n-2} & C_{n-1} & C_n^* \\ \cdots & D_{n-2} & D_{n-1} & D_n^* \\ & \vdots & \vdots & \vdots \\ \ddots & D_2 & D_3 & D_4^* \\ \ddots & D_1 & D_2 & D_3^* \\ \ddots & D_0 & D_1 & D_2^* \\ \cdots & 0 & D_0 & D_1^* \end{bmatrix}.$$

Equation (65) now yields

$$D_i^* = D_i + D_{i+1}^*(I - D_1^*)^{-1}D_0.$$

A derivation similar to the one given in connection with Lemma 1 indicates that

$$(I - D_1^*)^{-1}D_0 = G. \qquad (86)$$

Hence

$$D_i^* = D_i + D_{i+1}^* G.$$

The derivation given to obtain (82) from (80) can be used now to obtain

$$D_i^* = \sum_{j=0}^{\infty} D_{i+j} G^j.$$

For the C_i^*, a similar derivation yields

$$C_i^* = \sum_{j=0}^{\infty} C_{i+j} G^j.$$

Note that this equation implies equation (34) derived earlier. Once the D_i^* and C_i^* are given, (66) yields the following stable recursion for α_j (see also [Neuts, 1989, page 142]):

$$\alpha_j = \alpha_0 C_j^*(I - D_1^*)^{-1} + \sum_{i=1}^{j-1} \alpha_{j-i} D_{i+1}^*(I - D_1^*)^{-1}.$$

As in the case of the GI/M/1 paradigm, the knowledge of the D_i^* and C_i^* allows one to cover cases with irregular initial conditions by switching to block-elimination. For instance, in the case given by (6), one has for the matrix imbedded in levels 0 and 1

$$\begin{bmatrix} B_0 & C_1^* \\ B_1 & D_1^* \end{bmatrix}$$

from which the probabilities α_0 and α_1 can be derived as

$$\alpha_0 = \alpha_0(B_0 + C_1^*(I - D_1^*)^{-1}B_1)$$
$$\alpha_1 = \alpha_0 C_1^*(I - D_1^*)^{-1}$$

Instead of block-elimination, one can also use scalar elimination, but, as was pointed out earlier, the censored matrices will be the same. Hence, there is no big difference between scalar elimination and block elimination.

5.8 The GI/G/1 Paradigm

The great advantage of block-elimination is that it generalizes to the GI/G/1 paradigm whereas the GI/M/1 and the M/G/1 paradigms do not. As a starting point, we use the following matrix, which can be obtained by iterating an approximate matrix $P(\omega)$ long enough

$$\begin{matrix} \ddots & \ddots & & \ddots & & \ddots & & \ddots & & \ddots & & \ddots \\ \ddots & Q_{-1} & Q_0 & Q_1 & Q_2^{(n+1)} & Q_3^{(n+2)} & \ddots \\ \ddots & Q_{-2} & Q_{-1} & Q_0 & Q_1^{(n+1)} & Q_2^{(n+2)} & \ddots \\ \ddots & Q_{-3}^{(n+1)} & Q_{-2}^{(n+1)} & Q_{-1}^{(n+1)} & Q_0^{(n+1)} & Q_1^{(n+2)} & \ddots \\ \ddots & Q_{-4}^{(n+2)} & Q_{-3}^{(n+2)} & Q_{-2}^{(n+2)} & Q_{-1}^{(n+2)} & Q_0^{(n+2)} & \ddots \end{matrix}$$

Applying "Crout-elimination" as in (68) yields

$$Q_i^{(n)} = Q_i + \sum_{j=1}^{\infty} Q_{i+j}^{(n+j)}(I - Q_0^{(n+j)})^{-1} Q_{-j}^{(n+j)}, \quad i \geq 0 \tag{87}$$

$$Q_i^{(n)} = Q_i + \sum_{j=1}^{\infty} Q_j^{(n+j)}(I - Q_0^{(n+j)})^{-1} Q_{i-j}^{(n+j)}, \quad i \leq 0. \tag{88}$$

The $Q_i^{(n)}$ will converge to the values Q_i^*, and once converged, equations (87) and (88) become

$$Q_i^* = Q_i + \sum_{j=1}^{\infty} Q_{i+j}^*(I - Q_0^*)^{-1} Q_{-j}^*, \quad i \geq 0 \tag{89}$$

$$Q_i^* = Q_i + \sum_{j=1}^{\infty} Q_j^*(I - Q_0^*)^{-1} Q_{i-j}^*, \quad i \leq 0. \tag{90}$$

Equations (89) and (90) are duals of one another in the sense that if the subscripts are multiplied by -1, and if the order of the matrix multiplication is reversed, then (89) becomes (90) and vice-versa.

As before, we can use the most recent values of $Q_i^{(n)}$. This can be done by replacing all $Q_i^{(n+j)}$, $j > 0$ by $Q_i^{(n+1)}$, in which case (87) and (88) become

$$Q_i^{(n)} = Q_i + \sum_{j=1}^{\infty} Q_{i+j}^{(n+1)}(I - Q_0^{(n+1)})^{-1} Q_{-j}^{(n+1)}, \quad i \geq 0$$

$$Q_i^{(n)} = Q_i + \sum_{j=1}^{\infty} Q_j^{(n+1)}(I - Q_0^{(n+1)})^{-1} Q_{-i-j}^{(n+1)}, \quad i \leq 0.$$

Essentially, this amounts to successive substitution applied to equations (89) and (90). More elaborate methods are available if $Q_i = 0$ for $i < -g$ or $i > h$. In this case, $Q_{-g} = Q_{-g}^*$, and the knowledge of Q_{-g}^* can be used to find a new value for Q_{-g+1}^*, which can in turn used to find a new value for Q_{-g+2}^*, and so on. We will not discuss this further, but refer instead to [Grassmann and Jain, 1989], a paper in which this is worked out for the case where the Q_i are scalars.

Once the Q_k^* are known, all $P_{ij}^{(n)}$ can be calculated by using (68). In particular, we need

$$C_k^* = P_{0\,k}^{(0)} = P_{0\,k}^*, \quad \text{and} \quad B_k^* = P_{k\,0}^{(0)} = P_{k\,0}^*.$$

These become

$$C_k^* = C_k + \sum_{r=1}^{\infty} C_{k+r}^*(I - Q_0^*)^{-1} Q_{-r}^*, \quad k > 0 \qquad (91)$$

$$B_k^* = B_k + \sum_{r=1}^{\infty} Q_r^*(I - Q_0^*)^{-1} B_{k+r}^*, \quad k > 0. \qquad (92)$$

Finally, $P_{0\,0}^* = B_0^*$ is given as

$$B_0^* = B_0 + \sum_{r=1}^{\infty} C_r^*(I - Q_0^*)^{-1} B_r^*. \qquad (93)$$

Now, one can select one phase of level 0, say phase 1, and set $\alpha_{0,1} = 1$. Then one solves

$$\alpha_0 = \alpha_0 B_0^*. \qquad (94)$$

Because of (66), one has:

$$\alpha_n = \sum_{i=1}^{n-1} \alpha_{n-i} Q_i^*(I - Q_0^*)^{-1} + \alpha_0 C_n^*(I - Q_0^*)^{-1}, \quad n > 0. \qquad (95)$$

Hence, all vectors α_n can be obtained recursively.

It turns out to be convenient to define $R_i = Q_i^*(I - Q_0^*)^{-1}$, and write (95) as

$$\alpha_n = \sum_{i=1}^{n-1} \alpha_{n-i} R_i + \alpha_0 C_n^*(I - Q_0^*)^{-1}, \quad n > 0. \qquad (96)$$

Because all elements of R_i and C_n are non-negative, no subtractions are involved, and the recursion is numerically stable. Note that this equation is also applicable for GI/M/1 type processes, in which case $R_1 = R$, $R_i = C_i^* = 0$ for $i > 1$, and for M/G/1 type processes, in which case $R_i = D_{i+1}^*(I - D_1^*)^{-1}$, $i > 1$.

Equations (94) and (96) allow us to find all α_i, $i \geq 0$, but we really need the π_n. To find the π_n, we need the norming constant t, which is

$$t = \sum_{j=0}^{m_1} \alpha_{0j} + \sum_{n=1}^{\infty} \alpha_n e,$$

where e is the row vector of dimension m with all its elements equal to 1.

To find t, define the generating function

$$\alpha(z) = \sum_{n=1}^{\infty} \alpha_n z^n.$$

The sum starts with $n = 1$ because level zero has a different dimension. Multiplying (96) by z^n and adding from $n = 1$ to infinity yields

$$\sum_{n=1}^{\infty} \alpha_n z^j = \sum_{n=1}^{\infty} \sum_{i=1}^{n-1} \alpha_{n-i} z^{n-i} R_i z^i + \alpha_0 \sum_{n=1}^{\infty} C_n^* z^n (I - Q_0^*)^{-1}$$

$$= \sum_{n=1}^{\infty} \alpha_n z^n \sum_{i=1}^{\infty} R_i z^i + \alpha_0 \sum_{n=1}^{\infty} C_n^* z^n (I - Q_0^*)^{-1}.$$

Consequently

$$\alpha(z) = \alpha_0 C(z) + \alpha(z) R(z), \tag{97}$$

where

$$R(z) = \sum_{n=1}^{\infty} R_n z^n \quad \text{and} \quad C(z) = \sum_{n=1}^{\infty} C_n^* (I - Q_0^*)^{-1}.$$

After setting $z = 1$, (97) yields the following equation for $\alpha(1)$:

$$0 = \alpha_0 C(1) + \alpha(1)(R(1) - I). \tag{98}$$

This equation can be solved for $\alpha(1)$. The vector $\alpha(1)/\alpha(1)e$ represents the distribution of the phases, provided the system is not in level 0. Of course, $\alpha(1)e$ can be used to find t, and with it, $\pi_i = \alpha_i/t$.

The expected level is $\alpha'(1)e$, and it can be found by taking the derivative of (97) at $z = 1$. This yields:

$$0 = \alpha_0 C'(1) + \alpha(1) R'(1) + \alpha'(1)(R(1) - I).$$

Since this equation has the same coefficient matrix as (98), one can use the UL factorization obtain from solving (98) for finding $\alpha(1)$. Higher moments may be obtained in a similar fashion.

6 APPLICATION: THE BILINGUAL SERVER MODEL

As an illustration of a quasi-birth and death model with a boundary level of different dimension we present the following application. Often an organization provides service to customers with distinct service requirements. Many telephone administrations provide service in a bilingual environment, where each group is entitled to be served in its own language. Typically, in a given region one language will predominate, and it can usually be assumed that all servers hired would be required to speak the majority language of that region. Minority language customers still require service in their mother tongue. Bilingual servers are a premium resource in some sense (they are harder to find and/or they cost more). The problem is to determine the minimum number of bilingual servers needed in order to provide *satisfactory* service to the minority language group.

It is worth noting that this model can *also* be used to analyze other specialized services, where a certain fraction of the customers require a service that only some of the servers are able to provide.

Clearly, it would seem to make sense to direct specialized customers to fully qualified servers *immediately*, in order to eliminate wasted effort. This is becoming increasingly common with the advent of call centers. (An appropriate modification of the model described below readily accommodates the call center variant.) Nonetheless, many situations still exist where the necessity of specialized service can be discerned only at the start of the initial service attempt. This can still be the case in many telephone operator applications, where in the most common scenario all customers dial 0 to reach an operator, regardless of the language they speak. It can also be the case, for instance, in a large automobile repair shop, where initial repairs to a car reveal a specific problem that only some technicians are qualified to handle.

In what follows we initially assume that the need for specialized service can only be determined once a server attempts to serve a selected customer. The call center variant is described afterwards.

6.1 System Description

Service is provided by M bilingual and N unilingual servers, where the latter group only speak the majority language. Majority- and minority-language customers enter the system via the same queue (called the entry queue), and cannot be distinguished until the start of service.

All customers selected by the bilingual servers, and majority language customers selected by the unilingual ones, receive service in the usual fashion and then depart the system. However, whenever a minority language customer is selected by a unilingual server, a *language mismatch* occurs. In this case, the customer is placed in the transfer queue, which is served on a nonpreemptive priority basis in preference to the entry queue by the bilingual servers. Upon completion of their service, these transferred customers depart from the system.

It is assumed that all interarrival, service, and transfer times are exponentially distributed, with parameter values given by

f = fraction of customers who are of the minority language group
λ = arrival rate (total) of traffic
h = average service time, $\mu = 1/h$
h_T = average time to transfer a customer, $\theta = 1/h_T$.

In order to specify the transition rates in the system, we need to know three things: a) the number of customers waiting in the entry queue, b) the number currently waiting in the transfer queue, and c) the number of unilingual servers currently involved in language mismatches. (When the entry queue is empty, we also need to know the number of busy servers of each type.) The last item is bounded, however, the others are potentially unbounded. In reality, however, any *bilingual server* system would never tolerate a large number of customers in the transfer queue, so it is credible to cap this with an upper limit K.

This enables us to use a matrix-geometric model to analyze the system, whose structure exhibits exceptional boundary behavior as indicated in (4). While the level of a matrix-geometric model can be unbounded, the block size describing all secondary factors must be finite, except for very limited circumstances (see, for example, Miller [Miller, 1981]). In the current problem, the *level i* will be used to denote the number of customers waiting in the entry queue, while the *phase j* will be needed to describe simultaneously the number of mismatches n in progress, and the number k of people in the transfer queue. We set $j = k(N+1) + n$, so that the sub-states are ordered lexicographically:

nobody transferred, nobody mismatched
$\vdots$
nobody transferred, N mismatched
$\vdots$
K transferred, N mismatched

If $k < K$, transfers are permitted, but if $k = K$ then any further transfers are forced to leave once mismatched. Letting i and j respectively refer to the current level and sub-state, we define the entries in A_0, A_1, and A_2 below. All non-specified entries are 0.

Non-boundary transitions. Since A_0 corresponds to the arrival of a new customer, $A_0 = \lambda I$. A_1 corresponds to transitions in which the entry queue length does not change. In any given row of A_1, at most two entries are non-zero. The diagonal element is $-[(M + N - n)\mu + n\theta + \lambda]$. Whenever $k > 0$, bilingual servers select customers from the transfer queue at rate $M\mu$. Therefore $j \to j - (N+1)$ at rate $(1 - \Delta_k)M\mu$ where Δ_k is the Kronecker delta function, namely $\Delta_k = 1$ if $k = 0$, and 0 is otherwise.

A_2 corresponds to transitions in which the entry queue drops by one. Every service completion by a unilingual server, and those by bilingual ones when $k = 0$, will be followed by a selection from the entry queue if a customer is waiting. The following are the rates of sub-state changes:

1) $j \to j$ happens a) whenever a bilingual server selects from the entry queue, b) when a unilingual server completes service to one majority customer and immediately selects another, and c) when a unilingual server has two back-to-back minority customers, and the first one is forced to leave because $k = K$. The total of these rates is $M\mu\Delta_k + (N-n)\mu(1-f) + n\theta f \Delta_{(k-K)}$.

2) $j \to k(N+1)+n+1 = j+1$ occurs when a normal completion by a unilingual server is followed by a language mismatch. This happens at rate $(N-n)\mu f$.

3) $j \to (k+1)(N+1) + n - 1 = j + N$ arises when a language mismatch is followed by the selection of a majority language customer. The rate is $n\theta(1-f)(1 - \Delta_{(k-K)})$.

4) $j \to (k+1)(N+1) + n = j + N + 1$ happens when a unilingual server serves one mismatched customer right after another. Again, $k < K$. Here the rate is $n\theta f(1 - \Delta_{(k-K)})$.

5) $j \to k(N+1) + n - 1 = j - 1$: occurs when a unilingual server cannot complete a transfer since the transfer queue is full ($k = K$), and then selects a majority language customer for service. The rate is $n\theta(1-f)\Delta_{(k-K)}$.

The boundary transitions are even more complicated since one must also specify the number of busy bilingual and unilingual servers, and because the number of mismatches in progress is bounded by the latter of these. We conclude by observing that the matrix B is very large relative to A_0, A_1, and A_2, containing $(N + 1)(N + 2)(M + K + 1)/2$ rows. This makes state reduction an attractive solution method once R is known; in fact we even use state reduction to determine R. The states governed by B are arranged as follows:

$$B = \begin{bmatrix} A_{01} & A_{00} & 0 & 0 & \cdots & 0 \\ A_{12} & A_{11} & A_{10} & 0 & \cdots & 0 \\ 0 & A_{22} & A_{21} & A_{20} & \cdots & 0 \\ \vdots & \vdots & \vdots & \vdots & \ddots & \vdots \\ 0 & 0 & 0 & \cdots & A_{N\,2} & A_{N\,1} \end{bmatrix}$$

In this formulation, the submatrix $A_{m\,0}$ pertains to transitions from sub-level m to $m+1$, while $A_{m\,1}$ refers to transitions within sub-level m, and $A_{m\,2}$ refers to transitions from sub-level m to $m-1$. The most important of these matrices is $A_{N\,1}$, which is of size $(M + K + 1)(N + 1)$. The first M rows correspond to states where some number of bilingual severs are idle. We distinguish the lower-right $(K + 1)(N + 1)$ block of $A_{N\,1}$ by $A'_{N\,1}$; this block is the crossover point where all servers are busy yet no one is waiting in the entry queue. In a slight modification of the usual thinking, π_0 will be used to denote the probability of these states in what follows below.

When state reduction is employed to find the steady-state probabilities corresponding to B, the higher states are ties in by the following equivalent to (43):
$$Y = A'_{N\,1} + RA_2$$
where R is found in the usual way. The interested reader is referred to [Stanford and Grassmann, 1993] for the details.

Performance measures. The principal performance measures of interest are the average delays experienced by the two customer types. All customers pass through the entry queue. Mismatched minority customers pass through the transfer queue, so the average transfer queue delay is needed as well. The means to calculate both of these queue delays is to apply Little's Law: the average number in both the entry queue and the transfer queue is found, and the result is divided by the average arrival rate to the respective queue, yielding the average delay.

Since the level i of our matrix-geometric model is precisely the number waiting in the entry queue, the mean entry queue length is given by Neuts [13], p.36, equation (1.8.4). Thus the average entry queue delay is given by

$$W_E = \frac{1}{\lambda}\{\sum_{k=1}^{\infty} k\pi_k e\} = (\frac{1}{\lambda})\pi_0 R(I-R)^{-2} e. \tag{99}$$

Minority customers arrive to the entry queue at rate $f\lambda$, and only a fraction α of these are mismatched, so the average rate of entry to the transfer queue is $f\lambda\alpha$. Therefore the average transfer queue delay is

$$W_T = \frac{1}{f\lambda\alpha} N_T. \tag{100}$$

The overall average delay W_M experienced by minority customers is

$$W_M = W_E + \alpha\{h_T + W_T\}. \tag{101}$$

Since the only customers who enter the transfer queue are mismatched minority customers, the average rate into it must equal the average rate at which the unilingual servers are transferring minority customers. But the former of these rates is $f\lambda\alpha$, and the latter is $N_X\theta$, where N_X denotes the average number of (unilingual) servers handling mismatched customers. Thus

$$\alpha = N_X\theta/(f\lambda). \tag{102}$$

We obtain the following equivalent form to (5.4) after substituting for α and W_T:
$$W_M = W_E + \{N_X + N_T\}/(f\lambda). \tag{103}$$

Thus it is not necessary to explicitly calculate α in order to quantify the average delays. However, α is a performance measure in its own right, and (5.5) provides

Table 1 Delay Performance (in seconds) of 6 Server Systems with 25% Minority Traffic

Average Service Time = 25 Seconds
Average Transfer Time = 5 Seconds

Occupancy		(No. Bilingual, No. Unilingual)			
		(5,1)	(4,2)	(3,3)	(2,4)
.4	Entry Queue Delay	0.28	0.28	0.27	0.26
	Minority Delay	1.33	2.58	4.51	9.57
.5	Entry Queue Delay	0.85	0.85	0.85	0.82
	Minority Delay	2.04	3.57	6.11	13.14
.6	Entry Queue Delay	2.12	2.17	2.19	2.17
	Minority Delay	3.49	5.39	8.68	18.17
.7	Entry Queue Delay	4.89	5.09	5.27	5.34
	Minority Delay	6.47	8.90	13.14	25.87
.8	Entry Queue Delay	11.55	12.36	13.19	13.93
	Minority Delay	13.36	16.81	22.59	40.10
.9	Entry Queue Delay	35.07	40.37	47.09	55.88
	Minority Delay	37.13	45.50	58.13	89.19

a means to calculate it at the same time as the delays. Formulas for N_X and N_T are provided in [Stanford and Grassmann, 1993]; these merely sum terms over the applicable states.

Sample results. Table 1 presents results for a six server system assuming $f = 25\%$. As expected, the average delay for minority customers increases with both the utilization and the proportion of servers who are unilingual. However, for sufficiently low occupancies, the average entry queue delay decreases as the proportion of unilingual servers increases. This occurs because there are more mismatches taking place. Since it takes much less time to transfer than to actually serve a customer, the unilingual operators select new customers more frequently. At low load, this increased availability more than offsets the wasted effort by unilingual servers on minority language customers.

As a minimal requirement in these calculations, K was chosen to equal the number of unilingual servers plus 4, in order to render negligible the steady-state probability of transferred customers being blocked. In one case where the chances of a full transfer queue would be greater than 1 in 1000 under this rule (namely, the last column of Table 1), K has been increased until that target was met (specifically, to $K = 14$).

Bilingual Call Centers. In call center applications, customers can be assessed at the moment of entry to the system. Consequently, minority language customers can be routed directly to the pool of available bilingual servers, thereby eliminating language mismatches. This reduces the dimension of the blocks drastically: the call center variant has only $K+1$ phases rather than $(K+1)(N+1)$. Consequently, systems on the order of 100 servers can be dealt with.

The following would seem to be logical operating rules for a bilingual call center. The bilingual servers would check the minority language queue first, and would only select a waiting majority language customer if the minority queue were empty. Likewise, in order to balance the load as much as possible, arriving majority language customers would only select an idle bilingual server if all unilingual servers were busy.

Under these assumptions, the matrices A_0 A_1 and A_2 assume the following values: $A_0 = \lambda I$, $A_2 = \text{diag}((M+N)\mu, N\mu, \ldots, N\mu)$ and

$$A_1 = \lambda_2 S_1 + M\mu S_{-1} - \text{diag}(\lambda + \theta, \ldots, \lambda + \theta, \lambda_1 + \theta)$$

where $\theta = (M+N)\mu$.

One can similarly define new values for the $A_{m\,0}$, $A_{m\,1}$ and $A_{m\,2}$ matrices. The same methods can be applied as in the case of [Stanford and Grassmann, 1993], and results will appear shortly in [Stanford and Grassmann, 1999].

References

[Alfa et al., 1998] Alfa, A., Sengupta, B., and Takine, T. (1998). The use of non-linear programming in matrix analytic methods. *Stochastic Models*, 14(1&2):351–367.

[Alfa and Chakravarthy, 1998] Alfa, A. S. and Chakravarthy, S. R., editors (1998). *Advances in Matrix-Analytic Methods in Stochastic Models.* Notable Publications, Inc., New Jersey.

[Bini and Meini, 1996] Bini, D. and Meini, B. (1996). On the solution of a nonlinear matrix equation arising in queuing problems. *SIAM J. Matrix Anal. Apl.*, 17(4):906–926.

[Chakravarthy and Alfa, 1993] Chakravarthy, S. and Alfa, A. (1993). A multi-server queue with Markovian arrivals and group service with thresholds. *Naval Research Logistics*, 40:811–827.

[Chakravarthy and Alfa, 1997] Chakravarthy, S. R. and Alfa, A. S., editors (1997). *Matrix-Analytic Methods in Stochastic Models*, volume 183 of *Lecture Notes in Pure and Applied Mathematics*. Marcel Dekker, New York.

[Daigle and Lucantoni, 1991] Daigle, J. N. and Lucantoni, D. M. (1991). Queueing systems having phase-dependent arrival and service rates. In Stewart, W. J., editor, *Numerical Solution of Markov Chains*, pages 179–215. Marcel Dekker, New York, NY.

[Gail et al., 1998] Gail, H. R., Hantler, S. L., and Taylor, B. A. (1998). Matrix-geometric invariant measures for G/M/1 type Markov chains. *Stochastic Models*, 14(3):537–569.

[Grassmann, 1982] Grassmann, W. K. (1982). The GI/PH/1 queue. *INFOR*, 20:144–156.

[Grassmann, 1985] Grassmann, W. K. (1985). The factorization of queueing equations and their interpretation. *J. Opl. Res. Soc.*, 36:1041–1050.

[Grassmann, 1986] Grassmann, W. K. (1986). The $PH^X/M/c$ queue. In *Selecta Statistica Canadiana*, volume 7, pages 25–52.

[Grassmann, 1993] Grassmann, W. K. (1993). Rounding errors in certain algorithms involving Markov chains. *ACM Transactions on Math Software*, 19:496–508.

[Grassmann, 1998] Grassmann, W. K. (1998). Finding test data for Markov chains with repeating columns. In *Advances in Matrix Analytical Methods for Stochastic Models*, pages 261–278. Notable Publications, 1049 Hillcrest Drive, Neshanic Station, NJ 08853.

[Grassmann and Heyman, 1990] Grassmann, W. K. and Heyman, D. P. (1990). Equilibrium distribution of block-structured Markov chains with repeating rows. *Journal of Applied Probability*, 27:557–576.

[Grassmann and Heyman, 1993] Grassmann, W. K. and Heyman, D. P. (1993). Computation of steady-state probabilities for infinite-state Markov chains with repeating rows. *ORSA Journal on Computing*, 5:292–303.

[Grassmann and Jain, 1989] Grassmann, W. K. and Jain, J. L. (1989). Numerical solutions of the waiting time distribution and idle time distribution of the arithmetic GI/G/1 queue. *Operations Research*, 37:141–150.

[Grassmann et al., 1985] Grassmann, W. K., Taksar, M., and Heyman, D. P. (1985). Regenerative analysis and steady state distributions for Markov chains. *Operations Research*, 33:1107–1117.

[Gross and Harris, 1998] Gross, D. and Harris, C. M. (1998). *Fundamentals of Queueing Theory*. Wiley, New York, 3rd edition.

[Heyman, 1987] Heyman, D. P. (1987). Further comparisons of direct methods for computing stationary distributions of Markov chains. *SIAM J. Algebraic and Discrete Methods*, 8:226–232.

[Heyman, 1995] Heyman, D. P. (1995). Accurate computation of the fundamental matrix of a Markov chain. *SIAM Journal of Matrix Anal. Appl.*, 16:151–159.

[Heyman and O'Leary, 1998] Heyman, D. P. and O'Leary, D. P. (1998). Overcoming instability in computing the fundamental matrix for a Markov chain. *SIAM J. Matrix Anal. Appl.*, 19:534–540.

[Heyman and Reeves, 1989] Heyman, D. P. and Reeves, A. (1989). Numerical solutions of linear equations arising in Markov chain models. *ORSA Journal on Computing*, 1:52–60.

[Jain and Grassmann, 1988] Jain, J. L. and Grassmann, W. K. (1988). Numerical solution for the departure process from the GI/G/1 queue. *Comput. Opsn. Res.*, 15:293–296.

[Kao and Lin, 1989] Kao, E. and Lin, C. (1989). The M/M/1 queue with randomly varying arrival and service rates: a phase substitution solution. *Management Science*, 35:561–570.

[Kao and Lin, 1991] Kao, E. and Lin, C. (1991). Using state reduction for computing steady state probabilities of queues of GI/M/1 types. *ORSA Journal on Computing*, 3:231–240.

[Kemeny et al., 1966] Kemeny, J. G., Snell, J., and Knapp, A. W. (1966). *Denumerable Markov Chains*. Van Nostrand, Princeton, NJ.

[Kohlas, 1986] Kohlas, J. (1986). Numerical computation of mean passage times and absorption probabilities in Markov and semi-Markov models. *Zeitschrift für Operations Res.*, 30:A197–A207.

[Lal and Bhat, 1988] Lal, R. and Bhat, U. N. (1988). Reduced system algorithms for Markov chains. *Management Science*, 88:1202–1220.

[Latouche, 1987] Latouche, G. (1987). A note on two matrices occurring in the solution of quasi birth-and-death processes. *Stochastic Models*, 3:251–258.

[Latouche, 1992] Latouche, G. (1992). Algorithms for infinite Markov chains with repeating columns. In *Linear Algebra, Markov Chains, and Queueing Models*, volume 48 of *IMA Volumes in Mathematics and its Applications*, pages 231–265. Springer-Verlag, Heidelberg, Germany.

[Latouche and Ramaswami, 1993] Latouche, G. and Ramaswami, V. (1993). A logarithmic reduction algorithm for quasi-birth-death processes. *J. Appl. Prob.*, 30:650–674.

[Liu and Zhao, 1996] Liu, D. and Zhao, Y. Q. (1996). Determination of explicit solution for a general class of Markov processes. In Chakravarthy, S. and Alfa, A., editors, *Matrix-Analytic Methods in Stochastic Models*, pages 343–357. Marcel Dekker.

[Meini, 1998] Meini, B. (1998). Solving M/G/1 type Markov chains: Recent advances and applications. *Stochastic Models*, 14(1&2):479–496.

[Miller, 1981] Miller, D. R. (1981). Computation of steady-state probabilities for M/M/1 priority queues. *Operations Research*, 29:945–958.

[Mitrani and Chakka, 1995] Mitrani, I. and Chakka, R. (1995). Spectral expansion solution for a class of Markov models: Application and comparison with the matrix-geometric method. *Performance Evaluation*, 23:241–260.

[Neuts, 1981] Neuts, M. F. (1981). *Matrix-Geometric Solutions in Stochastic Models, An Algorithmic Approach*. Johns Hopkins University Press, Baltimore.

[Neuts, 1989] Neuts, M. F. (1989). *Structured Stochastic Matrices of M/G/1 Type and Their Applications*. Marcel Dekker, New York.

[O'Cinneide, 1993] O'Cinneide, C. O. (1993). Entrywise perturbation theory and error analysis for Markov chains. *Numerische Mathematik*, 65:109–120.

[Ramaswami, 1998] Ramaswami, V. (1998). The generality of the quasi birth-and-death process. In *Advances in Matrix Analytical Methods for Stochastic Models*, pages 93–113. Notable Publications, 1049 Hillcrest Drive, Neshanic Station, NJ 08853.

[Ramaswami and Latouche, 1986] Ramaswami, V. and Latouche, G. (1986). A general class of Markov processes with explicit matrix-geometric solutions. *OR Spectrum*, 8:209–218.

[Ramaswami and Lucantoni, 1985] Ramaswami, V. and Lucantoni, D. M. (1985). Stationary waiting time distribution in queues with phase-type service and quasi-birth-and-death processes. *Stoch. Models*, 1:125–136.

[Sheskin, 1985] Sheskin, T. J. (1985). A Markov partitioning algorithm for computing steady-state probabilities. *Operations Research*, 33:228–235.

[Sonin and Thornton, 1998] Sonin, I. and Thornton, J. (1998). Recursive computation of the fundamental/group inverse matrix of a Markov chain from an explicit formula. Technical Report 28223, Dept. of Mathematics, University of North Carolina at Charlotte.

[Sonin, 1999] Sonin, I. M. (1999). The state reduction and related algorithms and their application to the study of Markov chains, graph theory and the optimal stopping problem. Accepted in Advances in Math.

[Stanford and Grassmann, 1993] Stanford, D. A. and Grassmann, W. K. (1993). The bilingual server model: a queueing model featuring fully and partially qualified servers. *INFOR*, 31(4):261–277.

[Stanford and Grassmann, 1999] Stanford, D. A. and Grassmann, W. K. (1999). Bilingual call centres. Technical report, Dept. of Statistical and Actuarial Sciences, University of Western Ontario.

[Stewart, 1973] Stewart, G. W. (1973). *Introduction to Matrix Computations*. Academic Press, New York.

[Wallace, 1969] Wallace, V. (1969). *The Solution of Quasi Birth and Death Processes Arising from Multiple Access Computer Systems*. PhD thesis, Systems Engineering Laboratory, University of Michigan.

[Zhao et al., 1998] Zhao, Y., Li, W., and Braun, W. (1998). Infinite block-structured transition matrices and their properties. *Adv. Appl. Prob.*, 30:365–384.

[Zhao et al., 1999] Zhao, Y. Q., Braun, W. J., and Li, W. (1999). Northwest corner and banded matrix approximation to a countable Markov chain. *Naval Research Logistics*, 46:187–197.

[Zhao and Liu, 1996] Zhao, Y. Q. and Liu, D. (1996). The censored Markov chain and the best augmentation. *J. Appl. Prob*, 33:623–629.

Winfried Grassmann got his education in economics in Zurich, Switzerland. After his Masters, he joined the Operations Research Department of Swissair, the Swiss flag carrier. There, he developed a system for inventory control, and a system for dealing with rotating parts. Both systems were implemented with great success. While at Swissair, he also finished his Ph.D., which he defended in 1968 with summa cum laude. He then joined the Computer Science Department of the University of Saskatchewan, where he taught Operations Research and Computer Science. Dr. Grassmann was on the editorial boards of Naval Research Logistics and Operations Research, and he is presently associate editor of the INFORMS Journal on Computing. He has written a book on stochastic processes and, more recently, a book on logic and discrete mathematics. His main areas of research are queueing theory and Markov modeling, two areas in which he has published widely. His papers have appeared in Operations Research, Journal of Applied Probability, Interfaces, Naval Research Logistics, INFOR and other journals. For his lifetime achievements, he received the 1999 merit award of the Canadian Operational Research Society.

David Stanford was born in Montreal. He received his B. Sc. from Concordia University in 1976 and his M. Eng. and Ph.D. from Carleton University in 1978 and 1981 respectively. His doctoral thesis dealt with prediction of delays and queue lengths in queueing systems. In 1981 he joined Bell-Northern Research. Following two years at Concordia University, he moved to his current appointment in 1988 at the University of Western Ontario. Dr. Stanford was president of the Canadian Operational Research Society from 1995-96. He also acted as associate editor of INFOR. He has published on correlation in queues, departure-time characterizations, queueing in call centers, and risk theory. His papers have appeared in Operations Research, the Journal of Applied Probability, QUESTA, INFORS, Performance Evaluation and other journals.

7 USE OF CHARACTERISTIC ROOTS FOR SOLVING INFINITE STATE MARKOV CHAINS

H. Richard Gail[1], Sidney L. Hantler[2] and B. Alan Taylor[3]

[1] H. Richard Gail
IBM Thomas J. Watson Research Center
Yorktown Heights, NY 10598
rgail.us.ibm.com

[2] Sid L. Hantler IBM Thomas J. Watson Research Center
Yorktown Heights, NY 10598
shaantler.us.ibm.com

[3] B. Alan Taylor
Department of Mathemtics
The University of Michigan
Ann Arbor, MI 48109
taylor@math.lsa.umich.edu

1 INTRODUCTION

In this chapter, our interest is in determining the stationary distribution of an irreducible positive recurrent Markov chain with an infinite state space. In particular, we consider the solution of such chains using roots or zeros. A root of an equation $f(z) = 0$ is a zero of the function $f(z)$, and so for notational convenience we use the terms root and zero interchangeably. A natural class of chains that can be solved using roots are those with a transition matrix that has an almost Toeplitz structure. Specifically, the classes of M/G/1 type chains and G/M/1 type chains lend themselves to solution methods that utilize roots. In the M/G/1 case, it is natural to transform the stationary equations and solve for the stationary distribution using generating functions. However, in the G/M/1 case the stationary probability vector itself is given directly in terms of roots or zeros. Although our focus in this chapter is on the discrete-time case, we will show how the continuous-time case can be handled by the same techniques. The M/G/1 and G/M/1 classes can be solved using the matrix analytic method [Neuts, 1981, Neuts, 1989], and we will also discuss the relationship between the approach using roots and this method.

Consider an irreducible positive recurrent discrete-time Markov chain with an infinite number of states and with ($\infty \times \infty$) transition matrix $\mathcal{P} = [p_{i,j}]$, $i = 0, 1, \ldots$, $j = 0, 1, \ldots$. The chain has a unique stationary distribution $\pi = [\pi_i]$, $i = 0, 1, \ldots$, and it satisfies the matrix equation $\pi = \pi \mathcal{P}$ and the conservation of probability equation $\sum_{i=0}^{\infty} \pi_i = 1$. We wish to determine the $\infty \times 1$ vector π.

The classical transform approach to the problem is to multiply the kth equation of $\pi = \pi \mathcal{P}$ by z^k and sum on k. In general, transforming the equations $\pi = \pi \mathcal{P}$ will not yield a generating function equation that is amenable to solution. However, when the matrix $\mathcal{P}$ has a special Toeplitz-like structure, the transform method can be applied (see [Horn and Johnson, 1985] page 27 for the definition of a Toeplitz matrix). An example is the class of (scalar) M/G/1 type Markov chains, for which $p_{i,j} = a_{j-i+N}$ for $i = N, N+1, \ldots$, $j = i - N, i - N + 1, \ldots$, with $\sum_{k=0}^{\infty} a_k = 1$. When $\mathcal{P}$ has this special form, transforming yields an equation that expresses the generating function $\pi(z) = \sum_{k=0}^{\infty} \pi_k z^k$ in terms of a finite set of unknown boundary probabilities $\pi_0, \ldots, \pi_{N-1}$. In fact, $\pi(z)$ has the form $\pi(z) = [\sum_{i=0}^{N-1} \pi_i f_i(z)]/A(z)$, where $A(z) = z^N - \sum_{k=0}^{\infty} a_k z^k$. Since $\pi(1) = 1$, the function $\pi(z)$ is analytic in the open unit disk and continuous in the closed unit disk, and so a zero $|z^*| \leq 1$ of the function $A(z)$ must also be a zero of the numerator, i.e., $\sum_{i=0}^{N-1} \pi_i f_i(z^*) = 0$. Proceeding in this manner for all such zeros yields a set of linear equations involving the boundary probabilities $\pi_0, \ldots, \pi_{N-1}$. Once these probabilities are found, then the generating function $\pi(z)$ can be determined.

Transforming the equations $\pi = \pi \mathcal{P}$ is equivalent to multiplying by the $\infty \times 1$ vector $e(z)$, where the kth entry of $e(z)$ is z^k, $k = 0, 1, \ldots$. This yields the generating function equation $\pi[\mathcal{I} - \mathcal{P}]e(z) = 0$, where we have used the notation $\mathcal{I}$ for the $\infty \times \infty$ identity matrix. Thus transforming leads naturally to the study of the equation $[\mathcal{I} - \mathcal{P}]X = 0$ or $\mathcal{P}X = X$. In the M/G/1 case, except for a finite set of boundary equations, one obtains a discrete Wiener-Hopf system (convolution followed by projection), namely, $\sum_{k=0}^{\infty} a_k x_{k+l} = x_{N+l}$ for $l = 0, 1, \ldots$. It is known that solutions to these systems are given in terms of root vectors $e(\xi)$, where ξ is a zero of the function $A(z)$, and their derivatives if necessary (see [Gohberg and Fel'dman, 1974, Goldberg, 1958]). This is an alternative explanation for the use of roots in finding the vector π.

The class of G/M/1 type Markov chains can also be solved using roots or zeros. For this class, the transition matrix (in the scalar case) satisfies $p_{i,j} = a_{i-j+N}$ for $j = N, N+1, \ldots$, $i = j - N, j - N + 1, \ldots$, with $\sum_{k=0}^{\infty} a_k \leq 1$. The equations $\pi = \pi \mathcal{P}$ form a Wiener-Hopf system directly except for a finite set of boundary equations, and there is no need to use transforms. In fact, one obtains the set $\pi_{N+l} = \sum_{k=0}^{\infty} \pi_{k+l} a_k$ for $l = 0, 1, \ldots$. Thus the stationary probability vector π itself is given in terms of root vectors and derivatives. Since the entries of π sum to 1, only zeros $|\xi| < 1$ need to be considered.

For both the M/G/1 and G/M/1 classes, the analysis using roots can be extended to the case for which the a_k are finite matrices. Then singularities of $A(z)$ (equivalently, zeros of $\det A(z)$) and corresponding null vectors are required. That is, certain eigenvalues and eigenvectors related to $a(z) = \sum_{k=0}^{\infty} a_k z^k$ are used.

In the M/G/1 case, in order for the formula for $\pi(z)$ to make sense, the numerator must vanish on the zeros of $A(z)$ in the closed unit disk. Thus there must exist exactly the correct number of zeros of $A(z)$, and these must give rise to exactly the correct number of linearly independent equations to determine the boundary probabilities $\pi_0, \ldots, \pi_{N+1}$. In fact, this happens if the M/G/1 type chain is positive recurrent, and it never happens otherwise. Similarly, for G/M/1 there must exist enough zeros to determine a linear combination of root vectors (and derivatives) to yield the stationary distribution.

In both cases, questions involving the number and location of zeros of a certain analytic function must be addressed. Classical results from complex analysis, such as Rouché's Theorem and the Argument Principle, are used. However, these technical results often are applied incorrectly in the literature. Thus we discuss these theorems and their use, including examples to show their correct (and incorrect) application. In addition, proofs involving roots are sometimes rediscovered for specific instances of functions for which results are already known for a more general class. Thus we describe results on roots which have been proved under very general assumptions and for general classes of functions.

2 M/G/1 TYPE CHAINS

The classical transform method is a well-known approach that is often used when solving for the stationary distribution of an irreducible positive recurrent Markov chain with an infinite state space. Chains that are amenable to this approach have a Toeplitz-like structure to their transition probability matrix $\mathcal{P}$. One natural class of chains that can be solved using transforms is the set of M/G/1 type chains. Before introducing this class, let us consider an arbitrary discrete-time (time-homogeneous) irreducible positive recurrent Markov chain, and attempt to use transform analysis to find the corresponding stationary probability vector. The matrix of one-step transition probabilities is given by

$$\mathcal{P} = \begin{bmatrix} p_{0,0} & p_{0,1} & p_{0,2} & \cdots \\ p_{1,0} & p_{1,1} & p_{1,2} & \cdots \\ p_{2,0} & p_{2,1} & p_{2,2} & \cdots \\ \vdots & \vdots & \vdots & \ddots \end{bmatrix}. \quad (1)$$

Here the scalar $p_{i,j}$ represents the probability of a transition from state i to state j. Let $\pi = [\pi_0, \pi_1, \pi_2, \ldots]$ be the stationary probability vector of the chain, so that it satisfies $\pi = \pi\mathcal{P}$. This is equivalent to the infinite set of equations

$$\pi_j = \sum_{i=0}^{\infty} \pi_i p_{i,j}, \qquad j = 0, 1, \ldots. \tag{2}$$

If we multiply the jth equation by z^j and sum, we obtain the generating function equation

$$\pi(z) = \sum_{i=0}^{\infty} \pi_i p_i(z). \tag{3}$$

Here we have defined $\pi(z) = \sum_{j=0}^{\infty} \pi_j z^j$ and, for $i = 0, 1, \ldots$, we have also defined $p_i(z) = \sum_{j=0}^{\infty} p_{i,j} z^j$. Note that $\pi(1) = 1 = p_i(1)$ for all i. Unfortunately, trying to solve (2) by solving (3) instead does not seem to be a useful approach in general.

2.1 Scalar Case

If the transition matrix $\mathcal{P}$ has a particular structure that is almost Toeplitz, then the transform method can be utilized to determine π by finding zeros and exploiting the analyticity of probability generating functions in the unit disk. Suppose, for example, that the irreducible matrix $\mathcal{P}$ has the form

$$\mathcal{P} = \begin{bmatrix} b_0 & b_1 & b_2 & b_3 & \cdots \\ a_0 & a_1 & a_2 & a_3 & \cdots \\ 0 & a_0 & a_1 & a_2 & \cdots \\ 0 & 0 & a_0 & a_1 & \cdots \\ \vdots & \vdots & \vdots & \vdots & \ddots \end{bmatrix}. \tag{4}$$

In fact, this is exactly the form of the discrete-time Markov chain for number in system at departure instants for the classical M/G/1 queue [Kleinrock, 1975]. A chain with matrix $\mathcal{P}$ of the form (4) is called skip-free to the left, since transitions to lower numbered states can only take place from a state i to the neighboring state $i - 1$.

Define $a(z) = \sum_{j=0}^{\infty} a_j z^j$ and $b(z) = \sum_{j=0}^{\infty} b_j z^j$. We have $p_0(z) = b(z)$, and for $i = 1, 2, \ldots$, $p_i(z) = z^{i-1} a(z)$. Thus the generating function equation (3) becomes

$$\pi(z) = \pi_0 b(z) + \sum_{i=1}^{\infty} \pi_i z^{i-1} a(z). \tag{5}$$

Setting $\psi(z) = \sum_{i=1}^{\infty} \pi_i z^{i-1}$, so that $\pi(z) = \pi_0 + z\psi(z)$, equation (5) can also be written as

$$\pi_0[1 - b(z)] + \psi(z)[z - a(z)] = 0. \tag{6}$$

The unknown generating function $\psi(z)$ is given in terms of the scalar π_0, and to find it we simply use the fact that the probabilities sum to 1 as follows.

Differentiating (6) with respect to z and evaluating the result at $z = 1$ gives

$$\pi_0[-b'(1)] + \psi(1)[1 - a'(1)] = 0.$$

Now using the conservation of probability $\pi_0 + \psi(1) = 1$, we see that

$$\pi_0 = \frac{1 - a'(1)}{1 - a'(1) + b'(1)}.$$

Since $\mathcal{P}$ is irreducible, both $a'(1) > 0$ and $b'(1) > 0$ hold. As long as $a'(1) < 1$ and $b'(1) < +\infty$, then $0 < \pi_0 < 1$ (in fact, these are the conditions for the chain to be positive recurrent). Note that no appeal to the use of roots was needed for this simple case of one unknown constant.

Let us now consider a more general homogeneity property that $\mathcal{P}$ might possess, namely, suppose that N is a positive integer and

$$\mathcal{P} = \begin{bmatrix} b_{0,0} & b_{0,1} & b_{0,2} & b_{0,3} & \cdots \\ \vdots & \vdots & \vdots & \vdots & \ddots \\ b_{N-1,0} & b_{N-1,1} & b_{N-1,2} & b_{N-1,3} & \cdots \\ a_0 & a_1 & a_2 & a_3 & \cdots \\ 0 & a_0 & a_1 & a_2 & \cdots \\ 0 & 0 & a_0 & a_1 & \cdots \\ \vdots & \vdots & \vdots & \vdots & \ddots \end{bmatrix}. \quad (7)$$

A classical example of a chain with such a $\mathcal{P}$ is Bailey's bulk queue [Bailey, 1954]. The homogeneous part of the chain begins with state N, while states $0, \ldots, N - 1$ are boundary states. When $N = 1$ the chain is skip-free to the left, and we obtain the previous case with transition matrix (4). When $N > 1$ the chain is called non-skip-free.

Define $a(z) = \sum_{j=0}^{\infty} a_j z^j$ and, for $i = 0, \ldots, N - 1$, $b_i(z) = \sum_{j=0}^{\infty} b_{i,j} z^j$. We have $p_i(z) = b_i(z)$, for $i = 0, \ldots, N - 1$, and for $i = N, N + 1, \ldots$, $p_i(z) = z^{i-N} a(z)$. In this case, the generating function equation (3) is

$$\pi(z) = \sum_{i=0}^{N-1} \pi_i b_i(z) + \sum_{i=N}^{\infty} \pi_i z^{i-N} a(z). \quad (8)$$

Setting $\psi(z) = \sum_{i=N}^{\infty} \pi_i z^{i-N}$, so that $\pi(z) = \sum_{i=0}^{N-1} \pi_i z^i + z^N \psi(z)$, equation (5) is

$$\sum_{i=0}^{N-1} \pi_i [z^i - b_i(z)] + \psi(z)[z^N - a(z)] = 0. \quad (9)$$

Solving for $\psi(z)$, we obtain

$$\psi(z) = -\frac{\sum_{i=0}^{N-1} \pi_i [z^i - b_i(z)]}{z^N - a(z)}. \quad (10)$$

A more complicated formula occurs when solving for $\pi(z)$, namely,

$$\pi(z) = -\frac{\sum_{i=0}^{N-1} \pi_i[z^i a(z) - z^N b_i(z)]}{z^N - a(z)}, \tag{11}$$

and so it is advantageous to work with $\psi(z)$ using equations (9) and (10).

Since $\psi(z)$ is a probability generating function, it is analytic in the open unit disk and continuous in the closed unit disk. Thus if $z^N - a(z)$ has a zero, say ξ, that satisfies $|\xi| \leq 1$, then since $\psi(\xi)$ is finite we obtain the equation

$$\sum_{i=0}^{N-1} \pi_i[\xi^i - b_i(\xi)] = 0. \tag{12}$$

If ξ is a zero of $z^N - a(z)$ of multiplicity 2 in the closed unit disk (we will see later that this can only happen when $|\xi| < 1$ in the positive recurrent M/G/1 case), then the additional equation

$$\pi_0[-b_0'(\xi)] + \sum_{i=1}^{N-1} \pi_i[i\xi^{i-1} - b_i'(\xi)] = 0 \tag{13}$$

is obtained by differentiating (9). In general, a zero of multiplicity k will yield k linear equations in the unknown boundary probabilities π_i, $i = 0, \ldots, N-1$, through successive differentiation of (9), namely,

$$\sum_{i=0}^{l-1} \pi_i[-b_i^{(l)}(\xi)] + \sum_{i=l}^{N-1} \pi_i[i(i-1)\cdots(i-l+1)\xi^{i-l} - b_i^{(l)}(\xi)] = 0, \quad l = 0, \ldots, k-1. \tag{14}$$

Note that $z = 1$ is always a root since $a(1) = 1$, but it yields the identity $0 = 0$ in (9). However, evaluating (10) at $z = 1$ using l'Hôpital's rule and the conservation of probability gives an equation in $\pi_0, \ldots, \pi_{N-1}$, namely,

$$\sum_{i=0}^{N-1} \pi_i[i - b_i'(1)] + \left(1 - \sum_{i=0}^{N-1} \pi_i\right)[N - a'(1)] = 0. \tag{15}$$

If $z^N - a(z)$ has $N - 1$ zeros (other than $z = 1$) in the closed unit disk, then the resulting $N - 1$ equations from (9) and its derivatives (if necessary) plus the equation (15) from the zero $z = 1$ will enable the N unknown constants $\pi_0, \ldots, \pi_{N-1}$ to be determined. This illustrates the use of roots in solving for the stationary distribution π via the transform method.

For an irreducible chain with transition matrix of the form (7), at least one of the coefficients $a_0, \ldots, a_{N-1}$ must be nonzero. Otherwise, there would be no path from the homogeneous part of the chain to the boundary states. Let L be the smallest integer such that $a_L \neq 0$, so that $0 \leq L \leq N-1$ and

$$\mathcal{P} = \begin{bmatrix} b_{0,0} & \cdots & b_{0,L-1} & b_{0,L} & b_{0,L+1} & \cdots \\ \vdots & \ddots & \vdots & \vdots & \vdots & \ddots \\ b_{N-1,0} & \cdots & b_{N-1,L-1} & b_{N-1,L} & b_{N-1,L+1} & \cdots \\ 0 & \cdots & 0 & a_L & a_{L+1} & \cdots \\ 0 & \cdots & 0 & 0 & a_L & \cdots \\ \vdots & \ddots & \vdots & \vdots & \vdots & \ddots \end{bmatrix}. \qquad (16)$$

When $L = 0$, we transform all equations of $\pi = \pi\mathcal{P}$ and the analysis of (7) applies. However, if all equations are transformed when $L > 0$, then (9) takes the form

$$\sum_{i=0}^{N-1} \pi_i[z^i - b_i(z)] + \psi(z)z^L[z^{N-L} - a^*(z)] = 0, \qquad (17)$$

where $a^*(z) = \sum_{j=L}^{\infty} a_j z^{j-L}$. The function $z^N - a(z) = z^L[z^{N-L} - a^*(z)]$ has a zero of multiplicity L at $z = 0$. Further, the L equations obtained from (9) and its derivatives at this zero are

$$\pi_j(j!) - \sum_{i=0}^{N-1} \pi_i b_i^{(j)}(0) = 0, \qquad j = 0, \ldots, L-1. \qquad (18)$$

Recalling that the jth coefficient of the power series $b_i(z)$ is $b_i^{(j)}(0)/j!$, we see that (18) is simply the first L equations of $\pi = \pi\mathcal{P}$. Equations obtained from the other zeros of $z^N - a(z)$ (i.e., the zeros of $z^{N-L} - a^*(z)$) are unduly complicated, since simplifications clearly occur once the zeros at $z = 0$ are factored from the problem. Thus in the scalar case, zeros at the origin may indicate that the equations $\pi = \pi\mathcal{P}$ have not been transformed in an advantageous manner.

To solve for the stationary distribution using generating functions when $L > 0$, it is best to transform only the equations of $\pi = \pi\mathcal{P}$ which involve the homogeneous state probabilities a_k. That is, we do not transform the first L equations from $\pi = \pi\mathcal{P}$, namely,

$$\pi_j = \sum_{i=0}^{N-1} \pi_i b_{i,j} \qquad j = 0, \ldots, L-1. \qquad (19)$$

Instead, first define the generating function

$$\pi^*(z) = \sum_{j=L}^{\infty} \pi_j z^{j-L}. \qquad (20)$$

Multiplying the equation for π_j, $j = L, L+1, \ldots$, in $\pi = \pi\mathcal{P}$ by z^{j-L} and summing, we have similar to (3)

$$\pi^*(z) = \sum_{i=0}^{\infty} \pi_i p_i^*(z). \qquad (21)$$

Here, for $i = 0, \ldots, N - 1$, $p_i^*(z) = \sum_{j=L}^{\infty} b_{i,j} z^{j-L} = b_i^*(z)$, and for $i = N, N + 1, \ldots$, $p_i^*(z) = z^{i-N} \sum_{j=L}^{\infty} a_j z^{j-L} = z^{i-N} a^*(z)$.

Set $\psi(z) = \sum_{i=N}^{\infty} \pi_i z^{i-N}$ as before, so that $\pi^*(z) = \sum_{i=L}^{N-1} \pi_i z^{i-L} + z^{N-L} \psi(z)$. Then (21) becomes

$$\sum_{i=0}^{L-1} \pi_i [-b_i^*(z)] + \sum_{i=L}^{N-1} \pi_i [z^{i-L} - b_i^*(z)] + \psi(z)[z^{N-L} - a^*(z)] = 0. \qquad (22)$$

If $\xi \neq 1$ is a zero of $z^{N-L} - a^*(z)$ in the closed unit disk, then from (22) we obtain the equation

$$\sum_{i=0}^{L-1} \pi_i [-b_i^*(\xi)] + \sum_{i=L}^{N-1} \pi_i [\xi^{i-L} - b_i^*(\xi)] = 0. \qquad (23)$$

For a multiple zero of $z^{N-L} - a^*(z)$, (22) must be differentiated repeatedly to obtain the appropriate number of equations.

By summing the L equations (19), it can be seen that using the root $z = 1$ in (22) gives the equation $0 = 0$. So, as before, we differentiate (22), evaluate the result at $z = 1$, and use the conservation of probability to get

$$\sum_{i=0}^{L-1} \pi_i [-b_i^{*\prime}(1)] + \sum_{i=L}^{N-1} \pi_i [(i - L) - b_i^{*\prime}(1)] + \left(1 - \sum_{i=0}^{N-1} \pi_i\right) [(N - L) - a^{*\prime}(1)] = 0. \qquad (24)$$

Assuming that $z^{N-L} - a^*(z)$ has $N - L - 1$ zeros other than $z = 1$ in the closed disk, this procedure only gives $N - L$ equations in the N unknowns $\pi_0, \ldots, \pi_{N-1}$. However, additional equations are simply the first L from $\pi = \pi P$ as given in (19).

Characterization. Since our goal is to determine the stationary distribution of an irreducible positive recurrent Markov chain, characterizing the chain in terms of various input parameters is of interest. For the scalar case, the positive recurrence, null recurrence and transience of the M/G/1 type chain with transition matrix (7) is given in terms of

$$A'(1) = \frac{d}{dz}[z^N - a(z)]\bigg|_{z=1} = N - \sum_{k=0}^{\infty} k a_k. \qquad (25)$$

In fact, let d be the expected one-step drift from the homogeneous part of the chain. Then, the drift is $k - N$ with probability a_k, and so

$$d = \sum_{k=0}^{\infty} (k - N) a_k = -A'(1).$$

In particular, the drift is negative (toward the boundary) when $A'(1)$ is positive. In general, we have the following result.

Proposition 7.1 *Suppose that $\sum_k k b_{i,k} < +\infty$, $i = 0, \ldots, N-1$, $\sum_k k a_k < +\infty$, and $\sum_k k^2 a_k < +\infty$ when $A'(1) = 0$. Then the scalar M/G/1 type Markov chain with transition matrix of the form (7) is*

(i) *positive recurrent if and only if $A'(1) > 0$;*

(ii) *null recurrent if and only if $A'(1) = 0$;*

(iii) *transient if and only if $A'(1) < 0$.*

2.2 Matrix Case

In the above discussion, the elements a_k and $b_{i,k}$ of the transition matrix $\mathcal{P}$ of the Markov chain are scalars. However, determination of the stationary probabilities using transforms (and roots) can be applied when these elements are themselves finite matrices. This general class, called Markov chains of M/G/1 type, was introduced by Neuts [Neuts, 1989], who pioneered their analysis. His matrix analytic approach is based on probabilistic arguments and is amenable to numerical calculation. Also, as we shall see, it has connections to the classical transform method based on roots. Another solution approach, called the method of repeating rows, is due to Grassmann and Heyman [Grassmann and Heyman, 1990, Grassmann and Heyman, 1993] and is based on the notion of stochastic complement [Meyer, 1989]. We now describe a third approach, the use of transforms, for calculating the stationary distribution of an irreducible positive recurrent Markov chain of M/G/1 type.

Let $K \geq 0$, $N > 0$, $M > 0$ be integers. An M/G/1 type Markov chain has a transition matrix $\mathcal{P}$ of the form

$$\mathcal{P} = \begin{bmatrix} b_{-1,-1} & b_{-1,0} & b_{-1,1} & \cdots & b_{-1,N-1} & b_{-1,N} & b_{-1,N+1} & \cdots \\ b_{0,-1} & b_{0,0} & b_{0,1} & \cdots & b_{0,N-1} & b_{0,N} & b_{0,N+1} & \cdots \\ \vdots & \vdots & \vdots & \ddots & \vdots & \vdots & \vdots & \ddots \\ b_{N-1,-1} & b_{N-1,0} & b_{N-1,1} & \cdots & b_{N-1,N-1} & b_{N-1,N} & b_{N-1,N+1} & \cdots \\ 0 & a_0 & a_1 & \cdots & a_{N-1} & a_N & a_{N+1} & \cdots \\ 0 & 0 & a_0 & \cdots & a_{N-2} & a_{N-1} & a_N & \cdots \\ \vdots & \vdots & \vdots & \ddots & \vdots & \vdots & \vdots & \ddots \end{bmatrix}, \quad (26)$$

where $b_{-1,-1}$ is $K \times K$, $b_{-1,k}$ is $K \times M$ for $k = 0, 1, \ldots$, and $b_{i,-1}$ is $M \times K$ for $i = 0, \ldots, N - 1$. Also, for $i = 0, \ldots, N - 1$, $k = 0, 1, \ldots$, $b_{i,k}$ and a_k are $M \times M$ matrices. Thus when $M = 1$ we have the scalar case discussed previously. In fact, setting $K \to L$, $N + K \to N$ and $M \to 1$ gives a matrix as in (16). Here we assume that $a_0 \neq 0$, since otherwise we simply redefine the integer K so that this condition holds. The chain is called skip-free to the left for levels when $N = 1$, and non-skip-free when $N > 1$.

The chain can be thought of as having a two-dimensional state space

$$\mathcal{S} = \{(-1, j) : j = 0, \ldots, K - 1\} \bigcup \{(i, j) : i = 0, 1, \ldots, \; j = 0, \ldots, M - 1\}.$$

Corresponding to a state (i,j), the first coordinate i is called the level and the second coordinate j is called the phase. Thus, $b_{i,k}$ gives the transition probabilities from level i to level k, while a_k gives the transition probabilities from level i ($i \geq N$) to level $i - N + k$. The stationary distribution π is written to correspond to the subblocks of $\mathcal{P}$ as $\pi = [\pi_{-1}, \pi_0, \pi_1, \ldots]$. Here π_{-1} is $1 \times K$, while for $i = 0, 1, \ldots$, π_i is $1 \times M$. The equation $\pi = \pi \mathcal{P}$ then becomes

$$\pi_{-1} = \pi_{-1} b_{-1,-1} + \sum_{i=0}^{N-1} \pi_i b_{i,-1} \tag{27}$$

and

$$\pi_k = \pi_{-1} b_{-1,k} + \sum_{i=0}^{N-1} \pi_i b_{i,k} + \sum_{i=N}^{N+k} \pi_i a_{N+k-i}, \qquad k = 0, 1, \ldots. \tag{28}$$

The analysis of M/G/1 type chains with transition matrix of the form (26) can be reduced to the case $K = 0$. To see this, first transform (28) to obtain

$$\pi_{-1}[-b_{-1}(z)] + \sum_{i=0}^{N-1} \pi_i[z^i I_M - b_i(z)] + \psi(z)[z^N I_M - a(z)]. \tag{29}$$

Similar to the scalar case, $b_i(z) = \sum_{k=0}^{\infty} b_{i,k} z^k$, $i = -1, 0, \ldots, N-1$, $a(z) = \sum_{k=0}^{\infty} a_k z^k$, and $\psi(z) = \sum_{i=N}^{\infty} \pi_i z^{i-N}$ is a $1 \times M$ vector function. Since $\mathcal{P}$ is irreducible, the subblock $b_{-1,-1}$ must have spectral radius < 1, i.e., for each state in level -1 there must be a path with positive probability that leads outside that level. Therefore, the matrix $I_K - b_{-1,-1}$ is invertible, and $[I_K - b_{-1,-1}]^{-1} = \sum_{n=0}^{\infty} [b_{-1,-1}]^n$ (see [Horn and Johnson, 1985]). Then (27) can be solved for π_{-1} to yield

$$\pi_{-1} = \sum_{i=0}^{N-1} \pi_i b_{i,-1} [I_K - b_{-1,-1}]^{-1} = \sum_{i=0}^{N-1} \pi_i b_{i,-1} \sum_{n=0}^{\infty} [b_{-1,-1}]^n. \tag{30}$$

Substituting this expression into (29), we obtain

$$\sum_{i=0}^{N-1} \pi_i [z^i I_M - b_i^*(z)] + \psi(z)[z^N I_M - a(z)] = 0, \tag{31}$$

where

$$b_i^*(z) = b_i(z) + \sum_{n=0}^{\infty} b_{i,-1}[b_{-1,-1}]^n b_{-1}(z), \qquad i = 0, \ldots, N-1. \tag{32}$$

The $M \times M$ matrix functions $b_i^*(z)$ have nonnegative power series coefficients. Also, $b_i^*(1)$ are stochastic matrices from (32), since $b_{-1}(1) \mathbf{1}_M = [I_K - b_{-1,-1}] \mathbf{1}_K$. Thus, using the $b_i^*(z)$ and $a(z)$, we have found an equivalent problem for which $K = 0$. In fact, $b_i^*(z)$ and $a(z)$ correspond to the censored Markov chain obtained from the original M/G/1 type chain with transition matrix (26) by considering only those times when it is outside the K states of level -1.

Assuming that $K = 0$, we transform the equations $\pi = \pi\mathcal{P}$ by multiplying the kth block row by z^k and summing to obtain the matrix version of (9), namely,

$$\sum_{i=0}^{N-1} \pi_i B_i(z) + \psi(z)A(z) = 0. \tag{33}$$

Here $B_i(z) = z^i I_M - b_i(z)$, $i = 0, \ldots, N-1$, $A(z) = z^N I_M - a(z)$ are $M \times M$ matrix functions. Multiplying (33) on the right by $\mathrm{adj}\, A(z)$, where adj represents the classical adjoint matrix, gives the generating function equation

$$\sum_{i=0}^{N-1} \pi_i B_i(z) \,\mathrm{adj}\, A(z) + \psi(z)\det A(z) = 0. \tag{34}$$

Solving for $\psi(z)$, we find that

$$\psi(z) = -\frac{\sum_{i=0}^{N-1} \pi_i B_i(z) \,\mathrm{adj}\, A(z)}{\det A(z)}. \tag{35}$$

The above procedure can also be described in terms of the $\infty \times M$ root vector

$$e(z) = \begin{bmatrix} I_M \\ zI_M \\ z^2 I_M \\ \vdots \end{bmatrix}. \tag{36}$$

The Toeplitz-like structure of $\mathcal{P}$ gives

$$[\mathcal{I} - \mathcal{P}]e(z) = \begin{bmatrix} I_M - b_0(z) \\ \vdots \\ z^{N-1}I_M - b_{N-1}(z) \\ z^N I_M - a(z) \\ z^{N+1} I_M - za(z) \\ \vdots \end{bmatrix} = \begin{bmatrix} B_0(z) \\ \vdots \\ B_{N-1}(z) \\ A(z) \\ zA(z) \\ \vdots \end{bmatrix}. \tag{37}$$

Thus transforming yields the equation $\pi e(z) = \pi \mathcal{P} e(z)$, and from (37) we obtain (33). Furthermore, multiplying (37) on the right by $\mathrm{adj}\, A(z)$ gives

$$[\mathcal{I} - \mathcal{P}]e(z)\,\mathrm{adj}\, A(z) = \begin{bmatrix} B_0(z)\,\mathrm{adj}\, A(z) \\ \vdots \\ B_{N-1}(z)\,\mathrm{adj}\, A(z) \\ \det A(z) I_M \\ z \det A(z) I_M \\ \vdots \end{bmatrix}. \tag{38}$$

Multiplying on the left by π, we obtain (34).

As in the scalar case, since $\psi(z)$ is analytic in the open unit disk and continuous in the closed unit disk, if ξ is a zero of det $A(z)$ satisfying $|\xi| \le 1$, we have

$$\sum_{i=0}^{N-1} \pi_i B_i(\xi) \operatorname{adj} A(\xi) = 0. \qquad (39)$$

Further, if ξ is a zero of det $A(z)$ of multiplicity 2, then we also have the equation

$$\sum_{i=0}^{N-1} \pi_i D_i(\xi) = 0, \qquad (40)$$

where

$$D_i(z) = \frac{d}{dz}\{B_i(z) \operatorname{adj} A(z)\}. \qquad (41)$$

For zeros of greater multiplicity, successive differentiation of $B_i(z) \operatorname{adj} A(z)$ is utilized to obtain additional equations.

In the scalar case, zeros of $z^N - a(z)$ yield linear equations in the unknown boundary probabilities. In the matrix case, not only are zeros of det $A(z)$ required, but the matrix adj $A(z)$ evaluated at these zeros is needed, at least according to the equations (39). However, such an apparent requirement can be simplified as follows. First consider the case of a simple zero ξ of det $A(z)$. We claim that the $M \times M$ matrix $A(\xi)$ has rank $M-1$, while the rank of adj $A(\xi)$ is 1. To see this, evaluate the matrix equation $A(z) \operatorname{adj} A(z) = \det A(z) I_M$ and its derivative at the zero $z = \xi$ to get

$$A(\xi) \operatorname{adj} A(\xi) = 0, \qquad (42)$$

and

$$A(\xi) \frac{d}{dz} \operatorname{adj} A(z) \bigg|_{z=\xi} + A'(\xi) \operatorname{adj} A(\xi) = \gamma(\xi) I_M. \qquad (43)$$

Here $\gamma(\xi) = \frac{d}{dz} \det A(z)|_{z=\xi} \ne 0$, since ξ has multiplicity 1. Now rank$A(\xi) \le M - 1$, because $\det A(\xi) = 0$. If adj $A(\xi) = 0$, then from (43) it follows that $A(\xi)$ is invertible, a contradiction. So adj $A(\xi) \ne 0$, which means that some $(M-1) \times (M-1)$ submatrix of $A(\xi)$ has a nonzero determinant. We conclude that rank$A(\xi) = M - 1$. From (42) we have rank$A(\xi)$ + rank adj $A(\xi) \le M$ (see [Horn and Johnson, 1985] page 13), and so rank adj $A(\xi) = 1$. In particular, the columns of adj $A(\xi)$ are all multiples of the same vector. Thus any nonzero column (there must exist at least one), say $U(\xi)$, can be used in (39) to give the single equation

$$\sum_{i=0}^{N-1} \pi_i B_i(\xi) U(\xi) = 0. \qquad (44)$$

When $|\xi| < 1$ is a zero of det $A(z)$, at least one of the coefficients in (44) is nonzero. For, if $B_i(\xi)U(\xi) = 0$, $i = 0, \ldots, N-1$, then multiplying (37) on the right by $U(\xi)$ gives $\mathcal{P}X = X$, where $X = e(\xi)U(\xi)$. In the terminology of [Kemeny et al., 1966], X is a regular function (such a vector is also called harmonic). Let x_i, $i = 0, 1, \ldots$, be the ith entry of the $\infty \times 1$ vector X. Since $|\xi| < 1$, then $x_i \to 0$ as $i \to \infty$ from the definition of $e(\xi)$. Thus X has an entry of maximum modulus, i.e., there is x_j such that $|x_j| \geq |x_i|$ for all i. Since $\mathcal{P}$ is irreducible, given i there is $n = n(i)$ such that the (j,i) entry of $\mathcal{P}^n$, say $p_{j,i}^{(n)}$, is nonzero. Now $\mathcal{P}^n X = X$, and so by maximality of $|x_j|$

$$|x_j| \leq \sum_{l=0}^{\infty} p_{j,l}^{(n)} |x_l| \leq p_{j,i}^{(n)} |x_i| + \sum_{l \neq i} p_{j,l}^{(n)} |x_j| \leq |x_j|.$$

Since $p_{j,i}^{(n)} \neq 0$, we must have $|x_i| = |x_j|$. It follows that all entries of X have the same modulus. But the entries converge to 0, and so we conclude that $X = 0$. In particular, from the first block of M entries we have $U(\xi) = 0$, a contradiction. Therefore, (44) is not a degenerate equation.

Note that $U(\xi)$ is a right null vector for the matrix $A(\xi)$, i.e., $A(\xi)U(\xi) = 0$. Equivalently, $U(\xi)$ is a right eigenvector for the matrix $a(\xi)$ with corresponding eigenvalue ξ^N. Thus, eigenvalues and eigenvectors determine the linear equations for the boundary probabilities from simple zeros. From a numerical point of view, it is more beneficial to use eigenvalue-eigenvector packages instead of attempting to find zeros of the determinant function, which is usually poorly behaved.

When ξ is a multiple zero of $\det A(z)$, more complicated behavior occurs. For example, suppose that ξ is a zero of multiplicity 2. Then it is shown in [Gail et al., 1996] that $\operatorname{rank} \operatorname{adj} A(\xi) \leq 1$ and

$$\operatorname{rank} \begin{bmatrix} \operatorname{adj} A(\xi) & \frac{d}{dz} \operatorname{adj} A(z)\big|_{z=\xi} \\ 0 & \operatorname{adj} A(\xi) \end{bmatrix} = 2. \tag{45}$$

If $\operatorname{rank} \operatorname{adj} A(\xi) = 0$, then $\frac{d}{dz} \operatorname{adj} A(z)\big|_{z=\xi}$ has two linearly independent columns $U_0(\xi)$ and $U_1(\xi)$. Also, $\gamma(\xi) = 0$ in (43) since ξ is a multiple zero, and so $U_0(\xi)$ and $U_1(\xi)$ are both right null vectors of $A(\xi)$, i.e., they are both eigenvectors of $a(\xi)$ with eigenvalue ξ^N. These two vectors are used to obtain the equations for the boundary probabilities in (40). If $\operatorname{rank} \operatorname{adj} A(\xi) = 1$, let $U_0(\xi)$ be any nonzero column of $\operatorname{adj} A(\xi)$ and $U_1(\xi)$ be the corresponding column of $\frac{d}{dz} \operatorname{adj} A(z)\big|_{z=\xi}$. From (42) and (43) we see that $A(\xi)U_0(\xi) = 0$ and $A(\xi)U_1(\xi) + A'(\xi)U_0(\xi) = 0$. In this case, one equation from (44) and a second equation from (40) are obtained for the boundary probabilities. In general, for a zero of multiplicity k, one must consider (right) Jordan chains of $A(z)$ of length $l \leq k$, namely, vectors $U_0(\xi), \ldots, U_{l-1}(\xi)$ that satisfy (see [Gohberg et al., 1982])

$$\begin{aligned} A(\xi)U_0(\xi) &= 0 \\ &\vdots \\ A(\xi)U_{l-1}(\xi) + \cdots + \frac{A^{(l-1)}(\xi)}{(l-1)!} U_0(\xi) &= 0. \end{aligned} \tag{46}$$

Since $a(1)$ is stochastic, $z = 1$ is always a zero of $\det A(z)$. However, unlike the case of a zero $\xi \neq 1$, evaluating (39) with $z = 1$ yields $0 = 0$. This follows because each $b_i(1)$ is stochastic and $\operatorname{adj} A(1)$ has constant columns for $a(1)$ irreducible (see [Gantmacher, 1959]). Instead differentiate (39), evaluate the result at $z = 1$, multiply on the right by $\mathbf{1}_M$, and use the conservation of probability $\psi(1)\mathbf{1}_M + \sum_{i=0}^{N-1} \pi_i \mathbf{1}_M = 1$. The result is

$$\sum_{i=0}^{N-1} \pi_i [\gamma I_M - D_i(1)] \mathbf{1}_M = \gamma, \tag{47}$$

where

$$\gamma = \frac{d}{dz} \det A(z) \bigg|_{z=1}. \tag{48}$$

Assuming $\det A(z)$ has enough zeros in the closed unit disk (other than $z = 1$) to yield $MN - 1$ independent equations from (39) and its derivatives, they along with (47) will give MN equations in the N unknown $1 \times M$ vectors π_i, $i = 0, \ldots, N - 1$. Note that these unknown vectors represent the MN boundary probabilities, M probabilities for each of the N boundary levels. Thus the matrix case raises additional questions involving the roots. For example, how many independent equations does each zero yield, and how many total independent equations are obtained? We will discuss the case of simple zeros in detail later in the paper.

Characterization. In the scalar case, the characterization of an M/G/1 type Markov chain was given in terms of $A'(1)$. In the matrix case, there are two different functions whose derivatives at $z = 1$ characterize the chain. One function is $\det A(z)$, while the other function is $\rho(z)$, which is defined as follows. Recall that the spectral radius of a matrix F, denoted $\rho(F)$, is the maximum modulus of the eigenvalues of F (see [Horn and Johnson, 1985]). When $a(z)$ converges, we define $\rho(z)$ to be the spectral radius of $a(z)$, that is,

$$\rho(z) = \rho(a(z)). \tag{49}$$

For $a(1)$ irreducible, the positive recurrence, null recurrence and transience of the M/G/1 type chain is given in terms of the quantity γ defined in (48), and also in terms of $\rho'(1)$. In the reducible case, $a(1)$ may be decomposed into irreducible stochastic and nonstochastic subblocks (the normal form described on page 90 of [Gantmacher, 1959]), and the characterization of the chain is given in terms of these subblocks (see [Gail et al., 1996] for details). Concentrating on the irreducible case, we have the following result.

Proposition 7.2 *Suppose $a(1)$ is irreducible stochastic. Suppose also $\sum_k k b_{i,k} < +\infty$ for $i = 0, \ldots, N - 1$, $\sum_k k a_k < +\infty$, and $\sum_k k^2 a_k < +\infty$ when $\gamma = 0$.*

(i) *The M/G/1 type Markov chain is positive recurrent if and only if $\gamma > 0$ if and only if $\rho'(1) < N$.*

(ii) *The M/G/1 type Markov chain is null recurrent if and only if $\gamma = 0$ if and only if $\rho'(1) = N$.*

(iii) *The M/G/1 type Markov chain is transient if and only if $\gamma < 0$ if and only if $\rho'(1) > N$.*

Differentiating $\operatorname{adj} A(z) A(z) = \det A(z) I_M = A(z) \operatorname{adj} A(z)$ and evaluating the result at $z = 1$, it can be shown that

$$\gamma/c = \eta A'(1) \mathbf{1}_M = N - \eta a'(1) \mathbf{1}_M,$$

where $c > 0$ is a scalar and η is the stationary distribution of the irreducible stochastic matrix $a(1)$ (the entries of η give the stationary probabilities of the phases). Thus γ is a negative multiple of the expected one-step drift in the homogeneous part of the chain, which provides an intuitive interpretation for the results of the above proposition. Further, it was shown in [Keilson and Wishart, 1964] that $\rho'(1) = \eta a'(1) \mathbf{1}_M$, which explains the second part of the characterization.

2.3 Continuous-Time Case

The analysis of continuous-time M/G/1 type Markov chains can be easily reduced to the discrete-time case. Note that the analogue of (26) for continuous-time (with $K = 0$, and M, N positive integers) is the generator

$$Q = \begin{bmatrix} d_{0,0} & d_{0,1} & \cdots & d_{0,N-1} & d_{0,N} & d_{0,N+1} & \cdots \\ d_{1,0} & d_{1,1} & \cdots & d_{1,N-1} & d_{1,N} & d_{1,N+1} & \cdots \\ \vdots & \vdots & \ddots & \vdots & \vdots & \vdots & \ddots \\ d_{N-1,0} & d_{N-1,1} & \cdots & d_{N-1,N-1} & d_{N-1,N} & d_{N-1,N+1} & \cdots \\ c_0 & c_1 & \cdots & c_{N-1} & c_N & c_{N+1} & \cdots \\ 0 & c_0 & \cdots & c_{N-2} & c_{N-1} & c_N & \cdots \\ \vdots & \vdots & \ddots & \vdots & \vdots & \vdots & \ddots \end{bmatrix}. \quad (50)$$

As in the discrete-time case, the pair of integers (i, j), $i = 0, 1, \ldots$, $j = 0, \ldots, M - 1$, denotes a state of the irreducible continuous-time Markov chain of M/G/1 type. Levels and phases are defined as before, with i representing the level and j representing the phase. The $M \times M$ matrices $d_{i,k}$ and c_k, $i = 0, \ldots, N - 1$, $k = 0, 1, \ldots$, consist of transition rates between the levels of the continuous-time chain. Specifically, for $j = 0, \ldots, M - 1$, $l = 0, \ldots, M - 1$, we let $d_{i,k;j,l}$ be the (j, l) entry of the matrix $d_{i,k}$, with a similar definition for $c_{k;j,l}$. The off-diagonal entries of Q represent transition rates between states, namely, for $(i, j) \neq (k, l)$,

$$d_{i,k;j,l} = [\text{rate from state } (i, j) \text{ to state } (k, l)] \qquad i = 0, \ldots, N - 1, \quad (51)$$

and for $(i, j) \neq (k + i - N, l)$,

$$c_{k;j,l} = [\text{rate from state } (i, j) \text{ to state } (k + i - N, l)] \qquad i = N, N + 1, \ldots. \quad (52)$$

The diagonal entries of Q represent the negative of the rate out of a state, namely,

$$d_{i,i;j,j} = -[\text{rate out of state } (i,j)] \qquad i = 0, \ldots, N-1 \tag{53}$$

and

$$c_{N;j,j} = -[\text{rate out of state } (i,j)] \qquad i = N, N+1, \ldots. \tag{54}$$

Note that these diagonal entries are all nonzero, since otherwise the corresponding state would be absorbing.

The stationary probability vector π of an irreducible positive recurrent chain with generator Q satisfies the matrix equation $\pi Q = 0$. Here $\pi = [\pi_0, \pi_1, \ldots]$ in the M/G/1 case, where the $1 \times M$ vector π_i corresponds to level i of the chain. Specifically, if $\pi_{i,j}$ denotes the stationary probability of the chain being in state (i,j), then we have $\pi_i = [\pi_{i,0}, \ldots, \pi_{i,M-1}]$. Thus the family of equations $\pi Q = 0$ for the chain is

$$\sum_{i=0}^{N-1} \pi_i d_{i,k} + \sum_{i=N}^{N+k} \pi_i c_{N+k-i} = 0, \qquad k = 0, 1, \ldots. \tag{55}$$

Define $\widehat{\pi}_i = [\widehat{\pi}_{i,0}, \ldots, \widehat{\pi}_{i,M-1}]$, where

$$\widehat{\pi}_{i,j} = \begin{cases} \pi_{i,j} d_{i,i;j,j} & i = 0, \ldots, N-1, \; j = 0, \ldots, M-1 \\ \pi_{i,j} c_{N;j,j} & i = N, N+1, \ldots, \; j = 0, \ldots, M-1. \end{cases} \tag{56}$$

For $(i,j) \neq (k,l)$, $i = 0, \ldots, N-1$, define $\widehat{d}_{i,k;j,l} = d_{i,k;j,l}/d_{i,i;j,j}$ and set $\widehat{d}_{i,i;j,j} = 0$. For $(N,j) \neq (k,l)$, define $\widehat{c}_{k;j,l} = c_{k;j,l}/c_{N;j,j}$ and set $\widehat{c}_{N;j,j} = 0$. Then $\widehat{d}_{i,k}$ and $\widehat{c}_k$ yield a stochastic matrix $\widehat{Q}$, and the matrix equation $\pi Q = 0$ becomes $\widehat{\pi} = \widehat{\pi}\widehat{Q}$. That is, (55) becomes

$$\widehat{\pi}_k = \sum_{i=0}^{N-1} \widehat{\pi}_i \widehat{d}_{i,k} + \sum_{i=N}^{N+k} \widehat{\pi}_i \widehat{c}_{N+k-i}, \qquad k = 0, 1, \ldots. \tag{57}$$

Each equation has been normalized by dividing by the diagonal elements of Q. Transforming (57) we obtain

$$\sum_{i=0}^{N-1} \widehat{\pi}_i [z^i I_M - \widehat{d}_i(z)] + \widehat{\psi}(z)[z^N I_M - \widehat{c}(z)] = 0, \tag{58}$$

where $\widehat{d}_i(z)$, $\widehat{c}(z)$ and $\widehat{\psi}(z)$ are defined as for discrete-time. Note that $\widehat{c}(1)$ and $\widehat{d}_i(1)$, $i = 0, \ldots, N-1$, are stochastic matrices, and we have reduced the continuous-time problem to the discrete-time case.

Equation (58) is the fundamental relation for the continuous-time Markov chain of M/G/1 type. The stationary probabilities of the embedded discrete-time Markov chain defined at transition epochs may be obtained from $\widehat{\pi}_i$, $i = 0, \ldots, N-1$, and $\widehat{\psi}(z)$, while the stationary probabilities of the original continuous-time chain may be obtained by scaling the entries of $\widehat{\pi}_i$ and $\widehat{\psi}(z)$ by the appropriate rates $d_{i,i;j,j}$ and $c_{N;j,j}$. Note that if these rates are all the same, then (58) gives an equation for $\psi(z)$ and the π_i directly. Since the number of distinct rates is at most $M(N+1)$, this case can always be guaranteed, i.e., the chain is uniformizable [Grassmann, 1977a, Grassmann, 1977b, Jensen, 1953]. Thus the stationary probabilities of the continuous-time chain are identical to those of a discrete-time chain obtained by adding fictitious self-transitions to yield a uniform transition rate $q \geq \max_{i,j}\{d_{i,i;j,j}, c_{N;j,j}\}$. Note that the discrete-time uniformized chain obtained from a continuous-time M/G/1 type chain is itself of M/G/1 type, so that the previous discrete-time transform approach applies directly to it.

2.4 Relation to the Matrix Analytic Method

An approach for finding the stationary distribution of an irreducible positive recurrent Markov chain of M/G/1 type was developed by Neuts [Neuts, 1989]. Called the matrix analytic method, this approach is amenable to numerical calculation and is based on probabilistic arguments. Instead of using transforms, another way of finding equations for the boundary probabilities involves the use of censored Markov chains (or stochastic complement [Meyer, 1989]). Suppose we consider the given M/G/1 type Markov chain only when it is in the boundary states, and let this finite-state censored Markov chain have transition matrix $\widehat{\mathcal{P}}$. If we can find $\widehat{\mathcal{P}}$, then solving the system $\widehat{\pi} = \widehat{\pi}\widehat{\mathcal{P}}$ where $\widehat{\pi} = [\pi_0, \ldots, \pi_{N-1}]$ will yield the boundary probabilities.

First consider the skip-free to the left case $N = 1$ (recall we can always assume $K = 0$ in (26)). There is an $M \times M$ matrix G which is the key to the matrix analytic approach in the M/G/1 case. The elements of G^k, $k = 0, 1, \ldots$, give the transition probabilities, starting in level k, of absorption in the boundary level 0. It can be shown that

$$\widehat{\mathcal{P}} = b(G) = \sum_{k=0}^{\infty} b_k G^k, \tag{59}$$

and

$$G = a(G) = \sum_{k=0}^{\infty} a_k G^k. \tag{60}$$

Further, G is the minimal nonnegative solution of $G = a(G)$. Also, in the positive recurrent case, G is a stochastic matrix. Once G is determined (for example, using algorithms in [Akar and Sohraby, 1997a], [Bini and Meini, 1996], [Latouche and Ramaswami, 1993]), then the boundary probabilities can be found from the system $\pi_0 = \pi_0 b(G)$. Finally the probability vectors for the homogeneous states can be found using the method of Ramaswami [Ramaswami, 1988]. In the Ramaswami approach, the vector π_n for the nth level is found from the censored Markov chain obtained by considering the original M/G/1 chain only in levels $0, \ldots, n$ (see also [Meini, 1997]).

The following theorem describes the relationship between the zeros found in the transform method and the key matrix G of the matrix analytic method (recall that $A(z) = zI_M - a(z)$ in the skip-free to the left case).

Theorem 1 *The eigenvalues of G, with multiplicity, are the zeros of $\det A(z)$, with multiplicity, in the closed unit disk.*

To see this, suppose that ξ is an eigenvalue of G with right eigenvector $\mathbf{x}$. Then $G\mathbf{x} = \xi\mathbf{x}$, and so $G^k\mathbf{x} = \xi^k\mathbf{x}$ for $k = 0, 1, \ldots$. Therefore, using $G = a(G)$, we have

$$a(\xi)\mathbf{x} = \sum_{k=0}^{\infty} a_k \xi^k \mathbf{x} = \sum_{k=0}^{\infty} a_k G^k \mathbf{x} = a(G)\mathbf{x} = G\mathbf{x} = \xi\mathbf{x}.$$

Thus the matrix $A(\xi) = \xi I - a(\xi)$ is singular with right null vector $\mathbf{x}$, and so ξ is a zero of $\det A(z)$.

Similar results can be shown when G has generalized eigenvectors. Recall that a cycle of generalized (right) eigenvectors of length l corresponding to an eigenvalue ξ of G, say $\mathbf{x}_0, \ldots, \mathbf{x}_{l-1}$, is given by [Friedberg et al., 1989]

$$\begin{aligned} G\mathbf{x}_0 &= \xi\mathbf{x}_0 \\ G\mathbf{x}_j &= \xi\mathbf{x}_j + \mathbf{x}_{j-1}, \quad j = 1, \ldots, l-1. \end{aligned} \quad (61)$$

It can be shown, using $G = a(G)$, that $\mathbf{x}_0, \ldots, \mathbf{x}_{l-1}$, constitute a (right) Jordan chain of length l for $A(z)$, i.e., they satisfy the set of equations (45). This can be used to show that $\det A(z)$ has a zero at ξ of at least order l. In fact, using a dimension argument, it can be shown that G has an eigenvalue at ξ of order r if and only if $\det A(z)$ has a zero at ξ of order r. Since G is stochastic, all of its M eigenvalues lie in the closed unit disk. Since $\det A(z)$ has exactly M zeros in the closed unit disk in the positive recurrent case, these zeros are the eigenvalues of G. Further, the eigenvectors and generalized eigenvectors of G for an eigenvalue ξ correspond to Jordan chains of $A(z)$ for the zero ξ. Thus, we see that the zeros of the transform method are strongly related to the Jordan canonical form of G. This form is $G = V^{-1}\Lambda V$, where Λ consists of Jordan blocks corresponding to the eigenvalues of G, V^{-1} consists of generalized right eigenvectors of G, and V consists of generalized left eigenvectors of G.

In the non-skip-free case when $N > 1$ in (26)), we obtain the following generalization (see [Gail et al., 1997]).

Theorem 2 *The eigenvalues of G, with multiplicity, are Nth powers of the zeros of $\det A(z)$, with multiplicity, in the closed unit disk.*

It is also shown in [Gail et al., 1997] that G is the Nth power of the companion matrix corresponding to its first block row (its first M rows). Bini and Meini [Bini and Meini, 1998] have exploited this fact to develop an efficient algorithm to calculate G in the non-skip-free case.

3 G/M/1 TYPE CHAINS

In this section we describe the use of roots in solving for the stationary distribution of an irreducible positive recurrent G/M/1 type Markov chain. For chains in this class, the equations $\pi = \pi\mathcal{P}$ exhibit a structure that enables them to be solved directly without the use of transforms. Except for a finite number of boundary equations, this system is a set of Wiener-Hopf equations, unlike $\pi = \pi\mathcal{P}$ in the M/G/1 case. However, the equations $\mathcal{P}X = X$ do constitute a Wiener-Hopf system for M/G/1, and we have seen from (37) that solutions are given in terms of root vectors. Similarly, mathematical theory suggests that solutions of $\pi = \pi\mathcal{P}$ for G/M/1 should be a linear combination of root vectors and their derivatives if necessary (see [Gail et al., 1996, Gohberg and Fel'dman, 1974]). As in the M/G/1 case, we begin by discussing the scalar G/M/1 type Markov chain.

3.1 Scalar Case

Consider an irreducible Markov chain with transition matrix of the form

$$\mathcal{P} = \begin{bmatrix} b_0 & a_0 & 0 & 0 & \cdots \\ b_1 & a_1 & a_0 & 0 & \cdots \\ b_2 & a_2 & a_1 & a_0 & \cdots \\ b_3 & a_3 & a_2 & a_1 & \cdots \\ \vdots & \vdots & \vdots & \vdots & \ddots \end{bmatrix}. \tag{62}$$

Here b_k gives the probability of a transition from state k to state 0, while a_k gives the probability of a transition from state $k+l$ to state $l+1$. The Markov chain for number in system at arrival instants for the classical G/M/1 queue has a transition matrix of this form (see [Kleinrock, 1975]). The chain is skip-free to the right, since transitions to higher numbered states can only occur from a state i to the neighboring state $i+1$.

The system $\pi = \pi\mathcal{P}$ consists of the equations

$$\pi_0 = \sum_{k=0}^{\infty} \pi_k b_k \tag{63}$$

and

$$\pi_i = \sum_{k=i-1}^{\infty} \pi_k a_{k-i+1}, \quad i = 1, 2, \ldots. \tag{64}$$

It is natural to seek solutions of the system (64) in terms of root vectors (see [Gohberg and Fel'dman, 1974, Goldberg, 1958]). Specifically, define the $1 \times \infty$ vector function

$$e^T(z) = [1, z, z^2, z^3, \ldots]. \tag{65}$$

The result of the matrix equation $e^T(z) = e^T(z)\mathcal{P}$ is

$$1 = \sum_{k=0}^{\infty} z^k b_k = b(z) \tag{66}$$

and
$$z^i = \sum_{k=i-1}^{\infty} z^k a_{k-i+1} = z^{i-1} a(z), \qquad i = 1, 2, \ldots. \tag{67}$$

Now if $a(\sigma)$ converges and σ is a zero of $z - a(z)$, then the equations (67) immediately hold for $e^T(\sigma)$. Since a probability distribution π has entries that sum to 1, our interest is in summable solutions (an $\infty \times 1$ vector $\mathbf{v} = [v_0, v_1, \ldots]$ is summable if $\sum_{i=0}^{\infty} |v_i| < +\infty$). Thus σ must lie in the open unit disk, for otherwise $e^T(\sigma)$ is not a summable vector. Also note that $\sigma \neq 0$, since $a(0) = a_0 \neq 0$ by the irreducibility of $\mathcal{P}$. In fact, for a positive recurrent G/M/1 type chain, there is a unique zero of $z - a(z)$ in the open unit disk, and it satisfies $0 < \sigma < 1$ (see Section 4).

We claim that the initial equation (66) also holds for $e^T(\sigma)$. To see this, first note that the row sums of $\mathcal{P}$ satisfy

$$b_k + \sum_{i=0}^{k} a_i = 1, \qquad k = 0, 1, \ldots \tag{68}$$

Multiplying the kth equation of (68) by z^k, summing on k, and interchanging the order of summation gives

$$b(z) + a(z) \sum_{k=0}^{\infty} z^k = \sum_{k=0}^{\infty} z^k. \tag{69}$$

Evaluating (69) at $z = \sigma$ (recall that we require $|\sigma| < 1$) and using $\sigma = a(\sigma)$ yields $b(\sigma) = 1$, and so equation (66) is automatically satisfied. Any scalar multiple $Ce^T(\sigma)$ is also a solution of $\pi = \pi \mathcal{P}$. Since the sum of the π_i must be 1, setting $C = 1 - \sigma$ gives the stationary probability vector π with $\pi_i = \pi_0 \sigma^i = (1-\sigma)\sigma^i$. Therefore, we have determined the stationary distribution of the chain.

A more general scalar G/M/1 type Markov chain has transition matrix

$$\mathcal{P} = \begin{bmatrix} b_{0,0} & \cdots & b_{0,N-1} & a_0 & 0 & 0 & 0 & \cdots \\ b_{1,0} & \cdots & b_{1,N-1} & a_1 & a_0 & 0 & 0 & \cdots \\ b_{2,0} & \cdots & b_{2,N-1} & a_2 & a_1 & a_0 & 0 & \cdots \\ b_{3,0} & \cdots & b_{3,N-1} & a_3 & a_2 & a_1 & a_0 & \cdots \\ \vdots & \ddots & \vdots & \vdots & \vdots & \vdots & \vdots & \ddots \end{bmatrix}. \tag{70}$$

The chain is skip-free to the right if $N = 1$, and it reduces to the case of (62). When $N > 1$ the chain is called non-skip-free. The boundary equations of $\pi = \pi \mathcal{P}$ are

$$\pi_i = \sum_{k=0}^{\infty} \pi_k b_{k,i}, \qquad i = 0, \ldots, N-1 \tag{71}$$

while the homogeneous equations are

$$\pi_i = \sum_{k=i-N}^{\infty} \pi_k a_{k-i+N}, \qquad i = N, N+1, \ldots. \tag{72}$$

When $a(z) = \sum_{j=0}^{\infty} a_j z^j$ is a polynomial, the system (72) is a set of constant coefficient difference equations. Then it is well-known that solutions are given in terms of root vectors and their derivatives [Goldberg, 1958]. In the case of a general analytic function $a(z)$, this is no longer necessarily true. However, $a(z)$ and $A(z) = z^N - a(z)$ satisfy the additional properties that $a_k \geq 0$, $\sum_k k a_k < +\infty$ if $A'(1) \neq 0$, and $\sum_k k^2 a_k < +\infty$ if $A'(1) = 0$. Under these conditions, using results from [Gohberg and Fel'dman, 1974] it can be shown that summable solutions of $\pi = \pi \mathcal{P}$ are given in terms of the zeros of $a(z)$ in the open unit disk (see [Gail et al., 1996]).

From $e^T(z) = e^T(z)\mathcal{P}$, the system (71) becomes

$$z^i = \sum_{k=0}^{\infty} z^k b_{k,i} = b_i(z), \qquad i = 0, \ldots, N-1, \qquad (73)$$

while (72) is

$$z^i = \sum_{k=i-N}^{\infty} z^k a_{k-i+N} = z^{i-N} a(z), \qquad i = N, N+1, \ldots \qquad (74)$$

If $a(\sigma)$ converges and σ is a zero of $z^N - a(z)$, then the equations (74) hold.

For a zero of $z^N - a(z)$ of multiplicity 2, we also consider the derivative

$$e^{T'}(z) = [0, 1, 2z, 3z^2, \ldots]$$

of the root vector $e^T(z)$. Then $e^{T'}(z) = e^{T'}(z)\mathcal{P}$ yields the derivatives of (73) and (74), namely,

$$i z^{i-1} = b'_i(z), \qquad i = 0, \ldots, N-1 \qquad (75)$$

and

$$i z^{i-1} = z^{i-N} a'(z) + (i-N) z^{i-N-1} a(z), \qquad i = N, N+1, \ldots. \qquad (76)$$

Since $a(\sigma) = \sigma^N$ and $a'(\sigma) = N\sigma^{N-1}$ when σ is a zero of multiplicity 2, equations (76) are satisfied for $z = \sigma$. For a zero of multiplicity k, the vectors $e^{T^{(l)}}(z)$, $l = 0, \ldots, k-1$ must all be considered. A general solution of $\pi = \pi \mathcal{P}$ is a linear combination of root vectors and their derivatives (if necessary) evaluated at zeros of $z^N - a(z)$.

Suppose there are N distinct (simple) zeros of $z^N - a(z)$ in the open unit disk, say, $\sigma_1, \ldots, \sigma_N$. Any linear combination $\pi = \sum_{l=1}^{N} C_l e^T(\sigma_l)$ satisfies the homogeneous equations (72). However, such a vector does not necessarily satisfy the boundary equations, unless the constants C_l are chosen appropriately. For such π, the equations (71) are

$$\sum_{l=1}^{N} C_l \sigma_l^i = \sum_{l=1}^{N} C_l b_i(\sigma_l) \qquad i = 0, \ldots, N-1. \qquad (77)$$

Equivalently, we may write (77) in matrix form as

$$[C_1, C_2, \ldots, C_N] \begin{bmatrix} 1 - b_0(\sigma_1) & \sigma_1 - b_1(\sigma_1) & \cdots & \sigma_1^{N-1} - b_{N-1}(\sigma_1) \\ 1 - b_0(\sigma_2) & \sigma_2 - b_1(\sigma_2) & \cdots & \sigma_2^{N-1} - b_{N-1}(\sigma_2) \\ \vdots & \vdots & \ddots & \vdots \\ 1 - b_0(\sigma_N) & \sigma_N - b_1(\sigma_N) & \cdots & \sigma_N^{N-1} - b_{N-1}(\sigma_N) \end{bmatrix} = 0. \tag{78}$$

Although this gives N equations in the N unknowns C_l, they are not a linearly independent system. This is intuitively clear, since otherwise 0 is the only solution.

To show the dependence of the equations, first note that the row sums of $\mathcal{P}$ are

$$\sum_{i=0}^{N-1} b_{k,i} + \sum_{i=0}^{k} a_i = 1, \quad k = 0, 1, \ldots. \tag{79}$$

Multiply the kth equation of (79) by z^k and sum on k to yield

$$\sum_{i=0}^{N-1} b_i(z) + a(z) \sum_{k=0}^{\infty} z^k = \sum_{k=0}^{\infty} z^k. \tag{80}$$

Evaluating this at σ_l, $l = 1, \ldots, N$, (recall that $|\sigma_l| < 1$) and using $\sigma_l^N = a(\sigma_l)$ gives

$$\sum_{i=0}^{N-1} b_i(\sigma_l) = \sum_{i=0}^{N-1} \sigma_l^i, \quad l = 1, \ldots, N. \tag{81}$$

This shows that the matrix in (78) has row sums equal to 0, so that the equations in (77) are not linearly independent. In fact, this matrix has rank $N - 1$ to give a one-dimensional solution space of the C_l (see [Gail et al., 1996] for a proof in the general case of multiple zeros). However, an additional equation can be obtained using the conservation of probability. We have

$$\pi_i = \sum_{l=1}^{N} C_l \sigma_l^i, \quad i = 0, 1, \ldots,$$

and so using $\sum_{i=0}^{\infty} \pi_i = 1$, the final equation to find the C_l is ($|\sigma_l| < 1$)

$$\sum_{l=1}^{N} \frac{C_l}{1 - \sigma_l} = 1. \tag{82}$$

Similar to the scalar M/G/1 case, at least one of the coefficients $a_0, \ldots, a_{N-1}$ in (70) must be nonzero since $\mathcal{P}$ is irreducible. Otherwise there would be no path from the boundary states to the homogeneous part of the chain. Suppose that a_L is the first nonzero coefficient, so that $0 \leq L \leq N-1$ and

$$\mathcal{P} = \begin{bmatrix} b_{0,0} & \cdots & b_{0,N-1} & 0 & 0 & \cdots \\ \vdots & \ddots & \vdots & \vdots & \vdots & \ddots \\ b_{L-1,0} & \cdots & b_{L-1,N-1} & 0 & 0 & \cdots \\ b_{L,0} & \cdots & b_{L,N-1} & a_L & 0 & \cdots \\ b_{L+1,0} & \cdots & b_{L+1,N-1} & a_{L+1} & a_L & \cdots \\ \vdots & \ddots & \vdots & \vdots & \vdots & \ddots \end{bmatrix}. \tag{83}$$

An example of a Markov chain with a transition matrix of this form is the number in system at arrival instants of the classical G/M/m multiserver queue (see [Kleinrock, 1975]).

When $L > 0$, then $z^N - a(z) = z^L[z^{N-L} - a^*(z)]$, where $a^*(z) = \sum_{j=L}^{\infty} a_j z^{j-L}$. Thus $z^N - a(z)$ has a zero of multiplicity L at $z = 0$. Now note that

$$e^{T^{(j)}}(0) = [0, \ldots, 0, j!, 0, \ldots], \quad j = 0, 1, \ldots,$$

where $j!$ occurs in the jth entry. The solutions of $\pi = \pi \mathcal{P}$ contain linear combinations of the $e^{T^{(j)}}(0)$, $j = 0, \ldots, L-1$. As a consequence, solutions π have the form of L constant entries followed by entries corresponding to root vectors and their derivatives.

Specifically, suppose that $z^N - a(z)$ has $N - L$ distinct zeros in the open unit disk other than the zero at $z = 0$ of multiplicity L. That is, suppose that $z^{N-L} - a^*(z)$ has $N - L$ simple zeros in $|z| < 1$. By the above observation, a solution of the system $\pi = \pi \mathcal{P}$ for (83) has the form $\pi_i = \sum_{l=1}^{N-L} C_l \sigma_l^{i-L}$, $i = L, L+1, \ldots$, for certain scalars C_l and σ_l. Using this we obtain

$$\pi_i = \sum_{k=0}^{L-1} \pi_k b_{k,i} + \sum_{l=1}^{N-L} C_l b_i^*(\sigma_l), \quad i = 0, \ldots, L-1, \tag{84}$$

$$\sum_{l=1}^{N-L} C_l \sigma_l^{i-L} = \sum_{k=0}^{L-1} \pi_k b_{k,i} + \sum_{l=1}^{N-L} C_l b_i^*(\sigma_l), \quad i = L, \ldots, N-1, \tag{85}$$

$$\sum_{l=1}^{N-L} C_l \sigma_l^{i-L} = \sum_{l=1}^{N-L} C_l \sigma_l^{i-N} a^*(\sigma_l), \quad i = N, N+1, \ldots. \tag{86}$$

Here $b_i^*(z) = \sum_{k=L}^{\infty} b_{k,i} z^{k-L}$, $i = 0, \ldots, N-1$, and recall $a^*(z) = \sum_{k=L}^{\infty} a_k z^{k-L}$. The equations (86) are automatically satisfied when the σ_l are zeros of $z^{N-L} - a^*(z)$. Assuming this choice, from (84) and (85) we must solve N equations in the N unknowns π_i, $i = 0, \ldots, L-1$, and C_l, $l = 1, \ldots, N-L$.

Writing these N equations in matrix form in this case of distinct zeros gives

$$[\pi_0, \ldots, \pi_{L-1}, C_1, \ldots, C_{N-L}]$$

$$\begin{bmatrix} 1-b_{0,0} & \cdots & -b_{0,L-1} & -b_{0,L} & \cdots & -b_{0,N-1} \\ \vdots & \ddots & \vdots & \vdots & \ddots & \vdots \\ -b_{L-1,0} & \cdots & 1-b_{L-1,L-1} & -b_{L-1,L} & \cdots & -b_{L-1,N-1} \\ -b_0^*(\sigma_1) & \cdots & -b_{L-1}^*(\sigma_1) & 1-b_L^*(\sigma_1) & \cdots & \sigma_1^{N-1-L} - b_{N-1}^*(\sigma_1) \\ \vdots & \ddots & \vdots & \vdots & \ddots & \vdots \\ -b_0^*(\sigma_{N-L}) & \cdots & -b_{L-1}^*(\sigma_{N-L}) & 1-b_L^*(\sigma_{N-L}) & \cdots & \sigma_{N-L}^{N-1-L} - b_{N-1}^*(\sigma_{N-L}) \end{bmatrix} = 0.$$

(87)

Proceeding as before, one can show that the $N \times N$ matrix in (86) has row sums equal to 0, and so the equations (84) and (85) are not linearly independent. A final equation using the conservation of probability is (recall $|\sigma_l| < 1$)

$$\sum_{i=0}^{L-1} \pi_i + \sum_{l=1}^{N-L} \frac{C_l}{1-\sigma_l} = 1. \tag{88}$$

Characterization. In determining a summable solution π, there must exist a sufficient number of zeros of $z^N - a(z)$ in the open unit disk. In fact, as we shall see in Section 4, this occurs exactly when $a(z)$ corresponds to a positive recurrent G/M/1 type Markov chain. Clearly, if $a(1) < 1$ the chain is positive recurrent, since then there is probability of at least $1 - a(1)$ of a one-step transition to the boundary states from any state $i = N, N+1, \ldots$ in the homogeneous part of the chain. If $a(1) = 1$, the scalar G/M/1 type Markov chain is characterized in terms of

$$A'(1) = \frac{d}{dz}[z^N - a(z)]\bigg|_{z=1} = N - \sum_{k=0}^{\infty} ka_k.$$

We have the following general result.

Proposition 7.3 *Suppose that $a(1) < 1$.*

(i) *The scalar G/M/1 type Markov chain with transition matrix of the form (83) is positive recurrent.*

Suppose that $a(1) = 1$, $\sum_k ka_k < +\infty$, and $\sum_k k^2 a_k < +\infty$ when $A'(1) = 0$. Then the scalar G/M/1 type Markov chain with transition matrix of the form (83) is

(ii) *positive recurrent if and only if $A'(1) < 0$.*

(iii) *null recurrent if and only if $A'(1) = 0$.*

(iv) *transient if and only if $A'(1) > 0$.*

3.2 Matrix Case

Similar to the M/G/1 case, the scalar G/M/1 Markov chain can be generalized by considering the elements a_k and $b_{i,k}$ of the transition matrix $\mathcal{P}$ to be finite matrices. The resulting class, called Markov chains of G/M/1 type, was introduced and studied by Neuts [Neuts, 1981]. He showed that the stationary distribution of an irreducible positive recurrent G/M/1 type chain has a matrix-geometric form and developed a matrix analytic approach for the analysis of these chains. The method of repeating rows of Grasmann and Heyman [Grassmann and Heyman, 1990, Grassmann and Heyman, 1993] can also be used to study the G/M/1 class. We will review a third approach, namely, the use of roots to obtain the stationary distribution for such a chain.

Let $K \geq 0$, $N > 0$, $M > 0$ be integers. A G/M/1 type Markov chain has a transition matrix $\mathcal{P}$ of the form

$$\mathcal{P} = \begin{bmatrix} b_{-1,-1} & b_{-1,0} & b_{-1,1} & \cdots & b_{-1,N-1} & 0 & 0 & \cdots \\ b_{0,-1} & b_{0,0} & b_{0,1} & \cdots & b_{0,N-1} & a_0 & 0 & \cdots \\ b_{1,-1} & b_{1,0} & b_{1,1} & \cdots & b_{1,N-1} & a_1 & a_0 & \cdots \\ \vdots & \vdots & \vdots & \ddots & \vdots & \vdots & \vdots & \ddots \\ b_{N-1,-1} & b_{N-1,0} & b_{N-1,1} & \cdots & b_{N-1,N-1} & a_{N-1} & a_{N-2} & \cdots \\ b_{N,-1} & b_{N,0} & b_{N,1} & \cdots & b_{N,N-1} & a_N & a_{N-1} & \cdots \\ \vdots & \vdots & \vdots & \ddots & \vdots & \vdots & \vdots & \ddots \end{bmatrix},$$
(89)

where $b_{-1,-1}$ is $K \times K$, $b_{i,-1}$ is $M \times K$ for $i = 0, 1, \ldots$, and $b_{-1,k}$ is $K \times M$ for $k = 0, \ldots, N-1$. Also, for $i = 0, 1, \ldots$, $k = 0, \ldots, N-1$, $b_{i,k}$ and a_i are $M \times M$ matrices. A state of the chain has either the form $(-1, j)$, $j = 0, \ldots, K-1$, or the form (i, j), $i = 0, 1, \ldots$, $j = 0, \ldots, M-1$. The first coordinate i is called the level, and the second coordinate j is called the phase. Thus, $b_{i,k}$ gives the transition probabilities from level i to boundary level k, while a_k gives the transition probabilities from level $k + i$ ($i \geq 0$) to homogeneous level $N + i$. Note that, by redefining K if necessary, we can assume that $a_0 \neq 0$. When $N = 1$ the chain is skip-free to the right for levels, while when $N > 1$ it is non-skip-free.

We next rewrite the stationary distribution to correspond to the subblocks of $\mathcal{P}$ as $\pi = [\pi_{-1}, \pi_0, \pi_1, \ldots]$. Here π_{-1} is $1 \times K$, while for $i = 0, 1, \ldots$, π_i is $1 \times M$. The equations $\pi = \pi \mathcal{P}$ are

$$\pi_{-1} = \pi_{-1} b_{-1,-1} + \sum_{i=0}^{\infty} \pi_i b_{i,-1} \tag{90}$$

$$\pi_k = \pi_{-1} b_{-1,k} + \sum_{i=0}^{\infty} \pi_i b_{i,k} \quad k = 0, \ldots, N-1 \tag{91}$$

$$\pi_k = \sum_{i=k-N}^{\infty} \pi_i a_{i-k+N} \quad k = N, N+1, \ldots \tag{92}$$

Since $\mathcal{P}$ is irreducible, we have seen previously that $[I - b_{-1,-1}]^{-1}$ equals $\sum_{n=0}^{\infty} [b_{-1,-1}]^n$, and the case $K > 0$ can be reduced to $K = 0$ by eliminating π_{-1} from the equations. Then (91) becomes

$$\pi_k = \sum_{i=0}^{\infty} \pi_i b_{i,k}^* \qquad k = 0, \ldots, N-1 \qquad (93)$$

where

$$b_{i,k}^* = b_{i,k} + \sum_{n=0}^{\infty} b_{i,-1} [b_{-1,-1}]^n b_{-1,k}. \qquad (94)$$

The $b_{i,k}^*$ and a_k determine a transition matrix of the form (70), but with entries that are $M \times M$ subblocks. This corresponds to the censored Markov chain obtained from the original G/M/1 chain with transition matrix (89) by considering only those times when it is outside the K states of level -1. Thus, in the following discussion, we will assume that the equations of interest have the form (93) and (92), but we relabel the subblocks as $b_{i,k}$ and a_k.

To construct solutions of $\pi = \pi \mathcal{P}$ in the matrix case, we use the $M \times \infty$ vector

$$e^T(z) = [I_M, zI_M, z^2 I_M, z^3 I_M, \ldots], \qquad (95)$$

which reduces to the definition (65) when $M = 1$. Proceeding as in the scalar case, we consider the equation $e^T(z) = e^T(z)\mathcal{P}$ (see also [Gohberg and Fel'dman, 1974]). Note that

$$e^T(z)[\mathcal{I} - \mathcal{P}] = [B_0(z), \ldots, B_{N-1}(z), A(z), zA(z), \ldots], \qquad (96)$$

where $B_i(z) = z^i I_M - b_i(z)$ and $A(z) = z^N I_M - a(z)$. This is the analogue of the M/G/1 equation (37). Multiplying (96) on the left by adj $A(z)$ gives

$$\operatorname{adj} A(z) e^T(z)[\mathcal{I} - \mathcal{P}] = [\operatorname{adj} A(z) B_0(z), \ldots, \operatorname{adj} A(z) B_{N-1}(z), \det A(z) I_M, \ldots], \qquad (97)$$

which is the analogue of (38). Candidate solutions of $\pi = \pi \mathcal{P}$ that are summable are obtained by evaluating (97) at zeros of $\det A(z)$ in the open unit disk. Thus the result of $\operatorname{adj} A(z) e^T(z) = \operatorname{adj} A(z) e^T(z) \mathcal{P}$ is the system of equations

$$\operatorname{adj} A(z) B_i(z) = 0, \qquad i = 0, \ldots, N-1 \qquad (98)$$

and

$$z^{i-N} \det A(z) I_M = 0, \qquad i = N, N+1, \ldots. \qquad (99)$$

If σ is a zero of $\det A(z)$, then (99) automatically holds.

Taking the derivative $\frac{d}{dz}\{\operatorname{adj} A(z) e^T(z)\}[I - \mathcal{P}]$ in (97) yields the derivatives of (98) and (99), namely,

$$\frac{d}{dz}\{\operatorname{adj} A(z) B_i(z)\} = 0, \qquad i = 0, \ldots, N-1 \qquad (100)$$

and
$$\frac{d}{dz}\{z^{i-N}\det A(z)\}\mathbf{1}_M = 0, \quad i = N, N+1, \ldots. \tag{101}$$

If σ is a zero of $\det A(z)$ of multiplicity 2, then $\frac{d}{dz}\det A(z)|_{z=\sigma} = 0$. It follows that (101) is satisfied at $z = \sigma$, and so $\frac{d}{dz}\{\text{adj } A(z)e^T(z)\}|_{z=\sigma}$ is another candidate solution for $\pi = \pi \mathcal{P}$. For a zero of multiplicity k, the first k derivatives $\frac{d^l}{dz^l}\{\text{adj } A(z)e^T(z)\}$, $l = 0, \ldots, k-1$, are all considered. Thus we seek a linear combination of vectors of the form $\text{adj } A(\sigma)e^T(\sigma)$ and derivatives, if necessary, for zeros σ of $\det A(z)$ in the open unit disk. Such a linear combination automatically satisfies (99) (and derivatives), and we require it to also satisfy (98) (and derivatives).

Suppose that σ is a simple zero of $\det A(z)$ in the open unit disk. We have seen in Section 2.2 that the matrix $\text{adj } A(\sigma)$ has rank 1, and so all of its rows are multiples of the same vector. Let the $1 \times M$ vector $V(\sigma)$ be any nonzero row of $\text{adj } A(\sigma)$ (such a row must exist). Then $V(\sigma)$ is a left eigenvector of $a(\sigma)$ with eigenvalue σ^N (see also [Mitrani and Chakka, 1995, Mitrani and Mitra, 1992]). We may replace (98) by the equations $V(\sigma)B_i(\sigma) = 0$, $i = 0, \ldots, N-1$. It is this system that $V(\sigma)e^T(\sigma)$ must satisfy if it is to be a solution of $\pi = \pi\mathcal{P}$. This occurs automatically in the scalar skip-free case (i.e., when $M = 1$, $N = 1$), as we have seen in Section 3.1. For general M and N, suppose that $\det A(z)$ has MN distinct simple zeros in the open unit disk, say $\sigma_1, \ldots, \sigma_{MN}$. The linear combination $\pi = \sum_{l=1}^{MN} C_l V(\sigma_l)e^T(\sigma_l)$, where the C_l are scalars, yields MN equations in MN unknowns, namely,

$$[C_1, \ldots, C_{MN}] \begin{bmatrix} V(\sigma_1)B_0(\sigma_1) & V(\sigma_1)B_1(\sigma_1) & \cdots & V(\sigma_1)B_{N-1}(\sigma_1) \\ \vdots & \vdots & \ddots & \vdots \\ V(\sigma_{MN})B_0(\sigma_{MN}) & V(\sigma_{MN})B_1(\sigma_{MN}) & \cdots & V(\sigma_{MN})B_{N-1}(\sigma_{MN}) \end{bmatrix} = 0. \tag{102}$$

The row sums of the $MN \times MN$ matrix in (102) are 0, and so the equations are not linearly independent. To see this, similar to the derivation of (80) we have

$$\sum_{i=0}^{N-1} b_i(z)\mathbf{1}_M + a(z)\sum_{k=0}^{\infty} z^k \mathbf{1}_M = \sum_{k=0}^{\infty} z^k \mathbf{1}_M. \tag{103}$$

Evaluate (103) at σ_l and multiply the result on the left by $V(\sigma_l)$, $l = 1, \ldots, MN$. Then, using $V(\sigma_l)a(\sigma_l) = V(\sigma_l)\sigma_l^N$, we conclude that $V(\sigma_l)\sum_{i=0}^{N-1} B_i(\sigma_l)\mathbf{1}_M = 0$, i.e., the row sums are 0. In fact, the matrix in (102) has rank $MN - 1$ (see [Gail et al., 1996]). A final equation obtained using the conservation of probability $\sum_{i=0}^{\infty} \pi_i \mathbf{1}_M = 1$ is

$$\sum_{l=1}^{MN} \frac{C_l}{1 - \sigma_l} V(\sigma_l)\mathbf{1}_M = 1. \tag{104}$$

When det $A(z)$ has multiple zeros in the open unit disk, vectors involving the derivatives of adj $A(z)e^T(z)$ must be used in finding the stationary distribution π.

Characterization. The matrix G/M/1 type chain is characterized in terms of $\gamma = \frac{d}{dz}\det A(z)|_{z=1}$ (recall the definition (48) above), as well as in terms of $\rho'(1)$. We discuss the case for which $a(1)$ is irreducible, since the reducible case can be reduced to the consideration of irreducible subblocks by using the normal form of $a(1)$.

Proposition 7.4 *Suppose $a(1)$ is irreducible substochastic but not stochastic.*

(i) *The G/M/1 type chain is positive recurrent.*

Suppose $a(1)$ is irreducible stochastic, $\sum_k ka_k < +\infty$, and $\sum_k k^2 a_k < +\infty$ when $\gamma = 0$.

(ii) *The G/M/1 type Markov chain is positive recurrent if and only if $\gamma < 0$ if and only if $\rho'(1) > N$.*

(iii) *The G/M/1 type Markov chain is null recurrent if and only if $\gamma = 0$ if and only if $\rho'(1) = N$.*

(iv) *The G/M/1 type Markov chain is transient if and only if $\gamma > 0$ if and only if $\rho'(1) < N$.*

3.3 Continuous-Time Case

In this section we show that the analysis of continuous-time G/M/1 type Markov chains can be easily reduced to the discrete-time case. Consider a continuous-time Markov chain (not necessarily of G/M/1 type) with an irreducible generator

$$\mathcal{Q} = \begin{bmatrix} -q_0 & q_{0,1} & q_{0,2} & q_{0,3} & \cdots \\ q_{1,0} & -q_1 & q_{1,2} & q_{1,3} & \cdots \\ q_{2,0} & q_{2,1} & -q_2 & q_{2,3} & \cdots \\ \vdots & \vdots & \vdots & \vdots & \ddots \end{bmatrix}, \tag{105}$$

where $q_i = \sum_{j \neq i} q_{i,j}$, $i = 0, 1, \ldots$, $q_i \neq 0$, and $0 \leq q_{i,j} < +\infty$. In the positive recurrent case, the stationary distribution $\pi = [\pi_0, \pi_1, \ldots]$ satisfies $\pi \mathcal{Q} = 0$. Defining $p_i = \pi_i q_i$ for $i = 0, 1, \ldots$, we obtain the equivalent set of equations $\mathbf{p} = \mathbf{p}\mathcal{R}$, where the stochastic matrix $\mathcal{R}$ has off-diagonal elements $r_{i,j} = q_{i,j}/q_i$, $i \neq j$, and diagonal elements $r_{i,i} = 0$. This gives a discrete-time system (the original continuous-time chain at transition epochs) that can be solved to yield $\mathbf{p}$. Since $\mathcal{R}$ has a G/M/1 type structure if and only if $\mathcal{Q}$ does, one can, for example, solve for $\mathbf{p}$ by using roots. (Similarly, $\mathcal{R}$ has an M/G/1 type structure if and only if $\mathcal{Q}$ does, so one can use generating functions to find $\mathbf{p}$ in this case.) The stationary distribution π is then recovered by adjusting the p_i using the mean length of time $1/q_i$ spent in state i and the conservation of probability, namely, $\pi_i = (p_i/q_i)/\sum_{j=0}^{\infty}(p_j/q_j)$.

When the rates q_i are uniformly bounded, say $q_i \leq q$ for all i where $0 < q < +\infty$, then a technique called uniformization or randomization [Grassmann, 1977a, Grassmann, 1977b, Jensen, 1953] (also known as Jensen's method [Grassmann, 1991]) can be used to transform the continuous-time chain into a discrete-time chain subordinated to a Poisson process. By adding fictitious transitions from each state back to itself, transitions can be thought of as occurring at the constant rate q. In fact, define the matrix $\mathcal{P} = \mathcal{I} + \mathcal{Q}/q$, which is stochastic by choice of q. Then $\pi\mathcal{Q} = 0$ is equivalent to solving $\pi = \pi\mathcal{P}$, as $q \neq 0$.

To see that uniformization may be applied to continuous-time G/M/1 type Markov chains, note that the analogue of (89) is the generator

$$\mathcal{Q} = \begin{bmatrix} d_{-1,-1} & d_{-1,0} & d_{-1,1} & \cdots & d_{-1,N-1} & 0 & 0 & \cdots \\ d_{0,-1} & d_{0,0} & d_{0,1} & \cdots & d_{0,N-1} & c_0 & 0 & \cdots \\ d_{1,-1} & d_{1,0} & d_{1,1} & \cdots & d_{1,N-1} & c_1 & c_0 & \cdots \\ \vdots & \vdots & \vdots & \ddots & \vdots & \vdots & \vdots & \ddots \\ d_{N-1,-1} & d_{N-1,0} & d_{N-1,1} & \cdots & d_{N-1,N-1} & c_{N-1} & c_{N-2} & \cdots \\ d_{N,-1} & d_{N,0} & d_{N,1} & \cdots & d_{N,N-1} & c_N & c_{N-1} & \cdots \\ \vdots & \vdots & \vdots & \ddots & \vdots & \vdots & \vdots & \ddots \end{bmatrix}.$$

(106)

Here the matrices $d_{i,k}$ and c_k represent transition rates between the levels of the continuous-time Markov chain. Since there are only a finite number of distinct diagonal subblocks, namely, $d_{i,i}$, $i = -1, 0, \ldots, N-1$, and c_N, and since each of these subblocks is a finite matrix, $\mathcal{Q}$ has a finite number of distinct diagonal elements. Thus the G/M/1 type chain with generator $\mathcal{Q}$ is uniformizable, and so solving for the stationary probabilities can be reduced to the discrete-time case. The matrix $\mathcal{P} = \mathcal{I} + \mathcal{Q}/q$ is also of G/M/1 type, since $\mathcal{Q}$ has that form.

3.4 Relation to the Matrix Analytic Method

We have seen that the stationary distribution of an irreducible positive recurrent Markov chain of G/M/1 type is given in terms of a linear combination of root vectors in the case when all zeros of $\det A(z)$ in the open unit disk are simple, with suitable modifications when multiple zeros occur. A matrix analytic method based on probabilistic arguments was developed by Neuts [Neuts, 1981] for Markov chains of G/M/1 type. We now show how the approach using zeros is related to the matrix analytic method.

First consider the skip-free to the right case $N = 1$. For a Markov chain of G/M/1 type, Neuts showed there is an $M \times M$ matrix R such that π has the matrix-geometric form

$$\pi_i = \pi_0 R^i, \qquad i = 0, 1, \ldots. \tag{107}$$

Since π is a summable solution, the matrix R has spectral radius < 1 in the positive recurrent case, and so all of its M eigenvalues lie in the open unit disk. Further, it can be shown that

$$R = a[R] = \sum_{k=0}^{\infty} R^k a_k. \tag{108}$$

In fact, R is the minimal nonnegative solution of $R = a[R]$. Here we use the notation $a[\cdot]$ to indicate that multiplication by the terms a_k is on the right, as opposed to the notation $a(\cdot)$ when multiplication by the a_k is on the left. The boundary probability vector satisfies

$$\pi_0 = \pi_0 b[R] = \sum_{k=0}^{\infty} \pi_0 R^k b_k. \tag{109}$$

Similar to the M/G/1 case, the key matrix R can be determined from zeros of the function $\det A(z) = \det\{zI_M - a(z)\}$. In fact, the zeros in the open unit disk are the eigenvalues of the matrix R. Note that in the positive recurrent G/M/1 case, either $a(1)$ is not stochastic or $a(1)$ is stochastic and $\gamma < 0$.

Theorem 3 *The eigenvalues of R, with multiplicity, are the zeros of $\det A(z)$, with multiplicity, in the open unit disk.*

In this case, we use left eigenvectors instead of right eigenvectors. If ξ is an eigenvalue of R with left eigenvector $\mathbf{y}$, then $\mathbf{y}R^k = \mathbf{y}\xi^k$ for $k = 0, 1, \ldots$. Therefore, using $R = a[R]$, we have

$$\mathbf{y}a(\xi) = \mathbf{y}\sum_{k=0}^{\infty} \xi^k a_k = \mathbf{y}\sum_{k=0}^{\infty} R^k a_k = \mathbf{y}a[R] = \mathbf{y}R = \mathbf{y}\xi.$$

For multiple eigenvalues of R, the (right) Jordan chains in the M/G/1 case are replaced by (left) Jordan chains in the G/M/1 case.

Since R has spectral radius < 1 in the positive recurrent G/M/1 case, all of its M eigenvalues lie in the open unit disk. Since $\det A(z)$ has exactly M zeros in the open unit disk in this case, these zeros are the eigenvalues of R. Further, the eigenvectors and generalized eigenvectors of R for an eigenvalue ξ correspond to (left) Jordan chains of $A(z)$. Thus, we see that the zeros of $\det A(z)$ can be used to calculate the Jordan canonical form of R, i.e., the form $R = V^{-1}\Lambda V$.

In the non-skip-free case, for which $N > 1$, we have the following generalization (see [Gail et al., 1997] for details).

Theorem 4 *The eigenvalues of R, with multiplicity, are Nth powers of the zeros of $\det A(z)$, with multiplicity, in the open unit disk.*

4 SOLUTION METHOD

The procedures outlined in Sections 2 and 3 for finding the stationary distribution of an irreducible positive recurrent M/G/1 or G/M/1 type Markov chain raise several questions involving the zeros of certain types of analytic functions. For example, how many zeros of the function $z^N - a(z)$, or $\det\{z^N I_M - a(z)\}$ in the matrix case, lie in the closed unit disk, and where are they located? Note that, since $a(z)$ has real power series coefficients, complex zeros of these functions (if any) appear in conjugate pairs. Can multiple roots exist, and if so how are they accounted for in these approaches? Even if enough zeros can be found, are the corresponding equations from these roots linearly independent in the M/G/1 case? Can the roots be used to characterize the positive recurrence, null recurrence and transience of the given Markov chain?

4.1 Number and Location of Zeros

Zeros in the Scalar Case. We now consider the scalar case, and return to the matrix case later in the section. The most important question involves determining the number and location of the zeros of $z^N - a(z)$, N a nonnegative integer, in the closed unit disk. Here $a(z)$ has nonnegative power series coefficients with $a(1) \leq 1$. In order to find this information about the zeros, an appeal to Rouché's Theorem is commonly made.

Rouché's Theorem (Usual Version): *Let $f(z)$ and $g(z)$ be analytic inside and on a closed contour C with $|f(z) - g(z)| < |f(z)|$ on C. Then $f(z)$ and $g(z)$ have the same number of zeros inside C.*

The following stronger version of Rouché's Theorem implies the above result (see [Berenstein and Gay, 1991, Glicksberg, 1976]).

Rouché's Theorem (Strong Version): *Let $f(z)$ and $g(z)$ be analytic inside and on a closed contour C with $|f(z) - g(z)| < |f(z)| + |g(z)|$ on C. Then $f(z)$ and $g(z)$ have the same number of zeros inside C.*

The triangle inequality $|f(z) - g(z)| \leq |f(z)| + |g(z)|$ is always true. However, strict inequality must hold at every point of C in order to apply Rouché's Theorem. Note that if strict inequality holds, then $f(z) \neq 0$ and $g(z) \neq 0$ on C.

Let us try to find the zeros of $z^N - a(z)$ in the closed unit disk using Rouché's Theorem. Set $f(z) = z^N$ and $g(z) = z^N - a(z) = A(z)$, so that $f(z) - g(z) = a(z)$. To use the Usual Version of Rouché's Theorem directly, we must show that $|a(z)| < |z^N|$ on the unit circle C. If $a(1) < 1$, then this holds, since $|a(z)| \leq a(|z|)$ because $a(z)$ has nonnegative power series coefficients. Thus, in this case we have the following.

Theorem 5 *Suppose that $a(1) < 1$. Then $A(z)$ has N zeros in the open unit disk and no zeros on the unit circle.*

This simple application of Rouché's Theorem fails when $a(1) = 1$, specifically at the point $z = 1$, for then we have $|a(z)| = 1 = |z^N|$. Similarly, the strong version also fails at this point. Thus some type of limiting argument is usually used to find the zeros.

Assume that $a(z)$ is analytic in the closed unit disk, i.e., it is analytic in an open disk containing $|z| \leq 1$. Suppose that $a(1) = 1$ and $A'(1) > 0$ (the case of a positive recurrent M/G/1 type chain), so that $a'(1) < N$. Now $g(z) = z^N - a(z)$ satisfies $g(1) = 0$ and $g'(1) > 0$, so for small $\epsilon > 0$ we have $g(1 + \epsilon) > 0$. Since $a(z)$ has nonnegative power series coefficients, we have $|a(z)| \leq a(1 + \epsilon) < (1 + \epsilon)^N = |z|^N$ for z on the circle of radius $1 + \epsilon$. By the usual version of Rouché's Theorem, $z^N - a(z)$ has N zeros in the open disk $|z| < 1 + \epsilon$, since this is true for the function z^N. Letting $\epsilon \to 0$, we see that $a(z)$ has N zeros in the closed unit disk.

Suppose that $a(1) = 1$ and $A'(1) < 0$, i.e., $a'(1) > N$ (this corresponds to the case of a positive recurrent G/M/1 type chain). Now $g(z) = z^N - a(z)$ satisfies $g(1) = 0$ and $g'(1) < 0$, so for small $\epsilon > 0$ we have $g(1 - \epsilon) > 0$. Since $a(z)$ has nonnegative power series coefficients, we have $|a(z)| \leq a(|z|) = a(1 - \epsilon) < (1 - \epsilon)^N = |z|^N$ for z on the circle of radius $1 - \epsilon$. By the usual version of Rouché's Theorem, $z^N - a(z)$ has N zeros in the open disk $|z| < 1 - \epsilon$, since this is true for the function z^N. Letting $\epsilon \to 0$, we see that $a(z)$ has N zeros in the open unit disk.

We have shown for $a(1) = 1$ that $A(z)$ has N zeros in the closed unit disk when $A'(1) > 0$, and $A(z)$ has N zeros in the open unit disk when $A'(1) < 0$. The argument in the first case required that $A(z)$ be analytic in the closed disk, instead of simply analytic in the open disk and continuous in the closure. Also, no information about zeros on the unit circle was obtained (we do know that $z = 1$ is such a zero).

Example: It was assumed in the above argument that $a(z)$ is analytic in the closed unit disk. However, this may not hold as the following example illustrates. Let $0 < a_0 < 1$ and define, for $k = 1, 2, \ldots$, $a_k = (1 - a_0)4/k(k + 1)(k + 2)$. Then $a(1) = 1$, and so the a_k give a probability mass function. Also, $a'(1) = 2(1 - a_0)$, and $A'(1) > 0$ if and only if $2(1 - a_0) < N$. This condition is satisfied when $N > 1$ or when $N = 1$ and $a_0 > 1/2$, and it corresponds to the case of a positive recurrent M/G/1 type Markov chain. If $N = 1$ and $a_0 < 1/2$ then $A'(1) < 0$, which corresponds to a positive recurrent G/M/1 type Markov chain. The function $a(z)$ has radius of convergence 1, i.e., $a(z)$ is analytic in the open unit disk, but it is not analytic in any larger disk. Thus the above argument using Rouché's Theorem is not valid, and a more delicate analysis must be done for such functions.

Example: The N zeros identified when $A'(1) > 0$ were found to be in the closed unit disk, with at least one zero (namely $z = 1$) on the unit circle. Further, when $A'(1) < 0$ N zeros were found in the open unit disk, with another zero at $z = 1$. In fact, additional zeros that satisfy $|z| = 1$ may occur. Let $N = 2$ and consider $a(z)$ with $a_0 = p$, $a_4 = 1 - p$, where $0 < p < 1$ ($p > 1/2$ corresponds to a positive recurrent M/G/1 type chain, while $p < 1/2$ corresponds to a positive recurrent G/M/1 type chain). Then $z^N - a(z) = z^2 - p - (1-p)z^4$, which has zeros on the unit circle at $z = 1$ and $z = -1$. Note that $A(z) = F(z^2)$, where we define the analytic function $F(z) = z - p - (1-p)z^2$. The zero $z = 1$ of $F(z)$ leads to the two zeros of $A(z)$ on the unit circle. Thus, it is not necessarily true that $z^N - a(z)$ has $N - 1$ zeros such that $|z| < 1$ and a single zero for which $|z| = 1$ (at $z = 1$), even though this has been stated (erroneously) many times in the literature. In general a more complete analysis is required to determine which zeros lie in the open unit disk and which zeros are located on the unit circle.

Suppose that we alter the previous example slightly by letting $N = 3$ with $a_1 = p$ and $a_5 = 1-p$. Then $A(z) = z^3 - pz - (1-p)z^5$ still has zeros $z = 1$ and $z = -1$, but it no longer can be written in the form $F(z^2)$. However, once the zero at $z = 0$ is factored out, then the remaining function does have this form. In general, recall that $A(z) = z^L A^*(z) = z^L[z^{N-L} - \sum_{j=L}^{\infty} a_j z^{j-L}]$, where L is the smallest integer such that $a_L \neq 0$. Define

$$\kappa = \max\{k : A^*(z^{1/k}) \text{ is a (single-valued) function in } |z| \leq 1\}. \tag{110}$$

Then

$$\begin{aligned}\kappa &= \max\{k : k \text{ divides } N - L \text{ and } j - L \text{ for all } a_j \neq 0\} \\ &= \max\{k : k \text{ divides } j - N \text{ for all } a_j \neq 0\}.\end{aligned}$$

Thus, we also have

$$\kappa = \max\{k : z^{-N/k} A(z^{1/k}) \text{ is a (single-valued) function in } |z| \leq 1\}. \tag{111}$$

It can be shown that the zeros of $A(z)$ on the unit circle are κth roots of unity (see [Gail et al., 1996]). Furthermore, they all have the same multiplicity as the zero at $z = 1$.

When $A'(1) \neq 0$, then $z = 1$ is a simple zero of $A(z)$. However, when $A'(1) = 0$ (the null recurrent case for both M/G/1 and G/M/1), then $A(z)$ has a multiple zero at $z = 1$. It can be shown that this zero is of order 2 when $A(z) = z^N - a(z) \not\equiv 0$. To prove this fact, consider

$$A''(1) = \frac{d^2}{dz^2}[z^N - a(z)]\bigg|_{z=1} = N(N-1) - \sum_{k=0}^{\infty} k(k-1)a_k.$$

If $A'(1) = 0$, then $\sum_{k=0}^{\infty} ka_k = N$ and so $A''(1) = N^2 - \sum_{k=0}^{\infty} k^2 a_k$. Therefore,

$$\sum_{k=0}^{\infty}(k-N)^2 a_k = \sum_{k=0}^{\infty} k^2 a_k - 2N\sum_{k=0}^{\infty} ka_k + N^2 = \sum_{k=0}^{\infty} k^2 a_k - N^2 = -A''(1),$$

which shows that $A''(1) \leq 0$. If $A''(1) = 0$ (so that $z^N - a(z)$ has a zero at $z = 1$ of order greater than 2), then $\sum_{k=0}^{\infty}(k-N)^2 a_k = 0$. This implies that $a_k = 0$ for $k \neq N$, and $a_N = 1$, i.e., $a(z) = z^N$ and $z^N - a(z) \equiv 0$. This contradiction shows that the zero at $z = 1$ is of order 2. As an example, an M/G/1 or G/M/1 type Markov chain with $a_N = 1$ cannot be irreducible, and so the above argument applies to $a(z)$ from irreducible null recurrent chains of M/G/1 or G/M/1 type. We have also shown that $A''(1) < 0$ when $A'(1) = 0$.

The following theorem collects these results for the scalar case together and is proved in [Gail et al., 1996] under minimal assumptions (i.e., $a(z)$ analytic in the open unit disk and continuous in the closed unit disk).

Theorem 6 *Suppose that $a(1) = 1$, $\sum_k k a_k < +\infty$, and $\sum_k k^2 a_k < +\infty$ when $A'(1) = 0$.*

(i) *If $A'(1) > 0$, then $A(z)$ has N zeros (counting multiplicity) in the closed unit disk, with $N - \kappa$ zeros in the open unit disk and simple zeros on the unit circle at the κth roots of unity.*

(ii) *If $A'(1) = 0$, then $A(z)$ has $N + \kappa$ zeros (counting multiplicity) in the closed unit disk, with $N - \kappa$ zeros in the open unit disk and zeros of multiplicity 2 on the unit circle at the κth roots of unity.*

(iii) *If $A'(1) < 0$, then $A(z)$ has $N + \kappa$ zeros (counting multiplicity) in the closed unit disk, with N zeros in the open unit disk and simple zeros on the unit circle at the κth roots of unity.*

Zeros in the Matrix Case. In the matrix case, the function of interest is $\det A(z) = \det\{z^N I_M - a(z)\}$. One idea that has appeared in the literature to find the number and location of the zeros of $\det A(z)$ in the closed unit disk is to try to reduce the problem to the scalar case. That is, fix z and let the M eigenvalues of $a(z)$ be denoted as $\lambda_i(z)$, $i = 0, \ldots, M-1$. Then $\det A(z) = \prod_{i=0}^{M-1}[z^N - \lambda_i(z)]$, and we simply apply Rouché's Theorem to each factor. Of course, there is once again a problem with the point $z = 1$. However, the main reason for not using this approach is that it is mathematically incorrect. The functions $\lambda_i(z)$ are not necessarily analytic functions, and in fact they are not even continuous functions in general (see [Gail et al., 1994]). Thus Rouché's Theorem cannot be applied. We now provide a simple example to illustrate this behavior.

Example: Let $M = 2$ and N be arbitrary. Define $a(z) = e^{\mu(z-1)}\widetilde{a}(z)$, where $\mu > 0$ and $\widetilde{a}(z) = \begin{bmatrix} p & 1-p \\ z & 0 \end{bmatrix}$ with $0 < p < 1$. Then $\gamma = N(2-p) - (1-p) - \mu(2-p)$. Therefore, if $\mu < N - (1-p)/(2-p)$, then $a(z)$ corresponds to an irreducible positive recurrent M/G/1 type Markov chain, while if $\mu > N - (1-p)/(2-p)$, then $a(z)$ corresponds to an irreducible positive recurrent G/M/1 type Markov chain. The eigenvalues of $\widetilde{a}(z)$ are zeros of the function $\det\{\lambda I_2 - \widetilde{a}(z)\}$, namely,

$$\widetilde{\lambda}_0(z) = [p + \sqrt{p^2 + 4(1-p)z}]/2, \qquad \widetilde{\lambda}_1(z) = [p - \sqrt{p^2 + 4(1-p)z}]/2.$$

Note that $p^2 + 4(1-p)z = 0$ at the point $z^* = -p^2/4(1-p)$, and this branch point satisfies $|z^*| < 1$. Thus there is no way to define $\widetilde{\lambda}_i(z)$ as analytic functions (or even as continuous functions) in the open unit disk, and so this is also true for the eigenvalues of $a(z)$, namely, $\lambda_i(z) = e^{\mu(z-1)}\widetilde{\lambda}_i(z)$.

In the above example the eigenvalues $\lambda_0(z^*)$ and $\lambda_1(z^*)$ are equal. If the eigenvalues corresponding to a particular $a(z)$ happen to be distinct for all $|z| \le 1$, then the eigenvalue functions $\lambda_i(z)$ are analytic functions in the unit disk (see [Andrew et al., 1993]), and the procedure using Rouché's Theorem can be carried out (of course, there is still a problem at the point $z = 1$). However, it is usually difficult to verify this uniqueness condition, and as we have seen it does not hold in general. Arguments have appeared in the literature that advocate simply perturbing the input parameters in case multiple eigenvalues are encountered to obtain analytic eigenvalue functions. Note that there is no way to perturb the input parameters in the above example to yield analytic $\lambda_i(z)$. If a multiple eigenvalue occurs at a point z_1 in the unit disk, perturbing the parameters will only cause the multiple eigenvalue to appear at a different point z_2. Thus, in general, these perturbation arguments are also invalid.

A different approach that does not involve factoring a determinant as the product of eigenvalues is necessary for finding the number and location of the zeros of $\det A(z)$. Instead of using Rouché's Theorem, a proof based on the Argument Principle can be developed for the matrix case, and we state two versions of this principle below (for a different example of its use in queueing analysis see page 234 of [Coffman et al., 1983]).

Argument Principle: *Let $f(z)$ be analytic inside and on a closed contour C, and suppose that $f(z)$ has no zeros on C. Then*

$$\frac{1}{2\pi i} \int_C \frac{f'(z)}{f(z)} dz = \mathcal{Z}_f, \qquad (112)$$

where $\mathcal{Z}_f$ is the number of zeros of $f(z)$ inside C, counted according to multiplicity.

Interpreting the left hand side of equation (112) in terms of winding numbers yields the following equivalent statement (which also explains the name attached to the theorem).

Argument Principle: *Let $f(z)$ be analytic inside and on a closed contour C, and suppose that $f(z)$ has no zeros on C. Then the number of zeros of $f(z)$ inside C, counted according to multiplicity, is equal to $1/2\pi$ times the net change in the argument of $f(z)$ as z traverses C in the positive direction.*

A consequence of this result, which we now state in the form of a lemma, is useful in locating singularities for various types of analytic functions that arise in the analysis of Markov chains and queueing systems. In fact, we will apply it to the function $\det A(z)$ that appears in the analysis of M/G/1 and G/M/1 type chains.

Lemma 1 *Let $F(z,t)$ be a function analytic for z within and on a closed contour C, and continuous for t in some interval $\mathcal{J}$. If $F(z,t) \ne 0$ for $z \in C$ and $t \in \mathcal{J}$, then the number of zeros of $F(z,t)$ inside C is the same for all $t \in \mathcal{J}$.*

Proof: Since $F(z,t)$ is nonzero on C for all $t \in \mathcal{J}$, the number of times the image curve winds around the origin as z traverses this contour is the same for all $t \in \mathcal{J}$. That is, the winding number and hence the number of zeros remains the same for such t. Formally, let $\mathcal{Z}(t)$ be the number of zeros of $F(z,t)$ inside $\mathcal{C}$. By hypothesis, $F(z,t) \neq 0$ on $\mathcal{C}$ for all $t \in \mathcal{J}$. By the Argument Principle

$$\mathcal{Z}(t) = \frac{1}{2\pi i} \int_\mathcal{C} \frac{\frac{\partial}{\partial z} F(z,t)}{F(z,t)} dz,$$

and so it is a continuous integer valued function of t. Thus $\mathcal{Z}(t)$ is constant.

The idea in applying this lemma is to compare the zeros of a known function with those of the function of interest, similar to the application of Rouché's Theorem in the scalar case. A function $F(z,0)$ is identified for which the number of zeros is known, and a target function $F(z,1)$ is also defined. Then one function is deformed to the other through t in a way such that (hopefully) the image never hits the origin (i.e., the number of zeros remains constant). This lemma may be applied to $\det F(z,t)$, with $F(z,t)$ a matrix function, when $F(z,t)$ is nonsingular on the contour $\mathcal{C}$.

Corollary 1 *Let $F(z,t)$ be a matrix-valued function analytic for z within and on a closed contour $\mathcal{C}$, and continuous for t in some interval $\mathcal{J}$. If $F(z,t)$ is nonsingular for $z \in \mathcal{C}$ and $t \in \mathcal{J}$, then the number of zeros of $\det F(z,t)$ inside $\mathcal{C}$ is the same for all $t \in \mathcal{J}$.*

In the case of stochastic matrices (and infinitesimal generators), the nonsingularity can frequently be verified by using the notion of diagonal dominance when $\mathcal{C}$ is the unit circle. Recall that a square matrix $S = [s_{ij}]$ is (row) diagonally dominant if $|s_{ii}| \geq \sum_{j \neq i} |s_{ij}|$ for every row i. Note that if S is stochastic (or more generally substochastic), then $I - S$ is diagonally dominant. A matrix is called strictly diagonally dominant if the above inequalities are all strict. The relevance of diagonal dominance to the problem of verifying the nonsingularity of $F(z,t)$ on $|z| = 1$ is a consequence of the Levy-Desplanques Theorem [Horn and Johnson, 1985], which states that a strictly diagonally dominant matrix is nonsingular. A matrix is called irreducibly diagonally dominant if it is irreducible, diagonally dominant, and at least one of the above row inequalities is strict. Another well-known result states that an irreducibly diagonally dominant matrix is nonsingular (see [Horn and Johnson, 1985]). Thus we have the following.

Corollary 2 *Let $F(z,t)$ be a matrix function analytic in z in the closed unit disk, continuous in t (for t in some interval $\mathcal{J}$). If $F(z,t)$ is strictly diagonally dominant (or irreducibly diagonally dominant) for $|z| = 1$ and all t in $\mathcal{J}$, then the number of zeros of $\det F(z,t)$ in the open unit disk is the same for such t.*

Lemma 1 (and Corollary 2) may be used to determine the singularities of a matrix analytic function $F(z)$ using the following approach. Suppose that a decomposition of the form $F(z) = F_1(z) + F_2(z)$ is made, and define $F(z,t) = F_1(z) + tF_2(z)$ where t is a real parameter, $0 \leq t \leq 1$. Then one attempts to apply the above result to show that $F(z) = F(z,1)$ has the same (known) number of zeros as $F_1(z) = F(z,0)$. A plausible way of applying Lemma 1 (and Corollary 2), which was used in [Gail et al., 1992], is to decompose the matrix of interest $F(z)$ into diagonal and offdiagonal elements. That is, write $F(z) = D(z) + O(z)$, where $D(z)$ consists of the diagonal elements of $F(z)$ and $O(z)$ consists of the off-diagonal elements, and consider the family of functions $F(z,t) = D(z) + tO(z)$ for $0 \le t \le 1$.

In the case of the matrix $A(z) = z^N I_M - a(z)$, a more natural decomposition is readily apparent. Define the functions

$$A(z,t) = z^N I_M - ta(z),$$

where the real parameter t is in the closed interval $[0,1]$. We claim that $\det A(z,t)$ has MN zeros in $|z| < 1$ and none on $|z| = 1$ for $0 \le t < 1$. The claim clearly holds for $A(z,0) = z^N I_M$ (the determinant is simply z^{MN}), so we only need to show that $A(z,t)$ is nonsingular for $0 \le t < 1$, $|z| = 1$, and apply Lemma 1.

Let $\rho(z,t)$ be the spectral radius of $ta(z)$, and note that $\rho(z,1) = \rho(z)$, the spectral radius of $a(z)$. Since $a(1)$ is a substochastic matrix and the entries of $a(z)$ are power series with nonnegative coefficients, then $\rho(z,t) \le \rho(|z|,t) \le \rho(1,t) = t\rho(1)$ for $|z| \le 1$ and $0 \le t \le 1$. When $a(1)$ is irreducible substochastic but not stochastic, then $\rho(1) < 1$, and $A(z,t) = z^N[I_M - tz^{-N}a(z)]$ is nonsingular for z on the unit circle and $0 \le t \le 1$ (see [Horn and Johnson, 1985] page 301). When $a(1)$ is stochastic, then $\rho(1) = 1$, and $A(z,t)$ is nonsingular for $|z| = 1$ and $0 \le t < 1$. Thus, in all these cases, $\det A(z,t)$ has MN zeros in the open unit disk and no zeros on the unit circle. For scalar $a(z)$, this is an instance of a general result of Takács [Takács, 1962] (see also [Cohen and Down, 1996]).

When $t = 1$ and $a(1)$ is stochastic, the matrix function $A(z,1) = A(z)$ is clearly singular at $z = 1$, since it has zero row sums. The point $z = 1$ is a simple zero of $\det A(z)$ when $\gamma \ne 0$. Suppose first that $\gamma > 0$. Then $\det A(1-\epsilon) < 0$ for small $\epsilon > 0$. By continuity, there is small $\tau > 0$ so that $\det A(1-\epsilon, 1-\tau)$ is also negative. However, $\det A(1,0) = 1$ and $\det A(1,t) \ne 0$ for $0 \le t < 1$ as shown above. By continuity, $\det A(1,t) > 0$ for $0 \le t < 1$, so in particular, $\det A(1,1-\tau)$ is positive. Therefore, $\det A(1-\epsilon_1, 1-\tau) = 0$ for some $0 < \epsilon_1 < \epsilon$. The same argument holds for $\tau \to 0$, so the simple zero at $z = 1$ is the limit of zeros from inside the unit disk. Since this can also be shown for any additional zeros on the unit circle, it follows that $\det A(z)$ has MN zeros in the closed unit disk.

When $\gamma < 0$, we assume that $a(z)$ is analytic in the closed unit disk, i.e., it is analytic in an open disk containing the unit circle. Since $\gamma < 0$, then $\det A(1+\epsilon) < 0$ for small $\epsilon > 0$. By continuity, $\det A(1+\epsilon, 1-\tau)$ is also negative for small $\tau > 0$. But $\det A(1, 1-\tau) > 0$ as shown in the previous case, so there is $0 < \epsilon_1 < \epsilon$ such that $\det A(1+\epsilon_1, 1-\tau) = 0$. Thus the simple zero at $z = 1$ is the limit of zeros from outside the unit disk. In this case $\det A(z)$ has MN zeros in the open unit disk.

We now consider the case $\gamma = 0$. Define

$$\delta = \frac{d^2}{dz^2} \det A(z) \bigg|_{z=1}. \tag{113}$$

When $\gamma = 0$, it can be shown that $\delta < 0$ (see [Gail et al., 1996] for details). Therefore, $\det A(z)$ has a zero of order 2 at $z = 1$, similar to the null recurrent scalar case.

The function $\det A(z)$ may have zeros in addition to $z = 1$ on the unit circle. For $a(1)$ irreducible stochastic, the zeros of $\det A(z)$ with $|z| = 1$ are κth roots of unity, where

$$\kappa = \max\{k : z^{-MN/k} \det A(z^{1/k}) \text{ is a (single-valued) function in } |z| \leq 1\}. \tag{114}$$

Note that (114) reduces to (111) in the scalar case.

If $a(z)$ is analytic in the open unit disk but not in a larger disk, then the above arguments must be modified. However, the following theorem has been proved under a set of minimal assumptions for irreducible $a(1)$ (see [Gail et al., 1996]). Specifically, it is assumed that $a(z) = \sum_k a_k z^k$ with $a_k \geq 0$, and $a(z)$ is analytic in the open unit disk and continuous in the closed unit disk. The reducible case follows by considering the irreducible subblocks from the normal form of $a(1)$.

Theorem 7 *Suppose $a(1)$ is irreducible substochastic but not stochastic.*

(i) *Then $\det A(z)$ has MN zeros in the open unit disk and no zeros on the unit circle.*

Suppose $a(1)$ is irreducible stochastic, $\sum_k k a_k < +\infty$, and $\sum_k k^2 a_k < +\infty$ when $\gamma = 0$.

(ii) *If $\gamma > 0$, then $\det A(z)$ has MN zeros (counting multiplicity) in the closed unit disk, with $MN - \kappa$ zeros in the open unit disk, and simple zeros on the unit circle at the κth roots of unity.*

(iii) *If $\gamma = 0$, then $\det A(z)$ has $MN + \kappa$ zeros (counting multiplicity) in the closed unit disk, with $MN - \kappa$ zeros in the open unit disk, and zeros of multiplicity 2 on the unit circle at the κth roots of unity.*

(iv) *If $\gamma < 0$, then $\det A(z)$ has $MN + \kappa$ zeros (counting multiplicity) in the closed unit disk, with MN zeros in the open unit disk, and simple zeros on the unit circle at the κth roots of unity.*

4.2 Zeros Outside the Unit Circle

The formula (35) shows that the generating function $\pi(z) = \sum_{i=0}^{N-1} \pi_i z^i + \psi(z)$ for the stationary distribution of a positive recurrent M/G/1 type Markov chain is meromorphic (i.e., it is analytic except for poles) in the exterior of the closed unit disk whenever $a(z)$ and $b(z)$ are meromorphic in the complex plane. In particular, (35) shows that the poles of $\pi(z)$ outside of the unit disk are due to zeros of det $A(z)$ in $|z| > 1$ (or poles of $B(z)$ there). Thus, the rate that the stationary probabilities $\{\pi_j\}$ tend to 0 is (usually) determined by the smallest modulus of a zero of det $A(z)$ in the exterior of the unit disk.

In this section, we point out that the zero of det $A(z)$ of smallest modulus (if any) in $|z| > 1$ is necessarily real and simple. Therefore, when $z = 1$ is the only zero of det $A(z)$ on the unit circle, as is normally the case, $\pi(z)$ has the form

$$\pi(z) = \frac{\zeta}{\sigma - z} + f(z).$$

Here $\sigma > 1$ is the zero of det $A(z)$ of smallest modulus in the exterior of the unit disk, ζ is the residue of $\pi(z)$ at this point, and $f(z)$ is a function analytic in a neighborhood of $\{|z| \leq \sigma\}$. Thus, the principal asymptotic behavior of the coefficients $\{\pi_j\}$ is

$$\pi_j = \frac{\zeta}{\sigma^{j+1}} + o(\sigma^{-j}).$$

In case $a(z)$, $b(z)$ do not extend to be meromorphic in the exterior of the unit disk, some technical modifications of these remarks apply (see [Gail et al., 1998]). The fact that σ is real and simple is also of use, for example, in obtaining results concerning the convergence properties of algorithms for calculating the stationary probabilities (see [Bini and Meini, 1998]).

This analysis, as well as some other interesting properties of the chain, is made in terms of the spectral radius $\rho(z)$ of the $M \times M$ matrix function $a(z)$. The spectral radius is defined for $|z| < r_a$, the radius of convergence of $a(z)$, since $a(z)$ converges there (in some cases we also consider $|z| = r_a$). When $\rho(z)$ is restricted to the nonnegative real axis, we use the notation $\rho(r)$. The matrix $a(z)$ has M eigenvalues, which in general may be complex-valued. However, $\rho(z)$, the largest modulus of these eigenvalues, is nonnegative. For a particular z, there may be several eigenvalues that have modulus $\rho(z)$.

Since the eigenvalues of a matrix depend continuously on the entries of the matrix (see [Horn and Johnson, 1985] pages 539–540), it follows that $\rho(z)$ is a continuous function. This is in contrast to the eigenvalues of $a(z)$ themselves, for we have seen that there is no general way to select single-valued choices of the eigenvalues that are continuous functions in the unit disk, let alone analytic functions. However, when $a(1)$ is irreducible, by restricting our attention to the positive real axis, we find that $\rho(r)$ is, in fact, a real analytic function. In addition, it has a convexity property which we shall describe later in the section.

Since $a(z)$ has nonnegative power series coefficients, the nonzero entries of $a(r)$ for $r > 0$ are exactly the nonzero entries of $a(1)$. It follows that $a(r)$ is irreducible for $r > 0$, and so the spectral radius $\rho(r)$ is a simple eigenvalue of $a(r)$ (see page 508 of [Horn and Johnson, 1985]). By Theorem 2.1 of [Andrew et al., 1993], a neighborhood and an analytic eigenvalue function defined in that neighborhood that agrees with the simple eigenvalue at r can be found. Further, analytic right and left eigenvectors corresponding to this analytic eigenvalue can be defined in this neighborhood.

Lemma 2 Let $a(z) = \sum_{k=0}^{\infty} a_k z^k$, with $a_k \geq 0$, have radius of convergence r_a.

(i) Then $\rho(z)$ is a continuous function of z for $|z| < r_a$.

(ii) If $a(1)$ is irreducible, then $\rho(r)$ is a real analytic function for $0 < r < r_a$. Corresponding right and left eigenvectors $X(r)$ and $Y(r)$ can be chosen as real analytic functions for $0 < r < r_a$.

Proposition 7.5 Suppose that $a(1)$ is irreducible and stochastic, and that $\rho'(1) < N$ (i.e., $\gamma > 0$). If $\det A(z)$ has a zero outside the closed unit disk, then there is a smallest such zero σ in modulus and it is real, simple and positive. Furthermore, σ^N is the spectral radius of $a(\sigma)$.

Proof: Suppose that $|\xi| > 1$ is a zero of $\det A(z)$. Then $a(\xi)$ converges, and there is a vector $X \neq 0$ such that $a(\xi)X = \xi^N X$. It follows that $|\xi| \leq r_a$ and $|\xi|^N \leq \rho(\xi)$. We claim there is $1 < \hat{r} \leq |\xi|$ such that $a(\hat{r})$ converges and $\hat{r}^N \leq \rho(\hat{r})$. If $a(|\xi|)$ converges, then we choose $\hat{r} = |\xi|$. If $a(|\xi|)$ diverges, then $r_a = |\xi|$ and some entry $a_{i,j}(r)$ is unbounded as $r \uparrow r_a$. Let $f(r) \triangleq a(r)[I_M + a(r)]^{M-1} \geq a(r)[I_M + a(1)]^{M-1}$ for $1 \leq r < r_a$. Now $[I_M + a(1)]^{M-1}$ is a positive matrix, because $a(1)$ is irreducible. Therefore, all entries in the ith row of $f(r)$ are unbounded, and so each column of $f(r)$ has an unbounded entry. Thus all column sums of $f(r)$ are unbounded, and it follows that $\rho(f(r)) \uparrow +\infty$ as $r \uparrow r_a$ (see page 492 of [Horn and Johnson, 1985]). But the spectral radius of $f(r)$ is $\rho(r)[1 + \rho(r)]^{M-1}$, and so $\rho(r) \uparrow +\infty$ as $r \uparrow r_a$. Since $|\xi| = r_a$ in this case, the claim again holds. The derivative of $\rho(r) - r^N$ is negative at $r = 1$ by assumption, so there is $\epsilon > 0$ such that $\rho(1 + \epsilon) < (1 + \epsilon)^N$. By continuity, there is $1 + \epsilon < \sigma \leq \hat{r}$ which satisfies $\rho(\sigma) = \sigma^N$. In particular, $1 < \sigma \leq |\xi|$, which shows the result.

However, $\det A(z)$ may have no zeros outside the unit circle. As an example, this clearly is true if $a(z)$ has radius of convergence 1. Furthermore, given a constant $C \geq 1$, an $a(z)$ is constructed in [Gail et al., 1998] with radius of convergence C for which $\det A(z)$ has no zeros in $|z| > 1$. The zero σ studied in the previous proposition was identified as a root of $\rho(r) = r^N$. We will show that the number of zeros of $\det A(z)$ that satisfy this equation in $(0, r_a)$ is at most 2. This result follows, because the function $\rho(r)$ is not only real analytic, it has an interesting convexity property (see [Gail et al., 1998, Kingman, 1961]).

Proposition 7.6 Let $a(z) = \sum_{k=0}^{\infty} a_k z^k$, where $a_k \geq 0$, have radius of convergence r_a. For $0 < r < r_a$, $\log \rho(r)$ is a convex increasing function of $\log r$.

When characterizing M/G/1 and G/M/1 type Markov chains, we have seen that the derivatives of $\rho(r)$ at $r = 1$ are related to the derivatives of $\det A(z)$ at $z = 1$. This relationship is described in the next result (recall that η is the stationary distribution of $a(1)$, so that $\eta a(1) = \eta$ and the entries of η sum to 1).

Proposition 7.7 Suppose that $a(1)$ is irreducible stochastic, and $\sum_k k a_k < +\infty$.

(i) Then $\rho'(1) = \eta a'(1)\mathbf{1} = N - \gamma/c$, where c is a positive scalar.
Suppose further that $\sum_k k^2 a_k < +\infty$.
(ii) If $\rho'(1) = N$, then $\rho''(1) = N(N-1) - \delta/c > 0$.

Propositions 7.6 and 7.7 imply that $\rho(r) = r^N$ for at most two points in $(0, r_a)$. For, $\rho(r) = r^N$ if and only if $\log \rho(r) = N \log r$. The latter can happen for at most two points by convexity, unless $\rho(r) = r^N$ for all r in some subinterval. However, this implies that the real analytic function $\rho(r) - r^N$ is identically zero on $(0, r_a)$. This is impossible when $a(1)$ is not stochastic, since then $\rho(1) < 1$. When $a(1)$ is stochastic and $\gamma \neq 0$ (the positive recurrent and transient cases for M/G/1 and G/M/1), then the derivative at $r = 1$ satisfies $\rho'(1) - N = -\gamma/c \neq 0$. When $a(1)$ is stochastic and $\gamma = 0$ (the null recurrent case for M/G/1 and G/M/1), then the second derivative at $r = 1$ satisfies $\rho''(1) - N(N-1) = -\delta/c \neq 0$. Thus, in any case, $\rho(r) - r^N$ cannot be identically zero. However, the equation $\rho(r) = r^N$ may have three solutions (with an additional solution at 0) as the following example illustrates.

Example: Let $M = 2$, $N = 1$, and

$$a_0 = \begin{bmatrix} 0 & p \\ 0 & 0 \end{bmatrix}, \quad a_1 = \begin{bmatrix} 1-p-q & 0 \\ 1 & 0 \end{bmatrix}, \quad a_2 = \begin{bmatrix} 0 & q \\ 0 & 0 \end{bmatrix},$$

where $0 < p < 1$, $0 < q < 1$, $0 < p+q < 1$. Then $a(1)$ is irreducible. Further, $\rho(r) = r$ at the three points $r = 0$, $r = 1$, and $r = r^* = p/q$.

We close this section by mentioning a result involving roots outside the unit disk that holds for rational functions (i.e., for matrix-valued functions with entries that are ratios of polynomials). It was observed in [Akar and Sohraby, 1997b] that if $b_i(z)$, $i = 0, \ldots, N-1$, and $a(z)$ are rational functions corresponding to an irreducible positive recurrent M/G/1 type Markov chain, then the stationary distribution satisfies the "almost geometric form" $\pi_{k+N} = gF^k H$, where g is $1 \times S$, F is $S \times S$ and H is $S \times M$ for some positive integer S. This form holds for any rational function [Ball et al., 1990], and the following explanation was given in [H. R. Gail and Taylor, 1998]. Note that, in this rational case, the formula (35) shows that the generating function $\pi(z) = \sum_{i=0}^{N-1} \pi_i z^i + \psi(z)$ is also rational. Now any rational function can be expanded in partial fractions in terms of its poles. The generating function $\pi(z)$ can only have poles in the exterior of the closed unit disk, and these are (potentially) the zeros of $\det A(z)$ and the poles of $B_i(z)$ in $|z| > 1$. This partial fractions expansion is not advantageous from a numerical point of view. In fact, the minimal value of S is given in terms of the number of poles of $\pi(z)$, counting multiplicity. Simple scalar examples are presented in [H. R. Gail and Taylor, 1998] for which S is arbitrarily large.

4.3 Multiple Zeros and Simple Zeros

The M/G/1 and G/M/1 analysis proceeds without technical complications when only simple zeros of $\det A(z)$ exist in the closed unit disk. The equations for the boundary probabilities are much easier to derive and analyze. For multiple roots, Jordan chains must be considered instead of eigenvectors when determining the equations for the boundary probabilities (see [Gail et al., 1996] for details). But do multiple root cases exist? Are they easy to find and do they represent realistic systems? When are simple roots guaranteed to occur?

In the null recurrent case, we have seen that multiple zeros of det $A(z)$ exist on the unit circle. In fact, they also exist in the open unit disk for both positive recurrent M/G/1 and G/M/1 cases. For example, in a recent paper [Grassmann, 1998], Grassmann has presented a systematic way of producing cases with multiple zeros. Although he considers scalar continuous-time chains, his method applies equally well to the discrete-time case.

Dukhovny [Dukhovny, 1994, Dukhovny, 1998] has studied the question of multiple roots in the scalar case, and we now review his results. As one example, given an integer $n \geq 2$ and $0 < \xi < 1/n$, he constructs a scalar $a(z)$ with nonnegative power series coefficients which has a zero at $-\xi$ of multiplicity n as follows. Choose $\sigma \geq n\xi/(1-n\xi)$ and define

$$a(z) = z^{n+1} + \frac{1}{1+\sigma}(z+\xi)^n(z-\sigma)(z-1).$$

Then $a(z)$ is a polynomial of degree $n+2$ with coefficients $0 < a_k < 1$, $k = 0, \ldots, n+2$, such that $a(1) = 1$ (i.e., $a(z)$ is a generating function of a nonnegative random variable). Setting $N = n+1$, the function $A(z) = z^N - a(z)$ has a zero at $-\xi$ of multiplicity n. Note that, for $a(z)$ constructed in this way, $A'(1) > 0$, $A'(1) = 0$, $A'(1) < 0$ if and only if $\sigma > 1$, $\sigma = 1$, $\sigma < 1$, respectively. As a specific example, take $n = 2$, $\xi = 1/3$ and $\sigma = 2$ to yield

$$a(z) = \frac{1}{27}(9z^4 + 6z^3 + z^2 + 9z + 2).$$

Then (with $N = n+1 = 3$) it follows that $-1/3$ is a zero of $A(z) = z^N - a(z)$ of multiplicity 2. In this example, $A'(1) > 0$, and so $a(z)$ corresponds to a positive recurrent M/G/1 type Markov chain. In fact, $a(z)$ can be interpreted in terms of an M/M/1 queueing system with bulk service (bulk size at most 3) [Dukhovny, 1998]. The same construction yields $a(z)$ for positive recurrent G/M/1 chains for any integer $n \geq 2$ and any $0 < \xi < 1/2n$, since then the choice $n\xi/(1-n\xi) \leq \sigma < 1$ ensures that $A'(1) < 0$. Examples with multiple complex zeros of any multiplicity can also be constructed (see [Dukhovny, 1994, Grassmann, 1998]).

Cases for which simple zeros of $A(z)$ (scalar case $M = 1$) in the closed unit disk are guaranteed are also explored in [Dukhovny, 1994]. Suppose that $a(z) = g(z)^n$ where $g(z)$ is a generating function and $n \geq N$, the number of boundary states. Also suppose that $a(z)$ corresponds to a positive recurrent M/G/1 type chain, so that $a'(1) < N$. Then, assuming that $A(z)$ has a multiple zero $0 < |\xi| \leq 1$, we have $\xi^N = g(\xi)^n$ and (after some manipulations) $Ng(\xi)/\xi = ng'(\xi)$. Now

$$|ng'(\xi)| \leq ng'(1) = a'(1) < N,$$

and so $|g(\xi)/\xi| < 1$. Therefore, since $n \geq N$,

$$1 = |g(\xi)^n/\xi^n||\xi^{n-N}| < |\xi^{n-N}| \leq 1,$$

a contradiction. Here $a(z)$ is the generating function of a sum of $n \geq N$ independent and identically distributed random variables, each with generating function $g(z)$.

Another (scalar) case for which $A(z)$ has simple zeros in the closed unit disk occurs when $a(z)$ corresponds to an infinitely divisible distribution with $a'(1) < N$. That is, for every n there is a generating function $g_n(z)$ such that $a(z) = g_n(z)^n$, and so the result follows immediately from the above argument. An example of such a distribution is the Poisson, for which $a(z) = e^{\lambda(z-1)} = [e^{(\lambda/n)(z-1)}]^n$ for any n where λ is a given parameter. For other examples and a discussion of infinitely divisible distributions see [Feller, 1968, Feller, 1971].

Dukhovny mentions that simple zeros for $A(z) = z^N - a(z)$ are also guaranteed when $a(z) = \alpha(\lambda - \lambda g(z))$ with $a'(1) < N$, where $\alpha(s)$ is the Laplace-Stieltjes transform of a nonnegative continuous random variable, $\lambda > 0$, $g(z)$ is the generating function of a nonnegative integer random variable, and $|\alpha'(s)| \leq -\alpha'(0)|\alpha(s)|$ for $\text{Re}(s) \geq 0$. An example is presented in the analysis of Bailey's bulk queue [Bailey, 1954], for which $\alpha(s) = [\mu/(s+\mu)]^p$ and $g(z) = z$.

In the matrix case, simple roots for $\det A(z)$ can also be guaranteed to occur in some cases. One example, analyzed by Daigle and Lucantoni, was of a single server continuous-time queue with arrival and service rates that depend on a finite state birth-death process (with, say, M states) [Daigle and Lucantoni, 1991]. For this system, $a(z)$ is quadratic with a tridiagonal form for which $a_{j,j}(z) = \mu_j + \lambda_j z^2$, $a_{j,j-1}(z) = \alpha_j z$, and $a_{j,j+1}(z) = \beta_j z$. Here λ_j and μ_j represent the arrival and service rates when the birth-death process is in state j, while β_j and α_j represent the birth and death rates in state j (the rate of a transition to state $j+1$ and state $j-1$, respectively). The authors found that all zeros of $\det A(z)$ were simple and, in addition, real. These properties led them to develop efficient and stable methods for calculating the zeros. This particular example is a representative of a class of problems for which $\det A(z)$ has real and simple zeros in the closed unit disk. Other examples include queues with blocking [Konheim and Reiser, 1976, Konheim and Reiser, 1978], a multiserver queue [Mitrany and Avi-Itzhak, 1968], and a priority queue example [Gail et al., 1992]. In these cases the determinant of $A(z)$ can be expressed in terms of Sturm sequences of functions (see [Franklin, 1968] for more on Sturm sequences).

Multiple zeros of $\det A(z)$ also occur in the matrix case (see [Wuyts and Boel, 1998] for an example with multiple zeros at the origin even though $a_0 \neq 0$). For simple zeros, recall that right eigenvectors corresponding to $a(z)$ for M/G/1 and left eigenvectors for G/M/1 are required when finding equations for the boundary probabilities. For multiple zeros, this eigenvector case can again occur, i.e., if ξ is a zero of $\det A(z)$ of multiplicity k, then it is possible that $a(\xi)$ has k distinct eigenvectors corresponding to the eigenvalue ξ^N. However, cases with Jordan chains can also be encountered (recall the definition (45) of a Jordan chain).

4.4 Linear Independence of Root Equations

In the M/G/1 type Markov chain case, the classical transform method involves obtaining equations from zeros of the function $\det A(z)$ in the closed unit disk. It is shown in [Gail et al., 1996] that a zero of multiplicity k yields k independent equations. Another question of interest is whether these so-called root equations are linearly independent for all the different zeros. Even though there are enough zeros to generate the required number of equations, if the resulting set is not linearly independent, then the stationary probabilities cannot be determined from them. The first to consider this question was Bailey [Bailey, 1954], who studied a model now known as Bailey's bulk queue. He showed that the equations obtained from roots for this system corresponded to a Vandermonde matrix, which he then used to show independence for this particular scalar problem. In the analysis of Bailey's bulk queue, it was shown that simple zeros resulted, and the fact that $(N = 1)$ $b_k = a_k$ for all k was also exploited. In general, at least for the case of simple roots, we will outline the approach of [Gail et al., 1995] that can be followed to prove independence, even for matrix M/G/1 type chains. In the case of multiple zeros, complications arise involving possible occurrences of Jordan chains and the use of derivatives of root vectors when determining the root equations. But even in the multiple zero case, a dimension argument has been given to yield independence (see [Gail et al., 1996]).

For a positive recurrent M/G/1 type Markov chain, recall that $\gamma > 0$, and $\det A(z)$ has MN zeros in the closed unit disk, including a simple zero at $z = 1$. When $M = 1$, $N = 1$, the only zero satisfying $|z| \leq 1$ is $z = 1$, and as a result only the conservation of probability equation is obtained (see the analysis of (4)). Thus we assume that $MN > 1$ to avoid trivialities. We now sketch a proof of linear independence when $\det A(z)$ has all simple zeros in the closed unit disk, and $z = 1$ is the only zero on the unit circle. Let us denote these $MN - 1$ other zeros, which must lie in $|z| < 1$, by $\xi_1, \ldots, \xi_{MN-1}$. From (44), the equations from these simple zeros may be written in matrix form as

$$[\pi_0, \ldots, \pi_{N-1}]
\begin{bmatrix}
B_0(\xi_1)U(\xi_1) & \cdots & B_0(\xi_{MN-1})U(\xi_{MN-1}) & D_0(1)\mathbf{1}_M - \gamma\mathbf{1}_M \\
\vdots & \ddots & \vdots & \vdots \\
B_{N-1}(\xi_1)U(\xi_1) & \cdots & B_{N-1}(\xi_{MN-1})U(\xi_{MN-1}) & D_{N-1}(1)\mathbf{1}_M - \gamma\mathbf{1}_M
\end{bmatrix} = 0$$
(115)

We claim that the first $MN - 1$ columns of the above matrix are linearly independent.

Suppose there are (complex) scalars $\alpha_1, \ldots, \alpha_{MN-1}$ such that

$$\sum_{l=1}^{MN-1} \alpha_l \begin{bmatrix} B_0(\xi_l)U(\xi_l) \\ \vdots \\ B_{N-1}(\xi_l)U(\xi_l) \end{bmatrix} = 0.$$
(116)

We wish to show that $\alpha_l = 0$ for all l. From (44), we see that $\mathcal{P}X = X$, where

$$X = \sum_{l=1}^{MN-1} \alpha_l e(\xi_l) U(\xi_l). \tag{117}$$

Thus X is regular [Kemeny et al., 1966] (i.e., X is harmonic). Since all $|\xi_l| < 1$, the entries of X converge to 0, because $\xi_l^k \to 0$ as $k \to \infty$. Therefore, as shown in Section 2, these entries all have the same modulus. Necessarily, this common value is 0, and so $X = 0$.

From the definition (117) of X, concentrating on the first $MN - 1$ blocks (recall that $MN > 1$ by assumption), we have

$$\begin{bmatrix} I_M & \cdots & I_M \\ \vdots & \ddots & \vdots \\ \xi_1^{MN-2} I_M & \cdots & \xi_{MN-1}^{MN-2} I_M \end{bmatrix} \begin{bmatrix} \alpha_1 U(\xi_1) \\ \vdots \\ \alpha_{MN-1} U(\xi_{MN-1}) \end{bmatrix} = 0. \tag{118}$$

The $(MN-1) \times (MN-1)$ matrix in the above system is a block version of a Vandermonde matrix. Since all ξ_l are distinct, this matrix is invertible, and so $\alpha_l U(\xi_l) = 0$, $l = 1, \ldots, MN - 1$. Since each vector $U(\xi_l) \neq 0$, it follows that all scalars $\alpha_l = 0$. Thus the equations obtained from (simple) zeros in $|z| < 1$ are linearly independent.

We next wish to show that the final equation (47) from the conservation of probability is independent of the other root equations. In fact, we claim that the last column in the matrix of (114) cannot lie in the span of the first $MN - 1$ (linearly independent) columns. Suppose that the last column is a linear combination of the first $MN - 1$ columns, that is,

$$D_i(1)\mathbf{1}_M - \gamma \mathbf{1}_M = \sum_{l=1}^{MN-1} \beta_l B_i(\xi_l) U(\xi_l), \quad i = 0, \ldots, N-1, \tag{119}$$

where the β_l are (complex) scalars. Differentiating (38) and evaluating the result at $z = 1$ yields

$$[\mathcal{I} - \mathcal{P}] \frac{d}{dz} \{e(z) \, \text{adj}\, A(z)\} \bigg|_{z=1} = \begin{bmatrix} D_0(1) \\ \vdots \\ D_{N-1}(1) \\ \gamma I_M \\ \gamma I_M \\ \vdots \end{bmatrix}, \tag{120}$$

where $D_i(z)$ is defined in (41). From (44), (119) and (120) we have

$$[\mathcal{I} - \mathcal{P}] \frac{d}{dz} \{e(z) \, \text{adj}\, A(z)\} \bigg|_{z=1} \mathbf{1}_M - \gamma \mathbf{1}_\infty = [\mathcal{I} - \mathcal{P}] \sum_{l=1}^{MN-1} \beta_l e(\xi_l) U(\xi_l). \tag{121}$$

Then $Y = \mathcal{P}Y + \gamma \mathbf{1}_\infty$, where

$$Y = e'(1) \operatorname{adj} A(1) \mathbf{1}_M + e(1) \frac{d}{dz} \{\operatorname{adj} A(z)\}\bigg|_{z=1} \mathbf{1}_M - \sum_{l=1}^{MN-1} \beta_l e(\xi_l) U(\xi_l). \quad (122)$$

The imaginary part of Y is $\operatorname{Im} Y = -\operatorname{Im}\{\sum_{l=1}^{MN-1} \beta_l e(\xi_l) U(\xi_l)\}$, and it satisfies $\operatorname{Im} Y = \mathcal{P} \operatorname{Im} Y$ because γ is real. Since all $|\xi_l| < 1$, the entries of $\operatorname{Im} Y$ converge to 0, and by previous arguments $\operatorname{Im} Y = 0$, i.e., Y is real. The second and third terms on the left-hand side of (122) are bounded vectors, while the first term has nonnegative entries because $\operatorname{adj} A(1) \geq 0$ (see [Gantmacher, 1959]). It follows that $Y = \operatorname{Re} Y$ has entries that are bounded below. Thus, by adding a constant vector Y_c to Y if necessary (such a vector satisfies $Y_c = \mathcal{P} Y_c$), we may assume that Y is nonnegative. Since $\gamma > 0$ in the positive recurrent M/G/1 case, $Y \geq \mathcal{P}Y$ is superregular (see [Kemeny et al., 1966]) or superharmonic. Multiplying the equation $Y = \mathcal{P}Y + \gamma \mathbf{1}_\infty$ on the left by $\mathcal{P}^k$, $k = 0, \ldots, n-1$, and summing yields (since $\mathcal{P} \mathbf{1}_\infty = \mathbf{1}_\infty$)

$$Y = \mathcal{P}^n Y + n\gamma \mathbf{1}_\infty, \qquad n = 1, 2, \ldots.$$

Now $\mathcal{P}^n Y \geq 0$, so $Y \geq n\gamma \mathbf{1}_\infty$ for all n. Since $\gamma > 0$, we have $n\gamma \mathbf{1}_\infty \to +\infty$ as $n \to \infty$, which contradicts $Y < +\infty$.

We have shown that the root equations and the conservation of probability equation are linearly independent, at least in the case of simple zeros inside the unit circle. The general case, for which multiple zeros in $|z| < 1$ and zeros on $|z| = 1$ other than $z = 1$ are possible, requires a more complicated argument (see [Gail et al., 1996] for details).

References

[Akar and Sohraby, 1997a] Akar, N. and Sohraby, K. (1997a). An invariant subspace approach in M/G/1 and G/M/1 type Markov chains. *Communications in Statistics–Stochastic Models*, 13:381–416.

[Akar and Sohraby, 1997b] Akar, N. and Sohraby, K. (1997b). Matrix geometric solutions in M/G/1 type Markov chains with multiple boundaries: A generalized state-space approach. In Ramaswami, V. and Wirth, P. E., editors, *Teletraffic Contributions for the Information Age*, pages 487–496. Elsevier.

[Andrew et al., 1993] Andrew, A. L., Chu, K.-W. E., and Lancaster, P. (1993). Derivatives of eigenvalues and eigenvectors of matrix functions. *SIAM Journal on Matrix Analysis and Applications*, 14(4):903–926.

[Bailey, 1954] Bailey, N. T. J. (1954). On queueing processes with bulk service. *Journal of the Royal Statistical Society, Series B*, 16:80–87.

[Ball et al., 1990] Ball, J. A., Gohberg, I., and Rodman, L. (1990). *Interpolation of Rational Matrix Functions*. Birkhäuser.

[Berenstein and Gay, 1991] Berenstein, C. A. and Gay, R. (1991). *Complex Variables: An Introduction*. Springer-Verlag.

[Bini and Meini, 1996] Bini, D. and Meini, B. (1996). On the solution of a nonlinear matrix equation arising in queueing problems. *SIAM Journal on Matrix Analysis and Applications*, 17:906–926.

[Bini and Meini, 1998] Bini, D. and Meini, B. (1998). Using displacement structure for solving non-skip-free M/G/1 type Markov chains. In Alfa, A. S. and Chakravarthy, S. R., editors, *Advances in Matrix Analytic Methods for Stochastic Models*, pages 17–37. Notable Publications.

[Coffman et al., 1983] Coffman, Jr., E. G., Fayolle, G., and Mitrani, I. (1983). Two queues with alternating service periods. In Courtois, P. J. and Latouche, G., editors, *Performance 87*, pages 227–237. North-Holland.

[Cohen and Down, 1996] Cohen, J. W. and Down, D. G. (1996). On the role of Rouché's theorem in queueing analysis. *Queueing Systems*, 23:281–291.

[Daigle and Lucantoni, 1991] Daigle, J. N. and Lucantoni, D. M. (1991). Queueing systems having phase-dependent arrival and service rates. In Stewart, W. J., editor, *Numerical Solution of Markov Chains*, pages 161–202. Marcel Dekker.

[Dukhovny, 1994] Dukhovny, A. (1994). Multiple roots in some equations of queueing theory. *Communications in Statistics–Stochastic Models*, 10(2):519–524.

[Dukhovny, 1998] Dukhovny, A. (1998). Multiple roots in queueing equations: How realistic must examples get? *Communications in Statistics–Stochastic Models*, 14(3):763–765.

[Feller, 1968] Feller, W. (1968). *An Introduction to Probability Theory and its Applications*, volume I. John Wiley & Sons, 3rd edition.

[Feller, 1971] Feller, W. (1971). *An Introduction to Probability Theory and its Applications*, volume II. John Wiley & Sons, 2nd edition.

[Franklin, 1968] Franklin, J. N. (1968). *Matrix Theory*. Prentice-Hall.

[Friedberg et al., 1989] Friedberg, S. H., Insel, A. J., and Spence, L. E. (1989). *Linear Algebra*. Prentice-Hall, 2nd edition.

[Gail et al., 1994] Gail, H. R., Hantler, S. L., Konheim, A. G., and Taylor, B. A. (1994). An analysis of a class of telecommunications models. *Performance Evaluation*, 21:151–161.

[Gail et al., 1995] Gail, H. R., Hantler, S. L., Sidi, M., and Taylor, B. A. (1995). Linear independence of root equations for M/G/1 type Markov chains. *Queueing Systems*, 20:321–339.

[Gail et al., 1992] Gail, H. R., Hantler, S. L., and Taylor, B. A. (1992). On a preemptive Markovian queue with multiple servers and two priority classes. *Mathematics of Operations Research*, 17(2):365–391.

[Gail et al., 1996] Gail, H. R., Hantler, S. L., and Taylor, B. A. (1996). Spectral analysis of M/G/1 and G/M/1 type Markov chains. *Advances in Applied Probability*, 28:114–165.

[Gail et al., 1997] Gail, H. R., Hantler, S. L., and Taylor, B. A. (1997). Non-skip-free M/G/1 and G/M/1 type Markov chains. *Advances in Applied Probability*, 29:733–758.

[Gail et al., 1998] Gail, H. R., Hantler, S. L., and Taylor, B. A. (1998). Matrix-geometric invariant measures for G/M/1 type Markov chains. *Communications in Statistics–Stochastic Models*, 14(3):537–569.

[Gantmacher, 1959] Gantmacher, F. R. (1959). *Applications of the Theory of Matrices*. Interscience.

[Glicksberg, 1976] Glicksberg, I. (1976). A remark on Rouché's theorem. *The American Mathematical Monthly*, 83:186–187.

[Gohberg et al., 1982] Gohberg, I., Lancaster, P., and Rodman, L. (1982). *Matrix Polynomials*. Academic Press.

[Gohberg and Fel'dman, 1974] Gohberg, I. C. and Fel'dman, I. A. (1974). *Convolution Equations and Projection Methods for their Solution*. American Mathematical Society.

[Goldberg, 1958] Goldberg, S. (1958). *Introduction to Difference Equations*. John Wiley & Sons.

[Grassmann, 1977a] Grassmann, W. K. (1977a). Transient solutions in Markovian queueing systems. *Computers & Operations Research*, 4:47–53.

[Grassmann, 1977b] Grassmann, W. K. (1977b). Transient solutions in Markovian queues. *European Journal of Operational Research*, 1:396–402.

[Grassmann, 1991] Grassmann, W. K. (1991). Finding transient solutions in Markovian event systems through randomization. In Stewart, W. J., editor, *Numerical Solution of Markov Chains*, pages 357–371. Marcel Dekker.

[Grassmann, 1998] Grassmann, W. K. (1998). Finding test data for solving the equilibrium equations of Markov chains with repeating columns. In Alfa, A. S. and Chakravarthy, S. R., editors, *Advances in Matrix Analytic Methods for Stochastic Models*, pages 261–278. Notable Publications.

[Grassmann and Heyman, 1990] Grassmann, W. K. and Heyman, D. P. (1990). Equilibrium distribution of block-structured Markov chains with repeating rows. *Journal of Applied Probability*, 27:557–576.

[Grassmann and Heyman, 1993] Grassmann, W. K. and Heyman, D. P. (1993). Computation of steady-state probabilities for infinite-state Markov chains with repeating rows. *ORSA Journal on Computing*, 5:292–303.

[H. R. Gail and Taylor, 1998] H. R. Gail, S. L. H. and Taylor, B. A. (1998). M/G/1 type Markov chains with rational generating functions. In Alfa, A. S. and Chakravarthy, S. R., editors, *Advances in Matrix Analytic Methods for Stochastic Models*, pages 1–16. Notable Publications.

[Horn and Johnson, 1985] Horn, R. A. and Johnson, C. R. (1985). *Matrix Analysis*. Cambridge University Press.

[Jensen, 1953] Jensen, A. (1953). Markoff chains as an aid in the study of markoff processes. *Skandinavsk Aktuarietidskrift*, 36:87–91.

[Keilson and Wishart, 1964] Keilson, J. and Wishart, D. M. G. (1964). A central limit theorem for processes defined on a finite Markov chains. In *Proceedings of the Cambridge Philosophical Society*, volume 60, pages 547–567.

[Kemeny et al., 1966] Kemeny, J. G., Snell, J. L., and Knapp, A. W. (1966). *Denumerable Markov Chains*. Van Nostrand.

[Kingman, 1961] Kingman, J. F. C. (1961). A convexity property of positive matrices. *Quarterly Journal of Mathematics*, 12:283–284.

[Kleinrock, 1975] Kleinrock, L. (1975). *Queueing Systems Volume I: Theory*. John Wiley & Sons.

[Konheim and Reiser, 1976] Konheim, A. G. and Reiser, M. (1976). A queueing model with finite waiting room and blocking. *Journal of the ACM*, 23(2):328–341.

[Konheim and Reiser, 1978] Konheim, A. G. and Reiser, M. (1978). Finite capacity queueing systems with applications in computer modeling. *SIAM Journal on Computing*, 7(2):210–229.

[Latouche and Ramaswami, 1993] Latouche, G. and Ramaswami, V. (1993). A logarithmic reduction algorithm for quasi-birth-death processes. *Journal of Applied Probability*, 30:650–674.

[Meini, 1997] Meini, B. (1997). An improved FFT-based version of Ramaswami's formula. *Communications in Statistics–Stochastic Models*, 13:23–238.

[Meyer, 1989] Meyer, C. D. (1989). Stochastic complementation, uncoupling Markov chains, and the theory of nearly reducible systems. *SIAM Review*, 31(2):40–272.

[Mitrani and Chakka, 1995] Mitrani, I. and Chakka, R. (1995). Spectral expansion solution for a class of Markov models: Application and comparison with the matrix-geometric method. *Performance Evaluation*, 23(3):241–260.

[Mitrani and Mitra, 1992] Mitrani, I. and Mitra, D. (1992). A spectral expansion method for random walks on semi-infinite strips. In *Iterative Methods in Linear Algebra*, pages 141–149. North-Holland.

[Mitrany and Avi-Itzhak, 1968] Mitrany, I. L. and Avi-Itzhak, B. (1968). A many-server queue with service interruptions. *Operations Research*, 16(3):628–638.

[Neuts, 1981] Neuts, M. F. (1981). *Matrix-Geometric Solutions in Stochastic Models*. Johns Hopkins University Press.

[Neuts, 1989] Neuts, M. F. (1989). *Structured Stochastic Matrices of M/G/1 Type and Their Applications*. Marcel Dekker.

[Ramaswami, 1988] Ramaswami, V. (1988). A stable recursion for the steady state vector in Markov chains of M/G/1 type. *Communications in Statistics–Stochastic Models*, 4:183–188.

[Takács, 1962] Takács, L. (1962). *Introduction to the Theory of Queues*. Oxford University Press.

[Wuyts and Boel, 1998] Wuyts, K. and Boel, R. K. (1998). Efficient matrix geometric methods for B-ISDN by using the spectral decomposition of the rate matrices. In Alfa, A. S. and Chakravarthy, S. R., editors, *Advances in Matrix Analytic Methods for Stochastic Models*, pages 341–359. Notable Publications.

H. Richard Gail received the Ph.D. degree in Engineering from the University of California, Los Angeles. He is a research staff member at the IBM T. J. Watson Research Center in Yorktown Heights, NY. His research interests include queueing theory and computer systems modeling and analysis.

Sid L. Hantler received the Ph.D. degree in Mathematics from the University of Michigan. He is the manager of the Stochastic Analysis group at the IBM T. J. Watson Research Center in Yorktown Heights, NY. His research interests include queueing theory and network design and analysis.

B. A. Taylor received the Ph.D. degree in Mathematics from the University of Illinois. He is Professor of Mathematics and Chairman of the Mathematics Department at the University of Michigan. He is also Associate Treasurer of the American Mathematical Society. His research interests include complex analysis, pluripotential theory and queueing theory.

8 AN INTRODUCTION TO NUMERICAL TRANSFORM INVERSION AND ITS APPLICATION TO PROBABILITY MODELS

Joseph Abate[1], Gagan L. Choudhury[2] and Ward Whitt[3]

[1]900 Hammond Road
Ridgewood, NJ 07450-2908

[2]AT&T Labs, Room 1L-238
Holmdel, NJ 07733-3030
gagan@buckaroo.att.com

[3]AT&T Labs, Room A117
180 Park Avenue
Florham Park, NJ 07932-0971
wow@research.att.com

1 INTRODUCTION

Numerical transform inversion has an odd place in computational probability. Historically, transforms were exploited extensively for solving queueing and related probability models, but only rarely was numerical inversion attempted. The model descriptions were usually left in the form of transforms. Vivid examples are the queueing books by Takács [Takács, 1962] and Cohen [Cohen, 1982]. When possible, probability distributions were calculated analytically by inverting transforms, e.g., by using tables of transform pairs. Also, moments of probability distributions were computed analytically by differentiating the transforms and, occasionally, approximations were developed by applying asymptotic methods to transforms, but only rarely did anyone try to compute probability distributions by numerically inverting the available transforms. However, there were exceptions, such as the early paper by Gaver [Gaver, 1966]. (For more on the history of numerical transform inversion, see our earlier survey [Abate and Whitt, 1992a].) Hence, in the application of probability models to engineering, transforms became regarded more as mathematical toys than practical tools. Indeed, the conventional wisdom was that numerical transform inversion was very difficult. Even numerical analysts were often doubtful of the numerical stability of inversion algorithms. In queueing, both theorists and practitioners lamented about the "Laplace curtain" obscuring our understanding of system behavior.

Thus, the perceived difficulty of numerical transform inversion served as a motivation for much progress in computational probability. Instead of directly trying to invert available transforms, researchers primarily responded to this situation by developing new computational methods and new modeling approaches that avoided transforms. A good example is the collection of matrix-geometric methods in Neuts [Neuts, 1981]. In his preface, Neuts concludes that "the oft-lamented Laplacian curtain, which covers the solution and hides the structural properties of many interesting stochastic models, is now effectively lifted." Indeed, matrix-geometric methods and other alternative approaches are remarkably effective, as can be seen from Chapter 6 in this book.

However, since then, it has been recognized that the concerns about numerical transform inversion were misplaced. Contrary to previous impressions, it is not difficult to numerically compute probabilities and other quantities of interest in probability models by directly inverting the transforms. For example, all the transforms in the queueing books by Takács [Takács, 1962] and Cohen [Cohen, 1982] can be numerically inverted. To a large extent, the numerical inversion can be done by directly computing from the classical inversion formulas. However, there are complications, so that some additional thought is beneficial. But thought has been given, so that now there are a variety of effective inversion algorithms based on the classical inversion formulas.

The purpose of this chapter is to provide an introduction to numerical transform inversion and its application to compute probabilities in stochastic models. We focus on Laplace transforms and generating functions (z transforms), but similar inversion methods apply to other transforms such as characteristic functions (Fourier transforms). In Section 2 we present numerical inversion algorithms and in Sections 3–5 we present applications to queueing models with numerical examples. The queueing examples in Sections 3–5 include: computing transient characteristics in the M/M/c/0 Erlang loss model (Section 3), computing the steady-state waiting-time distribution in the general GI/GI/1 queue (Section 4) and computing steady-state characteristics of product-form closed queueing networks (Section 5). These three models are arguably the three most fundamental models in queueing theory. These three examples illustrate the extra work that often must be done to treat more difficult problems in which a simple closed-form transform is not available. Other examples can be found in the references.

Now we illustrate how transforms and numerical inversion can help by considering two simple examples. For these examples, the probability analysis is elementary or well known, but nevertheless calculation can be difficult without numerical inversion. Calculating cumulative distribution functions for these examples is rarely suggested in textbook discussions. On the other hand, calculation is elementary with numerical transform inversion, because in these examples a simple closed-form transform is readily available to be numerically inverted. Indeed, the numerical inversion approach to these problems is suitable for introductory textbooks. The introductory textbook on performance evaluation by Kobayashi [Kobayashi, 1978] included a brief discussion of numerical transform inversion (an early variant of the Fourier-series method for numerically inverting Laplace transforms to be discussed in Section 2), but it did not strongly influence practice. The recent introductory textbook on stochastic processes by Kao [Kao, 1997] uses the inversion algorithms for Laplace transforms and generating functions presented in Section 2 here.

Example 1 (Project Completion Time) Suppose that a project is composed of n independent tasks, one performed after another. We want to know the probability distribution of the time to complete the entire project when there is uncertainty about the time to complete each task. Let the time to complete task k be a random variable X_k with probability density function f_k (depending on k). Then the time to complete the project is the sum $S_n = X_1 + \ldots + X_n$, which has a density g_n that is the convolution of the n densities $f_1, \ldots, f_n$; i.e., g_n can be defined recursively by the convolution integral.

$$g_n(t) = \int_0^\infty g_{n-1}(t-x) f_n(x) dx$$

and its associated cumulative distribution function (cdf) is

$$G_n(t) = \int_0^t g_n(x) dx \ . \tag{1}$$

By the central limit theorem for independent non-identically distributed random variables, e.g., p. 262 of Feller [Feller, 1971], the sum S_n is approximately normally distributed with mean and variance equal to the sum of the component means and variances, provided that n is not too small and that the individual task times X_k have finite variances and are suitably small compared to the sum. However, if one or two task times are much larger than the others, then the normal approximation can be quite inaccurate. Then it may be much better to compute the exact distribution.

Unfortunately, however, except for very small n, the desired cdf $G_n(t)$ in (1) is difficult to compute directly because it involves an $(n-1)$-dimensional integral. However, it is usually easy to compute by numerical transform inversion provided that we know the Laplace transforms of the densities f_k, i.e.,

$$\hat{f}_k(s) \equiv \int_0^\infty e^{-st} f_k(t) dt \ .$$

If we only know the means and variances of the task times, then we might fit appropriate distributions, such as gamma distributions, to these moments, and then compute the distribution of the project completion time. A gamma density with shape parameter α and scale parameter λ has Laplace transform

$$\hat{f}(s) = \left(\frac{\lambda}{\lambda + s}\right)^\alpha . \tag{2}$$

The associated mean and variance are α/λ and α/λ^2. Hence, the parameters α and λ can easily be chosen to match the mean and variance.

To obtain the Laplace transform of the density of the project completion time, we use the basic transform law stating that the transform of a convolution is the product of the transforms. Thus the Laplace transform of the density g_n is

$$\hat{g}_n(s) \equiv \int_0^\infty e^{-st} g_n(t) dt = \prod_{k=1}^n \hat{f}_k(s) , \quad n \geq 1 . \tag{3}$$

Moreover, we use another basic transform law to get the Laplace transform of the cdf G_n. Since G_n is the integral of g_n,

$$\hat{G}_n(s) \equiv \int_0^\infty e^{-st} G_n(t) dt = \frac{\hat{g}_n(s)}{s} . \tag{4}$$

Combining (3) and (4), we see that the Laplace transform $\hat{G}_n$ of the cdf G_n is conveniently expressed in terms of the given transforms $\hat{f}_1, \ldots, \hat{f}_n$. Thus we can easily apply an inversion algorithm to calculate the cdf $G_n(t)$ for any desired t by numerically inverting the Laplace transform $\hat{G}_n$.

In the next section we describe inversion algorithms that can be used to calculate the cdf values $G_n(t)$ for any t given the Laplace transform $\hat{G}_n$. These algorithms are intended for the case in which the function to be calculated, here G_n, is relatively smooth. For example, there is no difficulty if each task time density has a gamma distribution as in (2). The major source of numerical difficulty in the inversion algorithm stems from discontinuities in the function or its derivatives. However, discrete probability distributions can be calculated easily by numerically inverting their generating functions.

In the numerical inversion of Laplace transforms, the required computation turns out to be remarkably light: For each value of t, we typically need to add about 50 transform terms and, by (3) and (4), we need to perform n multiplications for each.

Example 1 illustrated two basic laws for manipulating transforms. Unlike numerical transform inversion, these basic transform laws are quite well known, so we will not dwell upon them. Sources for additional background on transforms are the appendices of Kleinrock [Kleinrock, 1975], Giffin [Giffin, 1975], Doetsch [Doetsch, 1961], [Doetsch, 1974], Van Der Pol [Pol and Bremmer, 1987] and Poularikas [Poularikas, 1996].

Example 2 (The Renewal Function) Let $M(t)$ be the renewal function, recording the expected number of renewals in the interval $(0,t]$, associated with a renewal process having an interrenewal-time cdf F with density f, i.e.,

$$M(t) = \sum_{n=1}^{\infty} F^{n*}(t), \quad t \geq 0, \tag{5}$$

where F^{n*} is the n-fold convolution of the cdf F with itself. The renewal function is of considerable interest because it arises in many probability models, but because of the convolution in (5) it is rather difficult to calculate directly. However, it is elementary to compute using numerical inversion. Let $\hat{f}$ be the Laplace transform of f. Then the Laplace transform of M is

$$\hat{M}(s) \equiv \int_0^\infty e^{-st} M(t) dt = \sum_{n=1}^{\infty} \frac{\hat{f}(s)^n}{s} = \frac{\hat{f}(s)}{s(1-\hat{f}(s))}. \tag{6}$$

Given the transform $\hat{f}$, numerical inversion algorithms apply to easily calculate $M(t)$ for any desired t by inverting the transform $\hat{M}$ in (6).

Numerical inversion applies easily to the two examples given here, because the transform is available in closed form. The challenge in more difficult examples is to compute the transform values, because transforms often are not available in closed form. For example, for the busy period distribution in the M/G/1 queue, the transform is available implicitly via the Kendall functional equation. In that case, the transform values can be readily calculated by iteration; see Abate and Whitt [Abate and Whitt, 1992c]. The examples in Sections 3–5 illustrate how transform values can be obtained for numerical inversion in other more complicated queueing examples.

2 NUMERICAL INVERSION ALGORITHMS

In this section we present some numerical inversion algorithms. We primarily aim to explain how the algorithms work. Of course, in order to *apply* the inversion algorithms, it is not necessary to know how the algorithms work. Using the algorithms, transform expressions as in Example 1 and 2 above can be immediately applied to calculate the desired quantities. However, even if we only want to apply inversion algorithms, it is good to understand how they work.

We begin in Section 2.1 by presenting classical inversion formulas for Laplace transforms, which form the basis for the numerical inversion algorithms. Next, in Section 2.2 we present a variant of the Fourier-series algorithm for numerically inverting a Laplace transform. Afterwards, in Section 2.3 we provide further discussion of Euler summation to help explain why it is so effective in accelerating the convergence of the infinite series arising in the Fourier-series method. Finally, in Section 2.4 we present the Fourier-series algorithm for numerically inverting a generating function.

In this section we consider only one-dimensional transforms. For extensions of the inversion algorithms to multi-dimensional transforms; see Choudhury, Lucantoni and Whitt [Choudhury et al., 1994].

2.1 Inversion Formulas for Laplace Transforms

Given a real-valued function f on the nonnegative real line, its *Laplace transform* (LT) is defined by

$$\mathcal{L}(f)(s) \equiv \hat{f}(s) \equiv \int_0^\infty e^{-st} f(t) dt, \qquad (7)$$

where s is a complex variable with $Re(s) > 0$. (Let $Re(s)$ be the real part and $Im(s)$ the imaginary part of s.) Conditions ensuring that the integral in (7) exists appear in the literature; e.g., Doetsch [Doetsch, 1974].

For probability applications, it is useful to consider Laplace-Stieltjes transforms as well as ordinary Laplace transforms. Given a nonnegative *random variable* X with *cumulative distribution function* (cdf) F, i.e., $F(t) = P(X \leq t)$, $t \geq 0$, and *probability density function* (pdf) f when it exists, i.e., when $F(t) = \int_0^t f(u) du$, $t \geq 0$, we define its Laplace transform by

$$Ee^{-sX} \equiv \mathcal{L}(f)(s) \equiv \hat{f}(s) \equiv \int_0^\infty e^{-st} dF(t) = \int_0^\infty e^{-st} f(t) dt;$$

i.e., $\hat{f}$ is the Laplace transform of the pdf f as in (7), but $\hat{f}$ is the *Laplace-Stieltjes transform* (LST) of the cdf F. The associated LT of F is thus $\hat{F}(s) = \hat{f}(s)/s$. In probability applications, we are typically interested in the cdf, which can be computed by numerically inverting the LT $\hat{F}$. We are also often interested in the *complementary cdf* (ccdf) $F^c(t) \equiv 1 - F(t)$, which has LT $\hat{F}^c(s) = (1 - \hat{f}(s))/s$. For these probability transforms $\hat{f}, \hat{F}$ and $\hat{F}^c$, the integrals always exist. We will always apply the inversion to LTs instead of LSTs. We usually aim to calculate the ccdf F^c by numerically inverting its transform $\hat{F}^c$.

In this section we present a way to numerically invert the LT $\hat{f}$ in (7) in order to calculate $f(t)$ for any given t. We have assumed that f is real-valued, as when f is a pdf, cdf or ccdf, but inversion also applies when f is complex-valued. Indeed, if $f(t) = f_1(t) + i f_2(t)$, $t \geq 0$, for $i = \sqrt{-1}$, then $\hat{f}(s) = \hat{f}_1(s) + i \hat{f}_2(s)$, so it is clear that inversion extends easily to complex-valued functions. Complex-valued functions arise naturally in the iterated one-dimensional inversion of multidimensional transforms, as we illustrate here in Section 5. In that setting, the function to be calculated by the one-dimensional inversion will be complex-valued in all steps except the last, e.g., see [Choudhury et al., 1994].

A natural starting point for numerical inversion of Laplace transforms is the *Bromwich inversion integral*; e.g., see Theorem 24.4 on p. 257 of Doetsch [Doetsch, 1974].

Theorem 1 (*Bromwich inversion integral*) *Given the Laplace transform $\hat{f}$ in (7), the function value $f(t)$ can be recovered from the contour integral*

$$f(t) = \frac{1}{2\pi i} \int_{b-i\infty}^{b+i\infty} e^{st} \hat{f}(s) ds , \quad t > 0 , \qquad (8)$$

where b is a real number to the right of all singularities of $\hat{f}$, and the contour integral yields the value 0 for $t < 0$.

As usual with contour integrals, there is flexibility in choosing the contour provided that it is to the right of all singularities of $\hat{f}$. However, there is no need to be bothered by the complexities of complex variables. If we choose a specific contour and perform a change of variables, then we obtain an integral of a real-valued function of a real variable. First, by making the substitution $s = b + iu$ in (8), we obtain

$$f(t) = \frac{1}{2\pi} \int_{-\infty}^{\infty} e^{(b+iu)t} \hat{f}(b+iu) du . \qquad (9)$$

Then, since

$$e^{(b+iu)t} = e^{bt}(\cos ut + i \sin ut) ,$$

$\sin ut = -\sin(-ut)$, $\cos ut = \cos(-ut)$, $Im(\hat{f}(b+iu)) = -Im(\hat{f}(b-iu))$ and $Re(\hat{f}(b+iu)) = Re(\hat{f}(b-iu))$, and from the fact that the integral in (8) is 0 for $t < 0$, we obtain

$$f(t) = \frac{2e^{bt}}{\pi} \int_0^{\infty} Re(\hat{f}(b+iu)) \cos(ut) du \qquad (10)$$

and

$$f(t) = \frac{-2e^{bt}}{\pi} \int_0^{\infty} Im(\hat{f}(b+iu)) \sin(ut) du .$$

Theorem 1 implies that $f(t)$ can be calculated from the transform $\hat{f}$ by performing a numerical integration (quadrature). Since there are many numerical integration algorithms, e.g., see Davis and Rabinowitz [Davis and Rabinowitz, 1984], there are obviously many possible approaches to numerical transform inversion via the Bromwich inversion integral. In this context, the remaining goal is to exploit the special structure of the integrand in (10) in order to calculate the integral accurately and efficiently.

However, there also are quite different numerical inversion algorithms, because the Bromwich inversion integral is not the only inversion formula. To illustrate, we mention a few others. First, there is the Post-Widder inversion formula, which involves differentiation instead of integration. It is the basis for the Jagerman-Stehfest procedure in Section 8 of Abate and Whitt [Abate and Whitt, 1992a]. (See that source for further discussion.)

Theorem 2 (*Post-Widder inversion formula*) *Under regularity conditions,*

$$f(t) = \lim_{n\to\infty} \frac{(-1)^n}{n!} \left(\frac{n+1}{t}\right)^{n+1} \hat{f}^{(n)}((n+1)/t) , \qquad (11)$$

where $\hat{f}^{(n)}$ is the nth derivative of $\hat{f}$.

For small n, the terms on the right in (11) can serve as useful rough approximations, because they tend to inherit the structure of f; see Jagerman [Jagerman, 1978] and [Jagerman, 1982].

The next inversion formula is a discrete analog of the Post-Widder formula involving finite differences. It is the basis for the Gaver-Stehfest procedure in Section 8 of Abate and Whitt [Abate and Whitt, 1992a]. Let Δ be the difference operator, defined by $\Delta \hat{f}(n\alpha) = \hat{f}((n+1)\alpha) - \hat{f}(n\alpha)$ and let $\Delta^k = \Delta(\Delta^{k-1})$.

Theorem 3 (*discrete analog of Post-Widder formula*) *If f is a bounded real-valued function that is continuous at t, then*

$$f(t) = \lim_{n\to\infty} (-1)^n \frac{\ln 2}{t} \frac{(2n)!}{n!(n-1)!} \Delta^n \hat{f}(n \ln 2/t) .$$

Finally, we mention the Laguerre-series inversion formula, which is the basis for the Laguerre-series or Weeks' algorithm in Weeks [Weeks, 1966] and Abate, Choudhury and Whitt [Abate et al., 1996] and [Abate et al., 1997].

Theorem 4 (*Laguerre-series representation*) *Under regularity conditions,*

$$f(t) = \sum_{n=0}^{\infty} q_n l_n(t), \ t \geq 0, \qquad (12)$$

where

$$l_n(t) = e^{-t/2} L_n(t), \ t \geq 0, \qquad (13)$$

$$L_n(t) = \sum_{k=0}^{n} \binom{n}{k} \frac{(-t)^k}{k!}, \ t \geq 0, \qquad (14)$$

and

$$\hat{q}(z) \equiv \sum_{n=0}^{\infty} q_n z^n = (1-z)^{-1} \hat{f}((1+z)/2(1-z)) . \qquad (15)$$

The function L_n in (14) are the *Laguerre polynomials*, while l_n in (13) are the associated *Laguerre functions*. The scalars q_n in (12) are the *Laguerre coefficients* and $\hat{q}$ in (15) is the *Laguerre generating function* (the generating function of the Laguerre coefficients).

Now we consider a specific algorithm based on the Bromwich inversion integral.

2.2 The Fourier-Series Method for Laplace Transforms

In this section we develop one variant of the Fourier-series method. The specific algorithm is concisely summarized at the end of the section. Pointers to the literature appear there as well.

There are two natural ways to develop the Fourier-series method. One way starts with the Bromwich inversion integral. As indicated before, we can directly apply a standard numerical integration procedure to perform the integration in (10). Somewhat surprisingly, perhaps, one of the most naive approaches – the trapezoidal rule – proves to be remarkably effective in this context. If we use a step size h, then the trapezoidal rules gives

$$f(t) \approx f_h(t) \equiv \frac{he^{bt}}{\pi} Re(\hat{f}(b)) + \frac{2he^{bt}}{\pi} \sum_{k=1}^{\infty} Re(\hat{f}(b+ikh)) \cos(kht) , \qquad (16)$$

where $Re(\hat{f}(b)) = \hat{f}(b)$ since b is real. By letting $h = \pi/2t$, the cosine terms in (16) all become ± 1 or 0. Letting $h = \pi/2t$ and $b = A/2t$, we obtain the "nearly alternating" series

$$f_h(t) = f_A(t) \equiv \frac{e^{A/2}}{2t} Re(\hat{f}) \left(\frac{A}{2t}\right) + \frac{e^{A/2}}{t} \sum_{k=1}^{\infty} (-1)^k Re(\hat{f}) \left(\frac{A+2k\pi i}{2t}\right) . \qquad (17)$$

The remaining problem is to control the error associated with calculating $f(t)$ via this representation (17). There are three sources of error associated with the trapezoidal rule: first, the *discretization error* associated with approximating $f(t)$ by $f_h(t)$ in (16); second, the *truncation error* associated with approximately calculating the infinite series in (17) (which might not be by simple truncation); and, third, the *roundoff error* associated with addition and multiplication in calculating $f_A(t)$ in (17). The remaining discussion is devoted to showing how to control these three sources of error when we use the trapezoidal rule.

The Discretization Error

The standard theory of numerical integration, as on p. 32 of Davis and Rabinowitz [Davis and Rabinowitz, 1984], shows that the discretization error for the trapezoidal rule using step size h is bounded by a term of order $O(h^2)$, which is not so good. However, the special trigonometric structure here tends to yield a much better approximation, of order $O(e^{-c/h})$ for some constant c. The better error bound follows from a second way to develop the Fourier-series method, which does not rely on the Bromwich contour integral. We can anticipate the second approach, though, by looking at (16). From (16), we can recognize that f_h is a trigonometric series. We thus can ask if f_h is a Fourier series of some, necessarily periodic, function f_p related to f (expressed in terms of f instead of its LT $\hat{f}$). The discretization error is then $f_h - f = f_p - f$.

The second approach starts, without considering the Bromwich inversion integral, by directly constructing a periodic function f_p approximating f and then constructing the Fourier series of this periodic function. However, to facilitate the procedure, we first damp f by letting $g(t) = e^{-bt}f(t)$ and then extend it over the entire real line by letting $g(t) = 0$ for $t < 0$. We then form a periodic function approximating g by considering an infinite series of translates of the original function, i.e., we let

$$g_p(t) = \sum_{k=-\infty}^{\infty} g\left(t + \frac{2\pi k}{h}\right) . \tag{18}$$

The damping ensures that the series in (18) converges, and tends to make the terms in the tails relatively small. The construction in (18) is called *aliasing*. We then work with g and g_p, recovering f from $f(t) = e^{bt}g(t)$ at the end.

The idea now is to construct the Fourier series of the periodic function g_p. (We elaborate in the proof of Theorem 5 below.) The result is equivalent to the *Poisson summation formula*. This procedure yields what we want because the coefficients of the Fourier series can be expressed directly in terms of the Laplace transform values. Indeed, the Fourier series of the periodic function g_p in (18) is a minor modification of the function f_A in (17). This second approach yielding the discretization error in the trapezoidal rule explains the name *Fourier-series method*.

We summarize the main result in the following theorem.

Theorem 5 *Under regularity conditions, the discretization error for the trapezoidal rule approximation in (17) is*

$$e_A(t) \equiv f_A(t) - f(t) = \sum_{k=1}^{\infty} e^{-kA} f((2k+1)t) . \tag{19}$$

Proof. We start with the periodic function g_p in (18) with period $2\pi/h$, which we assume is well defined. (It clearly is when $|f(t)| \leq C$ for all t.) The term for $k = 0$ in the right side of (18) is clearly $g(t)$. Thus the other terms in the series make up the aliasing error for g.

We then represent the periodic function g_p by its *complex Fourier series*

$$g_p(t) = \sum_{k=-\infty}^{\infty} c_k e^{ikht}, \tag{20}$$

where c_k is the kth *Fourier coefficient* of g_p, i.e.,

$$c_k = \frac{h}{2\pi} \int_{-\pi/h}^{\pi/h} g_p(t) e^{-ikht} dt . \tag{21}$$

We *assume* that the Fourier-series representation (20) is valid. The large literature on Fourier series provides conditions; e.g., [Tolstov, 1976]. For practical purposes, it is important that t be a continuity point of the function f, and thus of g.

Now we substitute the series (18) for $g_p(t)$ into (21) to obtain

$$c_k = \frac{h}{2\pi} \int_{-\pi/h}^{\pi/h} \sum_{k=-\infty}^{\infty} g\left(t + \frac{2k\pi}{h}\right) e^{-ikht} dt = \frac{h}{2\pi} \int_{-\infty}^{\infty} g(t) e^{-ikht} dt$$

$$= \frac{h}{2\pi} \int_0^\infty e^{-bt} f(t) e^{-ikht} dt = \frac{h}{2\pi} \hat{f}(b + ikh) . \qquad (22)$$

Thus, we can write $g_p(t)$ in two different ways. First, from (18) and the relation between g and f, we have

$$g_p(t) = \sum_{k=-\infty}^{\infty} g\left(t + \frac{2\pi k}{h}\right) = \sum_{k=-\infty}^{\infty} f\left(t + \frac{2\pi k}{h}\right) e^{-b(t+2\pi k/h)} . \qquad (23)$$

Second, from (20) and (21), we have

$$g_p(t) = \sum_{k=-\infty}^{\infty} c_k e^{ikht} = \frac{h}{2\pi} \sum_{k=-\infty}^{\infty} \hat{f}(b + ikh) e^{ikht} . \qquad (24)$$

Combining the terms in (23) and (24) involving f and $\hat{f}$, we obtain a version of the *Poisson summation formula*

$$\sum_{k=-\infty}^{\infty} f\left(t + \frac{2\pi k}{h}\right) e^{-b(t+2\pi k/h)} = \frac{h}{2\pi} \sum_{k=-\infty}^{\infty} \hat{f}(b + ikh) e^{ikht} . \qquad (25)$$

We then obtain what we want by focusing on the single term for $k = 0$ on the left in (25); i.e.,

$$e^{-bt} f(t) = \frac{h}{2\pi} \sum_{k=-\infty}^{\infty} \hat{f}(b + ikh) e^{ikht} - \sum_{\substack{k=-\infty \\ k \neq 0}}^{\infty} f\left(t + \frac{2\pi k}{h}\right) e^{-b(t+2\pi k/h)} . \qquad (26)$$

Multiplying both sides of (26) by e^{bt} yields

$$f(t) = \frac{h}{2\pi} \sum_{k=-\infty}^{\infty} \hat{f}(b + ikh) e^{(b+ikh)t} - \sum_{\substack{k=-\infty \\ k \neq 0}}^{\infty} f\left(t + \frac{2\pi k}{h}\right) e^{-2\pi kb/h} . \qquad (27)$$

Letting $h = \pi/t$ and $b = A/2t$ in (27), we obtain

$$f(t) = \frac{e^{A/2}}{2t} \sum_{k=-\infty}^{\infty} (-1)^k \hat{f}\left(\frac{A + 2k\pi i}{2t}\right) - \sum_{k=1}^{\infty} e^{-kA} f((2k+1)t) = f_A(t) - e_A(t) . \qquad (28)$$

Because of the alternative route to approximation f_A in (17) via aliasing, the discretization error in (19) is also called the *aliasing error*. Depending on the functional form of f, it is easy to choose the parameter A to control the discretization error $e_A(t)$ in (19). (We discuss this further below.)

Summing the Infinite Series

The remaining task is to calculate $f_A(t)$ in (17). We will discuss measures to control the truncation and roundoff errors and then summarize the algorithm at the end of the section.

Summing the infinite series will clearly be an easy task (e.g., by simply truncating) if the transform $\hat{f}(u + iv)$, for u and v real and positive, decays rapidly as $v \to \infty$. Generally speaking, we understand when this will occur, because the tail behavior of the transform $\hat{f}$ depends on the smoothness of the function f. By the Riemann-Lebesgue lemma (e.g., pp. 513-514 of Feller [Feller, 1971]), if f has n integrable derivatives, then $\hat{f}(u+iv)$ is $o(v^{-n})$ as $v \to \infty$. Thus the inversion will be easy if f is very smooth, but may be difficult if f is not smooth. The ultimate lack of smoothness is a discontinuity at or near t, which will invariably cause a problem; see Section 14 of Abate and Whitt [Abate and Whitt, 1992a]. When there is initially not enough smoothness, it is often possible to introduce smoothness into the function before performing the inversion; see Section 6 of Abate and Whitt [Abate and Whitt, 1992a].

The infinite series in (17) can often be calculated by simply truncating, but a more efficient algorithm can be obtained by applying a summation acceleration method; e.g., see Wimp [Wimp, 1981]. An acceleration technique is especially important for coping with slowly decaying tails. As with numerical integration, there are many alternatives. An acceleration technique that has proven to be effective in our nearly-alternating series context is Euler summation. (Euler summation is intended for alternating series, in which successive summands alternate in sign; see Section 2.3 below for further discussion.)

Controlling Roundoff Error

The main source of roundoff error is the multiplication by the prefactor $e^{A/2}$ in (17). We can reduce this prefactor while still maintaining control of the discretization error by making a minor modification involving a parameter l, which is a positive integer.

We start with the basic inversion integral (9). The trapezoidal rule approximation and its discretization error are given in (27). We can rewrite the trapezoidal rule approximation in (27) as

$$f_h(t) = \frac{he^{bt}}{2\pi}f(b) + \frac{he^{bt}}{\pi}\sum_{k=1}^{\infty} Re[e^{ikht}\hat{f}(b+ikht)] . \qquad (29)$$

Now, setting $h = \pi/lt$ and $b = A/2lt$, we get

$$f_h(t) = f_{A,l}(t) \equiv \frac{e^{A/2l}}{2lt}\left[\hat{f}\left(\frac{A}{2lt}\right) + 2\sum_{k=1}^{\infty} Re\left[e^{ik\pi/l}\hat{f}\left(\frac{A}{2lt} + \frac{ik\pi}{lt}\right)\right]\right] . \qquad (30)$$

Next, by algebraic manipulation, we get

$$f_{A,l}(t) = \sum_{k=0}^{\infty}(-1)^k a_k(t), \qquad (31)$$

where
$$a_k(t) = \frac{e^{A/2\ell}}{2\ell t} b_k(t), \quad k \geq 0, \qquad (32)$$

$$b_0(t) = \hat{f}\left(\frac{A}{2\ell t}\right) + 2 \sum_{j=1}^{\ell} Re\left[\hat{f}\left(\frac{A}{2\ell t} + \frac{ij\pi}{\ell t}\right) e^{ij\pi/\ell}\right] \qquad (33)$$

and

$$b_k(t) = 2\sum_{j=1}^{\ell} Re\left[\hat{f}\left(\frac{A}{2\ell t} + \frac{ij\pi}{\ell t} + \frac{ik\pi}{t}\right) e^{ij\pi/\ell}\right], \quad k \geq 1. \qquad (34)$$

Notice that the prefactor in (32) becomes $(e^{A/2l})/2lt$, which is decreasing in l. Also the aliasing error in (27) becomes

$$e_h(t) \equiv e_{A,\ell}(t) = \sum_{j=1}^{\infty} e^{-Aj} f\left((1+2j\ell)t\right). \qquad (35)$$

Notice that $f_{A,l}(t)$ is (31) and $e_{A,l}(t)$ in (35) coincide with $f_A(t)$ in (17) and $e_A(t)$ in (19) when $l = 1$.

In (31)–(35) we have chosen the two parameters b and h to depend on the new parameters A and ℓ and the function argument t. This choice makes the period of the periodic function $2\pi/h = 2\ell t$. This choice makes the series in (31) correspond to an eventually alternating series if the real part of $\hat{f}(u+iv)$ is eventually of fixed sign as v gets large. This is what we mean by "nearly alternating."

If $|f(x)| \leq C$ for all $x \geq (1+2\ell)t$, as in probability applications (e.g., $C = 1$ when f is a cdf or ccdf), then the aliasing error in (35) is bounded by

$$|e_{A,\ell}(t)| \leq \frac{Ce^{-A}}{1-e^{-A}}, \qquad (36)$$

which is approximately equal to Ce^{-A} when e^{-A} is small. Note that the aliasing error bound in (36) is independent of the parameter ℓ. Hence, to have at most $10^{-\gamma}$ aliasing error when $C = 1$, we let $A = \gamma \log 10$. (With $A = 18.4$, the aliasing error bound is 10^{-8}; with $A = 25.3$, the aliasing error bound is 10^{-11}.)

Remark 8.1 It should be evident that formula (32) presents problems as t becomes very small, because the prefactor $(e^{A/2\ell})/2\ell t$ approaches infinity as $t \to 0$. Thus the variant of the algorithm in (31) should not be used for extremely small t, e.g., for $t < 10^{-2}$. We can of course approximate $f(t)$ by $f(0)$ for small t, and $f(0)$ can be calculated by taking limits using the initial value theorem, which states that if $f(t) \to f(0)$ as $t \to 0$, then $s\hat{f}(s) \to f(0)$ as $s \to \infty$. The small t problem can also be solved "automatically" by scaling; see Choudhury and Whitt [Choudhury and Whitt, 1997].

Remark 8.2 The form of the aliasing error in (35) has important implications for practical implementation. Note that the aliasing error is likely to be smaller for large t if we invert the transform $(1 - \hat{f}(s))/s$ of the ccdf $F^c(t) \equiv 1 - F(t)$ than if we invert the transform $\hat{f}(s)/s$ of the cdf $F(t)$. For the ccdf, the aliasing error is about $e^{-A}F^c((1+2\ell)t)$ when $F^c((k+1)t) \ll F^c(kt)$. Hence, for small ccdf values associated with large t, we can often achieve small *relative* aliasing error from (35).

We now indicate how to apply Euler summation to approximate the infinite series in (31). Euler summation can be very simply described as the weighted average of the last m partial sums by a binomial probability distribution with parameters m and $p = 1/2$. (It is not necessary to use $p = 1/2$, but it is usually not worthwhile to search for a better p.) In particular, let s_n be the approximation $f_{A,\ell}(t)$ in (31) with the infinite series truncated to n terms, i.e.,

$$s_n = \sum_{k=0}^{n} (-1)^k a_k, \qquad (37)$$

where t is suppressed in the notation and $a_k \equiv a_k(t)$ is given in (32)-(34). We apply Euler summation to m terms after an initial n, so that the Euler sum approximation to (31) is

$$E(m,n) \equiv E(m,n,t) \equiv \sum_{k=0}^{m} \binom{m}{k} 2^{-m} s_{n+k}, \qquad (38)$$

for s_n in (37). Hence, (38) is the binomial average of the terms $s_n, s_{n+1}, \ldots, s_{n+m}$. We typically use $m = 11$ and $n = 38$, increasing n as necessary. The overall computation is specified by (32)-(34), (37) and (38).

As in other contexts, the acceleration typically drastically reduces the required computation. The improvement in typical examples is very impressive; it should be tried on simple examples. We present additional supporting theory below. To quickly see how important this acceleration step is, note that the error in simple truncation to s_n can be estimated as the value of last term, a_n. For example, if $a_n = n^{-1}$, then we would need about $n = 10^8$ terms to achieve accuracy to 10^{-8}, whereas with Euler summation it typically suffices to have $n = 50$, as noted above.

In order to *estimate* the error associated with Euler summation, we suggest using the difference of successive terms, i.e., $E(m, n+1) - E(m, n)$. Unlike for the aliasing error, however, we have no simple general bound on the summation error associated with truncating the series at $n + m$ terms and applying Euler summation to average the last m terms. However, we discuss Euler summation further below. Experience indicates that the error estimate $E(m, n+1) - E(m, n)$ is usually accurate. Moreover, under regularity conditions, it is also an upper bound. However, it can happen that the error estimate is optimistic, as indicated in Section 11 of [Abate and Whitt, 1992a].

Practically, the accuracy can be verified by performing the same computation again with different parameters, e.g., changing, m, n, ℓ and A. Note that changing ℓ or A produces a very different computation, so that we obtain an accuracy check by doing the calculation for two different choices of the parameter pair (ℓ, A).

Balancing the Roundoff and Discretization Errors

In the setting of (36), we can obviously make the aliasing error arbitrarily small by choosing A sufficiently large. However, with limited precision (e.g., with double precision, usually 14–16 digits), we are constrained from choosing A too large, because the prefactor $(e^{A/2\ell})/2\ell t$ in (32) can then become very large, causing roundoff error. Given that high precision is often readily available, roundoff error should not to be regarded as a serious problem. From that perspective, our analysis below shows why higher precision may be needed. In most probability applications, an aliasing error of 10^{-8} is more than adequate, so that the required A does not produce too much roundoff error even with double precision.

We now indicate how to control both the roundoff and aliasing errors. The key to controlling the roundoff error is the fact that the aliasing error bound in (36) depends on A but not on ℓ. Suppose that we are able to compute the $b_k(t)$ in (34) with an error of about 10^{-m}. (This quantity is typically of the order of machine precision.) Then, after multiplying by the prefactor $(e^{A/2\ell})/2\ell t$ in (32), the final "roundoff" error in $f_{A,\ell}(t)$ from (31) will be about $10^{-m}(e^{A/2\ell})/2\ell t$. Since the roundoff error estimate is increasing in A, while the aliasing error estimate e^{-A} in (36) (assuming that $C = 1$) is decreasing in A, the maximum of the two error estimates is minimized where the two error estimates are equal. The estimated total error should thus be approximately minimized at this point. Thus we find an appropriate value for the parameter A by solving the equation

$$e^{-A} = 10^{-m}(e^{A/2\ell})/2\ell t, \tag{39}$$

which yields

$$A = \left(\frac{2\ell}{2\ell + 1}\right)(m \log 10 + \log 2\ell t). \tag{40}$$

Ignoring the final $\log 2\ell t$ term in (40), we see that (40) implies that the final error estimate is

$$e^{-A} \approx 10^{2\ell m/(2\ell+1)}, \tag{41}$$

which means that we lose a proportion of $1/(2\ell+1)$ of our m-digit precision due to multiplying by the prefactor. We obtain higher precision by increasing ℓ, but at a computational expense, because the computational effort is proportional to ℓ; see (33) and (34).

With $\ell = 1$, we lose about one third of the machine precision due to roundoff, and with $\ell = 2$, we lose about one fifth of the machine precision. If we take the typical double-precision value of $m = 14$ and let $t = 1$ and $\ell = 1$, then the error estimate (41) becomes about $10^{-9.3}$ and with $\ell = 2$ the error estimate is about $10^{-11.2}$.

From this analysis, we see that it is difficult to compute quantities that are smaller than or the same order as machine precision. This difficulty can be addressed by scaling; see [Choudhury and Whitt, 1997].

We now summarize the variant of the Fourier-series method described above.

Algorithm Summary

Based on the parameters A, ℓ, m and n (e.g., $A = 19$, $\ell = 1$, $m = 11$ and $n = 38$) and the function argument t, approximately compute $f(t)$ by (38), (37) and (32)–(34). The aliasing error is given by (35). If $|f(t)| \leq C$ for all $t \geq 0$, then the aliasing error is bounded by (36) and is approximately Ce^{-A}. The overall error (including the Euler summation error) is estimated by $E(m, n + 1) - E(m, n)$ using (38). The overall error is also estimated by performing the computation with two different parameter pairs (A, ℓ), e.g., $(18, 1)$ and $(19, 1)$ or $(18, 1)$ and $(18, 2)$.

We reiterate that we give no a priori error bound for the entire algorithm, primarily because we have no simple general bound on the error from Euler summation. (Under additional assumptions, bounds are possible, though; see below.) However, when the algorithm is applied, we invariably can see the achieved accuracy by comparing the results of computations with different parameters. If the achieved accuracy is not sufficient, then we suggest: first, increasing n; second, increasing the roundoff control parameter ℓ; and third, considering convolution smoothing, as described in Section 6 of [Abate and Whitt, 1992a].

This section is based on Abate and Whitt [Abate and Whitt, 1992a, Abate and Whitt, 1995] and Choudhury, Lucantoni and Whitt [Choudhury et al., 1994]. The Fourier-series method for numerically inverting Laplace transforms was proposed by Dubner and Abate [Dubner and Abate, 1968]. The use of the parameter ℓ for roundoff error control was proposed for the case $l = 2$ by Durbin [Durbin, 1974] and was also used more generally by Kwok and Barthez [Kwok and Barthez, 1989] and Choudhury, Lucantoni and Whitt [Choudhury et al., 1994]. The use of Euler summation in this context was proposed by Simon, Stroot and Weiss [Simon et al., 1972]. The Fourier-series method with Euler summation was later developed independently in Japan by Hosono [Hosono, 1979], [Hosono, 1981], [Hosono, 1984].

There is an extensive body of related literature; for further discussion see Abate and Whitt [Abate and Whitt, 1992a]. We have presented only one variant of the one method for numerically inverting Laplace transforms. There are other variants of the Fourier-series method and there are entirely different methods. The Jagerman-Stehfest and Gaver-Stehfest procedures in Section 8 of Abate and Whitt [Abate and Whitt, 1992a] are based on the Post-Widder formula and its discrete analog in Theorems 2 and 3. The Laguerre-series algorithm in Abate, Choudhury and Whitt [Abate et al., 1996] is based on the Laguerre-series representation in Theorem 4. See these sources for additional background. See Davies and Martin [Davies and Martin, 1979] for a (somewhat dated) comparison of alternative methods. See Choudhury and Whitt [Choudhury and Whitt, 1995] for a description of the Q^2 performance analysis tool based on numerical transform inversion.

2.3 More on Euler Summation

Since Euler summation proves to be so effective in accelerating convergence in the Fourier-series algorithm, it is interesting to examine it in more detail. This section is somewhat more technical than others, and so might be skipped. This section draws upon Johnsonbaugh [Johnsonbaugh, 1979], Section 6 of [Abate and Whitt, 1992a] and O'Cinneide [O'Cinneide, 1997]. For an advanced treatise on acceleration techniques, see Wimp [Wimp, 1981].

A good way to understand how Euler summation performs is to try it (apply (37) and (38)) with some simple examples, e.g., $a_k = k^{-p}$ for $p > 0$. Because of the alternating series structure, with simple truncation the maximum of the errors $|s_n - s_\infty|$ and $|s_{n-1} - s_\infty|$ must be at least $a_n/2$. On the other hand, the averaging associated with Euler summation can lead to amazing improvement.

We now analyze the performance more carefully. First note that if $a_n \geq a_{n+1} \geq 0$ for all n in (37), then

$$s_{2n+2} \geq s_{2n} \geq s_\infty \geq s_{2n+1} \geq s_{2n-1}$$

and

$$|s_\infty - s_n| \leq a_n \qquad (42)$$

for all $n \geq 1$.

Next observe that we can write $E(m,n)$ itself in the form of an alternating series by changing the order of summation, i.e.,

$$E(m,n) = \sum_{j=0}^{m} \binom{m}{j} 2^{-m} \sum_{k=0}^{n} (-1)^{k+j} a_{k+j} = \sum_{k=0}^{n} (-1)^k b_k,$$

where

$$b_k = \sum_{j=0}^{m} \binom{m}{j} 2^{-m} (-1)^j a_{k+j} . \qquad (43)$$

Of course, in general b_k in (43) need not be of constant sign. However, from (42), we see that if $b_k \geq b_{k+1} \geq 0$ for all k, then

$$|E(m,n) - s_\infty| \leq b_n = |E(m,n) - E(m,n-1)| \qquad (44)$$

and our error estimate $|E(m,n) - E(m,n-1)|$ becomes a bound on the actual error.

To see the quantitative benefits of Euler summation, it is useful to consider (37) and (38) with $a_n = (c+n)^{-k}$ for some constant c and positive integer k. (For any c, a_n is positive and decreasing for $n > -c$.) We can exploit the identity

$$\frac{1}{(c+n)^k} = \frac{1}{(k-1)!} \int_0^\infty e^{-(c+n)x} x^{k-1} dx,$$

which follows from the form of the gamma distribution, to obtain the following result. See O'Cinneide [O'Cinneide, 1997].

Theorem 6 *If $a_n = (c+n)^{-k}$, then the summands b_k in (43) indeed are positive and decreasing when $c + k > 0$, so that the bound (44) holds for n suitably large and*

$$|E(m,n) - E(m,n-1)| \leq \frac{(m+k-1)!}{2^m(k-1)!(c+n)^{m+k}} \ . \tag{45}$$

Hence, when $a_n = (c+n)^{-k}$, from (42) the error using s_n in (37) (direct truncation) is bounded by $(c+n)^{-k}$, whereas the error from Euler summation is bounded above by $C(c+n)^{-(m+k)}$ from (44) and (45). Asymptotically, as $n \to \infty$ for fixed m, the rate of convergence improves from n^{-k} to $n^{-(m+k)}$ when we use Euler summation.

We now show that in the inversion algorithm we actually have a_n asymptotically of the form $C(c+n)^{-k}$ for sufficiently large n, under mild smoothness conditions, so that the analysis above is directly applicable to the inversion problem. For this purpose, let

$$\text{Re}(f)(u+iv) = \int_0^\infty e^{-ut} \cos vt f(t)dt = \int_0^\infty \cos vt g(t)dt \ , \tag{46}$$

where g is the damped function $g(t) \equiv e^{-ut}f(t)$.

If f is twice continuously differentiable, then so is g. If $g(\infty) = g'(\infty) = 0$ (which is satisfied when $f(\infty) = f'(\infty) = 0$), then we can apply integration by parts in (46) to get

$$\text{Re}(\hat{f})(u+iv) = \frac{uf(0) - f'(0)}{v^2} - \frac{1}{v^2}\int_0^\infty \cos vt g''(t)dt \ . \tag{47}$$

If in addition g'' is integrable, then we can apply the Reimann-Lebesgue lemma to deduce that

$$\text{Re}(\hat{f})(u+iv) = \frac{uf(0) - f'(0)}{v^2} + o\left(\frac{1}{v^2}\right) \quad \text{as} \quad v \to \infty \ . \tag{48}$$

Similarly,

$$\begin{aligned}
\text{Im}(f)(u+iv) &= \int_0^\infty \sin vt g(t)dt \\
&= \frac{-f(0)}{v} + \frac{1}{v}\int_0^\infty \cos vt g'(t)dt \\
&= \frac{-f(0)}{v} + o\left(\frac{1}{v}\right) \quad \text{as} \quad v \to \infty \tag{49}
\end{aligned}$$

provided that g' is integrable.

From (48) and (48), we see that $\text{Re}(\hat{f})(u+iv)$ and $\text{Im}(\hat{f})(u+iv)$ will indeed be eventually of constant sign as v increases (including the cases in which $u = f'(0)$ and $f(0) = 0$, which require further analysis). From (34) we see that the summands a_n in the infinite series to be computed are asymptotically of the form $C(c+n)^k$ for $k=2$ (real part) or $k=1$ (imaginary part), where C and c are known constants. Thus the analysis above shows that Euler summation will be effective provided that the function f indeed has the smoothness and integrability properties required for (48) and (48).

Furthermore, with extra smoothness, we can repeat the integration by parts argument in (47) and obtain a more detailed analysis. Explicit error bounds were first determined by O'Cinneide [O'Cinneide, 1997]. The analysis shows that

$$|E(2m,n) - s_\infty| \sim Cn^{-(2m+2)} \text{ as } n \to \infty$$

if g has $2m+2$ derivatives with $g^{(k)}(\infty) = 0$ for $0 \le k \le 2m+1$ and $g^{(2m+2)}$ is integrable.

Example 3 To see that the desired properties supporting Euler summation do *not* hold in general, consider the ccdf of a unit probability point mass at x, with LT

$$\hat{F}^c(s) = \frac{1 - e^{-sx}}{s},$$

The ccdf F^c is constant except for the one jump at x. Note that

$$\text{Re}(\hat{F}^c)(u+iv) = \frac{u(1 - e^{-ux}\cos vx) + v\sin vx}{u^2 + v^2} \tag{50}$$

From (50), we see that, for every u, there is no v_0 such that $\text{Re}(\hat{F}^c)(u+iv)$ is of constant sign for $v \ge v_0$. Moreover, in this case Euler summation does not provide improvement over simple truncation; see Table 5 of Abate and Whitt [Abate and Whitt, 1992a].

Example 4 To see that Euler summation does not perform well on all alternating series, consider the series

$$\sum_{k=1}^{\infty} (-1)^k (\sin kx)^2 / k = \frac{\log(1/\cos x)}{2} \quad \text{for} \quad 0 < x < \pi/2.$$

The performance of Euler summation degrades as x increases. For example, the error in $E(11, 30)$ is of order 10^{-9}, 10^{-3} and 10^{-1} for $x = 0.1$, 1.0 and 1.5, respectively. If $x = 1.5$, s_{41} had an error of order 10^{-2}, which is better.

2.4 Generating Functions

The Fourier-series method also applies to invert generating functions. Suppose that $\{q_k : k \ge 0\}$ is a sequence of complex numbers with generating function

$$\mathcal{G}(q) \equiv \hat{q}(z) \equiv \sum_{k=0}^{\infty} q_k z^k, \tag{51}$$

where z is a complex number, and we want to calculate q_k.

At the outset, note that this problem should be regarded as easier than the Laplace transform inversion, because the coefficients q_k in (51) can be recovered from the generating function $\hat{q}$ by repeated differentiation and evaluation at $z = 0$. This means that numerical differentiation techniques and symbolic mathematical software are often readily applicable. However, it is often difficult to achieve desired accuracy with numerical differentiation techniques, especially for large n. It is also difficult to invoke symbolic mathematical software when the generating function is only expressed implicitly. Fortunately, in this setting numerical inversion is also a viable alternative. We will apply the numerical inversion algorithm for generating functions of complex numbers in Section 5 to calculate normalization constants in product-form closed queueing networks. In that context, it is important to allow q_k to be complex numbers in order to invert multidimensional transforms using iterative one-dimensional inversion.

Just as for Laplace transforms, numerical inversion of generating functions can be based on an inversion integral. The Cauchy contour integral plays the role for generating functions that the Bromwich contour integral plays for Laplace transforms. The following parallels Theorem 1.

Theorem 7 *Suppose that $\{q_k : k \geq 0\}$ is a sequence of complex numbers with $|q_k| \leq Kb^k$ for all $k \geq 1$, where K and b are positive constants. Then*

$$q_k = \frac{1}{2\pi i} \int_\Gamma \frac{\hat{q}(z)}{z^{n+1}} dz, \qquad (52)$$

$$= \frac{1}{2\pi r^k} \int_0^{2\pi} \hat{q}(re^{iu}) e^{-iku} du \ .$$

where the contour Γ is a circle centered at the origin with radius r, where r is less than the radius of convergence b^{-1}.

Paralleling the inversion of Laplace transforms, a good way to proceed is to apply the trapezoidal rule in the integral in (52). Just as in Section 2.1, we can apply the Fourier series method to determine the discretization error associated with the trapezoidal rule. The following result thus parallels Theorem 5. Note that $q_0 = \hat{q}(0)$, so it suffices to produce an algorithm to calculate q_k for $k \geq 1$.

Theorem 8 *Under the assumptions of Theorem 7, for $0 < r < b^{-1}$ and $k \geq 1$,*

$$q_k = q_k^a - e_a, \qquad (53)$$

where the trapezoidal-rule approximation to (52) is

$$q_k^a = \frac{1}{2k\ell r^k} \sum_{j=1}^{2k} (-1)^j a_j \qquad (54)$$

with

$$a_j \equiv a_j(k, \ell, r) = \sum_{j_1=0}^{\ell-1} e^{-\pi i j_1/\ell} \hat{q}(re^{\pi i(j_1+\ell j)/\ell k}), \quad 1 \leq j \leq 2k, \qquad (55)$$

TRANSFORM INVERSION

and the associated *discretization* or aliasing error *is*

$$e_a = \sum_{j=1}^{\infty} q_{k(1+2j\ell)} r^{2jk\ell} . \qquad (56)$$

Proof. To establish (54) and (56), we form the damped sequence $a_k = q_k r^k$ for $0 < r < b^{-1}$ and let $a_k = 0$ for $k < 0$. Then we apply the discrete Fourier transform to (calculate the discrete Fourier series of) the aliased sequence

$$a_k^p = \sum_{j=-\infty}^{\infty} a_{k+jm} . \qquad (57)$$

Note that $\{a_k^p\}$ is a periodic sequence with period m. The assumptions imply that $\sum_{k=-\infty}^{\infty} |a_k| < \infty$, so that the series in (57) converges.

In preparation for the next step, let the Fourier transform of the sequence $\{a_k\}$ be

$$\phi(u) = \sum_{k=-\infty}^{\infty} a_k e^{iku} = \hat{q}(re^{iu}) ,$$

which has inverse

$$a_k = \frac{1}{2\pi} \int_0^{2\pi} \phi(u) e^{-iku} du .$$

Now the *discrete Fourier transform* of the periodic sequence $\{a_k^p\}$ is

$$\begin{aligned}
\hat{a}_k^p &= \frac{1}{m} \sum_{j=0}^{m-1} a_j^p e^{i2\pi kj/m} \\
&= \frac{1}{m} \sum_{j=0}^{m-1} \sum_{\ell=-\infty}^{\infty} a_{j+\ell m} e^{i2\pi jk/m} \\
&= \frac{1}{m} \sum_{j=-\infty}^{\infty} a_j e^{i2\pi jk/m} = \frac{1}{m} \phi(2\pi k/m). \qquad (58)
\end{aligned}$$

Now apply the inversion formula for discrete Fourier transforms to obtain

$$a_k^p = \sum_{j=0}^{m-1} \hat{a}_j^p e^{-i2\pi jk/m} = \frac{1}{m} \sum_{j=0}^{m-1} \phi(2\pi j/m) e^{-i2\pi jk/m} . \qquad (59)$$

The combination of (57) and (59) is the *discrete Poisson summation formula*

$$\sum_{j=-\infty}^{\infty} a_{k+jm} = \frac{1}{m} \sum_{j=0}^{m-1} \phi(2\pi j/m) e^{-i2\pi jk/m} .$$

Putting the term for $k = 0$ on the left, we obtain for integers k and $m > 0$

$$a_k = \frac{1}{m} \sum_{j=0}^{m-1} \phi(2\pi j/m) e^{-i2\pi jk/m} - \sum_{\substack{j=-\infty \\ j \neq 0}}^{\infty} a_{k+jm} . \qquad (60)$$

Finally, we obtain (54) and (56) by setting $m = 2k\ell$ in (60). Note that $a_{k+jm} = 0$ for $j < 0$ with this choice. In particular,

$$q_k r^k \equiv a_k = \frac{1}{2k\ell} \sum_{j=0}^{2k\ell-1} \phi(\pi j/k\ell) e^{-i\pi j/\ell} - \sum_{j=1}^{\infty} q_{k(1+2j\ell)} r^{k+2jk\ell} ,$$

so that

$$q_k = \frac{1}{2k\ell r^k} \sum_{j=0}^{2k\ell-1} \hat{q}(re^{i\pi j/k\ell}) e^{-i\pi j/\ell} - \sum_{j=1}^{\infty} q_{k(1+2j\ell)} r^{2jk\ell}$$

which gives (53)–(56).

Note that if $|q_n| \leq C$ for all $n \geq (1+2\ell)k$, then

$$|e_a| \leq \sum_{j=1}^{\infty} Cr^{2jk\ell} \leq \frac{Cr^{2k\ell}}{1 - r^{2k\ell}}, \qquad (61)$$

which is approximately $Cr^{2k\ell}$ when $r^{2k\ell}$ is suitably small. To make the aliasing error about $10^{-\eta}$ when $C = 1$, we let $r = 10^{-\eta/2k\ell}$, where ℓ is chosen to control the roundoff error (the default value being $\ell = 1$). In typical applications we may choose $\eta = 8$. However, in some applications there is no bound on $|q_k|$. We then may need to apply an additional scaling algorithm, as illustrated in Section 5.

It is possible to reduce the computations by a factor of 2 if q_k is real-valued by using the fact that $\hat{q}(\bar{z}) = \overline{\hat{q}(z)}$. Then (54) can be replaced by

$$q_k^a = \frac{1}{2k\ell r^k} \sum_{j=1}^{2k} (-1)^j Re(a_j) \qquad (62)$$

$$= \frac{1}{2k\ell r^k} \left(a_0(k,\ell,r) + (-1)^k a_k(k,\ell,r) + 2 \sum_{j=1}^{k-1} (-1)^j Re\left(a_j(k,\ell,r)\right) \right)$$

for $a_j \equiv a_j(k,\ell,r)$ in (55).

Unlike for the Laplace transforms in Section 2.1, the series (54) is finite, so that there is no need to approximate an infinite series. Hence there is no need to apply Euler summation with generating functions. However, if the index k is very large, then it may still be advantageous to apply Euler summation to accelerate convergence of the finite sum (54), as we illustrate in the closed queueing network example in Section 5.4. The representation (54) has been chosen to be nearly an alternating series, so that it is directly in the form to apply Euler summation, just as in (38).

The roundoff control procedure is essentially the same as for Laplace transforms. If we can compute the sum in (54) without the prefactor $(2k\ell r^k)^{-1}$ with a precision (error estimate) of 10^{-m}, then the roundoff error after multiplying by the prefactor will be approximately $(2k\ell r^k)^{-1}10^{-m}$. Since the roundoff error estimate is decreasing in r as r approaches 1 from below, while the aliasing error estimate is increasing in r, the maximum of the two error estimates will be minimized when the two estimates are equal. Thus the total error should be approximately minimized at this point. This leads to the equation

$$r^{2k\ell} = (2k\ell r^k)^{-1}10^{-m}, \tag{63}$$

Assuming that we can ignore the term $2k\ell$ on the right in (63), we get $r^{(2\ell+1)k} \approx 10^{-m}$ or

$$r \approx 10^{-m/(2\ell+1)k} \quad \text{and} \quad r^{2\ell k} \approx 10^{-2\ell m/(2\ell+1)}.$$

As in Section 2.1, this analysis shows that we approximately lose a proportion $1/(2\ell + 1)$ of our precision due to roundoff error. If $m = 12$, then we can achieve an overall error estimate of about 10^{-8} by setting $\ell = 1$ and $r = 10^{-4/k}$. By increasing ℓ, we can get close to 10^{-m} but never below it. To accurately calculate smaller numbers than 10^{-m} we need to apply scaling; see Section 5.3.

We now summarize the algorithm.

Generating Function Algorithm Summary.. *Based on the desired sequence index k and the parameters ℓ and r (e.g., $\ell = 1$ and $r = 10^{-4/k}$ corresponding to $\eta = 8$), approximately compute q_k from its generating function $\hat{q}$ in (51) by (54). If q_k is real-valued, then replace (54) by (62). The aliasing error is (56). If $|q_k| \leq C$ for all k, then the aliasing error is bounded by (61) and approximated by $Cr^{2k\ell}$ (which is 10^{-8} when $C = 1$, $\ell = 1$ and $r = 10^{-4/k}$). If the index k is large, it may be advantageous to apply Euler summation using (38).*

The algorithm in this section is based on Abate and Whitt [Abate and Whitt, 1992a, Abate and Whitt, 1992b] and Choudhury, Lucantoni and Whitt [Choudhury et al., 1994], but as they indicate, there is a substantial body of related literature. Nevertheless, surprisingly, this algorithm was not well known.

3 TRANSIENT CHARACTERISTICS OF THE ERLANG LOSS MODEL

This section contains our first nontrivial example illustrating how the numerical inversion of Laplace transforms can be applied. We apply the Fourier-series algorithm in Section 2.1, but we could also apply other algorithms, such as the Laguerre-series algorithm based on Theorem 4.

Given that inversion algorithms are available, typically the major challenge in applications is efficiently computing the required Laplace transform values.

Fortunately, much previous work in applied probability has been devoted to deriving transforms of random quantities of interest. As indicated at the outset, excellent examples are the queueing books by Takács [Takács, 1962] and Cohen [Cohen, 1982]. Nevertheless, computing transform values can be a challenge. Sometimes transforms are only available as integrals, as in the Pollaczek contour integral expression for the GI/G/1 waiting time, to be discussed in Section 4. On other occasions, transforms are only available implicitly, as in Kendall functional equation for the M/G/1 busy period.

In this section we consider the classical Erlang loss model, i.e., the M/M/c/0 system with Poisson arrival process, exponential service times, c servers and no extra waiting space, where blocked calls are lost. We let the individual service rate be 1 and the arrival rate (which coincides with the offered load) be a. The way to compute steady-state characteristics for this model is very well known, but that is not the case for transient (time-dependent) characteristics. Transience arises by considering arbitrary fixed initial states. We show how to compute several transient characteristics by numerical transform inversion. This section draws on Abate and Whitt [Abate and Whitt, 1998].

Before starting, we mention other applications of numerical transform inversion to calculate transient characteristics of queueing models. The M/G/1 busy period distribution is treated in Abate and Whitt [Abate and Whitt, 1992c]. The time-dependent queue-length and workload processes in the M/G/1, BMAP/G/1 and $M_t/G_t/1$ queues are treated in Choudhury, Lucantoni and Whitt [Choudhury et al., 1994], Lucantoni, Choudhury and Whitt [Lucantoni et al., 1994], and Choudhury, Lucantoni and Whitt [Choudhury et al., 1997], respectively. Both steady-state and time-dependent distributions in polling models are calculated in Choudhury and Whitt [Choudhury and Whitt, 1996]. The time-dependent distributions of semi-Markov processes are calculated in Duffield and Whitt [Duffield and Whitt, 1998].

Here we develop algorithms for computing four quantities in the M/M/c/0 model: the time-dependent blocking probability starting at an arbitrary initial state i, i.e., the transition probability

$$P_{ic}(t) \equiv P(N(t) = c | N(0) = i) ,$$

where $N(t)$ is the number of busy servers at time t; the complementary cumulative distribution function (ccdf) $F_{ic}^c(t)$ of the time T_{ic} all servers first become busy starting at an arbitrary initial state i; i.e., where

$$F_{ic}^c(t) \equiv 1 - F_{ic}(t) \equiv P(T_{ic} > t) ,$$

and

$$T_{ic} \equiv \inf\{t \geq 0 : N(t) = c | N(0) = i\} ;$$

the time-dependent mean

$$M_i(t) \equiv E(N(t)|N(0) = i) ; \tag{64}$$

and the (stationary) covariance function

$$R(t) \equiv Cov(N_s(u), N_s(u+t)) \tag{65}$$
$$= E(N_s(u)N_s(u+t)) - EN_s(u)EN_s(u+t),$$

where $\{N_s(t) : t \geq 0\}$ is a stationary version of $\{N(t) : t \geq 0\}$, i.e., where $N_s(u)$ in (65) is distributed according to the steady-state distribution

$$\pi_j \equiv P(N_s(u) = j) = \frac{a^j/j!}{\sum_{k=0}^{c} a^k/k!}. \tag{66}$$

We also show how to compute these quantities for very large systems by performing computations for moderately sized systems and using scaling based on the established heavy-traffic limit in which $(N^{(a)}(t)-a)/\sqrt{a}$ converges to the reflected Ornstein-Uhlenbeck (ROU) process as $a \to \infty$ with $i(a) - a \sim \gamma_1\sqrt{a}$ and $c(a) - a \sim \gamma_2\sqrt{a}$, where $f(a) \sim g(a)$ means that $f(a)/g(a) \to 1$ as $a \to \infty$; see p. 177 of Borovkov [Borovkov, 1984] and Srikant and Whitt [Srikant and Whitt, 1996]. The ROU process is the ordinary OU process modified to have a reflecting upper barrier. The OU process is a diffusion process with constant diffusion coefficient and proportional state-dependent drift, i.e., with drift $-\delta x$ in state x. However, we will not focus on the ROU process; we will only use the scaling.

For example, suppose that we want to compute $P_{ic}(t)$ for some large a such as $a = 10^8$, where c and i are allowed to depend on a via $c(a) = \lfloor a + \sqrt{a} \rfloor$ and $i(a) = \lfloor a - 2\sqrt{a} \rfloor$, with $\lfloor x \rfloor$ being the greatest integer less than or equal to x. We will write $P_{i(a)c(a)}^{(a)}$ to indicate the dependence upon the offered load a. The heavy-traffic limit implies that $P_{i(a)c(a)}^{(a)}/B(c(a), a)$ should be approximately independent of a, where $B(c(a), a) \equiv P_{i(a)c(a)}^{(a)}(\infty) \equiv \pi_{c(a)}^{(a)}$ is the steady-state Erlang blocking probability, which is known to have the asymptotic relation

$$B(c(a), a) \sim \frac{1}{\sqrt{a}} \frac{\phi(\gamma)}{\Phi(-\gamma)} \text{ as } a \to \infty,$$

where ϕ is the density and Φ is the cdf of a standard (mean 0, variance 1) normal distribution and γ is the limit of $(a-c)/\sqrt{a}$; see Jagerman [Jagerman, 1974], Whitt [Whitt, 1984], and equation (15) of Srikant and Whitt [Srikant and Whitt, 1996]. Hence, we can compute $P_{i(10^8)c(10^8)}^{(10^8)}(t)$ approximately using results for $a = 400$ as follows:

$$P_{i(10^8)c(10^8)}^{(10^8)}(t) \approx \frac{B(10^8 + 10^4, 10^8)}{B(400 + 20, 400)} P_{i(400)c(400)}^{(400)}(t)$$
$$\approx \left(\frac{20}{10^4}\right) P_{i(400)c(400)}^{(400)}(t), \tag{67}$$

with $i(a) = \lfloor a - 2\sqrt{a} \rfloor$ and $c(a) = \lfloor a + \sqrt{a} \rfloor$ in each case; e.g., $i(a) = 360$ for $a = 400$ and $i(a) = 10^8 - 2(10^4)$ for $a = 10^8$. We will show the effectiveness of the scaling in numerical examples.

The algorithms here are based on computing the Laplace transforms of these quantities with respect to time and then applying the Fourier-series method. For the most part, algorithms for computing the transforms are available in the literature. In particular, an algorithm to calculate the Laplace transform of $P_{ij}(t)$ is given on pp. 81–84 of Riordan [Riordan, 1962], but it does not seem to be widely known. Formulas for the Laplace transform of the mean and the covariance are given in Beneš [Beneš, 1961], [Beneš, 1965] and Jagerman [Jagerman, 1978], but the formula for the covariance transform in (15) on p. 209 of [Beneš, 1965] and (15) on p. 136 of [Beneš, 1961] has a sign error. Abate and Whitt [Abate and Whitt, 1998] derived a new formula for the covariance transform, given below in Theorem 3.2.

The numerical inversion algorithm is an alternative to the spectral expansion described in Beneš [Beneš, 1961], [Beneš, 1965] and Riordan [Riordan, 1962]. The spectral expansion is efficient for computing values at many time points, because the eigenvalues and eigenvectors need only be computed once. However, the inversion algorithm is also fast, and remarkably simple.

The numerical inversion algorithm is also an alternative to the numerical solution of a system of ordinary differential equations (ODEs), which is discussed here in Chapter 3. Numerical solution of ODEs has the advantage that it applies to time-dependent models as well as the transient behavior of stationary models with nonstationary initial conditions. However, when the numerical inversion algorithm applies, it has the advantage that it can produce calculations at any desired t without having to compute the function over a large set of time points in the interval $[0, t]$.

Finally, asymptotic formulas can serve as alternatives to exact numerical algorithms in the appropriate asymptotic regimes. Such asymptotic formulas are given in Mitra and Weiss [Mitra and Weiss, 1989] and Knessl [Knessl, 1990]. These asymptotic formulas are very attractive when they are both simple and sufficiently accurate, but many of the asymptotic formulas are not simple. Then they properly should be viewed as alternatives to numerical algorithms. It appears that the numerical algorithm here is much more accurate than the asymptotic approximations.

3.1 Time-Dependent Blocking Probabilities

As shown on pp. 81–84 of Riordan [Riordan, 1962], the Laplace transform

$$\hat{P}_{ij}(s) \equiv \int_0^\infty e^{-st} P_{ij}(t) dt$$

is easily computed recursively, exploiting relations among the Poisson-Charlier polynomials. Since Riordan was not developing a numerical inversion algorithm, he was not interested in a numerical algorithm for computing the transform, so it is not highlighted, but it is there. The key relation is (8) on p. 84 of [Riordan, 1962] using the recursions (3) and (4). The determinant $|D|$ in (8) is evaluated in (6).

We will focus on $P_{ij}(t)$ only for $j = c$, but the general case can be computed as well. To express the result for $P_{ic}(t)$, let

$$d_n \equiv d_n(s, a) = (-1)^n C_n(-s, a) ,$$

where s is a complex variable and $C_n(s, a)$ are the Poisson-Charlier polynomials; i.e.,

$$d_n = \frac{1}{a^n} \sum_{k=0}^{n} \binom{n}{k} s(s+1) \ldots (s+k-1) a^{n-k} ; \qquad (68)$$

e.g.,

$$d_0 = 1, \quad d_1 = \frac{1}{a}(a + s) \qquad (69)$$

$$d_2 = \frac{1}{a^2}(a^2 + (2a+1)s + s^2) . \qquad (70)$$

We now specify the algorithm for computing $\hat{P}_{ic}(s)$ for any desired i, c and complex s. We use the polynomials d_n, but we do not compute them via (68); instead we compute them recursively. Our algorithm follows from the recursive relations in Riordan [Riordan, 1962].

Theorem 9 *The Laplace transform of the time-dependent blocking probability is*

$$\hat{P}_{ic}(s) = d_i \hat{P}_{0c}(s) , \qquad (71)$$

where

$$\hat{P}_{0c}(s) = \frac{1}{a(d_{c+1} - d_c)} ,$$

d_0 *and* d_1 *are given in (69) and*

$$d_{n+1} = (1 + \frac{n}{a} + \frac{s}{a}) d_n - \frac{n}{a} d_{n-1} , \quad n \geq 1 . \qquad (72)$$

Since $\{N_s(t) : t \geq 0\}$ is a stationary reversible process, e.g., see p. 26 of Keilson [Keilson, 1979], $\pi_i P_{ic}(t) = \pi_c P_{ci}(t)$. Hence, we can also calculate $P_{ci}(t)$ directly from $P_{ic}(t)$ by

$$P_{ci}(t) = (\pi_i / \pi_c) P_{ic}(t) = \frac{a^i c!}{a^c i!} P_{ic}(t) .$$

As indicated in the introduction, $P_{ic}^{(a)}(t)/B(c, a)$ should be approximately independent of a provided that $i \equiv i(a) \approx a + \gamma_1 \sqrt{a}$ and $c \equiv c(a) \approx a + \gamma_2 \sqrt{a}$ for arbitrary constants γ_1 and γ_2 (which we think of as being in the interval $[-5, 5]$). To calculate the Erlang blocking probability $B(c, a)$, we use the well known recurrence

$$B(c, a) = \frac{1}{1 + \frac{c}{aB(c-1,a)}} . \qquad (73)$$

The Erlang blocking probability B is related to the polynomial d_n by $d_n(1,a) = 1/B(n,a)$. The recurrence relation (73) itself follows directly from another recurrence relation for d_n, namely,

$$d_n(s,a) = d_n(s+1,a) - \frac{n}{a} d_{n-1}(s+1,a) \ ;$$

see Corollary 3 on p. 549 of Jagerman [Jagerman, 1974]. The polynomials d_n are related to the sigma functions used in Beneš [Beneš, 1965] and other early references by $\sigma_s(n) = a^n d_n(s,a)/n!$

We now illustrate the algorithm with a numerical example. We will consider five cases with five different values of a, ranging from $a = 100$ to $a = 10,000$, where $\gamma_1 = (i(a) - a)/\sqrt{a} = -3$ and $\gamma_2 = (c(a) - a)/\sqrt{a} = 2$. The five cases with steady-state performance measures are displayed in Table 1. Let M and V be the mean and variance of the steady-state number of busy servers, i.e., $M = a(1-B)$ and

$$V = M - aB(c - M) = M - aB(c - a) - (aB)^2 \ . \tag{74}$$

The effectiveness of the scaling is shown in Table 1 through the values of $\sqrt{a}B$ and V/a, which are nearly independent of a.

Table 1 The five cases ($\gamma_1 = -3$ and $\gamma_2 = 2$).

cases	c	a	i	B	M	V	$\sqrt{a}B$	V/a
I	120	100	70	.0056901	99.43	87.73	.056901	.877271
II	440	400	340	.0028060	398.88	352.72	.056120	.881806
III	960	900	810	.0018613	898.33	795.01	.055840	.883341
IV	2600	2500	2350	.0011122	2497.22	2211.45	.055608	.884579
V	10200	10000	9700	.0005543	9994.46	8855.13	.055430	.885513

Numerical values of $P^{(a)}_{i(a),c(a)}(t)/B(c(a),a)$ for nine time points are displayed in Table 2. The values of B are computed from (73), while the values of $P^{(a)}_{i(a)c(a)}(t)$ are computed by the Fourier-series method as in Section 2.1 after computing the transform values by the algorithm in Theorem 9. The inversion parameters were set so that the transform was computed at 40 values of complex s in each case. For the largest case, $a = 10^4$, the computation took about two minutes using UBASIC on a PC. (See [Abate and Whitt, 1992a] for more on UBASIC.) As in Table 1, the effectiveness of the scaling in Table 2 is evident in the similarity of values in each row.

3.2 Other Descriptive Characteristics

Let $f_{ij}(t)$ be the probability density function (pdf) of the first passage time T_{ij} from state i to state j in the M/M/c/0 model. Clearly,

$$P_{ij}(t) = f_{ij}(t) * P_{jj}(t)$$

Table 2 Values of $P^{(a)}_{i(a),c(a)}(t)/B(c(a),a)$ in the five cases of Table 1.

time	$I(a=100)$	$II(a=400)$	$III(a=900)$	$IV(a=2,500)$	$V(a=10,000)$
1.0	.038920	.040993	.041755	.042435	.042836
1.5	.220241	.225617	.227479	.227581	.230147
2.0	.459358	.464459	.466181	.467744	.468612
2.5	.657298	.660662	.661786	.662651	.663363
3.0	.792636	.794518	.795143	.795656	.796044
4.0	.928489	.928951	.929102	.929222	.929311
5.0	.976022	.976108	.976135	.976156	.976171
7.0	.9973498	.9973442	.9973420	.9973401	.9973386
10.0	.99990311	.99990208	.99990172	.99990141	.99990118

for all i and j, where $*$ denotes convolution. Hence, if

$$\hat{f}_{ij}(s) \equiv \int_0^\infty e^{-st} f_{ij}(t) dt ,$$

then

$$\hat{f}_{ij}(s) = \hat{P}_{ij}(s)/\hat{P}_{jj}(s) .$$

Since

$$\hat{F}^c_{ij}(s) = \frac{1 - \hat{f}_{ij}(s)}{s}$$

where

$$\hat{F}^c_{ij}(s) \equiv \int_0^\infty e^{-st} F^c_{ij}(t) dt$$

and $F^c_{ij}(t)$ is the ccdf of T_{ij}, we can calculate $F^c_{ij}(t)$ by numerical inversion too. In particular, given the algorithm for calculating $\hat{P}_{ic}(s)$ in Theorem 9, we can calculate $\hat{F}^c_{ic}(s)$ and $F^c_{ic}(t)$.

It is also possible to derive a recursion for the transform $\hat{f}_{i,i+1}(s)$ directly. Considering the possible times and locations of the first transition, we have $\hat{f}_{01}(s) = a/(a+s)$ and

$$\hat{f}_{i,i+1}(s) = \left(\frac{a+i}{a+i+s}\right) \left(\frac{a}{a+i} + \left(\frac{i}{a+i}\right) \hat{f}_{i-1,i}(s)\hat{f}_{i,i+1}(s)\right), \quad i \geq 1 . \quad (75)$$

From (75), we obtain for $i \geq 1$

$$\hat{f}_{i,i+1}(s) = \frac{a}{a+i+s - i\hat{f}_{i-1,i}(s)} . \quad (76)$$

On the other hand, we can derive (76) from (71) because

$$\hat{f}_{i,i+1}(s) = \frac{\hat{f}_{i,c}(s)}{\hat{f}_{i+1,c}(s)} = \frac{\hat{P}_{ic}(s)}{\hat{P}_{i+1,c}(s)} = \frac{d_i(s,a)}{d_{i+1}(s,a)} \quad (77)$$

and
$$\hat{f}_{0,i}(s) = 1/d_i(s,a) \ .$$

For example, the first relation in (77) holds because the first passage time from i to c is necessarily the sum of the independent first passage times from i to $i+1$ and from $i+1$ to c. The recursion (76) also follows from (72) and (77).

By the scaling for large a, the distribution of T_{ic} should be approximately independent of a when $c(a) = \lfloor a + \gamma_1 \sqrt{a} \rfloor$ and $i(a) = \lfloor a + \gamma_2 \sqrt{a} \rfloor$. Indeed, as $a \to \infty$ with $c(a) - a \sim \gamma_1 \sqrt{a}$ and $i(a) - a \sim \gamma_2 \sqrt{a}$, $T^{(a)}_{i(a)c(a)}$ converges in distribution to the first passage time τ_{γ_2,γ_1} of the Ornstein-Uhlenbeck (OU) diffusion process from γ_2 to γ_1; see Darling and Siegert [Darling and Siegert, 1953] and Keilson and Ross [Keilson and Ross, 1975].

We now give a numerical example. We compute the cdf $F_{ac}(t)$ for several values of t in the five cases given in Table 1. We let the initial state here be a instead of i; i.e., $\gamma_1 = 0$ instead of $\gamma_1 = -3$. The results are shown in Table 3.

Table 3 Values of the first-passage-time cdf $F_{ac(a)}(t)$ in the five cases given in Table 1 with $\gamma_1 = 0$ and $\gamma_2 = 2$.

time	$I(a=100)$	$II(a=400)$	$III(a=900)$	$IV(a=2,500)$	$V(a=10,000)$
2	.1755	.1694	.1674	.1657	.1644
4	.3318	.3230	.3199	.3175	.3156
6	.4564	.4461	.4426	.4397	.4375
8	.5576	.5467	.5429	.5398	.5375
10	.6400	.6291	.6252	.6221	.6197
20	.8715	.8638	.8611	.8588	.8571
30	.9541	.9500	.9485	.9473	.9463
40	.9836	.9817	.9809	.9803	.9798
80	.9997	.9997	.9996	.9996	.9996

We now turn to the time-dependent mean in (64). It has Laplace transform

$$\hat{M}_i(s) \equiv \int_0^\infty e^{-st} M_i(t) dt = \frac{i}{1+s} + \frac{a}{1+s}\left(\frac{1}{s} - \hat{P}_{ic}(s)\right) \ ;$$

see p. 215 of Beneš [Beneš, 1965]. Clearly $\hat{M}_i(s)$ is easily computed once we have $\hat{P}_{ic}(s)$.

Since $(N(t)-a)/\sqrt{a}$ converges to the ROU process as $a \to \infty$ with $i(a)-a \sim \gamma_1 \sqrt{a}$ and $c(a) - a \sim \gamma_2 \sqrt{a}$, we should have

$$m^{(a)}_{i(a)}(t) \equiv \frac{M^{(a)}_i(t) - a}{\sqrt{a}} \to m_i(t) \quad \text{as} \quad a \to \infty \ , \tag{78}$$

where $m_i(t)$ is the corresponding ROU mean function, provided that $i(a)$ and $c(a)$ are defined as above. We confirm the effectiveness of this scaling by computing the scaled mean $m^{(a)}_{i(a)}(t)$ in (78) for several different values of a. In particular, values of $-m_{i(a)}^{(a)}(t)$ are displayed in Table 4 for the same five cases as in Tables 1 and 2. Now we let $\gamma_1 = -3$ again, as in Tables 1 and 2.

Table 4 Values of the normalized mean $[a - M_{i(a)}^{(a)}(t)]/\sqrt{a}$ in the five cases given in Table 1 with $\gamma_1 = -3$.

time	$I(a = 100)$	$II(a = 400)$	$III(a = 900)$	$IV(a = 2,500)$	$V(a = 10,000)$
0.1	2.714512	2.714512	2.714512	2.714512	2.714512
0.5	1.819592	1.819592	1.819592	1.819592	1.819592
1.0	1.103903	1.103920	1.103925	1.103930	1.103638
1.5	.672385	.672445	.672466	.672483	.669390
2.0	.415669	.415718	.415733	.415743	.415751
3.0	.177146	.176943	.176865	.176800	.176748
5.0	.070190	.069547	.069316	.069124	.068976
7.0	.058365	.057607	.057335	.057111	.056938
10.0	.056954	.056174	.055895	.055664	.055486

We conclude this section by considering the covariance function in (65). We give two expressions for its Laplace transform derived in [Abate and Whitt, 1998].

Theorem 10 *The covariance function $R(t)$ has Laplace transform*

$$\hat{R}(s) \equiv \int_0^\infty e^{-st} R(t) dt$$

$$= \frac{V}{1+s} - \frac{(M-V)}{(1+s)^2} + \frac{(aB)^2}{(1+s)^2}\left(\frac{\hat{P}_{cc}(s)}{B} - \frac{1}{s}\right) \quad (79)$$

$$= \frac{V}{1+s} - \frac{(a-M)(\hat{M}_c(s) - (M/s))}{1+s}, \quad (80)$$

where $B \equiv B(c,a) \equiv \pi_c$ in (66), $M \equiv M_i(\infty) = a(1-B)$ and $V \equiv R(0)$ is given in (74).

We can apply (80) to obtain a useful direct expression for the covariance function.

Corollary. *The covariance can be expressed as*

$$R(t) = Ve^{-t} - (a-M)\int_0^t e^{-(t-u)}[M_c(u) - M]du \leq Ve^{-t}.$$

The Corollary to Theorem 10 yields a bound which is approached as $c \to \infty$; i.e., it is known that $R(t) = Ve^{-t}$ in the M/M/∞ model. Beneš proposes a simple approximation

$$R(t) \approx Ve^{-Mt/V}, \quad t \geq 0,$$

which is easy to compute and reasonably accurate; see p. 188 of [Beneš, 1965]. Since

$$Cov\left(\frac{N_s(u) - a}{\sqrt{a}}, \frac{N_s(u+t) - a}{\sqrt{a}}\right) = \frac{Cov(N_s(u), N_s(u+t))}{a}$$

we conclude that $C^{(a)}(t)/a$ should be approximately independent of a provided that $c(a) = a + \gamma\sqrt{a}$. We confirm this scaling in our numerical example below. In particular, values of the normalized covariance function $R(t)/a$ are displayed in Table 5. We use the same five cases (values of a) and same nine time points as in Table 4. From the evident convergence, it is clear that the values can be used to approximate the covariance function of the limiting ROU diffusion process as well.

Table 5 Values of the normalized covariance function $R(t)/a$ for the five cases in Table 1.

time	$I(a = 100)$	$II(a = 400)$	$III(a = 900)$	$IV(a = 2,500)$	$V(a = 10,000)$
0 .1	.784019	.788345	.789814	.791000	.791895
0 .5	.502346	.505750	.506913	.507853	.508564
1 .0	.288786	.291173	.291990	.292652	.293153
1 .5	.166203	.167816	.168370	.168819	.169159
2 .0	.095700	.096765	.097132	.097429	.097655
3 .0	.031748	.032192	.032345	.032469	.032564
5 .0	.003496	.003219	.003589	.003608	.003623
7 .0	.0003850	.0003948	.0003982	.0004010	.0004032
10 .0	.00001407	.00001455	.00001472	.00001486	.00001496

4 STEADY-STATE WAITING TIMES IN THE GI/G/1 QUEUE

This section contains a second nontrivial example illustrating how numerical inversion of Laplace transforms can be applied. In this section we consider the steady-state waiting-time distribution in the GI/G/1 queue, drawing upon Abate, Choudhury and Whitt [Abate et al., 1993], [Abate et al., 1994], [Abate et al., 1995b].

There is a single server with unlimited waiting space and the first-in first-out service discipline. The interarrival times and service times come from independent sequences of i.i.d. random variables. Let U and V be generic interarrival and service times with cdf's F and G, respectively. Let $\hat{f}$ and $\hat{g}$ be their Laplace Stieltjes transforms, e.g.,

$$\hat{f}(s) = \int_0^\infty e^{-st} dF(t) .$$

We assume that $EV < EU < \infty$, so that the system is stable; i.e., the steady-state waiting time, denoted by W, is well defined. (The steady-state limit exists and is proper; see [Asmussen, 1987] for details.) Without loss of generality, we assume that $EV = 1$ and $EU = 1/\rho$. Then $\rho < 1$ becomes the assumed stability condition.

We calculate the complementary cdf (ccdf) $P(W > x)$ by numerically inverting its Laplace transform

$$\hat{W}^c(s) \equiv \int_0^\infty e^{-st} P(W > t) dt = \frac{1 - \hat{w}(s)}{s}, \qquad (81)$$

where

$$\hat{w}(s) \equiv Ee^{-sW} = \int_0^\infty e^{-st} dP(W \le t) .$$

As in Section 3, the main challenge is computing the Laplace transform values $\hat{w}(s)$ for appropriate complex numbers s. The easiest special case is M/G/1 (when the interarrival time has an exponential distribution), in which the waiting-time Laplace transform $\hat{w}$ is available in closed form; see (82) below. In Section 4.1 we discuss a slightly more difficult case, in which the interarrival-time transform is rational. Then the waiting-time transform $\hat{w}$ is available once some roots of an equation have been found. Such roots can typically be found without difficulty; for further discussion see Chaudhry, Agarwal and Templeton [Chaudhry et al., 1992] and Chapter 10. The case in which the polynomial in the denominator has degree 2 is especially convenient; we discuss that case in detail.

In Section 4.2 we consider the general case. We apply Pollaczek's [Pollaczek, 1952] contour integral representation of the Laplace transform $\hat{w}$. In that case we must perform a numerical integration in order to calculate the transform values. This numerical integration approach applies directly when the service-time moment generating function is finite in a neighborhood of the origin. We show how to compute the required transform values more generally by this approach using exponential damping in Section 4.4.

We give numerical examples in Sections 4.3 and 4.4. In Section 4.3 we consider gamma distribution examples, i.e., $\Gamma_\alpha/\Gamma_\beta/1$ queues where Γ_α denotes the gamma distribution with shape parameter α. To include a case that is difficult for some algorithms (but not inversion), we consider the $E_k/E_k/1$ model (with Erlang distributions) with very high order k, namely, up to $k = 10^4$.

We conclude in Section 4.4 by considering long-tail service-time distributions. We show how exponential damping can be used together with numerical integration to get the Laplace transform values $\hat{w}(s)$ for general interarrival-time and service-time distributions (provided that Laplace transforms of these basic distributions are known). We also show that asymptotic results nicely complement the inversion algorithm by providing accurate values at very large arguments where the inversion gets difficult.

In this section we only consider single-server queues with renewal arrival processes. An inversion algorithm for single-server queues with a non-renewal arrival process (a batch Markovian arrival process) is described in Choudhury, Lucantoni and Whitt [Choudhury et al., 1996].

4.1 Rational Interarrival-Time Transform

The GI/G/1 model simplifies when one of the transforms $\hat{f}$ or $\hat{g}$ is rational, e.g., if $\hat{f} = \hat{\alpha}/\hat{\beta}$ where $\hat{\alpha}$ and $\hat{\beta}$ are polynomials. As shown by Smith [Smith, 1953], if the service-time transform $\hat{g}$ is rational, then the waiting-time transform $\hat{w}(s)$ itself is rational, and it is possible to obtain an explicit expression for the waiting-time ccdf $P(W > x)$; see p. 324 of Cohen [Cohen, 1982]. Hence, numerical inversion is especially attractive when the service-time transform is *not* rational, but it can be used for all service-time distribution.

The most familiar special case is the M/G/1 model, i.e., when $\hat{f}(s) = \rho/(\rho+s)$. Then the Laplace transform $\hat{w}$ is given by the Pollaczek-Khintchine formula

$$\hat{w}(s) = \frac{1-\rho}{1-\rho\hat{g}_e(s)}, \tag{82}$$

where

$$\hat{g}_e(s) = \int_0^\infty e^{-st} dG_e(t)$$

and G_e is the service-time stationary-excess cdf, defined by

$$G_e(t) = \frac{1}{EV} \int_0^t G^c(u) du, \quad t \geq 0. \tag{83}$$

Since $EV = 1$,

$$\hat{g}_e(s) = (1 - \hat{g}(s))/s. \tag{84}$$

If we can compute the transform values $\hat{g}(s)$, then the waiting-time ccdf $P(W > x)$ can be computed easily by inverting the transform $(1 - \hat{w}(s))/s$ for $\hat{w}$ in 82 and $\hat{g}_e$ in (84). Numerical examples are given in [Abate and Whitt, 1992a].

More generally, we can calculate the waiting-time ccdf $P(W > x)$ whenever the interarrival-time transform $\hat{f}$ is rational. Henceforth in this subsection we assume that $\hat{f} = \hat{\alpha}/\hat{\beta}$, where $\hat{\beta}$ is a polynomial of degree m and $\hat{\alpha}$ is a polynomial of degree at most $m - 1$. The model is then denoted $K_m/G/1$. In order to compute the transform, we must solve for the zeros of the equation

$$\hat{f}(s)\hat{g}(-s) = 1. \tag{85}$$

The following theorem comes from p. 329 of Cohen [Cohen, 1982].

Theorem 11 *Consider the $K_m/G/1$ queue with $\rho < 1$ in which the interarrival-time transform is rational, i.e., $\hat{f}(s) = \hat{\alpha}(s)/\hat{\beta}(s)$, where $\hat{\beta}(s)$ has degree m and $\hat{\alpha}(s)$ has degree at most $m-1$. Let the coefficient of s^m in $\hat{\beta}(s)$ be 1. Then equation (85) has m zeros with $\text{Re}(s) \leq 0$, exactly one of which is 0. Let $-\delta_i, 1 \leq i \leq m-1$, be the $m-1$ zeros with $\text{Re}(s) < 0$. Then the steady-state waiting time has Laplace transform.*

$$\hat{w}(s) = \frac{-\hat{\beta}(0)cs(1-\rho)}{\hat{\beta}(-s) - \hat{g}(s)\hat{\alpha}(-s)} \prod_{i=1}^{m-1} \frac{\delta_i - s}{\delta_i}, \tag{86}$$

where
$$c = \frac{\beta'(0) - \alpha'(0)}{\beta(0)}.$$

The mean waiting time and the probability of emptiness are

$$EW = \frac{\rho}{2(1-\rho)}\left\{EV^2 + EU^2 + 2EV\frac{\alpha'(0)}{\alpha(0)} - 2EU\frac{\beta'(0)}{\beta(0)}\right\} + \sum_{i=1}^{m-1}\delta_i^{-1}$$

and

$$P(W = 0) = (1-\rho)E[U]\beta(0)\prod_{i=1}^{m-1}\delta_i^{-1}.$$

The idle time within a busy cycle has Laplace transform

$$\hat{i}(s) = 1 - \frac{s}{\beta(s)}\prod_{i=1}^{m-1}(\delta_i + s).$$

When all the zeros with $Re(s) < 0$ of (85) can be found, which can usually be done without difficulty numerically, and which is easy in the case of $m = 2$, the waiting-time cdf can easily be calculated by numerically inverting the transform $(1 - \hat{w}(s))/s$ for $\hat{w}$ in (86).

We now give more explicit formulas for the case $m = 2$. In the case $m = 2$, equation (85) has precisely three roots: $\eta, 0$ and $-\delta$, where $\eta > 0$ and $\delta > 0$. Since the roots are all real, it is elementary to find them.

Let the interarrival time have transform

$$\hat{f}(s) = \frac{1 + (c_1 + c_2 - \rho^{-1})s}{(1 + c_1 s)(1 + c_2 s)} \quad (87)$$

for c_1 and c_2 real and positive with $c_1 \leq c_2$, so that the mean and squared coefficient of variation (SCV) are

$$f_1 = \rho^{-1} \text{ and } c_a^2 = 2(\rho c_1 + \rho c_2 - \rho^2 c_1 c_2) - 1.$$

Expanding (87) into partial fractions yields for $c_1 \neq c_2$:

$$\hat{f}(s) = \left(\frac{c_2 - \rho^{-1}}{c_2 - c_1}\right)(1 + c_1 s)^{-1} + \left(\frac{\rho^{-1} - c_1}{c_2 - c_1}\right)(1 + c_2 s)^{-1}. \quad (88)$$

We see that the pdf is hyperexponential (H_2) with $c_a^2 > 1$ if $c_2 > \rho^{-1} > c_1$; i.e.,

$$f(t) = p\lambda_1 e^{-\lambda_1 t} + (1-p)\lambda_2 e^{-\lambda_2 t}, \quad t \geq 0, \quad (89)$$

for $\lambda_i = 1/c_i$ and $p = (c_2 - \rho^{-1})/(c_2 - c_1)$ with $0 < p < 1$. On the other hand, if $\rho^{-1} > c_2 > c_1$, then (89) still holds but with $p < 0$. Then the pdf $f(t)$ in (89) is a difference of two exponentials and is called hypoexponential. Then the SCV satisfies $1/2 < c_a^2 < 1$.

For the special case of the hypoexponential with $c_1 + c_2 - \rho^{-1} = 0$, we can express the transform $\hat{f}(s)$ in (87) as

$$\hat{f}(s) = (1 + c_1 s)^{-1}(1 + c_2 s)^{-1} \, ,$$

so that the pdf $f(t)$ is the convolution of two exponential pdf's with means c_1 and c_2, which we refer refer to as generalized Erlang of order 2 (GE_2). When $c_1 = c_2$, we must have $c_1 = c_2 = 1/2\rho$ and the distribution becomes Erlang (E_2), i.e.,

$$\hat{f}(s) = (1 + [s/2\rho])^{-2} \, .$$

The degenerate exponential case is approached as $c_1 \to 0$ with $c_2 \to \rho^{-1}$.

The following is a direct consequence of Theorem 11.

Corollary. *Consider the $K_2/G/1$ queue having service-time transform $\hat{g}(s)$ with mean 1 and interarrival-time transform $\hat{f}(s)$ in (87) with mean ρ^{-1} for $0 < \rho < 1$. Then the steady-state waiting time has Laplace transform*

$$\hat{w}(s) = \frac{(1-\rho)(1-(s/\delta))}{(1-\rho\hat{g}_e(s)) + (\rho c_1 + \rho c_2 - 1)(1 - \hat{g}(s)) - \rho c_1 c_2 s} \, . \quad (90)$$

The mean waiting time and the emptiness probability are

$$EW = \frac{\rho(\rho^{-2}c_a^2 + c_s^2)}{2(1-\rho)} + \frac{1-\rho}{2\rho} + \frac{1}{\delta} - c_1 - c_2$$

and

$$P(W = 0) = \lim_{s \to \infty} \hat{w}(s) = \frac{1-\rho}{\rho c_1 c_2 \delta} \, .$$

The idle time transform is

$$\hat{i}(s) = \frac{1 + (c_1 + c_2 - \delta c_1 c_2)s}{(1 + c_1 s)(1 + c_2 s)} \, . \quad (91)$$

The first two idle-time moments are

$$i_1 = \delta c_1 c_2 \quad (92)$$

and

$$\frac{i_2}{2i_1} = c_1 + c_2 - \delta^{-1} \, .$$

Remark 8.3 It is of interest to compare the idle-time transform $\hat{i}(s)$ in (91) to $\hat{f}_e(s)$, the LST of the stationary-excess cdf F_e of the interarrival-time cdf F, defined as in (83). From (88), we see that

$$\hat{f}_e(s) = \frac{1 + \rho c_1 c_2 s}{(1 + c_1 s)(1 + c_2 s)} \, .$$

For any cdf H, let h_k be its kth moment. Then the first moment of the cdf F_e is

$$f_{e1} = \frac{f_2}{2f_1} = \frac{c_a^2 + 1}{2\rho} = c_1 + c_2 - \rho c_1 c_2 \qquad (93)$$

If we approximate i_1 by f_{e1}, we obtain from (92) and (93) an approximation for the root δ, namely,

$$\delta \approx \frac{1 + c_a^2}{2\rho c_1 c_2} \approx \frac{c_1 + c_2 - \rho c_1 c_2}{c_1 c_2},$$

which can be used as an initial guess when applying the Newton-Raphson root finding procedure. As indicated in [Abate et al., 1995b], a good initial guess for η is EW/ρ.

Example 5 (An $H_2/G/1$ Example) We conclude this subsection by giving a numerical example. Let the interarrival-time transform be as in (87) with $c_1 = 1/2\rho$ and $c_2 = 2/\rho$. Then the pdf is H_2, in particular,

$$f(t) = \frac{2}{3c_1} e^{-t/c_1} + \frac{1}{3c_2} e^{-t/c_2}, \quad t \geq 0,$$

so that $f_1 = 1/\rho, f_2 = 3/\rho^2$ and $f_3 = 33/2\rho^3$. Let the service-time pdf be gamma with mean 1 and shape parameter $1/2$ ($\Gamma_{1/2}$), i.e.,

$$g(t) = (2\pi t)^{-1/2} e^{-t/2}, \quad t \geq 0,$$

with $\hat{g}(s) = (1 + 2s)^{-1/2}$.

In Table 6 we display the root δ as a function of ρ. We also display several related quantities computable directly from δ, in particular, EW, $P(W = 0)$ and i_1. We compare EW to the heavy-traffic approximation $\rho(c_a^2 + c_s^2)/2(1-\rho)$ and we compare i_1 to f_{e1}. In this case the heavy-traffic approximation for the mean EW is quite good for all ρ. The mean stationary excess of an interarrival time f_{e1} consistently exceeds the idle-time mean i_1.

To do further analysis, we consider the case $\rho = 0.75$. Then we find that $\delta = 0.98115392$ and $P(W = 0) = 0.19110151$. (The high precision in $P(W = 0)$ would rarely be needed. On the other hand, the high precision in δ may be needed because it appears in the transform $\hat{w}(s)$ in (90) that we intend to invert. When the tail probability $P(W > x)$ is very small, we need very small absolute error to achieve reasonable relative error.) We compute the exact values of the ccdf $P(W > x)$ for several values of x in Table 7. We compare it to the Cramer-Lundberg asymptotic approximation

$$P(W > x) \sim \alpha e^{-\eta x} \quad \text{as } x \to \infty, \qquad (94)$$

where η is the positive real root of (85) with minimum real part and $f(x) \sim g(x)$ as $x \to \infty$ means that $f(x)/g(x) \to 1$ as $x \to \infty$; see p. 269 of Asmussen [Asmussen, 1987] and Abate, Choudhury and Whitt [Abate et al., 1995b].

Table 6 The root δ and other characteristics as functions of ρ in the $K_2/G/1$ model in Example 5.

ρ	δ	$P(W=0)$	EW	$\frac{\rho(c_a^2+c_s^2)}{2(1-\rho)}$	i_1	f_{e1}
.1	.105	.857	.25	.22	10.50	15.00
.2	.220	.729	.55	.50	5.49	7.50
.4	.475	.506	1.44	1.33	2.97	3.75
.5	.613	.408	2.13	2.00	2.45	3.00
.6	.757	.317	3.16	3.00	2.10	2.60
.8	1.058	.151	8.20	8.00	1.65	1.88
.9	1.214	.074	18.21	18.00	1.50	1.67
.99	1.357	.007	198.23	198.00	1.38	1.52

For the Cramér-Lundberg approximation with $\rho = 0.75$, we obtain

$$\eta^{-1} = 7.85645477 \text{ and } \alpha = 0.78472698 .$$

In this example, $EW = 6.185875$, so that the two rough estimates of η^{-1} are $EW/\rho = 8.2478$ and $(c_a^2 + c_s^2)/2(1-\rho) = 8.0$. Both are reasonable approximations that work well as initial guesses in the Newton-Raphson procedure.

From Table 7 we see that the Cramér-Lundberg approximation is excellent, even when x is not large. This numerical example illustrates a general phenomenon: When the Cramér-Lundberg approximation applies, it often serves as well as the exact values in applications. However, one should be careful about generalizing; asymptotic approximations do not always perform this well; see Section 4.4 below.

Finally, we remark that numerical inversion can also be used to calculate asymptotic parameters such as α and η in (94); e.g., see [Choudhury and Lucantoni, 1996] and [Abate et al., 1995a].

4.2 The Pollaczek Contour Integral

Pollaczek [Pollaczek, 1952] derived a contour-integral expression for the Laplace transform of the steady-state waiting-time distribution in the general GI/G/1 queue. Let H be the cumulative distribution (cdf) of $V - U$ and let ϕ be its moment generating function, defined by

$$\phi(z) = Ee^{z(V-U)} \equiv \int_{-\infty}^{\infty} e^{zt} dH(t) = \hat{f}(z)\hat{g}(-z) , \quad (95)$$

which we assume is analytic for complex z in the strip $|Re\ z| < \delta$ for some $\delta > 0$. A natural sufficient condition for this analyticity condition is for the service-time and interarrival-time distributions to have finite moment generating functions in a neighborhood of the origin, and thus moments of all orders,

Table 7 A comparison of the Cramér-Lundberg approximation with exact values obtained by numerical inversion for the waiting-time ccdf $P(W > x)$ for the queue $H_2/\Gamma_{1/2}/1$ with traffic intensity $\rho = 0.75$ in Example 5.

x	numerical transform inversion	Cramér-Lundberg approximation
10^{-8}	0.808898	0.785
.5	0.747832	0.736
1.0	0.697692	0.691
2.0	0.611125	0.608
4.0	0.472237	0.4716
8.0	0.283501	0.28346
16.0	0.1023909	0.1023905
30.0	0.01723289	same
50.0	0.00135140	same
70.0	0.00010598	same
80.0	0.00002968	same

but *neither the transform of the interarrival-time distribution nor the transform of the service-time distributions need be rational.*

Moreover, as noted on p. 40 of Pollaczek [Pollaczek, 1965] and in Section II.5.9 on p. 31 of Cohen [Cohen, 1982], it is possible to treat the case of more general service-time distributions by considering limits of service-time distributions that satisfy this analyticity condition. We discuss this extension here in Section 4.4. Now we assume that $\phi(z)$ in (95) is indeed analytic for complex z in the strip $|Re\ z| < \delta$ for some $\delta > 0$.

Here is Pollaczek's contour integral representation; see Chapter 5 of Cohen [Cohen, 1982].

Theorem 12 *In the GI/GI/1 model, the waiting-time Laplace transform is*

$$\hat{w}(s) \equiv Ee^{-sW} = \exp\left\{-\frac{1}{2\pi i}\int_C \frac{s}{z(s-z)}\log[1-\phi(-z)]dz\right\}, \quad (96)$$

where s is a complex number with $Re(s) \geq 0$, C is a contour to the left of, and parallel to, the imaginary axis, and to the right of any singularities of $\log[1 - \phi(-z)]$ in the left half plane, for ϕ in (95).

We have described algorithms for computing tail probabilities $P(W > x)$ by numerically inverting the Laplace transform $\hat{W}^c$ in (81). For example, the algorithm in Section 2.2 reduces to a finite weighted sum of terms $Re(\hat{W}^c(u + kvi))$ over integers k for appropriate real numbers u and v (the number of different k might be as low as 30.) To apply this algorithm, it suffices to compute Re $\hat{W}^c(s)$ for s of the required form $s = u + kvi$. For this purpose, it suffices to compute $\hat{w}(s)$ in (96) for s of this same form.

The standard expression for (96) has the contour just to the left of the imaginary axis, but this poses numerical difficulties because of the singularity in the first portion of the integrand, $s/z(s-z)$, and in the second portion, $\log[1 - \phi(-z)]$, at $z = 0$. However, this difficulty is easily avoided by moving the vertical contour of integration to the left, but still keeping it to the right of the singularity of $\log[1 - \phi(-z)]$ in the left halfplane closest to the origin, which we denote by $-\eta$. It turns out that this critical singularity of $\log[1 - \phi(-z)]$ also corresponds to the singularity of Ee^{-sW} in the left halfplane closest to the origin; i.e., the dominant singularity of (85) or

$$\eta = \sup\{s > 0 : Ee^{sW} < \infty\}. \tag{97}$$

Moreover, η is the asymptotic decay rate in the Cramér-Lundberg approximation in (94)

Given a reasonable estimate of η, we perform the integration (96) by putting the contour at $-\eta/2$. On this contour, $z = -\eta/2 + iy$ and y ranges from $-\infty$ to $+\infty$. Equation (96) becomes $Ee^{-sW} = \exp(-I)$, where

$$\begin{aligned} I &= \frac{1}{2\pi} \left(\int_{-\infty}^{0} \frac{s}{z(s-z)} \log[1 - \phi(z)] dy \int_{0}^{\infty} \frac{s}{z(s-z)} \log[1 - \phi(-z)] dy \right) \\ &= \frac{1}{2\pi} \int_{0}^{\infty} \left(\frac{s}{\bar{z}(s-\bar{z})} \log[1 - \phi(-\bar{z})] \frac{s}{z(s-z)} \log[1 - \phi(-z)] \right) dy \end{aligned} \tag{98}$$

with $\bar{z} = -\eta/2 - iy$. In general, I in (97) is complex; we compute its real and imaginary parts by integrating the real and imaginary parts of the integrand, respectively. However, if s is real, then so is I. In that case, the real parts of the two components of the integrand are the same, thereby simplifying the computation somewhat.

For the GI/G/1 queue, the desired parameter η in (97) can usually be easily found (by search algorithm) by solving the transform equation (85) for the positive real root with minimum real part. In order to find η, it suffices to restrict attention to the interval $(0, \eta_s)$, where

$$\eta_s = \sup\{s \geq 0 : Ee^{sV} < \infty\} \tag{99}$$

with V being a service time. (Of course η_s can be infinite, but that presents no major difficulty; in that case we start the search in the interval $(0,1)$. If the interval does not contain a root of (85), then we geometrically increase the upper limit until it contains the root.)

However, it can happen that transform equation (85) does not have a root even though the transform ϕ in (95) satisfies the analyticity condition; This means that $\eta = \eta_s > 0$ for η in (97) and η_s in (99), so that we can still put the vertical contour at $-\eta/2$.

A specific numerical integration procedure that can be used is fifth-order Romberg integration, as described in Section 4.3 of Press, Flannery, Teukolsky and Vetterling [Press et al., 1988]. First divide the integration interval $(0, \infty)$ in

(97) into a number of subintervals. If η is not too close to 0, then no special care is needed and it suffices to use the two subintervals $(0,1)$, and $(1,\infty)$ and then transform the infinite interval into $(0,1)$ using the transformation in (4.4.2) of [Press et al., 1988].

However, more care is required for less well behaved distributions (e.g., highly variable, nearly deterministic, or when η is close to 0). Then we examine the integrand more carefully and choose subintervals so that the ratio of the maximum to the minimum value within any subinterval is at most 10 or 100. This helps ensure that computational effort is expended where it is needed. Indeed, a version of the algorithm was developed to do this automatically. In this automatic procedure, the integration interval $(0,\infty)$ in (97) is divided into $m+1$ subintervals: $(0,b_1),(b_1,b_2),\ldots,(b_{m-1},b_m),(b_m,\infty)$. The last infinite subinterval (b_m,∞) is transformed into the finite interval $(0,b_m^{-1})$ using the transformation in (4.4.2) of [Press et al., 1988]. Within each subinterval, a fifth-order Romberg integration procedure is performed. An error tolerance of 10^{-12} is specified and the program generates successive partitions (going from n to $2n$ points) until the estimated improvement is no more than either the tolerance value itself or the product of the tolerance and the accumulated value of the integral so far (in the current subinterval as well as in earlier subintervals).

A specific procedure used for choosing the subintervals is as follows. If the integrand doesn't differ by more than a factor of 10 in the interval $(0,1)$ then b_1 is chosen as 1. Otherwise, b_1 is chosen such that the integrand roughly changes by a factor of 10 in the interval $(0,b_1)$. The endpoint b_1 is roughly determined by evaluating the integrand at 0 and at the points 10^{-n} with $n=10,9,\ldots,0$. For $2 \leq i \leq m$, the ratio b_i/b_{i-1} is assumed to be a constant K, where K is an input parameter. The number m is determined by looking at the ratio of the contribution from the subinterval (b_{i-1},b_i) to the total contribution so far. If this ratio is less than a constant ϵ, where ϵ is a second input parameter, then m is set to i, i.e., the next interval is made the last interval. A good choice of K and ϵ depends on the service-time and interarrival-time distributions. Typically less well behaved distributions require smaller K and/or ϵ. Our numerical experience indicates that $K=3$ and $\epsilon=10^{-4}$ works pretty well for most cases of interest.

The Laplace transform inversion algorithm also gives an estimate of the final error. If it is close to or below the 10^{-8} precision specified, we can be fairly confidence of a good computation.

4.3 Gamma Distribution Examples

In this subsection and the next we illustrate the numerical inversion algorithms for the GI/G/1 queue. In this subsection we consider $\Gamma_\alpha/\Gamma_\beta/1$ queues, where Γ denotes the gamma distribution, and α and β are the shape parameters of the interarrival-time and service-time distributions, respectively. The gamma distribution with scale parameter λ and shape parameter α has density

$$f(x) = \frac{1}{\Gamma(\alpha)}\lambda^\alpha x^{\alpha-1} e^{-\lambda x}, \quad x > 0,$$

mean α/λ, variance α/λ^2 and Laplace transform

$$Ee^{-sV} \equiv \int_0^\infty e^{-sx} f(x) dx = \left(\frac{\lambda}{\lambda+s}\right)^\alpha. \qquad (100)$$

The transform in (100) is rational if, and only if, the shape parameter α is a positive integer. When $\alpha = k$ for an integer k, the gamma distribution is also called Erlang of order k (E_k). Since convolutions of exponential distributions are smooth, we expect that this distribution will not be very difficult, at least when α is not too small; see Section 12 of Abate and Whitt [Abate and Whitt, 1992a].

We stipulate that the mean service time is 1 and that the arrival rate is ρ. The remaining two parameters α and β of the $\Gamma_\alpha/\Gamma_\beta/1$ queue are the shape parameters of the interarrival-time and service-time distribution. Since the squared coefficient of variation (SCV) is the reciprocal of the shape parameter, it suffices to specify the SCVs c_a^2 and c_s^2 of the interarrival-time and service-time distributions.

The algorithm can be checked against known results by considering the $E_k/\Gamma/1$ and $\Gamma/E_k/1$ special cases. These are special cases of the PH/G/1 and GI/PH/1 queues, for which there are alternative algorithms exploiting results for the M/G/1 and GI/M/1 paradigms in Neuts [Neuts, 1981, Neuts, 1989]. Another alternative for comparison is the root finding algorithm as in Chaudhry, Agarwal and Templeton [Chaudhry et al., 1992]; e.g., we found good agreement with results for the $E_{10}/E_{100}/1$ queue in Table 10 on p. 141 of Chaudhry, Agarwal and Templeton [Chaudhry et al., 1992].

Some other algorithms for $E_k/E_m/1$ queues get more difficult as k and m increase. Hence, we performed calculations for $E_k/E_k/1$ models with large k. Other $\Gamma_\alpha/\Gamma_\beta/1$ examples are given in [Abate et al., 1993].

Example 6 ($E_k/E_k/1$ Queues) We did calculations for the $E_k/E_k/1$ queue for $k = 10$, $k = 100$, $k = 1000$ and $k = 10,000$. In this case the transform equation in (85) for the asymptotic decay rate η becomes

$$\left(\frac{k}{k-\eta}\right)^k \left(\frac{k}{k+\eta/\rho}\right)^k = 1,$$

from which we easily obtain

$$\eta = k(1-\rho).$$

Since E_k is approaching a deterministic distribution as k increases, to avoid having negligible probabilities we let $\rho \equiv \rho_k$ increase with k. In particular, we

Table 8 Tail probabilities and cumulants of the steady-state waiting time in the $E_k/E_k/1$ model with traffic intensity $\rho = 1 - k^{-1}$, as a function of k. The case $k = \infty$ is an exponential with mean 1.

Congestion Measure	k				
	10	100	1,000	10,000	∞
$P(W > 0)$	0.7102575	0.9035808	0.9687712	0.9900406	1.0000000
$P(W > 1)$	0.2780070	0.3385844	0.3584117	0.3648607	0.3678794
$P(W > 3)$	0.0376169	0.0458224	0.0485057	0.0493785	0.0497871
$P(W > 5)$	0.0050909	0.0062014	0.0065645	0.0066825	0.0067379
$P(W > 7)$	0.0006890	0.0008393	0.0008884	0.0009044	0.0009119
$c_1(W)$	0.7484185	0.9195281	0.9741762	0.9917852	$0! = 1$
$c_2(W)$	0.9491038	0.9951389	0.9995064	0.9999502	$1! = 1$
$c_3(W)$	1.982543	1.995377	1.9999854	1.9999996	$2! = 2$
$c_4(W)$	5.992213	5.999948	5.999999	6.000000	$3! = 6$
$c_5(W)$	23.995966	23.999995	24.000000	24.000000	$4! = 24$
$c_6(W)$	119.997754	120.000000	120.000000	119.999993	$5! = 120$

let $\rho_k = 1 - k^{-1}$. With this choice, $\eta \equiv \eta_k = 1$ for all k. Also W_k, the steady-state waiting time in model k, converges to an exponential random variable with mean 1 as $k \to \infty$, as can be seen by applying the heavy-traffic argument of Kingman [Kingman, 1961] using (96).

Numerical values of some tail probabilities and cumulants are given for $E_k/E_k/1$ queues for these cases in Table 8. (The cumulants are calculated by other Pollaczek contour integrals; see [Abate et al., 1993].) The exponential limit is displayed as well under the heading $k = \infty$. None of these presented any numerical difficulties.

Interestingly, from Table 8, we see that for these cases W is quite well approximated by a mixture of an atom at 0 with probability $1/\sqrt{k} = \sqrt{1-\rho}$ and an exponential with mean 1 with probability $1 - 1/\sqrt{k}$.

4.4 Long-Tail Service-Time Distributions

Pollaczek's contour integral representation in (96) depends on an analyticity condition that is satisfied when the interarrival-time and service-time distributions have finite moment generating functions in a neighborhood of the origin; i.e., when $Ee^{sU} < \infty$ and $Ee^{sV} < \infty$ for some $s > 0$. However, it is possible to treat the general case by representing a general distribution as a limit of a sequence of distributions each of which satisfies this analyticity condition. It is known that the associated sequence of steady-state waiting-time distributions will converge to a proper limit provided that the distributions and their means also converge to proper limits; see p. 194 of Asmussen [Asmussen, 1987]. (The moment condition is actually on $(V - U)^+ = \max\{V - U, 0\}$.)

In fact, the long-tail interarrival-time distributions actually present no difficulty. It suffices to have $\phi(z)$ in (95) analytic in the strip $0 < \text{Re}(z) < \delta$ for some $\delta > 0$. However, the service-time distribution poses a real problem.

Hence, if $G^c(x)$ is the given service-time ccdf with Laplace transform $\hat{G}^c(s)$ and mean $m = \hat{G}^c(0)$, then it suffices to find an approximating sequence of service-time complementary cdf's $\{G_n^c(x) : n \geq 1\}$ with associated Laplace transforms $\{\hat{G}_n^c(s) : n \geq 1\}$ and means $\{m_n = \hat{G}_n^c(0) : n \geq 1\}$ such that $\hat{G}_n^c(s) \to \hat{G}^c(s)$ as $n \to \infty$ for all s. Then $G_n^c(x) \to G^c(x)$ as $n \to \infty$ for all x that are continuity points of the limiting complementary cdf G^c and $m_n \to m$ as $n \to \infty$.

A natural way to obtain a sequence of approximating service-time distributions with finite moment generating functions in some neighborhood of the origin when this condition is not satisfied originally is to introduce *exponential damping* in the Laplace-Stieltjes transform with respect to G. In particular, for any $\alpha > 0$ let the α-damped ccdf be

$$G_\alpha^c(x) = \int_x^\infty e^{-\alpha t} dG(t), \quad x \geq 0 . \tag{101}$$

Since we want a proper probability distribution, we put mass $1 - G_\alpha^c(0)$ at 0. If the original service-time distribution has mean 1 and we want the new service-time distribution also to have mean 1, then we also divide the random variable V_α with cdf G_α by the new mean m_α, i.e., we let the complimentary cdf be $G_\alpha^c(m_\alpha x)$.

The direct approximation in (101) makes the service-time distribution stochastically smaller than the original service-time distribution, which in turn makes the new steady-state waiting-time distribution stochastically smaller than the original one, which may be helpful for interpretation. However, keeping the same mean seems to give substantially better numbers. Here we keep the mean fixed at 1.

From (101), it is easy to see that, if $\hat{g}$ is the original Laplace-Stieltjes transform of G, then the Laplace-Stieltjes transform of $G_\alpha(m_\alpha x)$ with mean 1 is

$$\hat{g}_\alpha(s) = \hat{g}(\alpha + (s/m_\alpha)) + 1 - \hat{g}(\alpha) , \tag{102}$$

where

$$m_\alpha = -\hat{g}_\alpha'(0) = -\hat{g}'(\alpha) .$$

Thus, the Laplace transform $\hat{G}_\alpha^c$ of $G_\alpha^c(m_\alpha x)$ for G_α^c in (101) is

$$\hat{G}_\alpha^c(s) = \frac{1 - \hat{g}_\alpha(s)}{s}$$

for $\hat{g}_\alpha$ in (102). Hence, given $\hat{g}$, we can readily calculate values of $\hat{G}_\alpha^c$ for any $\alpha > 0$.

However, this approach is not trivial to implement, because it often requires a very small α before $G_\alpha(x)$ is a satisfactory approximation for $G(x)$, and a small α means a small η in (85). Indeed, $0 < \eta \leq \eta_s = \alpha$. In turn, a small η

means a relatively difficult computation, because the contour at $-\eta/2$ is near the singularity at 0. However, this can be handled by being careful with the numerical integration. Indeed, the algorithm employing an adaptive choice of integration subintervals was originally developed to handle this case.

Before considering an example illustrating exponential damping, we mention that it is also possible to approximate in other ways; e.g., see Asmussen, Nerman and Ollson [Asmussen et al., 1996], Feldmann and Whitt [Feldmann and Whitt, 1997] and Gaver and Jacobs [Gaver and Jacobs, 1998].

Example 7 (An M/G/1 Queue) To illustrate how the exponential damping approach works, we consider an M/G/1 queue with service-time density

$$g(x) = x^{-3}(1 - (1 + 2x + 2x^2)e^{-2x}), \quad x \geq 0 .$$

This distribution has first two moments $m_1 = 1$ and $m_2 = \infty$, so that there is a proper steady-state waiting-time distribution which has infinite mean; see pp. 181-184 of Asmussen [Asmussen, 1987]. It is easy to see that our service-time distribution has complementary cdf

$$G^c(x) = (2x^2)^{-1}(1 - (1 + 2x)e^{-2x}), \quad x \geq 0,$$

and Laplace transform

$$\hat{g}(s) = 1 - s + \frac{s^2}{2}\ln(1 + (2/s)) .$$

This service-time distribution is a *Pareto mixture of exponentials* (PME) distribution introduced in Abate, Choudhury and Whitt [Abate et al., 1994].

We use the M/G/1 queue to make it easier to compare our numerical results with other known results. In particular, we can also apply inversion directly to the Pollaczek-Khintchine (transform) formula (82). We also can compare the inversion results to a two-term asymptotic expansion derived by J. W. Cohen (personal communication). The two-term asymptotic expansion is

$$P(W > x) \sim \frac{\rho}{2(1-\rho)x}\left(1 + \frac{\rho}{(1-\rho)x}\right) \text{ as } x \to \infty, \qquad (103)$$

where $\gamma \equiv 0.5772\ldots$ is Euler's constant. The first term is just $P(W > x) \sim \rho/(1-\rho)x$. The inclusion of Euler's constant in equation (103) corrects an error in the second term of a conjecture on the bottom of p. 328 of [Abate et al., 1994]. The reasoning on p. 329 of [Abate et al., 1994] can be used, but the asymptotics for one term needs to be corrected. In particular, $\mathcal{L}^{-1}(s\log^2 s) \sim -2(\gamma - 1 + \log x)/x^2$ as $x \to \infty$.

In Table 9 we display the tail probabilities $P(W > x)$ for this M/G/1 example with $\rho = 0.8$ for five values of x: $x = 4$, $x = 20$, $x = 100$, $x = 500$ and $x = 2500$. The table shows the exact results (no damping, $\alpha = 0$) obtained from the Pollaczek-Khintchine formula and asymptotic approximations based

Table 9 A comparison of approximations for tail probabilities $P(W > x)$ with exact values in the M/G/1 model with $\rho = 0.8$ and the long-tail service-time distribution in Example 4.3.

Cases	x				
	4	20	100	500	2500
$\alpha = 0$ (exact)	0.4653	0.1558	0.02473	0.004221	0.000811
$\alpha = 10^{-2}$	0.4539	0.1175	0.0036	0.000002	0.000000
$\alpha = 10^{-3}$	0.4631	0.1474	0.0017	0.00096	0.000006
$\alpha = 10^{-4}$	0.4650	0.1544	0.02315	0.0032	0.00032
$\alpha = 10^{-6}$	0.4653	0.1557	0.02469	0.004193	0.000788
$\alpha = 10^{-8}$ (M/G/1)	0.4653	0.1558	0.02473	0.004221	0.000810
$\alpha = 10^{-8}$ (Pollaczek)	0.4653	0.1558	0.02473	0.004224	0.000815
asymptotics 1 term	0.50	0.10	0.020	0.0040	0.00080
2 terms	1.33	0.165	0.0239	0.00421	0.000810

on (103); as well as the approximations obtained from five values of the damping parameter α: $\alpha = 10^{-2}$, $\alpha = 10^{-3}$, $\alpha = 10^{-4}$, $\alpha = 10^{-6}$ and $\alpha = 10^{-8}$. The numerical results based on the algorithm here and the Pollaczek-Khintchine algorithm agreed to the stated precision for all values of α except $\alpha = 10^{-8}$, so only in the single case $\alpha = 10^{-8}$ are both numerical results given. Table 9 shows that the exact ($\alpha = 0$) results and the two algorithms with $\alpha = 10^{-8}$ are all quite close, even for $x = 2500$. Table 9 shows that the damping parameter α needs to be smaller and smaller as x increases in order for the calculations based on the approximating cdf G_α to be accurate. However, the calculation gets more difficult as x increases and α decreases. For the smaller α values reported, it was important to carefully choose the subintervals for the Romberg integration so that the integrand does not fluctuate too greatly within the subinterval. This was done by the automatic procedure described earlier.

In this example we are able to obtain a good calculation for all x because the asymptotics apply before the computation gets difficult. The relative percent error for the one-term (two-term) approximations at $x = 100$, $x = 500$ and

$x = 2,500$ are, respectively, 19%, 5.2% and 1.4% (3.2%, 0.2% and < 0.1%). This example illustrates a general phenomenon: Numerical methods work well together with asymptotics. One approach tends to work well when the other breaks down. They also serve as checks on each other.

5 CLOSED QUEUEING NETWORKS

This final section contains an example illustrating how the numerical inversion algorithm for generating functions of sequences of complex numbers in Section 2.4 can be applied. In this section we consider the product-form steady-state distribution of a closed (Markovian) queueing network (CQN), drawing upon Choudhury, Leung and Whitt [Choudhury et al., 1995b] and Choudhury and Whitt [Choudhury and Whitt, 1997]. This section also illustrates how inversion of multi-dimensional transforms can be applied; see Choudhury, Lucantoni and Whitt [Choudhury et al., 1994] for more on multi-dimensional inversion. Quantities of interest in stochastic loss networks can be calculated in the same way; see Choudhury, Leung and Whitt [Choudhury et al., 1995c], [Choudhury et al., 1995d], [Choudhury et al., 1995a].

It is known that the steady-state distribution of a CQN can be expressed in terms of a normalization constant. We show that steady-state characteristics of interest can be calculated by numerically inverting the multidimensional generating function of the normalization constant (regarded as a function of chain populations).

In Section 5.1 we introduce the CQN model and display the generating functions. In Section 5.2, we discuss the technique of dimension reduction for reducing the effective dimension of the inversion, which is important since the computational effort grows exponentially with the dimension. In Section 5.3 we discuss scaling to control the aliasing error, which is important since the normalization constants are not bounded. Finally, in Section 5.4 we illustrate by solving a challenging numerical example.

5.1 The Model and the Generating Functions

We start by describing the general class of probability distributions that we consider. We start somewhat abstractly, but below we will consider a special class of closed queueing networks. Let the state variable be a *job vector* $\mathbf{n} = (n_1, \ldots, n_L)$; n_l is the number of jobs of *type* ℓ; n_l might be the number of customers of a particular class at a particular queue. Let there be a specified *population vector* $\mathbf{K} = (K_1, \ldots, K_p)$; K_j is the population of *chain j*, a fixed quantity specified as part of the model data. The *state space* is the set of allowable job vectors, which depends on the population vector $\mathbf{K}$ and is denoted by $S(\mathbf{K})$. In this setting, the probability distributions that we consider have the form

$$p(\mathbf{n}) = g(\mathbf{K})^{-1} f(\mathbf{n}), \qquad (104)$$

where

$$g(\mathbf{K}) = \sum_{\mathbf{n} \in S(\mathbf{K})} f(\mathbf{n}) \qquad (105)$$

and f is a (known) nonnegative real-valued function on the L-fold product of the nonnegative integers. (For example, we might have $f(\mathbf{n}) = \prod_{\ell=1}^{L} f_\ell(n_\ell)$ with $f_\ell(n_\ell) = \rho_\ell^{n_\ell}$.) The term $g(\mathbf{K})$ in (104) and (105) is called the *normalization constant* or the *partition function*. For the closed queueing network models we will consider (and many other models), the state space has the special form

$$S(\mathbf{K}) = \left\{ \mathbf{n} \mid n_\ell \geq 0, \sum_{\ell \in C_j} n_\ell = K_j, \ 1 \leq j \leq p \right\}$$

for special sets C_j, $1 \leq j \leq p$.

Given a probability distribution as in (104), where the function f is relatively tractable, the major complication is determining the normalization constant $g(\mathbf{K})$ for the relevant population vector $\mathbf{K}$. In this setting, the convolution algorithm calculates $g(\mathbf{K})$ by expressing it in terms of values $g(\mathbf{K'})$ where $\mathbf{K'} < \mathbf{K}$ (i.e., $K'_\ell \leq K_\ell$ for all ℓ and $K'_\ell < K_\ell$ for at least one ℓ, e.g., see Conway and Georganas [Conway and Georganas, 1989] or Lavenberg [Lavenberg, 1983]. Other existing non-asymptotic algorithms proceed in a similar recursive manner. See Conway and Georganas [Conway and Georganas, 1989] for a unified view.

In contrast, we calculate $g(\mathbf{K}) \equiv g(K_1, \ldots, K_p)$ by numerically inverting its multi-dimensional generating function

$$\hat{g}(\mathbf{z}) \equiv \sum_{K_1=0}^{\infty} \cdots \sum_{K_p=0}^{\infty} g(\mathbf{K}) \prod_{j=1}^{p} z_j^{K_j} \qquad (106)$$

where $\mathbf{z} \equiv (z_1, \ldots, z_p)$ is a vector of complex variables. To quickly see the potential advantage of this approach, note that we can calculate $g(\mathbf{K})$ for one vector $\mathbf{K}$ without calculating $g(\mathbf{K'})$ for all the $\prod_{j=1}^{p} K_j$ nonnegative vectors $\mathbf{K'}$ less than or equal to $\mathbf{K}$, as is done with the convolution algorithm.

There are two obvious requirements for carrying out this program. First, we need to be able to compute the generating function values in (106) and, second, we need to be able to perform the numerical inversion. The first requirement often turns out to be surprisingly easy, because the generating function of a normalization constant often has a remarkably simple form. This has long been known in statistical mechanics. In that context, the normalization constant is usually referred to as the partition function and its generating function is referred to as the *grand partition function*; e.g., see pp. 213 and 347 of Reif [Reif, 1965]. Reiser and Kobayashi [Reiser and Kobayashi, 1975] used generating functions of normalization constants to derive their convolution algorithm. For more on generating functions of normalization constants in CQNs, see Bertozzi and McKenna [Bertozzi and McKenna, 1993].

We invert the p-dimensional generating function $\hat{g}(\mathbf{z})$ in (106) by recursively performing p one-dimensional inversions, using the algorithm in Section 2.4. To represent the recursive inversion, we define *partial generating functions* by

$$g^{(j)}(\mathbf{z}_j, \mathbf{K}_{j+1}) = \sum_{K_1=0}^{\infty} \cdots \sum_{K_j=0}^{\infty} g(\mathbf{K}) \prod_{i=1}^{j} z_i^{K_i} \quad \text{for} \quad 0 \leq j \leq p, \qquad (107)$$

where $\mathbf{z}_j = (z_1, z_2, \ldots, z_j)$ and $\mathbf{K}_j = (K_j, K_{j+1}, \ldots, K_p)$ for $1 \leq j \leq p$. Let $\mathbf{z}_0$ and $\mathbf{K}_{p+1}$ be null vectors. Clearly, $\mathbf{K} = \mathbf{K}_1, \mathbf{z} = \mathbf{z}_p$, $g^{(p)}(\mathbf{z}_p, \mathbf{K}_{p+1}) = \hat{g}(\mathbf{z})$ and $g^{(0)}(\mathbf{z}_0, \mathbf{K}_1) = g(\mathbf{K})$.

Let I_j represent inversion with respect to z_j. Then the step-by-step nested inversion approach is

$$g^{(j-1)}(\mathbf{z}_{j-1}, \mathbf{K}_j) = I_j \left[g^{(j)}(\mathbf{z}_j, \mathbf{K}_{j+1}) \right], \quad 1 \leq j \leq p, \qquad (108)$$

starting with $j = p$ and decreasing j by 1 each step. In the actual program implementation, we attempt the inversion shown in (108) for $j = 1$. In order to compute the righthand side we need another inversion with $j = 2$. This process goes on until at step p the function on the righthand side becomes the p-dimensional generating function and is explicitly computable. By simply relabeling the p transform variables, we see that the scheme above can be applied to the p variables in any order. In all steps except the last we have a sequence of complex numbers, as in Section 2.4. For further discussion of the multi-dimensional inversion, see [Choudhury et al., 1994] and [Choudhury et al., 1995b].

We now consider multi-chain closed queueing networks with only single-server queues (service centers with load-independent service rates) and (optionally) infinite-server queues. In this model all jobs are divided into *classes*. The combination of a class r job at queue i, (r, i) is called a *stage*. Two classes r and s communicate with each other if for some i and j, stage (s, j) can be reached from stage (r, i) in a finite number of steps (transitions) and vice versa. With respect to the relation of communication, all the classes can be divided into mutually disjoint equivalence classes called (closed) chains (ergodic sets in Markov chain theory). All classes within a chain communicate. No two classes belonging to different chains communicate. Since we are considering the steady-state distribution of a model with only closed chains, we do not need to consider any transient stages, i.e., stages (r, i) that will not be reached infinitely often. We now introduce further notation and give additional details about the model. For additional background, see Conway and Georganas [Conway and Georganas, 1989], Lavenberg [Lavenberg, 1983] or Chapter 11. We introduce the following features and notation:

- p = number of closed chains
- M = number of job classes ($M \geq p$).
- N = number of queues (service centers). Queues $1, \ldots, q$ are assumed to be of the single-server type and queues $q+1, \ldots, N$ are assumed to be

of the infinite-sever (IS) type. As usual, for the single-server queues, the service discipline may be first-come first-served (FCFS), last-come first-served preemptive-resume (LCFSPR) or processor sharing (PS). In the case of FCFS, the service times of all job classes at a queue are assumed to be exponential with the same mean.

- $R_{ri,sj}$ = routing matrix entry, probability that a class r job completing service at queue i will next proceed to queue j as a class s job for $1 \leq i, j \leq N$, $1 \leq r, s \leq M$ (i.e., class hopping is allowed). The pair (r, i) is referred to as a stage in the network.

- K_j = number of jobs in the jth closed chain, $1 \leq j \leq p$, which is fixed.

- $\mathbf{K} = (K_1, \ldots, K_p)$, the population vector, specified as part of the model data.

- n_{ri} = number of jobs of class r in queue i, $1 \leq r \leq M$, $1 \leq i \leq N$.

- n_i = number of jobs in queue i, i.e., $n_i = \sum_{r=1}^{M} n_{ri}$, $1 \leq i \leq N$.

- $\mathbf{n} = (n_{ri})$, $1 \leq r \leq M$, $1 \leq i \leq N$.

- C_j = set of stages in the jth closed chain. Clearly, $\sum_{(r,i) \in C_j} n_{ri} = K_j$, $1 \leq j \leq p$.

- $q_{ji} = \sum_{r:(r,i) \in C_j} n_{ri}$, number of jobs from chain j at queue i.

- $S(\mathbf{K})$ = state space of allowable job vectors or queue lengths (including those in service), i.e.,

$$S(\mathbf{K}) = \left\{ \mathbf{n} : n_{ri} \in \mathbf{Z}^+ \quad \text{and} \quad \sum_{(r,i) \in C_j} n_{ri} = K_j, \quad 1 \leq j \leq p \right\},$$

where $\mathbf{Z}^+$ is the set of nonnegative integers.

- e_{ri} = visit ratio, i.e., solution of the traffic rate equation

$$\sum_{(r,i) \in C_k} e_{ri} R_{ri,sj} = e_{sj} \quad \text{for all} \quad (s,j) \in C_k \quad \text{and} \quad 1 \leq k \leq p. \quad (109)$$

For each chain there is one degree of freedom in (109). Hence, for each chain j, the visit ratios $\{e_{ri} : (r, i) \in C_j\}$ are specified up to a constant multiplier.

- t_{ri} = the mean service time for class r at queue i.

- $\rho'_{ri} = t_{ri} e_{ri}$, $1 \leq r \leq M$, $1 \leq i \leq N$, the relative traffic intensities.

- $\rho_{j0} = \sum_{i=q+1}^{N} \sum_{(r,i) \in C_j} \rho'_{ri}$ and $\rho_{ji} = \sum_{(r,i) \in C_j} \rho'_{ri}$ for $i = 1, 2, \ldots, q$, the aggregate relative traffic intensities.

For this model, the steady-state distribution is given by (104) and the partition function is given by (105), where

$$f(\mathbf{n}) = \left[\prod_{i=1}^{q} n_i! \prod_{r=1}^{M} \frac{\rho_{ri}'^{n_{ri}}}{n_{ri}!}\right] \left[\prod_{i=q+1}^{N} \prod_{r=1}^{M} \frac{\rho_{ri}'^{n_{ri}}}{n_{ri}!}\right]. \quad (110)$$

The generating function $\hat{g}(\mathbf{z})$ is given by (106), using (105) and (110). By changing the order of summation, it can be seen that $\hat{g}(\mathbf{z})$ can be expressed remarkably simply as

$$\hat{g}(\mathbf{z}) = \frac{\exp\left(\sum_{j=1}^{p} \rho_{j0} z_j\right)}{\prod_{i=1}^{q} \left(1 - \sum_{j=1}^{p} \rho_{ji} z_j\right)}. \quad (111)$$

In general, there may be multiplicity in the denominator factors of (111) if two or more queues are identical with respect to visits by customers of all classes. In such a situation (111) becomes

$$\hat{g}(\mathbf{z}) = \frac{\exp\left(\sum_{j=1}^{p} \rho_{j0} z_j\right)}{\prod_{i=1}^{q'} \left(1 - \sum_{j=1}^{p} \rho_{ji} z_j\right)^{m_i}}, \quad (112)$$

where

$$\sum_{i=1}^{q'} m_i = q.$$

Here (112) is the preferred form, not (111); i.e., the key parameters are p and q'. The inversion algorithm simplifies by having different queues with identical single-server parameters.

Given the normalization constant $g(\mathbf{K})$ in (105) and (110), we can directly compute the steady-state probability mass function $p(\mathbf{n})$ in (104). Moreover, several important performance measures can be computed directly from ratios of normalization constants. For example, the throughput of class r jobs at queue i is

$$\theta_{ri} = e_{ri} \frac{g(\mathbf{K} - \mathbf{1}_j)}{g(\mathbf{K})} \quad \text{for} \quad (r, i) \in C_j, \quad (113)$$

where $\mathbf{1}_j$ is the vector with a 1 in the jth place and 0's elsewhere.

From (113) we see that we will often need to compute normalization constants $g(\mathbf{K})$ for several closely related vector arguments $\mathbf{K}$. When the population vector $\mathbf{K}$ is large, it is possible to calculate such closely related normalization constants efficiently by exploiting shared computation; see [Choudhury et al., 1995b]

The means $E[n_{ri}]$ and $E[q_{ji}]$ and higher moments $E[n_{ri}^k]$ and $E[q_{ji}^k]$ can also be computed directly from the normalization constants, but the standard formulas involve more than two normalization constant values. Choudhury,

Leung and Whitt [Choudhury et al., 1995b] developed an improved algorithm for means and higher moments via generating functions, which we now describe.

Given the steady-state probability mass function, we can calculate moments. Without loss of generality, let $(r,i) \in C_1$. We start with a standard expression for the probability mass function of q_{1i}, the number of chain 1 customers at queue i, namely,

$$P(q_{1i} = k) = \frac{\rho_{1i}^k(g(\mathbf{K} - k\mathbf{1}_1) - \rho_{1i}g(\mathbf{K} - (k+1)\mathbf{1}_1))}{g(\mathbf{K})}, \qquad (114)$$

see (3.257) on p. 147 of Lavenberg [Lavenberg, 1983]. (A similar expression holds for the mass function of n_{ri}. It involves ρ'_{ri} instead of ρ_{1i}.)

From the telescoping property of (114), we can write the tail probabilities as

$$P(q_{1i} \geq k) = \frac{\rho_{1i}^k g(\mathbf{K} - k\mathbf{1}_1)}{g(\mathbf{K})}. \qquad (115)$$

From (115) we obtain the standard formula for the mean,

$$E[q_{1i}] = \sum_{k=1}^{\infty} P(q_{1i} \geq k) = \sum_{k=1}^{K_1} \rho_{1i}^k \frac{g(\mathbf{K} - k\mathbf{1}_1)}{g(\mathbf{K})}; \qquad (116)$$

e.g., see (3.258) on p. 147 of Lavenberg [Lavenberg, 1983]. Unfortunately, formula (116) is not too convenient, because it requires $K_1 + 1$ normalization function calculations and thus $K_1 + 1$ numerical inversions. We now show how this mean can be calculated by *two* inversions.

For this purpose, we rewrite (116) as

$$E[q_{1i}] = \frac{\rho_{1i}^{K_1} h(\mathbf{K})}{g(\mathbf{K})} - 1, \qquad (117)$$

where

$$h(\mathbf{K}) = \sum_{k=0}^{K_1} \rho_{1i}^{-k} g(k, \mathbf{K}_2),$$

with $\mathbf{K}_2$ defined after (107). Let $\hat{h}_1(z_1)$ be the generating function of $h(\mathbf{K})$ with respect to K_1. Then

$$\hat{h}_1(z_1) = \sum_{m=0}^{\infty} z_1^m h(m, \mathbf{K}_2) = \sum_{m=0}^{\infty} z_1^m \sum_{k=0}^{m} \rho_{1i}^{-k} g(k, \mathbf{K}_2)$$

$$= \sum_{k=0}^{\infty} \rho_{1i}^{-k} g(k, \mathbf{K}_2) \sum_{m=k}^{\infty} z_1^m = \frac{g^{(1)}(z_1/\rho_{1i}, \mathbf{K}_2)}{1 - z_1}, \qquad (118)$$

where $g^{(1)}(\mathbf{z}_1, \mathbf{K}_2)$ is the partial generating function in (107). Now, if $\hat{h}(\mathbf{z})$ represents the full generating function of $h(\mathbf{K})$, then from (118) it is clear that

$$\hat{h}(\mathbf{z}) = \frac{\hat{g}(z_1/\rho_{1i}, z_2, \ldots, z_p)}{1 - z_1}.$$

Since $\hat{h}(\mathbf{z})$ is of the same form as $\hat{g}(\mathbf{z})$ it may be inverted by the established inversion procedure. Hence, we can obtain the mean $E[q_{1i}]$ using two inversions from (117). We invert $\hat{g}(\mathbf{z})$ and $\hat{h}(\mathbf{z})$, respectively, to obtain $g(\mathbf{K})$ and $h(\mathbf{K})$.

By the same approach, we can also calculate higher moments, see [Choudhury et al., 1995b].

5.2 Dimension Reduction by Decomposition

In general, the inversion of a p-dimensional generating function $\hat{g}(\mathbf{z})$ represents a p-dimensional inversion, whether it is done directly or by our proposed recursive technique. This presents a major problem because the computational complexity of the algorithm is exponential in the dimension. Fortunately, however, it is often possible to reduce the dimension significantly by exploiting special structure. To see the key idea, note that if $\hat{g}(\mathbf{z})$ can be written as a product of factors, where no two factors have common variables, then the inversion of $\hat{g}(\mathbf{z})$ can be carried out by inverting the factors separately and the dimension of the inversion is thus reduced. For example, if $\hat{g}(z_1, z_2, z_3) = \hat{g}_1(z_1)\hat{g}_2(z_2, z_3)$, then $\hat{g}$ can be treated as one two-dimensional problem plus one one-dimensional problem, which is essentially a two-dimensional problem, instead of a three-dimensional problem.

We call the direct factorization of the generating function $\hat{g}$ an *ideal decomposition*. It obviously provides reduction of computational complexity, but we do not really expect to be able to exploit it, because it essentially amounts to having two or more completely separate models, which we would not have with proper model construction. We would treat them separately to begin with.

Even though ideal decomposition will virtually *never* occur, key model elements (e.g., closed chains) are often only *weakly coupled*, so that we can still exploit a certain degree of decomposition to reduce the inversion dimensionality, often dramatically. The idea is to look for *conditional decomposition*. The possibility of conditional decomposition stems from the fact that when we perform the $(j-1)^{\text{st}}$ inversion in (108), the outer variables $z_1, \ldots, z_{j-1}$ are fixed. Hence, for the $(j-1)^{\text{st}}$ inversion it suffices to look for decomposition in the generating functions regarded as a function of the remaining $p - j + 1$ variables. For example, if $\hat{g}(z_1, z_2, z_3) = \hat{g}_1(z_1, z_2)\hat{g}_2(z_1, z_3)$, then for each fixed z_1, the transform $\hat{g}$ as a function of (z_2, z_3) factors into the product of two functions of a single variable. Hence $\hat{g}$ can be treated as two two-dimensional problems instead of one three-dimensional problems.

More generally, we select d variables that we are committed to invert. We then look at the generating function with these d variables fixed and see if the remaining function of $p - d$ variables can be factored. Indeed, we write the function of the remaining $p - d$ variables as a product of factors, where no two factors have any variables in common. The maximum dimension of the additional inversion required beyond the designated d variables is equal to the maximum number of the $p - d$ remaining variables appearing in one of the factors, say m. The overall inversion can then be regarded as being of dimension $d + m$. The idea, then, is to select an appropriate d variables, so that the resulting dimension $d + m$ is small.

This dimension reduction can be done whenever a multidimensional transform can be written as a product of factors. From (112), we see this structure always occurs with closed queueing networks. For closed queueing networks there is a factor for each queue in the network, and the variable z_j appears in the factor for queue i if and only if chain j visits queue i. Thus, conditional decomposition tends to occur when chains tend to visit relatively few queues. This property is called sparseness of routing chains in Lam and Lien [Lam and Lien, 1983]. As noted by Lam and Lien, this sparseness property is likely to be present in large communication networks and distributed systems.

To carry out this dimension reduction, we exploit the representation of the generating function $\hat{g}(\mathbf{z})$ as a product of separate factors, i.e.,

$$\hat{g}(\mathbf{z}) = \prod_{i=1}^{m} \hat{g}_i(\hat{\mathbf{z}}_i) \qquad (119)$$

where $m \geq 2$ and $\hat{\mathbf{z}}_i$ is a subset of $\{z_1, z_2, \ldots, z_p\}$. We assume that each $\hat{g}_i(\hat{\mathbf{z}}_i)$ cannot be further factorized into multiple factors, unless at least one of the latter is a function of all variables in the set $\hat{\mathbf{z}}_i$.

We now represent the conditional decomposition problem as a graph problem. We construct a graph, called an *interdependence graph*, to represent the interdependence of the variables z_k in the factors. We let each variable z_k be represented by a node in the graph. For each factor $\hat{g}_i(\hat{\mathbf{z}}_i)$ in (119), form a fully connected subgraph Γ_i by connecting all nodes (variables) in the set $\hat{\mathbf{z}}_i$. Then let $\Gamma = \bigcup_{i=1}^{m} \Gamma_i$.

Now for any subset D of Γ, we identify the *maximal connected subsets* $S_i(D)$ of $\Gamma - D$; i.e., $S_i(D)$ is connected for each i, $S_i(D) \cap S_j(D) = \emptyset$ when $i \neq j$ and $\bigcup_i S_i(D) = \Gamma - D$. Let $|A|$ be the cardinality of the set A. Then the dimension of the inversion resulting from the selected subset D is

$$\text{inversion dimension } = |D| + \max_i \{|S_i(D)|\} .$$

It is natural to consider the problem of minimizing the overall dimension of the inversion. This is achieved by finding the subset D to achieve the following minimum:

$$\text{minimal inversion dimension } = \min_{D \subseteq \Gamma} \{|D| + \max_i \{|S_i(D)|\} . \qquad (120)$$

In general, it seems difficult to develop an effective algorithm to solve this graph optimization problem; it seems to be an interesting research problem. However, for the small-to-moderate number of variables that we typically encounter, we can solve (120) by inspection or by enumeration of the subsets of Γ in increasing order of cardinality. Since our overall algorithm is likely to have difficulty if the reduced dimension is not relatively small (e.g., ≤ 10), it is not necessary to consider large sets D in (120). This dimension reduction is illustrated in the example in Section 5.4.

Even though it is not at first obvious, it turns out that the approach to dimension reduction developed by Choudhury, Leung and Whitt [Choudhury et al., 1995b] is essentially equivalent to the tree algorithm of Lam and Lien [Lam and Lien, 1983] used with the convolution algorithm. The connection can be seen by noting that convolution of normalization constants corresponds to multiplication of the generating functions. Dimension reduction may be easier to understand with generating functions, because multiplication is a more elementary operation than convolution.

5.3 Scaling

In this subsection we discuss scaling to control the aliasing error. This additional step is needed here because, unlike probabilities, the normalization constants are not bounded.

To make the motivation clear, we specify the inversion in (108) in more detail. Letting $g_j(K_j) = g^{(j-1)}(\mathbf{z}_{j-1}, \mathbf{K}_j)$ and $\hat{g}_j(z_j) = g^{(j)}(\mathbf{z}_j, \mathbf{K}_{j+1})$, we can apply Theorem 8 to express the jth step of the inversion in (108) as

$$g_j(K_j) = g_j^a(K_j) - e_j ,$$

where

$$g_j^a(K_j) = \frac{1}{2\ell_j K_j r_j^{K_j}} \sum_{k_1=1}^{l_j} e^{-\frac{\pi i k_1}{l_j}} \sum_{k_1=0}^{l_j-1} e^{-\frac{\pi i k_1}{l_j}} \sum_{k=-K_j}^{K_j-1} (-1)^k \hat{g}_j\left(r_j e^{\frac{\pi i (k_1+l_j k)}{l_j K_j}}\right), \quad (121)$$

l_j is a positive integer, r_j is a suitably small positive real number and e_j represents the *aliasing error*, which is given by

$$e_j = \sum_{n=1}^{\infty} g_j(K_j + 2nl_j K_j) r_j^{2nl_j K_j} . \quad (122)$$

Note that, for $j = 1$, $g_1(K_1) = g(\mathbf{K})$ is real, so that $\hat{g}_1(\bar{z}_1) = \overline{\hat{g}_1(z_1)}$. This enables us to cut the computation in (121) by about one half. For $j = 1$, we replace (121) by

$$g_1^a(K_1) = \frac{1}{2\ell_1 K_1 r_1^{K_1}} \left[\hat{g}_1(r_1) - (-1)^{K_1} \hat{g}_1(-r_1) + 2 \sum_{k_1=1}^{l_1} e^{\frac{-\pi i k_1}{l_1}} \right.$$

$$\left. \times \sum_{k=0}^{K_1-1} \hat{g}_1\left(r_1 e^{\pi i (k_1+l_1 k)/l_1 K_1}\right) \right] . \quad (123)$$

To control the aliasing error (122), we choose

$$r_j = 10^{-\frac{\gamma_j}{2\ell_j K_j}} . \quad (124)$$

Inserting (124) into (122), we get

$$e_j = \sum_{n=1}^{\infty} g_j(K_j + 2nl_jK_j)10^{-\gamma_j n} . \qquad (125)$$

This choice of r_j enables us to more easily control the aliasing error e_j using the parameter γ_j. For instance, if g_j were bounded above by 1, as is the case with probabilities, then the aliasing error would be bounded above by $10^{-\gamma_j}/(1 - 10^{-\gamma_j}) \approx 10^{-\gamma_j}$.

As is clear from (125), a bigger γ_j decreases the aliasing error. However, since $r_j^{-K_j} = 10^{\gamma_j/2\ell_j}$, the factor $r_j^{-K_j}$ in (121) increases sharply with γ_j and thus can cause roundoff error problems. Since the parameter l_j does not appear in the aliasing error term (125), it can be used to control the growth of $r_j^{-K_j}$ without altering the aliasing error. As indicated in Section 2.2, bigger values of l_j yield less roundoff error, but more computation because the number of terms in (121) is proportional to l_j.

Since the normalization constants can be arbitrarily large, the aliasing error e_j in (125) can also be arbitrarily large. Thus, in order to control errors, we scale the generating function in each step by defining a *scaled generating function* as

$$\hat{g}_j^*(z_j) = \alpha_{0j}\hat{g}_j(\alpha_j z_j),$$

where α_{0j} and α_j are positive real numbers. We invert this scaled generating function after choosing α_{0j} and α_j so that the errors are suitably controlled. Let $\overline{g}_j(K_j)$ represent the inverse function of $\hat{g}_j^*(z_j)$. The desired inverse function $g_j(K_j)$ may then be recovered from $\overline{g}_j(K_j)$ by

$$g_j(K_j) = \alpha_{0j}^{-1}\alpha_j^{-K_j}\overline{g}_j(K_j) .$$

A way to choose the parameters $\alpha_0, \alpha_1, \ldots, \alpha_p$ especially designed for CQNs was developed in Choudhury, Leung and Whitt [Choudhury et al., 1995b]. Here we present a more general approach from Choudhury and Whitt [Choudhury and Whitt, 1997] that can also be used in other contexts. For a p-dimensional inversion, our general strategy is to choose $p + 1$ parameters $\alpha_0, \alpha_1, \ldots, \alpha_p$ so that

$$g_\alpha(\mathbf{K}) = \alpha_0 \prod_{j=1}^{p} \alpha_j^{K_j} g(\mathbf{K}) \qquad (126)$$

is approximately a probability mass function (pmf) with $\mathbf{K}$ being approximately its mean vector. Then we should be able to control the discretization error in computing $g_\alpha(\mathbf{K})$ by acting as if it is bounded in $\mathbf{K}$. If we succeed in calculating $g_\alpha(\mathbf{K})$ with small discretization error, by (126), we succeed in calculating $g(\mathbf{K})$ with small relative discretization error.

The scaling algorithm is based on the following theorem.

Theorem 13 *Suppose that $\hat{g}(\mathbf{z})$ is the p-dimensional generating function of a nonnegative function $g(\mathbf{K})$. For any vector $(m_1, \ldots, m_p)$ of positive numbers, if the set of p equations*

$$z_i \frac{\partial}{\partial z_i} \log \hat{g}(z_1, \ldots, z_p)|_{z_j = \alpha_j} \text{ for all } j = m_i, \quad 1 \le i \le p, \quad (127)$$

has a solution $(\alpha_1, \ldots, \alpha_p)$ with $\hat{g}(\alpha_1, \ldots, \alpha_p) < \infty$, then $g_\alpha(\mathbf{K})$ in (126) with

$$\alpha_0 = \frac{1}{\hat{g}(\alpha_1, \ldots, \alpha_p)}$$

is a pmf with mean vector $(m_1, \ldots, m_p)$.

Proof. Note that

$$z_i \frac{\partial}{\partial z_i} \log \hat{g}(\mathbf{z})|_{\mathbf{z}=\boldsymbol{\alpha}} = \frac{z_i \frac{\partial}{\partial z_i} \hat{g}(\mathbf{z})|_{\mathbf{z}=\boldsymbol{\alpha}}}{\hat{g}(\boldsymbol{\alpha})}$$

$$= \frac{\sum_{K_1=0}^{\infty} \cdots \sum_{K_p=0}^{\infty} K_i \prod_{j=1}^{p} \alpha_j^{K_j} g(\mathbf{K})}{\sum_{K_1=0}^{\infty} \cdots \sum_{K_p=0}^{\infty} \prod_{j=1}^{p} \alpha_j^{K_j} g(\mathbf{K})} = m_i .$$

It remains to solve the p equations (127). Choudhury and Whitt [Choudhury and Whitt, 1997] suggest using an iterative procedure, fixing all parameters except α_0 and α_i for some i and the solving the single equation in (127) for i. Solving a single equation is relatively easy because the left side of (127) is strictly increasing in α_i. (See Theorem 7.1 of [Choudhury and Whitt, 1997].) To speed up convergence, after a few initial steps, a Newton-Raphson root finding algorithm can be used.

To illustrate, we consider a CQN with two chains. Equation (111) then yields the generating function

$$\hat{g}(z_1, z_2) = \frac{\exp(\rho_{10} z_1 + \rho_{20} z_2)}{(1 - \rho_{11} z_1 - \rho_{21} z_2)(1 - \rho_{12} z_1 - \rho_{22} z_2)} . \quad (128)$$

We work with the scaled generating function

$$\hat{g}_\alpha(z_1, z_2) = \alpha_0 \hat{g}(\alpha_1 z_1, \alpha_2 z_2) .$$

The scaling parameters α_1 and α_2 are obtained from the two equations

$$n_i = z_i \frac{\partial}{\partial z_i} \log \hat{g}(z_1, z_2)|_{z_1 = \alpha_1, z_2 = \alpha_2}$$

$$= \alpha_i \left[\rho_{i0} + \sum_{j=1}^{q} \frac{\rho_{ij}}{1 - \sum_{k=1}^{2} \rho_{kj} \alpha_k} \right] \text{ for } i = 1, 2 . \quad (129)$$

We must solve the pair of nonlinear equations in (128). As suggested above, we can fix α_2 and search for the value α_1 that satisfies the equation for $i = 1$. Next, fixing α_1 at the value obtained, we search for the value α_2 that satisfies the equation for $i = 2$. We do this repeatedly until convergence is achieved based on some prescribed error criterion. We observed that this procedure indeed converges, but the rate of convergence becomes slow as n_1 and n_2 increases. By contrast, the two-dimensional Newton-Raphson method (see Press et al. [Press et al., 1988], Chapter 9) converges very fast (less than 10 steps), provided that we start not too far from the root. So we initially use the search procedure a few times and then the Newton-Raphson method.

Here is a concrete with generating function (128). It corresponds to a CQN with two single-server queues, one infinite-server queue and two chains. Given (128), we need not specify all parameters. The relevant parameters are:

$$\rho_{1,0} = 1, \quad \rho_{2,0} = 1, \quad \rho_{1,1} = 1, \quad \rho_{2,1} = 2,$$
$$\rho_{1,2} = 2, \quad \rho_{2,2} = 3.$$

Numerical results for several values of the chain populations n_1 and n_2 are displayed in Table 10. The accurate computation in the last case would be challenging by alternative algorithms. For that case, we use Euler summation in each dimension and it took only seconds. Accuracy was checked by performing two independent computations with two sets of inversion parameters.

Table 10 Numerical results for the normalization constant g_{n_1,n_2} in a closed queueing network with two chains.

n_1	n_2	g_{n_1,n_2}	α_1	α_2	α_0
3	2	0.243883E+04	0.240070	0.104428	0.806627E-01
30	20	0.627741E+33	0.294397	0.130311	0.589928E-02
300	200	0.973460E+331	0.299451	0.133032	0.564701E-03
3000	2000	0.235196E+3318	0.299945	0.133303	0.562179E-04

5.4 A Challenging Example

We conclude this section and this chapter by giving a challenging numerical example. We calculate the normalization constant $g(\mathbf{K})$ given by (105) and using (110) for specified population vectors $\mathbf{K}$ from the generating function $\hat{g}(\mathbf{z})$ in (112). Thus the parameters are the number of chains, p, the number of distinct single-server queues, q', the multiplicities m_i, the aggregate relative traffic intensities ρ_{ji}, $1 \leq j \leq p$, $0 \leq i \leq q'$, and the desired population vector $\mathbf{K}$.

From (112), note that the normalization constant $g(\mathbf{K})$ only depends on these parameters p, q', m_i, ρ_{ji} and $\mathbf{K}$. Hence, we do not fully specify the models below. In particular, we do not give the routing probabilities $R_{ri,sj}$ or the mean service times t_{ri}. Thus, there are many detailed models consistent with our partial model specifications. One possible routing matrix consistent with the data that we provide is a cyclic routing matrix, all of whose entries are 0's and 1's, which yields visit ratios $e_{ri} = 1$ for all stages (r, i) from (109). If we consider this case, then $t_{ri} = \rho'_{ri}$ and the throughputs θ_{ri} in (113) coincide with the normalization constant ratios $g(\mathbf{K} - \mathbf{1}_j)/g(\mathbf{K})$. We also display these ratios along with the values of $g(\mathbf{K})$ in our numerical results below. We note that the throughputs for any more detailed model can be found by solving (109) for the visit ratios e_{ri} and then applying (113).

The particular example that we consider has 11 closed chains and 1000 queues. Specifically, $p = 11$, $q = 1000$, $q' = 10$ and $m_i = 100$, $1 \leq i \leq 10$. A crucial step is dimension reduction, which reduces the effective dimension from 11 to 2. We consider three cases. First, let the chain populations be $K_j = 200$ for $2 \leq j \leq 11$. We obtain four subcases by considering four different values for K_1. Our numerical results are given below in Table 11.

Table 11   Numerical results for Case 1 of the 11-chain example.

chain populations		normalization	ratio
K_j for $2 \leq j \leq 11$	K_1	constant $g(\mathbf{K})$	$g(\mathbf{K} - \mathbf{1}_1)/g(\mathbf{K})$
200	20	1.232036e278	4.582983e-3
200	200	2.941740e579	4.281094e-2
200	2000	3.399948e2037	2.585489e-1
200	20,000	9.07177e8575	4.846930e-1

The accuracy was checked by performing the calculations twice, once with $l_1 = 1$ and once with $l_1 = 2$.

The results in Table 11 were obtained in less than one minute by exploiting Euler summation with 51 terms. This example would seem to be out of the range of many other algorithms. For example, convolution would require a storage of size $200^{10} \times 2 \times 10^4 = 2.5 \times 10^{27}$ for the last case of Table 11. The last case would appear to be difficult even for the tree algorithm [Lam and Lien, 1983].

Some asymptotic methods require that there be an IS queue and that each chain visit this queue or that *all* chain populations be large. To show that our algorithm does not have such limitations, we consider two modifications of Case 1. Case 2 has classes 1 and 2 with small populations, while the other class populations remain large. In particular, we let $K_2 = 2$ and $K_3 = 5$. Numerical results for Case 2 appear below in Table 12.

Case 3 is a modification of Case 2 in which we remove all the IS nodes, i.e., we set $\rho_{j0} = 0$ for all j. Numerical results for Case 3 appear below in Table 13. As for Case 1, Cases 2 and 3 required about a minute on the SUN SPARC-2 workstation.

Before closing this section, we point out that if the model is such that we cannot take advantage of any of available speed-up techniques (namely, dimension reduction, fast summation of large sums and large queue multiplicities), then

Table 12   Numerical results for Case 2 of the 11-chain example.

chain populations K_j for $2 \leq j \leq 11$	K_1	normalization constant $g(\mathbf{K})$	ratio $g(\mathbf{K} - \mathbf{1}_1)/g(\mathbf{K})$
in all cases:	2	3.842031e407	5.128582e-4
$K_2 = 2$,	20	1.484823e454	5.087018e-3
$K_3 = 5$ and	200	6.003231e747	4.706783e-2
$K_j = 200, 4 \leq j \leq 11$	2000	5.442693e2154	2.705391e-1
	20,000	2.494765e8617	4.852817e-1

Table 13   Numerical results for Case 3 of the 11-chain example.

chain populations K_j for $2 \leq j \leq 11$	K_1	normalization constant $g(\mathbf{K})$	ratio $g(\mathbf{K} - \mathbf{1}_1)/g(\mathbf{K})$
in all cases:	2	9.959619e313	4.762073e-4
$K_2 = 2$,	20	1.447107e361	4.728591e-3
$K_3 = 5$ and	200	1.222889e660	4.417444e-2
$K_j = 200, 4 \leq j \leq 11$	2000	2.948943e2096	2.645993e-1
	20,000	4.210541e8588	4.851015e-1

the inversion algorithm will be slower than the convolution algorithm, as indicated in [Choudhury et al., 1995b]. Similarly, if dimension reduction is possible but none of the other speed-ups, then the inversion algorithm will be slower than the tree convolution algorithm, which also does dimension reduction in the time domain.

To illustrate, Lam and Lien [Lam and Lien, 1983] analyze an example with 64 queues and 32 chains, requiring about 3×10^7 operations. Choudhury, Leung and Whitt [Choudhury et al., 1995b] analyzed this model and observed that the effective dimension can be reduced from 32 to 9. However, all chain populations are between 1 and 5 and so the speed-up technique based on Euler summation does not apply. Also there are no multiplicities. Choudhury et al. estimated that the operation count for this example would be about 10^{12}, so that the inversion algorithm is considerably slower than the tree convolution algorithm, even though the inversion algorithm is faster than the pure convolution algorithm, which has an operation count of about 10^{23}. It appears that the inversion algorithm nicely complements the tree algorithm, because the tree algorithm will be faster if the effective dimension after dimension reduction remains large but all chain populations are small. In contrast, the inversion algorithm will be faster if the effective dimension after dimension reduction is small (typically 5 or less) but some of the chain populations are large.

References

[Abate et al., 1995a] Abate, J., Choudhury, G. L., Lucantoni, D. M., and Whitt, W. (1995a). Asymptotic analysis of tail probabilities based on the computation of moments. *Ann. Appl. Prob.*, 5:983–1007.

[Abate et al., 1993] Abate, J., Choudhury, G. L., and Whitt, W. (1993). Calculation of the GI/G/1 waiting-time distribution and its cumulants from Pollaczek's formulas. *Archiv für Elektronik und Übertragungstechnik*, 47:311–321.

[Abate et al., 1994] Abate, J., Choudhury, G. L., and Whitt, W. (1994). Waiting-time tail probabilities in queues with long-tail service-time distributions. *Queueing Systems*, 16:311–338.

[Abate et al., 1995b] Abate, J., Choudhury, G. L., and Whitt, W. (1995b). Exponential approximations for tail probabilities in queues, I: waiting times. *Oper. Res.*, 43:885–901.

[Abate et al., 1996] Abate, J., Choudhury, G. L., and Whitt, W. (1996). On the Laguerre method for numerically inverting Laplace transforms. *INFORMS Journal on Computing*, 8:413–427.

[Abate et al., 1997] Abate, J., Choudhury, G. L., and Whitt, W. (1997). Numerical inversion of multidimensional Laplace transforms by the Laguerre method. *Performance Evaluation*, 31:229–243.

[Abate and Whitt, 1992a] Abate, J. and Whitt, W. (1992a). The Fourier-series method for inverting transforms of probability distributions. *Queueing Systems*, 10:5–88.

[Abate and Whitt, 1992b] Abate, J. and Whitt, W. (1992b). Numerical inversion of probability generating functions. *Oper. Res. Letters*, 12:245–251.

[Abate and Whitt, 1992c] Abate, J. and Whitt, W. (1992c). Solving probability transform functional equations for numerical inversion. *Oper. Res. Letters*, 12:275–281.

[Abate and Whitt, 1995] Abate, J. and Whitt, W. (1995). Numerical inversion of Laplace transforms of probability distributions. *ORSA J. on Computing*, 7:36–43.

[Abate and Whitt, 1998] Abate, J. and Whitt, W. (1998). Calculating transient characteristics of the Erlang loss model by numerical transform inversion. *Stochastic Models*, 14.

[Asmussen, 1987] Asmussen, S. (1987). *Applied Probability and Queues*. Wiley, New York.

[Asmussen et al., 1996] Asmussen, S., Nerman, O., and Olsson, M. (1996). Fitting phase type distributions via the EM algorithm. *Scand. J. Statist.*, 23:419–441.

[Beneš, 1961] Beneš, V. E. (1961). The covariance function of a simple trunk group with applications to traffic measurements. *Bell System Tech. J.*, 40:117–148.

[Beneš, 1965] Beneš, V. E. (1965). *Mathematical Theory of Connecting Networks and Telephone Traffic.* Academic Press, New York.

[Bertozzi and McKenna, 1993] Bertozzi, A. and McKenna, J. (1993). Multidimensional residues, generating functions, and their application to queueing networks. *SIAM Review,* 35:239–268.

[Borovkov, 1984] Borovkov, A. A. (1984). *Asymptotics Methods in Queueing Theory.* Wiley, New York.

[Chaudhry et al., 1992] Chaudhry, M. L., Agarwal, M., and Templeton, J. G. C. (1992). Exact and approximate numerical solutions of steady-state distributions arising in the GI/G/1 queue. *Queueing Systems,* 10:105–152.

[Choudhury et al., 1995a] Choudhury, G. L., Leung, K. K., and Whitt, W. (1995a). An algorithm to compute blocking probabilities in multi-rate multi-class multi-resource loss models. *Adv. Appl. Prob.,* 27:1104–1143.

[Choudhury et al., 1995b] Choudhury, G. L., Leung, K. K., and Whitt, W. (1995b). Calculating normalization constants of closed queueing networks by numerically inverting their generating functions. *J. ACM,* 42:935–970.

[Choudhury et al., 1995c] Choudhury, G. L., Leung, K. K., and Whitt, W. (1995c). Efficiently providing multiple grades of service with protection against overloads in shared resources. *AT&T Tech. J.,* 74:50–63.

[Choudhury et al., 1995d] Choudhury, G. L., Leung, K. K., and Whitt, W. (1995d). An inversion algorithm to compute blocking probabilities in loss networks with state-dependent rates. *IEEE/ACM Trans. Networking,* 3:585–601.

[Choudhury and Lucantoni, 1996] Choudhury, G. L. and Lucantoni, D. M. (1996). Numerical computation of the moments of a probability distribution from its transform. *Oper. Res.,* 44:368–381.

[Choudhury et al., 1994] Choudhury, G. L., Lucantoni, D. M., and Whitt, W. (1994). Multidimensional transform inversion with applications to the transient M/G/1 queue. *Ann. Appl. Prob.,* 4:719–740.

[Choudhury et al., 1996] Choudhury, G. L., Lucantoni, D. M., and Whitt, W. (1996). Squeezing the most out of ATM. *IEEE Trans. Commun.,* 44:203–217.

[Choudhury et al., 1997] Choudhury, G. L., Lucantoni, D. M., and Whitt, W. (1997). Numerical solution of $M_t/G_t/1$ queues. *Oper. Res.,* 45:451–463.

[Choudhury and Whitt, 1995] Choudhury, G. L. and Whitt, W. (1995). Q^2: A new performance analysis tool exploiting numerical transform inversion. In *Proc. Third Int. Workshop on Modeling, Analysis and Simul. of Computer and Telecomm. Systems,* pages 411–415, Durham, NC. (MASCOTS '95).

[Choudhury and Whitt, 1996] Choudhury, G. L. and Whitt, W. (1996). Computing distributions and moments in polling models by numerical transform inversion. *Performance Evaluation,* 25:267–292.

[Choudhury and Whitt, 1997] Choudhury, G. L. and Whitt, W. (1997). Probabilistic scaling for the numerical inversion of non-probability transforms. *INFORMS J. Computing,* 9:175–184.

[Cohen, 1982] Cohen, J. W. (1982). *The Single Server Queue.* North-Holland, Amsterdam, second edition.

[Conway and Georganas, 1989] Conway, A. E. and Georganas, N. D. (1989). *Queueing Networks – Exact Computational Algorithms: A Unified Theory Based on Decomposition and Aggregation.* MIT Press, Cambridge, MA.

[Darling and Siegert, 1953] Darling, D. A. and Siegert, J. F. (1953). The first passage problem for a continuous Markov process. *Ann. Math. Statist.*, 24:624–639.

[Davies and Martin, 1979] Davies, B. and Martin, B. L. (1979). Numerical inversion of the Laplace transform: A survey and comparison of methods. *J. Comp. Phys.*, 33:1–32.

[Davis and Rabinowitz, 1984] Davis, P. J. and Rabinowitz, P. (1984). *Methods of Numerical Integration.* Academic Press, New York, second edition.

[Doetsch, 1961] Doetsch, G. (1961). *Guide to Applications of Laplace Transforms.* Van Nostrand, London.

[Doetsch, 1974] Doetsch, G. (1974). *Introduction to the Theory and Application of the Laplace Transformation.* Springer-Verlag, New York.

[Dubner and Abate, 1968] Dubner, H. and Abate, J. (1968). Numerical inversion of Laplace transforms by relating them to the finite Fourier cosine transform. *J. ACM*, 15:115–123.

[Duffield and Whitt, 1998] Duffield, N. G. and Whitt, W. (1998). A source traffic model and its transient analysis for network control. *Stochastic Models*, 14.

[Durbin, 1974] Durbin, F. (1974). Numerical inversion of Laplace transforms: an efficient improvement to Dubner and Abate's method. *Comput. J.*, 17:371–376.

[Feldmann and Whitt, 1997] Feldmann, A. and Whitt, W. (1997). Fitting mixtures of exponentials to long-tail distributions to analyze network performance models. *Performance Evaluation*, 31:245–279.

[Feller, 1971] Feller, W. (1971). *An Introduction to Probability Theory and its Applications*, volume II. Wiley, New York, second edition.

[Gaver, 1966] Gaver, D. P. (1966). Observing stochastic processes and approximate transform inversion. *Operations Research*, 14:444–459.

[Gaver and Jacobs, 1998] Gaver, D. P. and Jacobs, P. A. (1998). Waiting times when service times are stable laws: tamed and wild. In Shanthikumar, J. G. and (eds.), U. S., editors, *Recent Contributions in Applied Probability and Stochastic Processes, Festschrift for Julian Keilson.* Kluwer, Boston.

[Giffin, 1975] Giffin, W. C. (1975). *Transform Techniques for Probability Modeling.* Academic Press, New York.

[Hosono, 1979] Hosono, T. (1979). Numerical inversion of Laplace transform. *J. Inst. Elec. Eng. Jpn.*, pages A54–A64. 494 (In Japanese).

[Hosono, 1981] Hosono, T. (1981). Numerical inversion of Laplace transform and some applications to wave optics. *Radio Sci.*, 16:1015–1019.

[Hosono, 1984] Hosono, T. (1984). *Fast Inversion of Laplace Transform by BASIC*. Kyoritsu Publishers, Japan. (In Japanese).

[Jagerman, 1974] Jagerman, D. L. (1974). Some properties of the Erlang loss function. *Bell System Tech. J.*, 53:525–551.

[Jagerman, 1978] Jagerman, D. L. (1978). An inversion technique for the Laplace transform with applications. *Bell System Tech. J.*, 57:669–710.

[Jagerman, 1982] Jagerman, D. L. (1982). An inversion technique for the Laplace transform. *Bell Sys. Tech. J.*, 61:1995–2002.

[Johnsonbaugh, 1979] Johnsonbaugh, R. (1979). Summing an alternating series. *Amer. Math. Monthly*, 86:637–648.

[Kao, 1997] Kao, E. P. C. (1997). *An Introduction to Stochastic Processes*. Duxbury Press, New York.

[Keilson, 1979] Keilson, J. (1979). *Markov Chain Models – Rarity and Exponentiality*. Springer-Verlag, New York.

[Keilson and Ross, 1975] Keilson, J. and Ross, H. F. (1975). Passage time distributions for Gaussian Markov (Ornstein-Uhlenbech) statistical processes. *Selected Tables in Mathematical Statistics*, 3:233–327.

[Kingman, 1961] Kingman, J. F. C. (1961). The single server queue in heavy traffic. *Proc. Camb. Phil. Soc.*, 57:902–904.

[Kleinrock, 1975] Kleinrock, L. (1975). *Queueing Systems, Volume I: Theory*. Wiley, New York.

[Knessl, 1990] Knessl, C. (1990). On the transient behavior of the M/M/m/m loss model. *Stochastic Models*, 6:749–776.

[Kobayashi, 1978] Kobayashi, H. (1978). *Modeling and Analysis: An Introduction to System Performance Evaluation Methodology*. Addison-Wesley, Reading, MA.

[Kwok and Barthez, 1989] Kwok, Y. K. and Barthez, D. (1989). An algorithm for the numerical inversion of the Laplace transform. *Inverse Problems*, 5:1089–1095.

[Lam and Lien, 1983] Lam, S. S. and Lien, Y. L. (1983). A tree convolution algorithm for the solution of queueing networks. *Commun. ACM*, 26:203–215.

[Lavenberg, 1983] Lavenberg, S. S., editor (1983). *Computer Performance Modeling Handbook*. Academic Press, Orlando, FL.

[Lucantoni et al., 1994] Lucantoni, D. M., Choudhury, G. L., and Whitt, W. (1994). The transient BMAP/G/1 queue. *Stochastic Models*, 10:145–182.

[Mitra and Weiss, 1989] Mitra, D. and Weiss, A. (1989). The transient behavior in Erlang's model for large trunk groups and various traffic conditions. In *Teletraffic Science for new cost-Effective Systems, Networks and Services*, pages 1367–1374, Amsterdam. Elsevier-Science.

[Neuts, 1981] Neuts, M. F. (1981). *Matrix-Geometric Solutions in Stochastic Models*. The Johns Hopkins University Press, Baltimore.

[Neuts, 1989] Neuts, M. F. (1989). *Structured Stochastic Matrices of M/G/1 Type and Their Applications*. Marcel Dekker, New York.

[O'Cinneide, 1997] O'Cinneide, C. A. (1997). Euler summation for Fourier series and Laplace transform inversion. *Stochastic Models*, to appear.

[Pol and Bremmer, 1987] Pol, B. V. D. and Bremmer, H. (1987). *Operational Calculus*. Cambridge Press, reprinted Chelsea Press, New York.

[Pollaczek, 1952] Pollaczek, F. (1952). Fonctions caractéristiques de certaines répartitions définies au money de la notion d'ordre. Application à la théorie de attentes. *C. R. Acad. Sci. Paris*, 234:2334–2336.

[Pollaczek, 1965] Pollaczek, F. (1965). Concerning an analytic method for the treatment of queueing problems. In Smith, W. L. and Wilkinson, W. E., editors, *Proceedings of the Symposium on Congestion Theory*, pages 1–25 and 34–42, Chapel Hill. The University of North Carolina Press.

[Poularikas, 1996] Poularikas, A. D. (1996). *The Transforms and Applications Handbook*. CRC Press, Boca Raton, FL.

[Press et al., 1988] Press, W. H., Flannery, B. P., Teukolsky, S. A., and Vetterling, W. T. (1988). *Numerical Recipes, FORTRAN Version*. Cambridge University Press, Cambridge, England.

[Reif, 1965] Reif, F. (1965). *Fundamentals of Statistical and Thermal Physics*. McGraw-Hill, New York.

[Reiser and Kobayashi, 1975] Reiser, M. and Kobayashi, H. (1975). Queueing networks with multiple closed chains: theory and computational algorithms. *IBM J. Res. Dev.*, 19:283–294.

[Riordan, 1962] Riordan, J. (1962). *Stochastic Service Systems*. Wiley, New York.

[Simon et al., 1972] Simon, R. M., Stroot, M. T., and Weiss, G. H. (1972). Numerical inversion of Laplace transforms with applications to percentage labeled experiments. *Comput. Biomed. Res.*, 6:596–607.

[Smith, 1953] Smith, W. L. (1953). On the distribution of queueing times. *Proc. Camb. Phil. Soc.*, 49:449–461.

[Srikant and Whitt, 1996] Srikant, R. and Whitt, W. (1996). Simulation run lengths to estimate blocking probabilities. *ACM J. TOMACS*, 6:7–52.

[Takács, 1962] Takács, L. (1962). *Introduction to the Theory of Queues*. Oxford University Press, New York.

[Tolstov, 1976] Tolstov, G. P. (1976). *Fourier Series*. Dover, New York.

[Weeks, 1966] Weeks, W. T. (1966). Numerical inversion of Laplace transforms using Laguerre functions. *J. ACM*, 13:419–426.

[Whitt, 1984] Whitt, W. (1984). Heavy-traffic approximations for service systems with blocking. *AT&T Bell Lab. Tech. J.*, 63:689–708.

[Wimp, 1981] Wimp, J. (1981). *Sequence Transformations and Their Applications*. Academic Press, New York.

Joseph Abate received the B.S. degree from the City College of New York in 1961, and the Ph.D. degree in Mathematical Physics from New York University in 1967. From 1966 to 1970 he was employed by Computer Applications Corporation where he specialized in evaluating the performance of real-time computer systems. In 1971, he joined AT&T Bell Laboratories and retired in 1990. He has since served as a consultant to AT&T and Lucent Technologies. For most of his career, he worked on the design and analysis of transaction processing systems used in support of telephone company operations. For many years he has had a great passion for the use of Laplace transforms in queueing problems.

Gagan L. Choudhury received the B. Tech. degree in Radio Physics and Electronics from the University of Calcutta, India in 1979 and the MS and Ph.D. degrees in Electrical Engineering from the State University of New York (SUNY) at Stony Brook in 1981 and 1982, respectively. Currently he is a Technical Manager at the Teletraffic Theory and System Performance Department in AT&T Laboratories, Holmdel, New Jersey, USA. His main research interest is in the development of multi-dimensional numerical transform inversion algorithms and their application to the performance analysis of telecommunication and computer systems.

Ward Whitt received the A.B. degree in Mathematics from Dartmouth College, Hanover, NH, USA, in 1964 and the Ph.D. degree in Operations Research from Cornell University, Ithaca, NY, USA, in 1969. He was on the faculty of Stanford University and Yale University before joining AT&T Laboratories in 1977. He is currently a member of the Network Mathematics Research Department in AT&T Labs-Research in Florham Park, NJ, USA. His research has focused on probability theory, queueing models, performance analysis and numerical transform inversion.

9 OPTIMAL CONTROL OF MARKOV CHAINS

Shaler Stidham, Jr.

Department of Operations Research
University of North Carolina at Chapel Hill
CB #3180, Smith Building
Chapel Hill, NC 27599-3180, USA

sandy@or.unc.edu

1 INTRODUCTION

Add costs and decisions to a Markov chain and you have a *Markov decision chain* (*MDC*), the subject of this chapter. Roughly speaking, the problem is how to control a Markov chain to achieve an economic objective. Control is exercised by taking a sequence of actions, each of which may depend on the currently observed state and may influence both the immediate cost and the next state transition.

In the next section we introduce our general (discrete-time) Markov decision model. Since the purpose of this chapter is to discuss algorithms, the exposition is somewhat informal, but we do point out the technical difficulties and suggest how they can be resolved. Costs and transition probabilities may be time dependent. The objective is to minimize the expected total (discounted) cost over an infinite horizon, from each starting state. We derive the (Bellman) optimality equation satisfied by the optimal value function (that is, the minimal expected total discounted cost, considered as a function of the starting state.) We show how several commonly encountered problems emerge as applications of our general model. In the following section we consider at length the infinite-horizon discounted problem with time-homogeneous costs and transition probabilities. The emphasis is on algorithms for numerical computation or approximation of the optimal value function and an optimal (stationary) policy. (The material in this section draws extensively from the excellent text on Markov decision processes by Puterman [Puterman, 1994], to which the reader is referred for more details.) Our final section comments on extensions (e.g., to semi-Markov decision processes and the average-cost criterion) and applications to problems in the optimal control of queues.

2 DISCRETE-TIME MARKOV DECISION CHAIN: A GENERAL MODEL

At discrete time points or *stages*, $t = 0, 1, \ldots$, a decision maker observes the *state* X_t of a system and takes an *action* A_t. The state X_t belongs to a countable set $S = \{0, 1, \ldots\}$, called the *state space*. The action A_t belongs to a finite set $A = \{1, 2, \ldots, k\}$, called the *action space*. We assume that the system has the following Markovian property:

$$P\{X_{t+1} = j | X_t = i, A_t = a, X_{t-1}, A_{t-1}, \ldots, X_0, A_0\}$$
$$= P\{X_{t+1} = j | X_t = i, A_t = a\} =: p_{tij}(a) . \quad (1)$$

That is, the next state is determined (probabilistically) by the current state and action. In other words, given X_t and A_t (the present state and action), X_{t+1} (the immediate future state) is independent of $(X_0, A_0, \ldots, X_{t-1}, A_{t-1})$ (the past). Note that in this formulation the transition probabilities, $p_{tij}(a)$, are allowed to depend on the stage t. That is, they are not required to be time homogeneous.

The bivariate stochastic process, $\{(X_t, A_t), t = 0, 1, \ldots\}$, is called a *Markov decision chain (MDC)*. The formulation of $\{(X_t, A_t)\}$ as a stochastic process is incomplete, however, until we specify how actions are selected.

2.1 Definition of Policies

Define $D(A)$ is the set of all probability mass functions on A, that is

$$D(A) := \{(p_1, \ldots, p_k) : p_a \geq 0, a = 1, \ldots, k; \sum_{a=1}^{k} p_a = 1\} .$$

For each stage $t = 0, 1, \ldots$, let $H_t := (X_0, A_0, X_1, A_1, \ldots, X_{t-1}, A_{t-1}, X_t)$ be the *history* of the process at t. A function $\pi_t : (S \times A)^t \times S \to D(A)$ is called a *decision rule* for stage t, $t = 0, 1, \ldots$. Thus, the decision rule π_t assigns to each possible observed history, $h_t = (x_0, a_0, x_1, a_1, \ldots, x_{t-1}, a_{t-1}, x_t) \in (S \times A)^t \times S$, a probability mass function over the action space A, so that

$$P\{A_t = a | h_t\} = \pi_{t,a}(h_t) ,$$

where $\pi_t(h_t) = (\pi_{t,1}(h_t), \ldots, \pi_{t,k}(h_t))$. A sequence of decision rules

$$\pi = (\pi_0, \pi_1, \ldots, \pi_t, \ldots) ,$$

one for each stage t, is called a *policy*.

Now, for a fixed policy π, the bivariate sequence $\{(X_t, A_t), t = 0, 1, \ldots\}$ constitutes a well-defined stochastic process. (It is not necessarily Markov, since the action A_{t+1} is allowed to depend on the past, $(X_0, A_0, X_1, A_1, \ldots, X_{t-1}, A_{t-1})$, as well as the present, (X_t, A_t).) Informally, $\{(X_t, A_t)\}$ evolves as follows. At stage 0, we observe $X_0 = x_0$ and take action $A_0 = a_0$ with probability $\pi_{0,a_0}(x_0)$. The process moves to a new state $X_1 = x_1$ with probability $p_{0,x_0,x_1}(a_0)$. At stage 1, based on the observed history $h_1 = (x_0, a_0, x_1)$, we take action $A_1 = a_1$ with probability $\pi_{1,a_1}(h_1)$, whereupon the process moves to a new state $X_2 = x_2$ with probability $p_{1,x_1,x_2}(a_1)$, and so forth. More formally, it is a straightforward exercise to show that all the finite-dimensional distributions of (X_t, A_t) are completely determined by the initial distribution, $q(i), i \in S$, the transition probabilities, $p_{t,i,j}(a)$, and the policy π. As an example, note that

$$P_\pi\{X_1 = x_1\} = \sum_{x_0 \in S} q(x_0) \sum_{a_0 = A} \pi_{0,a_0}(x_0) p_{0,x_0,x_1}(a_0) ,$$

where we use $P_\pi\{\cdot\}$ to denote the probability measure (on events defined by (X_t, A_t)) induced by π.

Let Δ denote the set of all policies π.

Definition. A policy $\pi \in \Delta$ is called *Markov* if, for all $t = 0, 1, \ldots, i \in S$, $a \in A$,

$$P_\pi\{A_t = a | X_t = i, A_{t-1}, X_{t-1}, \ldots, A_0, X_0\} = P_\pi\{A_t = a | X_t = i\} .$$

The set of all Markov policies will be denoted by Δ_M.

Thus, for a Markov policy $\pi = (\pi_0, \pi_1, \ldots, \pi_t, \ldots)$, each decision rule π_t depends on the history, H_t, only through the current state, X_t, and thus can be represented (with slight abuse of notation) as a mapping $\pi_t : S \to D(A)$, so that, given $X_t = i$, action a is taken with probability $\pi_{t,a}(i)$.

Definition. A (Markov) policy π is called *stationary* if, for all $t = 0, 1, \ldots$, $i \in S, a \in A$,

$$P_\pi\{A_t = a | X_t = i, A_{t-1}, X_{t-1}, \ldots, A_0, X_0\} = P_\pi\{A_0 = a | X_0 = i\} .$$

The set of all stationary (Markov) policies will be denoted by Δ_{SM}.

A stationary policy π is completely determined by the single decision rule $\pi_0 : S \to D(A)$, since stationarity implies that $\pi_t = \pi_0$, for all t. Hence, for $\pi \in \Delta_{SM}$, we have $\pi = (\pi_0, \pi_0, \ldots, \pi_0, \ldots)$.

Definition. A Markov policy π is called *deterministic* if there are functions $f_t : S \to A$ such that

$$P_\pi\{A_t = f_t(i) | X_t = i\} = 1 ,$$

for all $t = 0, 1, \ldots, i \in S$. The set of all deterministic Markov policies will be denoted by Δ_{DM}.

In other words, under a deterministic Markov policy, each decision rule, $\pi_{t,\cdot}(i)$, is degenerate, putting all its mass on the single action $a = f_t(i)$.

The set of deterministic stationary (Markov) policies is denoted by Δ_{DSM}. A policy $\pi \in \Delta_{DSM}$ is completely specified by a single function $f : S \to A$, such that

$$P_\pi\{A_t = f(i)|X_t = i, A_{t-1}, X_{t-1}, \ldots, A_0, X_0\} = 1,$$

for all $t = 0, 1, \ldots$, $i \in S$. Such a policy π dictates that we choose action $a = f(i)$ (with probability one) whenever the process is in state i. We sometimes write $\pi = f^\infty$, or simply $\pi = f$ (again abusing notation), as shorthand for $\pi = (f, f, \ldots, f, \ldots)$, for a policy $\pi \in \Delta_{DSM}$.

Costs and Rewards

Let $c_t(i, a)$ denote the (expected) cost incurred in stage t, if the current state $X_t = i$ and action $A_t = a$ is taken. Negative costs represent rewards (benefits). Note that costs can be non stationary (that is, time dependent), but that they are "Markovian", in the sense that they depend on the observed history, $h_t = (x_0, a_0, x_1, a_1, \ldots, x_{t-1}, a_{t-1}, x_t)$ only through x_t. It is the Markovian nature of the costs, $c_t(i, a)$, and transition probabilities, $p_{tij}(a)$, that suggest that we should be able to restrict our attention to Markov policies without loss of optimality.

Let $\phi^\pi(i)$ denote the possibly infinite expected total cost (over the infinite horizon of stages, $t = 0, 1, \ldots$) of following policy π, if the system starts in state $X_0 = i$. The function $\phi^\pi(\cdot)$ is called the *value function* associated with policy π. Thus, for $\pi \in \Delta$, we have

$$\phi^\pi(i) := E_\pi \left[\sum_{t=0}^\infty c_t(X_t, A_t) | X_0 = i \right], \; i \in S, \quad (2)$$

where $E_\pi[\cdot]$ is the expectation operator associated with the probability measure $P_\pi\{\cdot\}$. Thus, for example,

$$E_\pi[c_t(X_t, A_t)|X_0 = i] = \sum_{j \in S} \sum_{a \in A} c_t(j, a) P_\pi\{X_t = j, A_t = a|X_0 = i\}. \quad (3)$$

Remark. We must be careful in applying the definition of $\phi(i)$ to make sure that the expectation in (2) is well defined. Indeed, the infinite series inside the expectation may not converge with probability one, unless we put some restrictions on the cost functions, c_t, and/or the transition probabilities, p_t. We shall do so carefully in each application. For example, if the costs are all non-negative, then the series converges (it may equal ∞), the expectation is well defined, and we have

$$\phi^\pi(i) := \sum_{t=0}^\infty E_\pi\left[c_t(X_t, f_t(X_t))|X_0 = i\right], \; i \in S,$$

where the interchange of expectation and summation is justified by Tonelli's theorem. (Note that each term in the sum can now be evaluated, at least in principle, from (3).)

Define for each $i \in S$ the *optimal value function* $v^*(i)$ by

$$v^*(i) := \inf_{\pi \in \Delta} \phi^\pi(i), \tag{4}$$

so that, for each $i \in S$, $v^*(i) \leq \phi^\pi(i)$, for all $\pi \in \Delta$, and, for any $\epsilon > 0$, there exists a policy $\pi \in \Delta$ such that

$$\phi^\pi(i) \leq v^*(i) + \epsilon. \tag{5}$$

(Note that such a π may depend on both i and ϵ.)

Definition. A policy $\pi \in \Delta$ is called *optimal* if $\phi^\pi(i) = v^*(i)$ for all $i \in S$.

Since the set of all policies, Δ, is uncountable, it is not clear that the infimum in the definition of $v^*(i)$ will be attained at each $i \in S$, let alone simultaneously for all $i \in S$ by the same policy π. Thus, an optimal policy need not exist.

Definition. Let $\epsilon > 0$ be fixed. A policy $\pi \in \Delta$ is called ϵ-*optimal* if

$$\phi^\pi(i) \leq v^*(i) + \epsilon, \; i \in S.$$

Note that not even the existence of an ϵ-optimal policy is assured, since thedefinition requires that the *same* policy satisfy (5) for all $i \in S$.

2.2 Examples

1. **Finite-Horizon Problem.** Consider a Markov decision process with state space $S = \{0, 1, \ldots\}$, action space $A = \{1, 2, \ldots, k\}$, one-stage cost function $c_t(i, a)$, and transition probabilities $p_{tij}(a)$. Suppose the system operates over a finite horizon of length T. That is, the system is observed at stages $t = 0, 1, \ldots, T-1, T$, and actions are taken at stages $t = 0, 1, \ldots, T-1$. At stage T, no action is taken and the system terminates in state X_T, incurring a *terminal cost* $g(X_T)$. This can be seen to be a special case of our general Markov decision model, by setting $c_T(i, a) = g(i)$ and $c_t(i, a) = 0$, for all $t \geq T+1$ ($i \in S$, $a \in A$). Then

$$\begin{aligned} \phi^\pi(i) &= E_\pi \left[\sum_{t=0}^\infty c_t(X_t, A_t) | X_0 = i \right] \\ &= E_\pi \left[\sum_{t=0}^{T-1} c_t(X_t, A_t) + g(X_T) | X_0 = i \right] \end{aligned}$$

is the expected total cost over the finite horizon, $t = 0, 1, \ldots, T-1, T$, following policy π and starting in state i.

2. **Discounted-Cost Problem.** Consider a Markov decision process with state space $S = \{0, 1, \ldots\}$, action space $A = \{1, 2, \ldots, k\}$, and time-homogeneous one-stage cost function $c(i, a)$ and transition probabilities $p_{ij}(a)$, operating over an infinite horizon. Suppose costs incurred in stage t are discounted back to time 0 by a discount factor β^t, $0 < \beta < 1$. (Here $\beta = (1 + \alpha)^{-1}$ is the one-stage discount factor, where $\alpha > 0$ is the one-stage interest rate.) Thus, for each $t \geq 0$, $\beta^t c(X_t, A_t)$ is the *present value* (at time 0) of the cost incurred at time t. This can be seen to be a special case of our general Markov decision model, by setting $c_t(i, a) = \beta^t c(i, a)$ ($i \in S$, $a \in A$), and $p_{tij}(a) = p_{ij}(a)$ ($i, j \in S$, $a \in A$), for all $t \geq 0$. Then

$$\phi^\pi(i) = E_\pi \left[\sum_{t=0}^{\infty} c_t(X_t, A_t) | X_0 = i \right]$$

$$= E_\pi \left[\sum_{t=0}^{\infty} \beta^t c(X_t, A_t) | X_0 = i \right]$$

is the expected total discounted cost over an infinite horizon, following policy π and starting in state i.

3. **Problems with an Absorbing State.** Consider a Markov decision process with state space $S = \{0, 1, \ldots\}$, action space $A = \{1, 2, \ldots, k\}$, and time-homogeneous one-stage cost function $c(i, a)$ and transition probabilities $p_{ij}(a)$. Suppose the system terminates at the first entrance time, τ, into a particular state. Without loss of generality assume this is state 0. This can be seen to be a special case of our general Markov decision model, by setting $c_t(i, a) = c(i, a)$, $p_{tij}(a) = p_{ij}(a)$, $i, j \in S$, $a \in A$, $i \neq 0$, and $c_t(0, a) = 0$, $p_{t00}(a) = 1$, $a \in A$, for all $t \geq 0$. Then

$$\phi^\pi(i) = E_\pi \left[\sum_{t=0}^{\infty} c_t(X_t, A_t) | X_0 = i \right]$$

$$= E_\pi \left[\sum_{t=0}^{\tau-1} c(X_t, A_t) | X_0 = i \right]$$

is the expected total cost incurred until the first entrance into state 0, following policy π and starting in state i.

4. **Optimal Stopping Problems.** Consider a $DTMC$ with state space $S = \{0, 1, \ldots\}$ and time-homogeneous transition probabilities p_{ij}. Upon observing state $X_t = i$ at time (stage) t, the system controller chooses whether to stop ($a = 0$) or continue ($a = 1$). If the choice is to stop, then a one-stage cost $c(i, 0)$ is incurred and the system terminates, with no future costs or rewards. If the choice is to continue, then a one-stage cost $c(i, 1)$ is incurred and the system moves to a new state $X_{t+1} = j$ at the next stage with probability p_{ij}. This problem is equivalent to a problem

with an absorbing state: append a new state $i = \infty$ (say) to the state space S, creating a new state space $\overline{S} := S \cup \{\infty\}$, and set $p_{ij}(1) = p_{ij}$, $p_{i,\infty}(0) = 1$, for all $i, j \in S$, and $p_{\infty,\infty}(a) = 1$, $c(\infty, a) = 0$, $a = 0, 1$. Then

$$\phi^\pi(i) = E_\pi \left[\sum_{t=0}^{\infty} c_t(X_t, A_t) | X_0 = i \right]$$

$$= E_\pi \left[\sum_{t=0}^{\tau-2} c(X_t, 1) + c(X_{\tau-1}, 0) | X_0 = i \right]$$

is the expected total cost incurred until the system stops, following policy π and starting in state i.

5. **Reduction of Discounted-Cost Problem to an Absorbing-State Problem.** The problem with discounted costs can be shown to be equivalent to the following problem with an absorbing state. Append an absorbing state, $i = \infty$, say, to the state space S, creating a new state space $\overline{S} := S \cup \{\infty\}$, and set $p'_{ij}(a) := \beta p_{ij}(a)$, $c'(i, a) := c(i, a)$, and $p'_{i,\infty}(a) := 1 - \beta$, $i, j \in S$, $a \in A$, and $p'_{\infty,\infty}(a) = 1$, $c'(\infty, a) = 0$, $a \in A$. Then

$$\phi^\pi(i) = E_\pi \left[\sum_{t=0}^{\infty} \beta^t c(X_t, A_t) | X_0 = i \right]$$

$$= E'_\pi \left[\sum_{t=0}^{\tau-1} c'(X_t, A_t) | X_0 = i \right], \quad (6)$$

where $E'_\pi[\cdot]$ is the expectation operator associated with the probability measure $P'_\pi\{\cdot\}$ induced by the transition probabilities $p'_{i,j}(a)$ on the augmented state space $\overline{S}$, and τ is the first entrance time into the absorbing state ∞. To motivate this result, note that the expected discounted cost in stage 1 is given by

$$E_\pi[\beta c(X_1, A_1) | X_0 = i]$$

$$= \sum_{a_0 \in A} \pi_{0,a_0}(i) \sum_{x_1 \in S} p_{i,x_1}(a_0) \sum_{a_1 \in A} \pi_{1,a_1}(i, a_0, x_1) \beta c(x_1, a_1)$$

$$= \sum_{a_0 \in A} \pi_{0,a_0}(i) \left[\sum_{x_1 \in S} \beta p_{i,x_1}(a_0) \sum_{a_1 \in A} \pi_{1,a_1}(i, a_0, x_1) c(x_1, a_1) \right.$$

$$\left. + (1-\beta) \cdot 0 \right]$$

$$= \sum_{a_0 \in A} \pi_{0,a_0}(i) \left[\sum_{x_1 \in S} p'_{i,x_1}(a_0) \sum_{a_1 \in A} \pi_{1,a_1}(i, a_0, x_1) c'(x_1, a_1) \right.$$

$$\left. + p'_{i,\infty}(a_0) \sum_{a_1 \in A} \pi_{1,a_1}(i, a_0, \infty) c'(\infty, a_1) \right]$$

$$= \sum_{a_0 \in A} \pi_{0,a_0}(i) \left[\sum_{x_1 \in \overline{S}} p'_{i,x_1}(a_0) \sum_{a_1 \in A} \pi_{1,a_1}(i, a_0, x_1) c'(x_1, a_1) \right]$$
$$= E'_\pi \left[c'(X_1, A_1) | X_0 = i \right].$$

The same reasoning can be extended to the other stages to prove the equivalence in (6).

For clarity we have presented these examples separately, but, of course, one may combine elements from more than one of them. For example, a finite-horizon problem with discounted costs or an optimal stopping problem with time-dependent costs and transition probabilities would also be special cases of our general *MDP* model.

2.3 Optimality Equation

Now we return to the general Markov decision model. Our goal is to derive the *dynamic-programming* optimality equation satisfied by the optimal value function, v^*. To this end, we must first extend our definition of the value function associated with a particular policy and the optimal value function to allow for different starting stages. For $\pi \in \Delta$ and $0 \leq t < \infty$, define

$$\phi_t^\pi(i) := E_\pi \left[\sum_{k=t}^{\infty} c_t(X_t, A_t) | X_t = i \right], \quad i \in S. \tag{7}$$

In words, $\phi_t^\pi(i)$ is the expected total cost incurred under policy π over an infinite horizon, starting at stage t in state i. Define the optimal value function, starting at stage t, by

$$v_t^*(i) := \inf_{\pi \in \Delta} \phi_t^\pi(i), \quad i \in S. \tag{8}$$

Note that $\phi_0^\pi(\cdot) \equiv \phi^\pi(\cdot)$ and $v_0^*(\cdot) \equiv v^*(\cdot)$, the value functions defined previously.

Theorem 1 *The functions $v_t^*(i)$, $i \in S$, $t = 0, 1, \ldots$, satisfy the following dynamic-programming optimality equation ($t = 0, 1, \ldots$):*

$$v_t^*(i) = \min_{a \in A} \left\{ c_t(i, a) + \sum_{j \in S} p_{tij}(a) v_{t+1}^*(j) \right\}, \quad i \in S. \tag{9}$$

Proof. Fix a state $i \in S$. First we show that

$$v_t^*(i) \geq \min_{a \in A} \left\{ c_t(i, a) + \sum_{j \in S} p_{tij}(a) v_{t+1}^*(j) \right\}.$$

Let $\pi \in \Delta$ be an arbitrary policy and let $h_t = (x_0, a_0, x_1, a_1, \ldots, x_{t-1}, a_{t-1}, i)$ be an arbitrary observed history in which $x_t = i$. Let $p(a) := \pi_{t,a}(h_t)$, the probability that policy takes action a at stage t under policy π, given the history h_t. Then

$$\begin{aligned}
\phi_t^\pi(i) &= E_\pi \left[\sum_{k=t}^\infty c_k(X_k, A_k) | X_t = i \right] \\
&= \sum_{a \in A} p(a) \Bigg(c_t(i,a) + \\
&\qquad \sum_{j \in S} p_{tij}(a) E_\pi \left[\sum_{k=t+1}^\infty c_k(X_k, A_k) | X_t = i, A_t = a, X_{t+1} = j \right] \Bigg) \\
&\geq \sum_{a \in A} p(a) \left(c_t(i,a) + \sum_{j \in S} p_{tij}(a) v_{t+1}^*(j) \right) \\
&\geq \min_{a \in A} \left\{ c_t(i,a) + \sum_{j \in S} p_{tij}(a) v_{t+1}^*(j) \right\}
\end{aligned}$$

But, since this inequality holds for all policies $\pi \in \Delta$, it follows that

$$\begin{aligned}
v_t^*(i) &= \inf_{\pi \in \Delta} \phi_t^\pi(i) \\
&\geq \min_{a \in A} \left\{ c_t(i,a) + \sum_{j \in S} p_{tij}(a) v_{t+1}^*(j) \right\}.
\end{aligned}$$

It remains to show that

$$v_t^*(i) \leq \min_{a \in A} \left\{ c_t(i,a) + \sum_{j \in S} p_{tij}(a) v_{t+1}^*(j) \right\}.$$

Fix an arbitrary $\epsilon > 0$ and, for each $j \in S$, choose a policy $\pi(j, \epsilon) \in \Delta$ such that

$$\phi_{t+1}^\pi(j) \leq v_{t+1}^*(j) + \epsilon.$$

Now, choose an action $a^*(i) \in A$ that attains the minimum in the r.h.s. of (9) and let $\pi \in \Delta$ be a policy that selects action $a^*(i)$ in state i at stage t and then follows policy $\pi(j, \epsilon)$ over stages $t+1, t+2, \ldots$, if $X_{t+1} = j$, $j \in S$. Then

$$\begin{aligned}
v_t^*(i) &\leq \phi_t^\pi(i) \\
&= E_\pi \left[c_t(X_t, A_t) + \sum_{k=t+1}^\infty c_k(X_k, A_k) | X_t = i \right] \\
&= c_t(i, a^*(i)) + \sum_{j \in S} p_{tij}(a^*(i)) \phi_{t+1}^{\pi(j,\epsilon)}(j)
\end{aligned}$$

$$\leq c_t(i, a^*(i)) + \sum_{j \in S} p_{tij}(a^*(i)) v_{t+1}^*(j) + \epsilon$$

$$= \min_{a \in A} \left\{ c_t(i, a) + \sum_{j \in S} p_{tij}(a) v_{t+1}^*(j) \right\} + \epsilon \,.$$

But, since ϵ was arbitrary, it follows that

$$v_t^*(i) \leq \min_{a \in A} \left\{ c_t(i, a) + \sum_{j \in S} p_{tij}(a) v_{t+1}^*(j) \right\}.$$

This completes the proof of the theorem.

2.4 Applications

Finite-Horizon Problem. Consider a T-stage problem as described in Example 1 above. To indicate the dependence on the horizon length T, we shall write $v_{t,T}^*(i)$ instead of $v_t^*(i)$ to denote the optimal value function associated with starting in state i at stage t. In this case we have $v_{T,T}^*(i) = g(i)$, for all $i \in S$, where g is the given terminal cost function. It follows that $v_{T-1,T}^*, v_{T-2,T}^*, \ldots, v_{0,T}^*$ can be calculated by backwards induction from the dynamic-programming optimality equation (9). When the state space S is finite, this gives a finite algorithm, with complexity of order $T \times |S| \times |A| \times |S|$. We also have the following theorem, which is easily proved by backwards induction on $t = T, T - 1, \ldots, 0$.

Theorem 2 *The optimal value functions, $v_{t,T}^*$, are given by the recursive optimality equations ($t = T - 1, T - 2, \ldots, 0$)*

$$v_{t,T}^*(i) = \min_{a \in A} \left\{ c_t(i, a) + \sum_{j \in S} p_{tij}(a) v_{t+1,T}^*(j) \right\}, \quad i \in S,$$

with $v_{T,T} \equiv g$. For each $0 \leq t < T$, define the decision rule $f_t : S \to A$ by

$$c_t(i, f_t(i)) + \sum_{j \in S} p_{tij}(f_t(i)) v_{t+1,T}^*(j) = \min_{a \in A} \left\{ c_t(i, a) + \sum_{j \in S} p_{tij}(a) v_{t+1,T}^*(j) \right\}, i \in S.$$

Then the deterministic Markov policy $\pi = (f_0, f_1, \ldots, f_{T-1})$ is optimal for the T-stage problem.

Now consider the effect of increasing the horizon length from T to $T + 1$, with the same given terminal cost function g and the same (non-time-homogeneous) one-stage costs, $c_t(i, a)$, and transition probabilities, $p_{tij}(a)$. In the T-stage problem, the first step of the backwards induction involves computing the function $v^*_{T-1,T}$ from

$$v^*_{T-1,T}(i) = \min_{a \in A} \left\{ c_{T-1}(i,a) + \sum_{j \in S} p_{T-1,i,j}(a) g(j) \right\},$$

whereas in the $T+1$-stage problem the first step involves computing the function $v^*_{T,T+1}$ from

$$v^*_{T,T+1}(i) = \min_{a \in A} \left\{ c_T(i,a) + \sum_{j \in S} p_{T,i,j}(a) g(j) \right\}.$$

Although they share the same terminal cost function, the r.h.s.'s of these two equations differ because of the time dependence of the one-stage costs and transition probabilities. This same difference persists at each step of the backwards induction. Thus, when the horizon length is increased from T to $T+1$, all the optimal value functions must be recomputed.

Time-Homogeneous Costs and Transition Probabilities. When the one-stage costs and transition probabilities are time homogeneous, this distinction disappears and we are able to utilize the computations already performed for the T-stage problem when we extend the horizon length to $T+1$. For in this case we have

$$v^*_{T-1,T}(i) = v^*_{T,T+1}(i) = \min_{a \in A} \left\{ c(i,a) + \sum_{j \in S} p_{i,j}(a) g(j) \right\};$$

that is, $v^*_{T-1,T}(i)$ is independent of the horizon length T and simply represents the optimal value in state i when one stage remains in the horizon. Similarly, for any $0 \leq t < T$, the value function $v^*_{t,T}$ depends only on $n = T - t$, that is, the number of stages remaining in the horizon. For this reason, in the time-homogeneous case it is convenient to introduce a new notation for the optimal value function. Let $v_n(i)$ denote the minimum expected total cost when the current state is i and n stages remain in the horizon. It follows that the optimal value functions v_n are uniquely determined by the recursive equations ($n = 1, 2, \ldots$)

$$v_n(i) = \min_{a \in A} \left\{ c(i,a) + \sum_{j \in S} p_{ij}(a) v_{n-1}(j) \right\}, \quad i \in S, \quad (10)$$

with $v_0 \equiv g$. Note also that $v_T \equiv v^*_{0,T}$. That is, v_T coincides with the optimal value function for stages $0, \ldots, T$ in the T-stage problem.

One-Stage Problem; Dynamic-Programming Algorithm

Let us consider the case $T = 1$. In this case we have

$$v_1(i) = \min_{a \in A} \left\{ c(i, a) + \sum_{j \in S} p_{ij}(a) v_0(j) \right\}, \quad i \in S. \tag{11}$$

Since $v_0(\cdot)$, $c(\cdot, \cdot)$, and $p_{\cdot,\cdot}(\cdot)$ are known functions, and A is a finite set, this equation can be evaluated provided S is a finite set. The complexity for doing this is obviously is $O(|S| \cdot |A| \cdot |S|)$.

Remark 1. If the quantity in braces in (10) has nice properties as a function of a (e.g., monotonicity, convexity), then enumeration of all $a \in A$ may not be necessary.

Now consider the T-stage problem for $T \geq 2$. Our problem is to calculate the optimal value function

$$v_T(i) = \min_{\pi} \phi_T^{\pi}(i), \quad i \in S.$$

We would like to construct an optimal policy for this problem. We can now see how to solve the T-stage problem by a backwards-recursion algorithm. At step 0 of the algorithm, solve a one-stage problem using equation (11). The resulting decision rule is the optimal decision rule, f_{T-1}, for stage $T - 1$. At step n of the algorithm ($n > 1$), solve the one-stage problem,

$$v_n(i) = \min_{a \in A} \left\{ c(i, a) + \sum_{j \in S} p_{ij}(a) v_{n-1}(j) \right\}, \quad i \in S. \tag{12}$$

The resulting decision rule is the optimal decision rule, f_{T-n}, for stage $T - n$. Stop after step $n = T$.

Discounted-Cost Infinite-Horizon Problem. Consider the problem in Example 2 above. That is, $c_t(i, a) = \beta^t c(i, a)$ and $p_{tij}(a) = p_{ij}(a)$, for $t = 0, 1, \ldots$, $(i, j \in S, a \in A)$. In this case the optimality equation (9) is equivalent to

$$v_t^*(i) = \min_{a \in A} \left\{ \beta^t c(i, a) + \sum_{j \in S} p_{ij}(a) v_{t+1}^*(j) \right\}, \quad i \in S,$$

$t = 0, 1, \ldots$ But, with stationary costs and transition probabilities and an infinite horizon, it is clear that the problem from stage t onward, given $X_t = i$, has exactly the same structure as the problem from stage $t + 1$ onward, given $X_{t+1} = i$, so that $v_{t+1}^*(i) = \beta v_t^*(i) = \beta^{t+1} v_0^*(i)$. That is, the only dependence of the optimal value function v_t^* on t arises from the discounting of all costs back to the fixed stage 0. Let $v_\beta(i)$ denote the minimal expected discounted cost over an infinite horizon, starting at an arbitrary stage t in state i, where costs are discounted back to stage t, rather than to stage 0 (as is the case with $v_t^*(i)$). Then it follows that $v_t^*(i) = \beta^t v_\beta(i)$ for all $t = 0, 1, \ldots$, and the new optimal value function, v_β, satisfies the optimality equation

$$v_\beta(i) = \min_{a \in A} \left\{ c(i,a) + \beta \sum_{j \in S} p_{ij}(a) v_\beta(j) \right\}, \; i \in S. \tag{13}$$

Note that (13) is a functional equation in which the same function, v_β, appears on both sides. Thus, unlike the recursive equation optimality for the finite-horizon problem, it cannot be used directly to compute the optimal value function.

Discounted-Cost Finite-Horizon Problem. Now consider a problem combining the assumptions of Examples 1 and 2 above. That is, $c_t(i,a) = \beta^t c(i,a)$ and $p_{tij}(a) = p_{ij}(a)$, for $t = 0, 1, \ldots, T-1$, $c_T(i,a) = g(i)$, and $c_t(i,a) = 0$, for $t = T+1, T+2, \ldots$ ($i, j \in S$, $a \in A$). In this case the optimality equation (9) is equivalent to ($i \in S$)

$$v_{t,T}^*(i) = \min_{a \in A} \left\{ \beta^t c(i,a) + \sum_{j \in S} p_{ij}(a) v_{t+1,T}^*(j) \right\}, \; t = 0, 1, \ldots, T-1,$$

$$v_{T,T}^*(i) = \beta^T v_{\beta,0}(i),$$

where $v_{\beta,0}(i) := \beta^{-T} g(i)$. In particular, for $t = T-1$, we have

$$\begin{aligned} v_{T-1,T}^*(i) &= \min_{a \in A} \left\{ \beta^{T-1} c(i,a) + \sum_{j \in S} p_{ij}(a) \beta^T v_{\beta,0}(j) \right\} \\ &= \beta^{T-1} \min_{a \in A} \left\{ c(i,a) + \beta \sum_{j \in S} p_{ij}(a) v_{\beta,0}(j) \right\} \\ &= \beta^{T-1} v_{\beta,1}(i), \end{aligned}$$

where $v_{\beta,1}(\cdot)$ satisfies

$$v_{\beta,1}(i) = \min_{a \in A} \left\{ c(i,a) + \beta \sum_{j \in S} p_{ij}(a) v_{\beta,0}(j) \right\}, \; i \in S.$$

As an induction hypothesis, let $1 \leq n \leq T-1$ and suppose $v_{T-n,T}^*(i) = \beta^{T-n} v_{\beta,n}(i)$, where $v_{\beta,n}(\cdot)$ satisfies

$$v_{\beta,n}(i) = \min_{a \in A} \left\{ c(i,a) + \beta \sum_{j \in S} p_{ij}(a) v_{\beta,n-1}(j) \right\}, \; i \in S. \tag{14}$$

Then

$$v^*_{T-n-1}(i) = \min_{a \in A}\left\{\beta^{T-n-1}c(i,a) + \sum_{j \in S} p_{ij}(a)\beta^{T-n}v_{\beta,n}(j)\right\}$$

$$= \beta^{T-n-1}\min_{a \in A}\left\{c(i,a) + \beta \sum_{j \in S} p_{ij}(a)v_{\beta,n}(j)\right\}$$

$$= \beta^{T-n-1}v_{\beta,n+1}(j),$$

where $v_{\beta,n+1}(\cdot)$ satisfies (14) with n replaced by $n+1$. It follows that the discounted-cost, finite-horizon problem with stationary costs and transition probabilities can be solved by recursively calculating the optimal value functions, $v_{\beta,n}$, from the optimality equations (14) for $n = 1, 2, \ldots$, starting with the known (undiscounted) terminal-cost function, $v_{\beta,0}(\cdot)$.

Note that $v_{\beta,n}(i)$ can be interpreted as the minimal expected discounted cost over stages $t, t+1, \ldots, T$, given that $X_t = i$ and the total horizon length is $T = t + n$, where costs are discounted back to stage t, rather than to stage 0 (as is the case with $v^*_t(i)$). As in the undiscounted finite-horizon model with stationary costs and transition probabilities, we see that the optimal value functions, expressed in this way, depend only on the number of stages remaining and not separately on the current stage index or the total horizon length. Thus an optimal (deterministic Markov) policy can be determined for the T-stage problem by first solving the $T-1$-stage problem, and then doing one more iteration of the recursion (14), rather than having to solve T new recursions of the form (9).

3 DISCOUNTED MARKOV DECISION CHAIN: ALGORITHMS

In this section we focus attention on a infinite-horizon MDC with time-homogeneous costs and transition probabilities and discounting.

We consider a Markov decision process with countable state space $S = \{0, 1, \ldots\}$ and action space $A = \{1, \ldots, k\}$. The one-period cost function is $c(i, a)$ ($i \in S$, $a \in A$); the transition probabilities are $p_{ij}(a) = P\{X_{t+1} = j | X_t = i, A_t = a\}$ ($i, j \in S$, $a \in A$); the one-period discount factor is β, $0 < \beta < 1$. Note that in our original notation, $c_t(i,a) = \beta^t c(i,a)$. For a policy π, define the value function ϕ^π by

$$\phi^\pi(i) := E_\pi\left[\sum_{t=0}^{\infty} c_t(X_t, A_t) | X_0 = i\right]$$

$$= E_\pi\left[\sum_{t=0}^{\infty} \beta^t c(X_t, A_t) | X_0 = i\right].$$

Define the optimal value function v_β by

$$v_\beta(i) := \inf_{\pi \in \Delta} \phi^\pi(i).$$

OPTIMAL CONTROL OF MARKOV CHAINS 339

We assume the one-stage cost functions are uniformly bounded; that is, there exists a constant $M < \infty$ such that $|c(i,a)| \leq M$, for all $i \in S$, $a \in A$. Hence, $\sum_{t=0}^{\infty} \beta^t |c(X_t, A_t)| \leq M(1-\beta)^{-1} \leq M' < \infty$ for all realizations $\{(X_t, A_t), t = 0, 1, \ldots\}$ of the process. Hence, for every policy π the expectation in the definition of $\phi^\pi(i)$ is well defined and finite, and

$$|\phi_\pi(i)| \leq E_\pi \left[\sum_{t=0}^{\infty} \beta^t |c(X_t, A_t)| \mid X_0 = i \right] \leq M' < \infty, \text{ for all } i \in S,$$

and thus
$$|v_\beta(i)| \leq M' < \infty, \text{ for all } i \in S.$$

That is, both ϕ^π and v_β belong to $\mathcal{B}(S)$, the set of all bounded functions $v : S \to \text{Re}$.

Theorem 3 *The optimal value function $\{v_\beta(i), i \in S\}$ is the unique solution in $\mathcal{B}(S)$ to the functional (optimality) equation:*

$$v_\beta(i) = \min_{a \in A} \left\{ c(i,a) + \beta \sum_{j \in S} p_{ij}(a) v_\beta(j) \right\}, \ i \in S. \tag{15}$$

Moreover, if for each $i \in S$, $f(i)$ is an action $a \in A$ that achieves the above minimum, then the deterministic stationary policy $\pi = f^{(\infty)}$ is optimal:

$$\phi^f(i) = v_\beta(i), i \in S.$$

Proof. It follows from Theorem 1 that $\{v_\beta(i), i \in S\}$ is a solution to the optimality equation (15). To show that it is unique, let $u \in \mathcal{B}(S)$ be another solution to (15). We shall show that $u \equiv v_\beta$. First observe that, by the definition of f,

$$v_\beta(i) = c(i, f(i)) + \beta \sum_{j \in S} p_{ij}(f(i)) v_\beta(j)$$

$$u(i) = \min_{a \in A} \left\{ c(i,a) + \beta \sum_{j \in S} p_{ij}(a) u(j) \right\}$$

$$\leq c(i, f(i)) + \beta \sum_{j \in S} p_{ij}(f(i)) u(j).$$

Therefore, for all $i \in S$,

$$u(i) - v_\beta(i) \leq c(i, f(i)) + \beta \sum_{j \in S} p_{ij}(f(i)) u(j)$$

$$- c(i, f(i)) - \beta \sum_{j \in S} p_{ij}(f(i)) v_\beta(j)$$

$$= \beta \sum_{j \in S} p_{ij}(f(i))[u(j) - v_\beta(j)]$$

$$\leq \beta \sum_{j \in S} p_{ij}(f(i)) \| u - v_\beta \|$$

$$= \beta \| u - v_\beta \|,$$

where $\| u - v_\beta \| := \max_{i \in S} |u(i) - v_\beta(i)| < \infty$. Reversing the roles of u and v_β and repeating this argument, we have $v_\beta(i) - u(i) \leq \beta \| u - v_\beta \|$, for all $i \in S$. Hence

$$-\beta \| u - v_\beta \| \leq u(i) - v_\beta(i) \leq \beta \| u - v_\beta \|,$$

which implies that $|u(i) - v_\beta(i)| \leq \beta \| u - v_\beta \|$ for all $i \in S$, and hence $\| u - v_\beta \| \leq \beta \| u - v_\beta \|$. But, since $\| u - v_\beta \| < \infty$ and $\beta < 1$, it must be the case that $u \equiv v_\beta$. Therefore, v_β is the unique solution to (15). ∎

This theorem tells us that the problem of finding an optimal policy for the infinite-horizon β-discounted problem can be solved if we can find a solution, $\{u(i), i \in S\}$, to the optimality equations (15). That solution must be v_β by the uniqueness, and then $\pi = f^\infty$ will be an optimal policy.

Remark. Note that the optimality equations (15) are *non-linear* equations, since the parameters $c(i, a)$ and $p_{ij}(a)$ depend on which action a achieves the minimum, which in turn depends on the variables $u(j), j \in S$. If, instead, we were using a fixed, prespecified action $a = f(i)$, say, for state i, then we would have the linear system,

$$u(i) - \beta \sum_{j \in S} p_{ij}(f(i))u(j) = c(i, f(i)), \; i \in S, \tag{16}$$

which can be solved by standard numerical methods (e.g., Gaussian elimination). In fact, as we shall see presently, this is the linear system satisfied by the value function $\phi^f(i), i \in S$, for the fixed policy $\pi = f^\infty$.

There are two main methods for solving (15): *value iteration* (successive approximations in function space) and *policy iteration* (successive approximations in policy space).

3.1 Value Iteration

Set $v_{\beta,0}(i), i \in S$, be an initial guess of the optimal value function, $v_\beta(i), i \in S$, and for $n \geq 1$, define the function $v_{\beta,n}$, the nth approximation of v_β, by

$$v_{\beta,n}(i) = \min_{a \in A} \left\{ c(i, a) + \beta \sum_{j \in S} p_{ij}(a) v_{\beta, n-1}(j) \right\}, \; i \in S. \tag{17}$$

Then it can be shown (the argument is similar to the proof of the above theorem) that $v_{\beta,n}(i) \to v_\beta(i)$ as $n \to \infty$, for all $i \in S$. Moreover, the convergence is uniform and geometric at rate β; that is, if $||a||$ denotes the norm of vector a,

$$\| v_\beta - v_{\beta,n} \| \le \beta^n \| v_\beta - v_{\beta,0} \| . \tag{18}$$

Note that the equations (17) are identical to the recursive optimality equations for the optimal value function for the finite-horizon discounted problem (cf. equation (14). Thus, the nth approximation, $v_{\beta,n}(i)$, of $v_\beta(i)$ can also be interpreted as the minimum expected total discounted cost in a finite-horizon problem with terminal cost function $g \equiv v_{\beta,0}$, when n stages remain and the current state is i.

While establishing that value iteration converges at geometric rate, the convergence bound (18) does not give us an implementable stopping criterion, since the right-hand side involves the unknown function v_β, which we are trying to approximate. The next theorem provides such a stopping criterion. First, we shall introduce matrix-vector and operator notation, in order to facilitate its statement and proof.

Let $f : S \to A$ be a given decision rule. Define the vector $c(f)$ and the matrix $P(f)$ by

$$(c(f))_i = c(i, f(i)), \ (P(f))_{ij} = p_{ij}(f(i)), \ i,j \in S .$$

It follows from the definition of $\phi^f(i)$ that $(i \in S)$

$$\begin{aligned}
\phi^f(i) &= E_f[\sum_{t=0}^\infty \beta^t c(X_t, A_t) | X_0 = i] \\
&= c(i, f(i)) + \beta \sum_{j \in S} p_{ij}(f(i)) E_f[\sum_{t=1}^\infty \beta^{t-1} c(X_t, A_t) | X_1 = j] \\
&= c(i, f(i)) + \beta \sum_{j \in S} p_{ij}(f(i)) E_f[\sum_{t=0}^\infty \beta^t c(X_t, A_t) | X_0 = i] \\
&= c(i, f(i)) + \beta \sum_{j \in S} p_{ij}(f(i)) \phi^f(j) ,
\end{aligned}$$

where the second equality follows from the Markov property and the third from the time homogeneity of the costs and transition probabilities. Thus we have shown that ϕ^f satisfies the linear system (16), as promised. This system of equations can be written equivalently in matrix-vector form as

$$\phi^f = c(f) + \beta P(f)\phi^f = L(f)\phi^f , \tag{19}$$

where the operator $L(f) : \mathcal{B}(S) \to \mathcal{B}(S)$ is defined by

$$L(f)u := c(f) + \beta P(f)u , \ u \in \mathcal{B}(S) .$$

In other words, ϕ^f is a *fixed point* of the operator $L(f)$. By an argument exactly like the proof of Theorem 3, it can be shown that ϕ^f is the *unique* fixed point of $L(f)$ in $\mathcal{B}(S)$.

The optimality equation (15) can also be written in matrix-vector form:
$$v_\beta = \min_f \{c(f) + \beta P(f)v_\beta\} = \min_f L(f)v_\beta ,$$
or
$$v_\beta = Lv_\beta , \qquad (20)$$
where the operator $L : \mathcal{B}(S) \to \mathcal{B}(S)$ is defined by
$$Lu := \min_f L(f)u , \ u \in \mathcal{B}(S) .$$

Thus, we can restate Theorem 0.1 in equivalent form: v_β is the (unique) fixed point of the operator L. Similarly, (17) says that the functions $v_{\beta,n}$, $n \geq 1$, satisfy the recursive system,
$$v_{\beta,n} = Lv_{\beta,n-1} . \qquad (21)$$
Note that, for any decision rule f and functions $u, v \in \mathcal{B}(S)$,
$$\| L(f)u - L(f)v \| \leq \beta \| u - v \| , \qquad (22)$$
$$\| Lu - Lv \| \leq \beta \| u - v \| \qquad (23)$$

(cf. proof of Theorem 3). That is, both $L(f)$ and L are *contraction mappings* on $\mathcal{B}(S)$, with modulus β. We are now ready to state our next theorem.

Theorem 4 *Let $\epsilon > 0$ be given and suppose $n \geq 1$ is chosen so that*
$$\| v_{\beta,n} - v_{\beta,n-1} \| \leq \frac{\epsilon(1-\beta)}{2\beta} . \qquad (24)$$

For each $i \in S$, let $f_\epsilon(i)$ be the policy which minimizes
$$\min_{a \in A} \left\{ c(i,a) + \beta \sum_{j \in S} p_{ij}(a)v_{\beta,n}(j) \right\} . \qquad (25)$$

Then
  a. $\| v_{\beta,n} - v_\beta \| \to 0$, as $n \to \infty$;
  b. *there exists a finite integer N such that (24) holds for all $n \geq N$;*
  c. $\| v_{\beta,n} - v_\beta \| < \epsilon/2$ *whenever (24) holds;*
  d. *the stationary policy $\pi = f_\epsilon^{(\infty)}$ is ϵ-optimal.*

Proof. Part (a) follows from Theorem 3. The proof of (b) follows from (a) because of the definition of a limit. To prove (c) and (d), first note that
$$\| \phi^{f_\epsilon} - v_\beta \| \leq \| \phi^{f_\epsilon} - v_{\beta,n} \| + \| v_{\beta,n} - v_\beta \| . \qquad (26)$$
It follows from (19), the definition of f_ϵ, (21), (22), and (23) that
$$\| \phi^{f_\epsilon} - v_{\beta,n} \| \leq \| L(f_\epsilon)\phi^{f_\epsilon} - Lv_{\beta,n} \| + \| Lv_{\beta,n} - v_{\beta,n} \|$$
$$= \| L(f_\epsilon)\phi^{f_\epsilon} - L(f_\epsilon)v_{\beta,n} \| + \| Lv_{\beta,n} - Lv_{\beta,n-1} \|$$
$$\leq \beta \| \phi^{f_\epsilon} - v_{\beta,n} \| + \beta \| v_{\beta,n} - v_{\beta,n-1} \| .$$

Therefore,

$$\| \phi^{f_\epsilon} - v_{\beta,n} \| \leq \left(\frac{\beta}{1-\beta}\right) \| v_{\beta,n} - v_{\beta,n-1} \| \leq \frac{\epsilon}{2},$$

whenever (24) holds. By a similar argument, one can show that

$$\| v_{\beta,n} - v_\beta \| \leq \left(\frac{\beta}{1-\beta}\right) \| v_{\beta,n} - v_{\beta,n-1} \| \leq \frac{\epsilon}{2},$$

whenever (24) holds, thus establishing (c). Combining these two inequalities, we conclude from (26) that

$$\| \phi^{f_\epsilon} - v_\beta \| \leq \left(\frac{2\beta}{1-\beta}\right) \| v_{\beta,n} - v_{\beta,n-1} \| \leq \epsilon,$$

whenever (24) holds, thus establishing (d).

3.2 Policy Iteration

Let f be a fixed stationary, deterministic policy. Recall that its value function, $\phi^f(i)$, $i \in S$, is the unique solution to the linear equations (16), or, equivalently, in matrix-vector form,

$$\phi^f = L(f)\phi^f. \qquad (27)$$

The basic idea of policy iteration is to start with a guess, $f = f_0$, say, of the optimal policy, to use (27) to find its value function, ϕ^{f_0}, and then to determine if we can improve the policy. We do this by checking to see if we can do better than f_0 from starting state i by choosing a different action in stage $t = 0$, and then following policy f_0 for stages $t = 1, 2, \ldots$. To this end, let a new policy, f_1, be defined by choosing, in each state i, an action $a = f_1(i)$ achieving the minimum in

$$w(i) := \min_{a \in A} \left\{ c(i,a) + \beta \sum_{j \in S} p_{ij}(a) \phi^{f_0}(j) \right\}.$$

(Equivalently, $w = L(f_1)\phi^{f_0} = \min_f L(f)\phi^{f_0}$. Note that $w \leq \phi^{f_0}$, since $\phi^{f_0} = L(f_0)\phi^{f_0}$.)

Lemma 1 *For all $i \in S$, $\phi^{f_1}(i) \leq \phi^{f_0}(i)$. If $w(i) = \phi^{f_0}(i)$ for all $i \in S$, then f_0 is optimal ($\phi^{f_0}(i) = v_\beta(i), i \in S$). If $w(i) < \phi^{f_0}(i)$ for at least one i, then $\phi^{f_1}(i) < \phi^{f_0}(i)$, in which case we say that policy f_1 is an improvement on policy f_0.*

Proof. To show that $\phi^{f_1}(i) \leq \phi^{f_0}(i)$ for all $i \in S$, observe that, in matrix-vector notation,

$$\begin{aligned} w &= c(f_1) + \beta P(f_1)\phi^{f_0} \\ &\leq c(f_0) + \beta P(f_0)\phi^{f_0} = \phi^{f_0}, \end{aligned}$$

which implies that
$$c(f_1) + \beta P(f_1)(c(f_1) + \beta P(f_1)\phi^{f_0}) \leq c(f_1) + \beta P(f_1)\phi^{f_0} \leq \phi^{f_0},$$
that is, $c(f_1) + \beta P(f_1)c(f_1) + \beta^2 P^2(f_1)\phi^{f_0} \leq \phi^{f_0}$. Iterating n times on this inequality, we obtain
$$\sum_{t=0}^{n-1} \beta^t (P(f_1))^t c(f_1) + \beta^n (P(f_1))^n \phi^{f_0} \leq \phi^{f_0}.$$

Now let $n \to \infty$ and use the fact that $\beta^n (P(f_1))^n \phi^{f_0} \to 0$ (since $0 < \beta < 1$ and ϕ^{f_0} is bounded) to conclude that

$$\begin{aligned}
\phi^{f_1} &= E_{f_1}[\sum_{t=0}^{\infty} \beta^t c(X_t, A_t) | X_0] \\
&= \sum_{t=0}^{\infty} \beta^t E_{f_1}[c(X_t, A_t) | X_0] \\
&= \sum_{t=0}^{\infty} \beta^t (P(f_1))^t c(f_1) \\
&\leq \phi^{f_0}.
\end{aligned}$$

The rest of the proof is similar.

This lemma forms the basis for the *policy-iteration* algorithm.

Step 0. Choose a policy f_0; set $n = 0$.

Step n. (i) (*Value Determination.*) Solve the linear equations
$$\phi^{f_n} = c(f_n) + \beta P(f_n)\phi^{f_n}.$$
for ϕ^{f_n}.

(ii) (*Policy Improvement.*) For each $i \in S$, compute
$$w(i) := \min_{a \in A} \left\{ c(i, a) + \beta \sum_{j \in S} p_{ij}(a)\phi^{f_n}(j) \right\}.$$

If $w(i) = \phi^{f_n}(i)$ for all $i \in S$, **stop**: f_n is optimal. If $w(i) < \phi^{f_n}(i)$ for at least one i, define a new policy f_{n+1} by

$$\begin{aligned}
w(i) &= c(i, f_{n+1}(i)) + \beta \sum_{j \in S} p_{ij}(f_{n+1}(i))\phi^{f_n}(j) \\
&= \min_{a \in A} \left\{ c(i, a) + \beta \sum_{j \in S} p_{ij}(a)\phi^{f_n}(j) \right\},
\end{aligned}$$

$i \in S$. Set $n \leftarrow n + 1$ and repeat Step n.

3.3 Linear Programming

For this section assume S is finite. We shall show how to formulate the problem of solving the optimality equation (15) as a linear program. First observe that a function $v: S \to \mathrm{Re}$ satisfies (15) if and only if the vector v satisfies the linear constraints,

$$v(i) - \beta \sum_{j \in S} p_{ij}(a) v(j) \leq c(i,a) , \ a \in A ,$$

for each $i \in S$, with equality for at least one $a \in A$. To ensure that the latter condition holds, let $\alpha(j)$, $j \in S$, be arbitrary positive constants such that $\sum_{j \in S} \alpha(j) = 1$ and consider the following linear program (P):

$$\max_v \sum_{j \in S} \alpha(j) v(j)$$

$$\text{s.t. } v(i) - \beta \sum_{j \in S} p_{ij}(a) v(j) \leq c(i,a) , \ i \in S , \ a \in A$$

$$v(i) \text{ unconstrained}, \ i \in S .$$

Theorem 5 *An optimal solution, $\{v^*(i), i \in S\}$, to problem (P) satisfies the optimality equation (15); hence $v^* \equiv v_\beta$.*

Proof. By the above remarks and the uniqueness of the solution to (15), it suffices to show that, for each $i \in S$, $v(i) = v^*(i)$ satisfies

$$v(i) - \beta \sum_{j \in S} p_{ij}(a) v(j) = c(i,a) ,$$

for at least one $a \in A$. Suppose we have a feasible solution $\{v(i), i \in S\}$ to (P) for which this is not the case. Then there exists an $i \in S$ and an $\epsilon > 0$ such that

$$v(i) + \epsilon \leq c(i,a) + \beta \sum_{j \in S} p_{ij}(a) v(j) , \ a \in A . \tag{28}$$

To show that $\{v(i), i \in S\}$ cannot be optimal for Problem (P), define $v'(i)$, $i \in S$, as follows:

$$v'(i) := v(i) + \epsilon ; \ v'(j) := v(j) , \ j \neq i .$$

Then it follows from (28) that

$$\begin{aligned}
v'(i) &= v(i) + \epsilon \\
&\leq c(i,a) + \beta \sum_{j \in S} p_{ij}(a) v(j) \\
&\leq c(i,a) + \beta \sum_{j \in S} p_{ij}(a) v'(j) ,
\end{aligned}$$

for all $a \in A$. For $j \neq i$, it follows from the fact that v is feasible for (P) that

$$\begin{aligned} v'(j) &= v(j) \\ &\leq c(j,a) + \beta \sum_{k \in S} p_{jk}(a) v(k) \\ &\leq c(j,a) + \beta \sum_{k \in S} p_{jk}(a) v'(k) \,, \end{aligned}$$

for all $a \in A$. Thus, $v'(i)$, $i \in S$, is also feasible for (P). But

$$\sum_{j \in S} \alpha(j) v'(j) = \sum_{j \in S} \alpha(j) v(j) + \alpha(i)\epsilon \,,$$

which implies that $\{v(i), i \in S\}$ cannot be optimal for (P).

Thus we can find the optimal value function v_β and thereby identify an optimal policy for the discounted problem by solving the linear program (P). Note that this problem has $|S|$ variables and $|S| \times |A|$ constraints. The large number of constraints relative to the number of variables suggests that it may be more efficient to solve the dual problem (D):

$$\min_x \sum_{j \in S} \sum_{a \in A} c(j,a) x(j,a)$$

s.t. $\sum_{a \in A} x(j,a) - \beta \sum_{i \in S} \sum_{a \in A} x(i,a) p_{ij}(a) = \alpha(j) \,,\ j \in S$

$x(j,a) \geq 0 \,,\ j \in S,\ a \in A \,.$

Interpretation of (P) and (D). The vector, $\{\alpha(j), j \in S$, can be interpreted as an initial probability distribution:

$$\alpha(j) = P\{X_0 = j\} \,,\ j \in S \,.$$

Thus the objective function, $\sum_{j \in S} \alpha(j) v(j)$, of Problem (P) can be interpreted as the unconditional expectation of the infinite-horizon expected discounted cost, taken with respect to this initial distribution. Since solving (P) yields the optimal value function, v_β, and thence a stationary policy that is optimal from *each* starting state $i \in S$, we deduce *a fortiori* that an optimal solution to the linear program (P) is independent of objective-function coefficients, $\alpha(j), j \in S$, as long as they are all positive. This is a remarkable (and unusual) property that is characteristic of a class of linear programs called *Leontief substitution systems*, to which Problem (P) belongs.

Consider Problem (D). Let π be a randomized Markov policy and define $x^\pi(i,a)$ $(i \in S, a \in A)$ by

$$x^\pi(i,a) := \sum_{j \in S} \alpha(j) \sum_{t=0}^{\infty} \beta^t P_\pi\{X_t = i, A_t = a | X_0 = j\} \,. \qquad (29)$$

Then it can easily be verified that $x^\pi(i, a)$, $i \in S$, $a \in A$, is a feasible solution to Problem (D). Note that $x^\pi(i, a)$ can be interpreted as the expected discounted joint frequency that the system is in state i and that action a is taken, following policy π. Conversely, suppose we have a feasible solution, $x(i, a)$, $i \in S$, $a \in A$, to Problem (D). Define the randomized stationary policy π by

$$d(i,a) := P\{A_t = a | X_t = i\} = \frac{x(i,a)}{\sum_{a' \in A} x(i,a')}, \ i \in S, \ a \in A. \tag{30}$$

Then $x^\pi(i,a)$ as defined by (29) for this policy is a feasible solution to (D) and $x^\pi(i,a) = x(i,a)$, for all $i \in S$, $a \in A$.

These observations provide an alternative explanation for why it suffices to consider deterministic stationary policies when solving the infinite-horizon discounted MDC: the objective function to be minimized in fact only depends on the discounted *state-action frequencies*, $x^\pi(i,a)$, not on the policy π itself. Hence we can, without loss of optimality, choose a *stationary* policy with the specified state-action frequencies. Note also that any feasible solution to (D) must have $\sum_{a \in A} x(i,a) > 0$ for each $i \in S$. Hence, at least one $x(i,a)$ must be positive for each $i \in S$. But, since basic feasible solutions have at most $|S|$ positive variables, at most one $x(i,a)$ is positive for each $i \in S$ in an optimal basic feasible solution. It follows that the corresponding optimal stationary policy defined by (30) will be deterministic, with $d(i,a) = 1$ (0) if $x(i,a) > 0$ ($= 0$).

3.4 Variants of Value Iteration and Policy Iteration

In this section we again assume S is finite. Number the states $i = 1, 2, \ldots, m$, where $m = |S|$. In the nth step of value iteration, suppose the value function $v_{\beta,n}(i)$ is evaluated from the recursive optimality equations (17) in the order $i = 1, 2, \ldots, m$. In this case, when the time comes to evaluate $v_{\beta,n}(i)$, we will already have evaluated $v_{\beta,n}(k)$, for $k = 1, 2, \ldots, i-1$. Presumably $v_{\beta,n}(k)$ is a better estimate (of $v_\beta(k)$) than $v_{\beta,n-1}(k)$ is, so why not plug the former rather than the latter into the right-hand side of (17) for $k < i$? This idea leads to the following *Gauss-Seidel* variant of value iteration:

$$v_{\beta,n}(i) = \min_{a \in A} \left\{ c(i,a) + \beta \sum_{j<i} p_{ij}(a) v_{\beta,n}(j) + \beta \sum_{j \geq i} p_{ij}(a) v_{\beta,n-1}(j) \right\}, \ i \in S. \tag{31}$$

Let f' be a decision rule such that $f'(i)$ attains the minimum on the right-hand side of (31) for each $i \in S$. Then

$$v_{\beta,n} = c(f') + \beta P^L(f') v_{\beta,n} + \beta P^U(f') v_{\beta,n-1}, \tag{32}$$

where the matrices $P_L(f')$ and $P^U(f')$ are defined as

$$P^L(f') = \begin{bmatrix} 0 & 0 & 0 & \cdots & 0 & 0 \\ p_{21} & 0 & 0 & \cdots & 0 & 0 \\ p_{31} & p_{32} & 0 & \cdots & 0 & 0 \\ \vdots & \vdots & \vdots & & \vdots & \vdots \\ p_{m1} & p_{m2} & p_{m3} & \cdots & p_{m,m-1} & 0 \end{bmatrix},$$

$$P^U(f') = \begin{bmatrix} p_{11} & p_{12} & p_{13} & \cdots & p_{1,m-1} & p_{1m} \\ 0 & p_{22} & p_{23} & \cdots & p_{2,m-1} & p_{2,m} \\ 0 & 0 & p_{33} & \cdots & p_{3,m-1} & p_{3,m} \\ \vdots & \vdots & \vdots & & \vdots & \vdots \\ 0 & 0 & 0 & \cdots & 0 & p_{mm} \end{bmatrix},$$

where we have written p_{ij} instead of $p_{ij}(f'(i))$ to simplify the notation. This decomposition, $P(f') = P^L(f') + P^U(f')$, into lower and upper triangular matrices is sometimes called a *splitting*. (See chapter 4 for more details.) It follows from (31) and (32) that

$$v_{\beta,n} \leq c(f) + \beta P^L(f) v_{\beta,n} + \beta P^U(f) v_{\beta,n-1}, \text{ and hence}$$
$$(I - \beta P^L(f)) v_{\beta,n} \leq c(f) + \beta P^U(f) v_{\beta,n-1},$$

for all decision rules f, with equality for $f = f'$. Since $0 < \beta < 1$ and $P^L(f)$ is substochastic, $\beta^k (P^L(f))^k \to 0$ as $k \to \infty$. It follows (the proof is left as an exercise) that $I - \beta P^L(f)$ has an inverse, which is given by

$$(I - \beta P^L(f))^{-1} = \lim_{K \to \infty} \sum_{k=0}^{K} \beta^k (P^L(f))^k \geq 0.$$

Therefore,

$$v_{\beta,n} \leq (I - \beta P^L(f))^{-1} c(f) + (I - \beta P^L(f))^{-1} \beta P^U(f) v_{\beta,n-1},$$

for all decision rules f, with equality for $f = f'$. Equivalently,

$$v_{\beta,n} = \min_f \left\{ (I - \beta P^L(f))^{-1} c(f) + (I - \beta P^L(f))^{-1} \beta P^U(f) v_{\beta,n-1} \right\}. \quad (33)$$

It remains to be seen whether the $v_{\beta,n}$ defined by this *Gauss-Seidel* recursion converges to v_β and if so at what rate. We answer this question in a more general setting.

A splitting, $I - \beta P(f) = Q(f) - R(f)$, of the matrix $I - \beta P(f)$ is called a *regular splitting* if $Q(f)^{-1} \geq 0$ and $R(f) \geq 0$. Let $M = (m_{ij})_{i,j \in S}$ be an $m \times m$ matrix of real numbers. Define the *norm* of M, denoted $\| M \|$, by

$$\| M \| := \max_{i \in S} \sum_{j \in S} |m_{ij}|.$$

Theorem 6 *Suppose that, for each $f \in F$, $(Q(f), R(f))$ is a regular splitting of $I - \beta P(f)$ and that*

$$\gamma := \sup_{f \in F} \| Q(f)^{-1} R(f) \| < 1 . \qquad (34)$$

Then:
  a. for all $v_{\beta,0} \in \mathcal{B}(S)$, the iterative scheme

$$v_{\beta,n} = \min_{f \in F} \{Q(f)^{-1} c(f) + Q(f)^{-1} R(f) v_{\beta,n-1}\} = T v_{\beta,n-1}$$

converges to the optimal value function v_β;
  b. v_β is the unique fixed point in $\mathcal{B}(S)$ of the operator T;
  c. the convergence of $v_{\beta,n}$ to v_β is uniform at rate less than or equal to γ.

The proof proceeds by showing that

$$\| Tu - Tv \| \le \gamma \| u - v \|$$

for all $u, v \in \mathcal{B}(S)$. Hence, T is a contraction mapping on $\mathcal{B}(S)$ with modulus γ and has a unique fixed point, which can be shown to coincide with v_β. To apply this theorem to Gauss-Seidel value iteration, let $Q(f) = I - \beta P^L(f)$, $R(f) = \beta P^U(f)$, and note that

$$\| Q(f)^{-1} R(f) \| = \| (I - \beta P^L(f))^{-1} \beta P^U(f) \| \le \| \beta P(f) \| = \beta < 1$$

(see Puterman [Puterman, 1994], Theorem 6.3.7, p. 169). Hence (34) holds and the theorem applies.

3.5 Modified Policy Iteration

At each value-determination step of Policy Iteration, we must solve the system of m linear equations,

$$\phi^f = c(f) + \beta P(f) \phi^f = L(f) \phi^f \qquad (35)$$

for the value function, ϕ^f, associated with the stationary policy f currently being evaluated. This can be done by a direct method, such as Gauss elimination, which requires on the order of m^3 elementary operations (additions, multiplications, or comparisons). An alternative is to solve (35) by an iterative method (successive approximations). For example, let $v \in \mathcal{B}(S)$ and define $\phi_0^f = v$ and, for $n \ge 1$,

$$\phi_n^f = c(f) + \beta P(f) \phi_{n-1}^f = L(f) \phi_{n-1}^f . \qquad (36)$$

Note that these equations are identical to those of value iteration, except that they are for a fixed policy and hence do not involve minimization at each iteration. Since $L(f)$ is a contraction mapping (cf. (22)), it follows that $\phi_n^f \to \phi^f$ (uniformly and geometrically at rate β) as $n \to \infty$.

An advantage of this iterative method is that it is numerically stable and, if $P(f)$ is sparse, the matrix multiplications required at each iteration can be performed efficiently, whereas in Gauss elimination the matrix is subject to fill in. Moreover, we may not require a high degree of accuracy in our approximation ϕ_n^f of ϕ^f, since our purpose in calculating ϕ^f (at least in the early stages of Policy Iteration) is to yield a basis for determining an improved policy. It is plausible that the policy improvement step may not be very sensitive to the accuracy of this approximation. These observations suggest setting a limit, l_n, on the number of iterations of (36) to be carried out at each value-determination step. In practice, l_n should increase with n, as it is more important to have an accurate approximation of ϕ^f towards the end of the algorithm, when the policy under examination is more likely to be close to optimal.

We are now ready to state the Modified Policy Iteration algorithm. Let f_n denote the stationary policy under evaluation at step n and let v^n denote our approximation of ϕ^{f_n}.

Step 0. Set $v^0 \equiv 0$ and $n = 0$. Go to Step n.

Step n.

(i) (Policy Improvement.) Set $k = 0$ and

$$u_0^n := \min_f \{c(f) + \beta P(f)v^n\}$$

and let f_{n+1} be a stationary policy that attains the minimum for each state. If

$$\| u_0^n - v^n \| \leq \frac{\epsilon(1-\beta)}{2\beta},$$

then **stop**: the policy f_{n+1} is ϵ-optimal. Otherwise:

(ii) (Value Determination.) Compute

$$u_{k+1}^n = c(f_{n+1}) + \beta P(f_{n+1})u_k^n$$

and increment k until $k = l_n$. Then set $v^{n+1} = u_{l_n}^n$ and $n \leftarrow n+1$ and repeat Step n.

Note that v^{n+1} represents the expected total discounted cost of a (nonstationary) Markov policy that uses f_{n+1} for l_n stages, then f_n for l_{n-1} stages, etc., in an $(l_n + l_{n-1} + \ldots + l_0)$-stage problem with terminal cost function v^0. Another way of viewing Modified Policy Iteration is as a variant of value iteration, in which we only intermittently perform a minimization. Specifically, after identifying an improving policy, f, we perform a number of "cheap" iterations, in which we simply iterate with the fixed operator, $L(f)$, instead of minimizing. This saves computational effort, relative to that required by value iteration. Note that value iteration may be viewed as the special case of Modified Policy Iteration in which $l_n = 0$ for all n, whereas ordinary policy iteration corresponds to the limiting case in which $l_n \to \infty$ for all n.

3.6 Spans, Bounds, Accelerated Convergence of Algorithms

Now we consider additional techniques for accelerating the convergence of value iteration and policy iteration. First we introduce some new notation and definitions. Then we motivate our approach by reviewing some of our previous results from a slightly different point of view.

Define the operator $B : \mathcal{B}(S) \to \mathcal{B}(S)$ by:

$$Bv := \min_f \{c(f) + \beta P(f)v\} - v = Lv - v, \; v \in \mathcal{B}(S).$$

That is, Bv measures the difference between terminating now with terminal-cost function v and choosing a decision rule f that minimizes the sum of the immediate cost $c(f)$ and the expected discounted cost incurred if we terminate one stage from now with terminal-cost function v. Note that $v = Lv$ if and only if $Bv = 0$. It follows that the optimal value function v_β is the unique function $v \in \mathcal{B}(S)$ such that $Bv = 0$.

Definition. Let $v \in \mathcal{B}(S)$. A decision rule f is called v-*improving* if $L(f)v = Lv$.

Example. At step n of value iteration, let $f_n(i)$ attain the minimum in

$$v_{\beta,n}(i) = \min_{a \in A} \left\{ c(i,a) + \beta \sum_{j \in S} p_{ij}(a) v_{\beta,n-1}(j) \right\},$$

for each $i \in S$. Then $v_{\beta,n} = Lv_{\beta,n-1} = L(f_n)v_{\beta,n-1}$ and hence f_n is $v_{\beta,n-1}$-improving.

Example. At step n of policy iteration, let $f_n(i)$ attain the minimum in

$$w(i) = \min_{a \in A} \left\{ c(i,a) + \beta \sum_{j \in S} p_{ij}(a) \phi^{f_{n-1}}(j) \right\},$$

for each $i \in S$. Then $w = L\phi^{f_{n-1}} = L(f_n)\phi^{f_{n-1}}$ and hence f_n is $\phi^{f_{n-1}}$-improving.

Compare the following lemma with Theorem 4, which gave stopping criteria for value iteration.

Lemma 2 *Let $v \in \mathcal{B}(S)$. Let $\epsilon > 0$ and suppose $\| Bv \| < [(1-\beta)/2\beta]\epsilon$. Then*

$$\| Lv - v_\beta \| < \frac{\epsilon}{2}. \tag{37}$$

Now suppose in addition that f is an Lv-improving decision rule. (That is, $L^2v = L(f)Lv$.) Then

$$\| \phi^f - v_\beta \| < \epsilon. \tag{38}$$

Proof. We have

$$\|Lv - v_\beta\| \le \|Lv - L^2v\| + \|L^2v - Lv_\beta\|$$
$$\le \beta\|v - Lv\| + \beta\|Lv - v_\beta\|.$$

Therefore,

$$(1-\beta)\|Lv - v_\beta\| \le \beta\|v - Lv\|,$$

so that

$$\|Lv - v_\beta\| \le \left(\frac{\beta}{1-\beta}\right)\|Bv\| \le \frac{\epsilon}{2},$$

thus establishing (37). To prove (38), first note that

$$\|\phi^f - Lv\| \le \|\phi^f - L^2v\| + \|L^2v - Lv\|$$
$$= \|L(f)\phi^f - L(f)Lv\| + \|L^2v - Lv\|$$
$$\le \beta\|\phi^f - Lv\| + \beta\|Lv - v\|.$$

Therefore,

$$(1-\beta)\|\phi^f - Lv\| \le \beta\|Lv - v\|,$$

so that

$$\|\phi^f - Lv\| \le \left(\frac{\beta}{1-\beta}\right)\|Bv\| \le \frac{\epsilon}{2}. \tag{39}$$

Since $\|\phi^f - v_\beta\| \le \|\phi^f - Lv\| + \|Lv - v_\beta\|$, (38) follows from (37) and (39).

Now we shall show how to apply a similar approach, using the concept of the *span* of a vector, rather than the norm. This leads to improved estimates of the optimal value function when value iteration terminates.

For $v \in \mathcal{B}(S)$, let $l(v) := \min_{i \in S} v(i)$ and $u(v) := \max_{i \in S} v(i)$. Define $s(v)$, the *span* of v, by

$$s(v) := u(v) - l(v).$$

Lemma 3 *Let $v \in \mathcal{B}(S)$ and let f be a v-improving decision rule. Then, for any integer $m \ge -1$,*

$$G_m(v) := v + \sum_{k=0}^{m} \beta^k (P(f))^k Bv + \left(\frac{\beta^{m+1}}{1-\beta}\right) l(Bv)e$$
$$\le v_\beta \le \phi^f$$
$$\le v + \sum_{k=0}^{m} \beta^k (P(f))^k Bv + \left(\frac{\beta^{m+1}}{1-\beta}\right) u(Bv)e =: G^m(v),$$

where e denotes an $|S|$-vector of ones.

Proof. First note that, for any $u, u' \in \mathcal{B}(S)$ and u'-improving decision rule f,

$$\begin{aligned}
Bu = Lu - u &\leq c(f) + \beta P(f)u - u \\
&= c(f) + (\beta P(f) - I)u \\
Bu' = Lu' - u' &= c(f) + \beta P(f)u' - u' \\
&= c(f) + (\beta P(f) - I)u',
\end{aligned}$$

so that

$$Bu \leq Bu' + (\beta P(f) - I)(u - u').$$

Applying this inequality with $u = v_\beta$ and $u' = v$, we have

$$0 = Bv_\beta \leq Bv + (\beta P(f) - I)(v_\beta - v).$$

Multiplying both sides of this inequality by $(I - \beta P(f))^{-1}(\geq 0)$ yields the inequality

$$0 \leq (I - \beta P(f))^{-1} Bv - (v_\beta - v),$$

from which it follows that

$$\begin{aligned}
v_\beta &\leq v + (I - \beta P(f))^{-1} Bv \\
&= v + \sum_{k=0}^{\infty} \beta^k (P(f))^k Bv \\
&\leq v + \sum_{k=0}^{m} \beta^k (P(f))^k Bv + \left(\frac{\beta^{m+1}}{1-\beta}\right) u(Bv)e.
\end{aligned}$$

The rest of the proof is similar.

It is easy to verify that $G_m(v)$ is non-decreasing in m and $G^m(v)$ is non-increasing in m. Taking $m = -1$ and $m = 0$ therefore yields the following corollary:

Corollary 1 *Let $v \in \mathcal{B}(S)$ and let f be a v-improving decision rule. Then*

$$\begin{aligned}
v + \frac{l(Bv)e}{1-\beta} &\leq v + Bv + \left(\frac{\beta}{1-\beta}\right) l(Bv)e \\
&\leq v_\beta \leq \phi^f \\
&\leq v + Bv + \left(\frac{\beta}{1-\beta}\right) u(Bv)e \leq v + \frac{u(Bv)e}{1-\beta}.
\end{aligned}$$

Example. Consider the case $v \equiv 0$. Then

$$Bv(i) = \min_{a \in A} c(i, a) =: c^*(i).$$

Let f_0 be a v-improving decision rule. Then $f_0(i)$ attains the minimum in the above expression. We call f_0 a *greedy* decision rule, since it minimizes the immediate costs, with no regard for the future. Applying the corollary, we have

$$\frac{l(c^*)e}{1-\beta} \leq c^* + \left(\frac{\beta}{1-\beta}\right) l(c^*)e$$
$$\leq v_\beta \leq \phi^{f_0}$$
$$\leq c^* + \left(\frac{\beta}{1-\beta}\right) u(c^*)e \leq \frac{u(c^*)e}{1-\beta}.$$

Now we come to our main theorem, which will be the basis for improved stopping criteria for iterative algorithms such as value iteration and modified policy iteration. (Compare this theorem to Lemma 2.)

Theorem 7 *Let* $v \in \mathcal{B}(S)$. *Let* $\epsilon > 0$ *and suppose* $s(Bv) < [(1-\beta)/\beta]\,\epsilon$. *Then*

$$\| Lv + \left(\frac{\beta}{1-\beta}\right) u(Bv)e - v_\beta \| < \epsilon. \tag{40}$$

Now suppose in addition that f *is an v-improving decision rule. (That is,* $Lv = L(f)v$.) *Then*

$$\| \phi^f - v_\beta \| < \epsilon. \tag{41}$$

Proof. From Corollary 1, we have

$$0 \leq v + Bv + \left(\frac{\beta}{1-\beta}\right) u(Bv)e - v_\beta$$
$$\leq v + Bv + \left(\frac{\beta}{1-\beta}\right) u(Bv)e - v - Bv - \left(\frac{\beta}{1-\beta}\right) l(Bv)e$$
$$= \left(\frac{\beta}{1-\beta}\right) s(Bv)e.$$

Taking norms and using the fact that $v + Bv = Lv$, we obtain (40). The proof of (41) is similar.

To apply this result to value iteration, take $v = v_{\beta,n}$. The stopping criterion then becomes

$$sp(Bv_{\beta,n}) = sp(Lv_{\beta,n} - v_{\beta,n}) = sp(v_{\beta,n+1} - v_{\beta,n}) < \left(\frac{1-\beta}{\beta}\right)\epsilon,$$

which should be compared to the original stopping criterion,

$$\| v_{\beta,n} - v_{\beta,n-1} \| \leq \frac{\epsilon(1-\beta)}{2\beta}.$$

Note that when we stop we now use the *upper-bound extrapolation* $v_{\beta,n+1} + \left(\frac{\beta}{1-\beta}\right) u(v_{\beta,n+1} - v_{\beta,n})e$, rather than $v_{\beta,n+1}$, as an approximation of v_β. (This comes from (40) applied to $v = v_{\beta,n}$.)

It can be shown (cf. Puterman [Puterman, 1994], pp. 202-203) that $sp(Bv_{\beta,n})$ converges to zero faster than $\| Bv_{\beta,n} \|$.

Elimination of Non-Optimal Actions

Now we show how to use the bounds to identify actions that cannot appear in an optimal policy.

For $i \in S$, $a \in A$, $v \in \mathcal{B}(S)$, define

$$B(i,a)v := c(i,a) + \beta \sum_{j \in S} p_{ij}(a)v(j) - v(i) .$$

Note the similarity to the definition of the operator B. In particular, note that $Bv(i) = \min_{a \in A} B(i,a)v$.

Lemma 4 *If $B(i,a')v_\beta > 0$, then a' does not attain the minimum in the r.h.s of the optimality equation,*

$$v_\beta(i) = \min_{a \in A} \left\{ c(i,a) + \beta \sum_{j \in S} p_{ij}(a)v_\beta(j) \right\} ,$$

and any stationary policy that uses action a' in state i is not optimal.

Proof. Suppose $B(i,a')v_\beta > 0$. Then

$$c(i,a') + \beta \sum_{j \in S} p_{ij}(a')v_\beta(j) > v_\beta(i) = \min_{a \in A} \left\{ c(i,a) + \beta \sum_{j \in S} p_{ij}(a)v_\beta(j) \right\} ;$$

that is, a' does *not* attain the minimum in the r.h.s of the optimality equation. Now let f be a decision rule such that $f(i) = a'$. Then

$$v_\beta(k) \leq c(k,f(k)) + \beta \sum_{j \in S} p_{kj}(f(k))v_\beta(j) , \text{ for all } k \in S, \text{ with}$$

$$v_\beta(i) < c(i,f(i)) + \beta \sum_{j \in S} p_{ij}(f(i))v_\beta(j) .$$

That is, $v_\beta L(f) < v_\beta$ (where we write $u < v$ to mean that $u(k) \leq v(k)$ for all $k \in S$, with $u(k) < v(k)$ for at least one k). It follows that

$$\begin{aligned} v_\beta &< L(f)v_\beta \\ &= c(f) + \beta P(f)v_\beta \\ &\leq c(f) + \beta P(f)(c(f) + \beta P(f)v_\beta) \\ &= c(f) + \beta P(f)c(f) + \beta^2 (P(f))^2 v_\beta \\ &\leq \sum_{t=0}^{n-1} \beta^t (P(f))^t c(f) + \beta^n (P(f))^n v_\beta , \end{aligned}$$

for all $n \geq 1$. Letting $n \to \infty$ and using the fact that $\beta < 1$ and v_β is uniformly bounded, we conclude that

$$v_\beta < c(f) + \beta P(f) v_\beta \leq \sum_{t=0}^{n-1} \beta^t (P(f))^t c(f) = \phi^f .$$

Thus, the stationary policy f^∞ is not optimal.

Since we do not know v_β *a priori*, we use upper and lower bounds on v_β in numerical implementations of this result. The following corollary is immediate.

Corollary 2 *Suppose there exist functions $v^L, v^U \in B(S)$ such that $v^L \leq v_\beta \leq v^U$. Then, if*

$$c(i,a') + \beta \sum_{j \in S} p_{ij}(a') v^L(j) > v^U(i) .$$

Then any stationary policy that uses action a' in state i is not optimal.

We know show how to apply this result to value iteration, policy iteration, and modified policy iteration. Let v_n denote the value function evaluated' at the nth step of the algorithm. Then

$$v_{n+1} = \begin{cases} L v_n & \text{, for value interation} \\ \phi^{f_{n+1}} = (I - \beta P(f_{n+1})^{-1} c(f) & \text{, for policy iteration} \\ v_n + \sum_{k=0}^{l_n} \beta^k (P(f_{n+1}))^k B v_n & \text{, for modified policy iteration ,} \end{cases} \tag{42}$$

where f_{n+1} achieves the minimum in $\min_f \{c(f) + \beta P(f) v_n\}$. The expression for v_n in the case of modified policy iteration may be verified as follows. We have

$$\begin{aligned} u_0^n &= L(f_{n+1}) v_n = \min_f \{c(f) + \beta P(f) v_n\} \\ u_{k+1}^n &= L(f_{n+1}) u_k^n , \ k = 0, 1, \ldots, l_n \\ v_{n+1} &= u_{l_n+1} , \end{aligned}$$

from which it follows by recursive substitution that

$$v_{n+1} = \sum_{l=0}^{l_n} \beta^l (P(f_{n+1}))^l c(f_{n+1}) + \beta^{l_n+1} (P(f_{n+1}))^{l_n+1} v_n . \tag{43}$$

Now $B v_n = L v_n - v_n = c(f_{n+1}) + \beta P(f_{n+1}) v_n - v_n$, so that $c(f_{n+1}) = B v_n - \beta P(f_{n+1}) v_n + v_n$. Substituting this expression for $c(f_{n+1})$ in (43) yields the desired result,

$$v_{n+1} = v_n + \sum_{k=0}^{l_n} \beta^k (P(f_{n+1}))^k B v_n ,$$

after some algebra.

The following theorem applies the previously derived upper and lower bounds for v_β to modified policy iteration.

Theorem 8 *Suppose that at iteration n of modified policy iteration, we have*

$$c(i,a') + \beta \sum_{j \in S} p_{ij}(a')v_n(j) + \left(\frac{\beta}{1-\beta}\right) l(Bv_n) > G^{l_n}(v_n)(i).$$

Then action a' is non-optimal in state i.

Note that the l.h.s in the above inequality is just

$$c(i,a') + \beta \sum_{j \in S} p_{ij}(a')G_{-1}(v_n).$$

4 RELATED TOPICS

We now briefly discuss other models and issues related to optimal control of Markov chains and more general Markov processes.

4.1 Finite-State Approximations to Infinite-State Problems

In our discussion of algorithms for the discounted-cost problem, we developed much of theory in the context of an infinite state space with uniformly bounded costs. Since the implementation of an algorithm requires that the state space be truncated to finite dimensions, the reader may wonder why we did not just assume a finite state space from the beginning.

One reason for preferring the infinite-state-space setting (with bounded costs) is that convergence rates and criteria for ϵ-optimality do not depend on our choice of truncation parameters. Another reason is that, in applications to problems with special structure (e.g., problems in the optimal control of queues), it is often known or conjectured that an optimal policy has a simple form, provided the state space is infinite. For example, the optimal admission policy for many queueing systems has a threshold form: admit if the state is below the threshold, otherwise reject. (See Stidham [Stidham, 1985], [Stidham, 1988] for surveys of control models of this type.) Truncating the state space may corrupt this property because of boundary effects caused by the finite capacity. Thus, when one solves such a problem numerically in order to test the conjectured property of optimal policies, one may not be sure whether the test failed because the property is not true or because of the truncation effect. Thus one needs to understand not only the properties of the infinite-state-space formulation of the problem, but how closely optimal policies for finite-state truncations approximate the optimal policy for the infinite-state problem. For results of this type, see Puterman [Puterman, 1994], Section 6.10.2.

4.2 Unbounded Rewards

The assumption of uniformly bounded costs is not always realistic in the context of infinite-state models. In queueing control, for example, the one stage-cost function may take the form, $c(i,a) = c(a)+h(i)$, where $h(i)$ is the (expected discounted) holding cost incurred during one stage when the number of customers present at the beginning of the stage is i. Typically, $h(i) \to \infty$ as $i \to \infty$. Fortunately, there are several ways of relaxing the bounded-cost assumption.

Under considerably weaker assumptions (e.g., $\beta < 1$ and $c(i,a) \geq -M$ for all $i \in S$, $a \in A$, where $M < \infty$), it can be shown that:

1. $v_\beta(i)$ is well defined (it may equal $+\infty$) for all $i \in S$;

2. v_β is a solution (not necessarily unique) to the optimality equation (15);

3. $v_{\beta,n}(i) \to v_\beta(i)$ as $n \to \infty$, for all $i \in S$, provided $v_{\beta,0} \equiv 0$.

(See Schäl [Schäl, 1975], Puterman [Puterman, 1994], Chapter 7; Bertsekas [Bertsekas, 1995], Chapter 3.) Thus we have pointwise convergence of the value-iteration algorithm when the starting function is identically zero. (See van Hee et al [van Hee et al., 1977], Whittle [Whittle, 1979], [Whittle, 1980], [Whittle, 1983], Stidham [Stidham, 1981], [Stidham and van Nunen, 1983], [Stidham, 1994] for extensions to non-zero starting functions.)

While establishing pointwise convergence, these results do not provide rates of convergence nor stopping criteria for value iteration. An alternative technique for handling unbounded rewards employs weighted supremum norms (cf. van der Wal [van der Wal, 1974], Puterman [Puterman, 1994], Section 6.10.2), so that the theory of contraction mappings can still be employed. This approach requires rate-of-growth assumptions about the one-stage cost and the transition probabilities and leads to stronger results on the rate of convergence than the pointwise approach. A related approach (Stidham and van Nunen [Stidham and van Nunen, 1983], [Stidham, 1994], establishes uniform convergence of value iteration for the ordinary supremum norm for starting functions $v_{\beta,0}$ in a restricted class, provided certain regularity conditions are satisfied. (These conditions hold in many queueing-control problems.) The difference between this approach and the approach based on weighted supremum norms is that, instead of dividing the cost function by a function that increases at a faster rate, one subtracts a "reference" function from the value functions. For example, the reference function might be the (easily computed) value function for a particular policy, such as the policy that admits no customers (in the case of optimal admission to a queue).

4.3 Semi-Markov and Continuous Time Markov Decision Chains

Consider the following extension of our discounted-cost Markov decision chain model. The time interval between observation points is now a random variable, T, with a distribution that may depend on the observed current state, $X_t = i$, and action, $A_t = a$ (but not on the history of past states, actions, or times between observations). Future costs are discounted continuously at rate $\alpha > 0$, so that the present value of a cost c incurred τ time units in the future is $c \cdot \exp\{-\alpha\tau\}$. Let $c(i,a)$ denote the expected α-discounted cost incurred during stage t (i.e., until the next observation point), given $X_t = i$ and $A_t = a$. Define discounted transition probabilities,

$$p'_{ij}(a) := E[e^{-\alpha T}\mathbf{1}\{X_1 = j\}|X_0 = i, A_0 = a], \ i,j \in S, \ a \in A.$$

Then, by introducing an absorbing state in much the same way as we did in Example 5, one can convert this problem to an equivalent problem in our general format (with stationary costs and transition probabilities), in which the objective is to minimize the expected total cost over an infinite horizon. Many of our results concerning algorithmic convergence can easily be extended to this class of problems. (See, e.g., van der Wal [van der Wal, 1974], Stidham [Stidham, 1994], Puterman [Puterman, 1994], Chapter 11.)

In the special case in which T has an exponential distribution, the infinite-horizon discounted-cost problem has an equivalent representation as a problem in continuous-time control (a continuous-time Markov decision chain). Specifically, suppose the system is observed continuously through time and let X_τ and A_τ now represent the state and action, respectively, at time $\tau \in [0, \infty)$. Costs are incurred continuously at rate $d(i, a)$ while $X_\tau = i$ and $A_\tau = a$. Future costs are discounted at rate $\alpha > 0$. Let $p_{ij}(a)$ denote the transition probabilities at the jump points of X_τ ($p_{ii}(a) = 0$). Assume

$$P\{T \leq x|X_0 = i, A_0 = a\} = \exp\{-q(i,a)x\}, \ x \geq 0, \ i \in S, \ a \in A.$$

Care must be taken in defining admissible policies for such a problem, but if one makes the plausible assumption that an optimal policy for the infinite-horizon problem will not change actions between changes in state, then it is easy to show that the problem is equivalent to a semi-Markov decision chain of the form given above, with one-stage cost function

$$c(i,a) = \frac{w(i,a)}{\alpha + q(i,a)}, \ i \in S, \ a \in A,$$

and discounted transition probabilities

$$p'_{ij}(a) = \left(\frac{q(i,a)}{\alpha + q(i,a)}\right) p_{ij}(a), \ i,j \in S, \ a \in A.$$

Thus the optimality equation for this problem is

$$v_\beta(i) = \min_{a \in A} \left\{ \frac{w(i,a)}{\alpha + q(i,a)} + \left(\frac{q(i,a)}{\alpha + q(i,a)}\right) \sum_{j \neq i} p_{ij}(a) v_\beta(j) \right\}, \ i \in S.$$

Setting $q_{ij}(a) = q(i,a)p_{ij}(a)$, $i \neq j$, and $q_{ii}(a) = q(i,a)$, it can easily be shown that an equivalent optimality equation is

$$\alpha v_\beta(i) = \min_{a \in A} \left\{ w(i,a) + \sum_{j \in S} q(i,a) v_\beta(j) \right\}, \ i \in S. \tag{44}$$

Alternatively, if $q(i,a) \leq q < \infty$ for all $i \in S$, $a \in A$, then one may uniformize (Lippman [Lippman, 1975], Serfozo [Serfozo, 1979]) the process by introducing "null events", which do not change the state, at rate $q - q(i,a)$. The semi-Markov decision chain in which one observes the system at each change of state and at each null event has one-stage cost function

$$c(i,a) = \frac{w(i,a)}{\alpha + q} \, , \, i \in S \, , \, a \in A \, ,$$

and discounted transition probabilities

$$p'_{ij}(a) = \frac{q_{ij}(a)}{\alpha + q} \, , \, i,j \in S \, , \, i \neq j \, , \, a \in A \, ;$$

$$p'_{ii}(a) = \frac{q - q(i,a)}{\alpha + q} \, , \, i \in S \, , \, a \in A \, .$$

For this formulation one has the optimality equation

$$v_\beta(i) = \left(\frac{1}{\alpha + q}\right) \min_{a \in A} \left\{ w(i,a) + \sum_{j \neq i} q_{ij}(a) v_\beta(j) + (q - q(i,a)) v_\beta(i) \right\} \, , \, i \in S \, . \quad (45)$$

This formulation has many advantages. For one thing, it is formally equivalent to the optimality equation (15) for the discrete-time MDC (take $c(i,a) = w(i,a)/(\alpha+q)$, $\beta = q/(\alpha+q)$, $p_{ij}(a) = q_{ij}(a)/q$, $i \neq j$, $p_{ii}(a) = (q - q_{ii}(a))/q$). Thus, all the algorithms and convergence results discussed in the previous sections can be applied to this problem. Another advantage is that this formulation is amenable to the use of inductive arguments to prove that the n-stage optimal value functions, $v_{\beta,n}$, have certain properties (e.g., convexity), which then can be used to show that an optimal policy has a certain structure (e.g., monotonicity). (For early applications of this approach to problems in queueing control, see Lippman [Lippman, 1975], Lippman and Stidham [Lippman and Stidham, 1977]. For a survey, see Stidham [Stidham, 1985], [Stidham, 1988], Stidham and Weber [Stidham, 1994].) Note that the optimality equation (45) may also be derived from (44) by elementary algebraic manipulations.

5 AVERAGE-COST CRITERION

In some applications discounting of future costs may not be appropriate. This is often the case, for example, in applications to communication systems, in which decisions and changes of state take place on an extremely short time scale, during which the effects of discounting are negligible. More to the point, numerical computation using the iterative algorithms that we have discussed becomes inefficient and/or unstable when the discount factor is very close to one. In such cases, a more appropriate objective is to minimize the long-run average expected cost,

$$\psi_\pi(i) := \limsup_{n \to \infty} n^{-1} E_\pi \left[\sum_{t=0}^{n-1} c(X_t, A_t) | X_0 = i \right] \, , \, i \in S \, , \, \pi \in \Delta \, .$$

Under certain regularity conditions (cf. Weber and Stidham [Weber and Stidham, 1987], Sennott [Sennott, 1989], [Sennott, 1998], Puterman [Puterman, 1994], Chapter 8), it can be shown that the minimal long-run average expected cost is a constant, g^* (independent of the state i) and satisfies the following optimality equation:

$$g^* + h^*(i) = \min_{a \in A} \left\{ c(i,a) + \sum_{j \in S} p_{ij}(a) h^*(j) \right\}, \; i \in S.$$

A stationary policy that achieves the minimum can then found by selecting in each state i an action a achieving the minimum on the right-hand side of this equation.

The function $h^*(\cdot)$ is the *relative-cost function* and has several interpretations. In particular, $h^*(i) = \lim_{m \to \infty} (v_{\beta_m}(i) - v_{\beta_m}(0))$, where the limit is taken through a sequence of discount factors $\beta_m \to 1$. The conditions developed in [Weber and Stidham, 1987] and [Sennott, 1989] are particular well suited to problems in queueing control.

Variants of value iteration and policy iteration are also available for iterative solution of the average-cost optimality equation. For details see Puterman [Puterman, 1994] and Sennott [Sennott, 1998].

References

[Bertsekas, 1995] Bertsekas, D. (1995). *Dynamic Programming and Optimal Control, Vol. II*. Athena Scientific, Belmont, Massachusetts.

[Lippman and Stidham, 1977] Lippman, S. and Stidham, S. (1977). Individual versus social optimization in exponential congestion systems. *Operations Research*, 25:233–247.

[Lippman, 1975] Lippman, S. A. (1975). Applying a new device in the optimization of exponential queuing systems. *Operations Research*, 23:687–710.

[Puterman, 1994] Puterman, M. (1994). *Markov Decision Processes: Discrete Stochastic Dynamic Programming*. Wiley, New York.

[Schäl, 1975] Schäl, M. (1975). Conditions for optimality in dynamic programming and for the limit of n-stage optimal policies to be optimal. *Z. Wahrscheinlichkeitstheorie verw. Geb.*, 32:179–196.

[Sennott, 1989] Sennott, L. (1989). Average cost optimal stationary policies in finite state Markov decision processes with unbounded costs. *Operations Research*, 37:626–633.

[Sennott, 1998] Sennott, L. (1998). *Stochastic Dynamic Programming and the Control of Queueing Systems*. Wiley, New York. (forthcoming).

[Serfozo, 1979] Serfozo, R. (1979). An equivalence between continuous and discrete-time Markov decision processes. *Operations Research*, 27:616–620.

[Stidham, 1981] Stidham, S. (1981). On the convergence of successive approximations in dynamic programming with non-zero terminal reward. *Z. Operations Res.*, 25:57–77.

[Stidham, 1985] Stidham, S. (1985). Optimal control of admission to a queueing system. *IEEE Transactions on Automatic Control*, 30:705–713.

[Stidham, 1988] Stidham, S. (1988). Scheduling, routing, and flow control in stochastic networks. *Stochastic Differential Systems, Stochastic Control Theory and Applications*, IMA-10:529–561. W. Fleming and P.L. Lions, eds.

[Stidham, 1994] Stidham, S. (1994). Successive approximations for Markovian decision processes with unbounded rewards: a review. In Kelly, F., editor, *Probability, Statistics, and Optimisation*, pages 467–483, London. John Wiley and Sons Ltd.

[Stidham and van Nunen, 1983] Stidham, S. and van Nunen, J. (1983). The shift-function approach for Markov decision processes with unbounded returns. Research Report, Program in Operations Research, N.C. State University, Raleigh.

[van der Wal, 1974] van der Wal, J. (1974). *Stochastic Dynamic Programming. Math. Centre Tract 139.* Mathematisch Centrum, Amsterdam.

[van Hee et al., 1977] van Hee, K., Hordijk, A., and van der Wal, J. (1977). Successive approximations for convergent dynamic programming. In Tijms, H. and Wessels, J., editors, *Markov Decision Theory. Math. Centre Tract 93*, pages 183–211, Amsterdam. Mathematisch Centrum.

[Weber and Stidham, 1987] Weber, R. and Stidham, S. (1987). Control of service rates in networks of queues. *Advances in Applied Probability*, 24:202–218.

[Whittle, 1979] Whittle, P. (1979). A simple condition for regularity in negative programming. *J. Appl. Prob.*, 16:305–318.

[Whittle, 1980] Whittle, P. (1980). Stability and characterisation conditions in negative programming. *J. Appl. Prob.*, 17:635–645.

[Whittle, 1983] Whittle, P. (1983). *Optimization over Time: Dynamic Programming and Stochastic Control, Vol. II.* John Wiley, New York.

Shaler Stidham, Jr received the B.A. degree in mathematics from Harvard University in 1963, the M.S. degree in operations research and computing from Case Institute of Technology in 1964, and the Ph.D. degree in operations research from Stanford University in 1968.

He is currently a professor in the Department of Operations Research at University of North Carolina at Chapel Hill. Previously he held faculty positions at Cornell University and N.C. State University. He has held visiting positions at Stanford University, University of Aarhus, the Technical University of Denmark, Bell Laboratories, University of Cambridge, and INRIA, Sophia Antipolis. He is an Overseas Fellow of Churchill College, Cambridge.

Dr. Stidham's research and publications are primarily in the design and control of queueing systems, with an emphasis on applications to communications. He has been an associate editor of *Management Science* and *Operations Research*, Area Editor for Stochastic Processes for *Operations Research*, and is currently an associate editor of *QUESTA*. He was program co-chair of the Conference on Applied Probability (Chapel Hill, 1988) and the TIMS XXIX International Meeting (Osaka, 1989). In 1990-91 he served as chair of the TIMS/ORSA College/Technical Section on Applied Probability.

10 ON NUMERICAL COMPUTATIONS OF SOME DISCRETE-TIME QUEUES

Mohan L. Chaudhry

Department of Mathematics and Computer Science
Royal Military College of Canada
P.O. Box 17000, STN Forces
Kingston, Ontario K7K 7B4 Canada
chaudhry-ml@rmc.ca

1 INTRODUCTION

Since Erlang did pioneering work in the application of queues to telephony, a lot has been written about queues in continuous time (see, for example [Asmussen, 1987, Bacelli and Bremaud, 1994, Bhat and Basawa, 1992, Boxma and Syski, 1988, Bunday, 1986, Bunday, 1996, Chaudhry and Templeton, 1983, Cohen, 1982, Cooper, 1981, Daigle, 1992, Gnedenko and Kovalenko, 1989, Gross and Harris, 1985, Kalashnikov, 1994, Kashyap and Chaudhry, 1988, Kleinrock, 1975, Lipsky, 1992, Medhi, 1991, Prabhu, 1965, Robertazzi, 1990, Srivastava and Kashyap, 1982, Tijms, 1986, White et al., 1975]). In comparison to that large body of literature, not much has been written about queues in discrete time.

Though the well-known model M/D/c and the *imbedded Markov chain* results may be considered discrete-time queues, the real work on discrete-time queues was initiated by Meisling [Meisling, 1958]. In the early 1970s, much impetus was given to this topic by Neuts and his collaborators [Neuts, 1973, Klimko and Neuts, 1973, Neuts and Klimko, 1973, Heimann and Neuts, 1973]. Recently, the interest in discrete-time queues has increased because of their applicability in computer systems and communication [Bruneel and Kim, 1993, Kouvatsos, 1995, Kouvatsos, 1996, Woodward, 1994]. Transmission of data, in particular, is done in well defined *slots*. Since every model in continuous time has its analog in discrete time, there is a large number of different models. Here, we can only describe a small number of these models. For further details on discrete queues, we refer to the ample literature [Bruneel and Kim, 1993, Hunter, 1983, Miyazawa and Takagi, 1994, Takagi, 1993, Tran-Gia et al., 1995, Woodward, 1994] and references given in this chapter.

The notation of discrete-time queues is similar to the continuous-time queues. In particular, for one and multiserver queues, the usual Kendall's notation, in its extended form as given in Chaudhry and Templeton [Chaudhry and Templeton, 1983], is used with the exception that M (exponential) is replaced by Geom (geometric). The models discussed in this chapter include the Geom/G/1 queue and the GI/G/1 queue, both in steady state. In all cases, generating functions are derived, which are then used to obtain expectations and variances. The queue-length and waiting-time distributions are then obtained by inverting the transforms numerically by using characteristic roots. The reader should be alerted to the fact that there are other methods to find the distributions in question, including the methods discussed in [Alfa, 1998, Alfa et al., 1995, Alfa and Neuts, 1995, Alfa and Chakravarthy, 1994, Grassmann and Jain, 1989, Lipsky, 1992], as well as in Chapter 6.

2 PROBLEMS ARISING WHEN TIME IS DISCRETE

In this chapter, we consider discrete-time queues, that is, queues where the times between arrivals, and the service times are both integers. Discrete-time queues arise in a natural way in communication, where packets are "slotted". For simplicity, we assume that the width of all slots is equal to 1, and that only one message is transmitted per slot. Slots may also arise in production, particularly in cases where parts are placed into containers which move at a constant speed, with some containers being empty. One can also approximate continuous-time queues by discrete-time queues. In this case, one replaces the continuous times by values of the type $k\delta$, where k is an integer. By choosing δ as the time unit, these times become integers.

If interarrival times and service times are all integers, the number in the system only changes at times $0, 1, 2, ..., m, ...$ Of interest is the number in the system at time m. At this point, the question arises as to how to count arrivals and departures that occur at m. To clarify this issue, we use a slightly different approach. We assume that arrivals and departures occur around m, that is in the intervals $(m-, m)$ or $(m, m+)$. Now two cases arise:

(i) Arrivals may occur in $(m-, m)$ and departures in $(m, m+)$.

(ii) Departures may occur in $(m-, m)$ and arrivals in $(m, m+)$.

One may note that in case (i) arrivals have precedence over departures and in case (ii) departures have precedence over arrivals. Case (i) is referred to as **arrival-first** (AF) policy and case (ii) as **departure-first** (DF) policy. These two conventions were used by Gravey and Hébuterne [Gravey and Hébuterne, 1992].

There is another way of looking at these two cases. Since in case (i) arrivals occur in the interval $(m-, m)$ and departures take place in the interval $(m, m+)$, it is called a *late arrival system* (LAS) as arrivals occur late in the the mth slot, see Figure 1. Similarly, if we consider case (ii) where departures occur in the interval $(m-, m)$ and arrivals take place in the interval $(m, m+)$, it is called an *early arrival system* (EAS), see Figure 2.

Consider now arrivals to the empty system under LAS. The natural case is that an element with a service time S arriving around m will leave around $m + S$. In this case, one talks about *delayed access* (DA). There is, however, an alternative to DA, and this is *immediate access* (IA): under the IA policy, arrivals to the empty system receive one unit of service right away, that is, if the system is idle at m, and if an arrival with a service requirement of S occurs at this time, it will leave around $m + S - 1$. However, if the system is busy at $m-$, an element starting service around time m with service time S will leave around $m + S$ rather than around $m + S - 1$.

In summary, there are three systems to consider, namely EAS, LAS-IA and LAS-DA. We now compare these systems under the assumption that all interarrival times and service times have the same (integer) length. To simplify the analysis, we assume that the system starts empty. Furthermore, let A_n be the time of the nth arrival, or, to be more specific, assume that the nth arrival occurs between A_n- and A_n+. Moreover, assume that S_n is the service time of the element arriving around A_n. We first compare the systems EAS and LAS-DA. In both systems, the first busy period starts around A_1. Hence, until A_1-, both systems are idle. and both systems contain one element at A_1+. If $A_2 < A_1 + S_1$, the busy period continues, and under both EAS and LAS-DA, the number in the system is 1 from A_1+ to A_2-, and 2 at A_2+. This continues until the busy period ends: the number in the system is always the same for EAS and LAS-DA, except possibly for the intervals $(1-, 1+)$, $(2-, 2+)$, $(3-, 3+)$, ..., until the end of the first busy period. However, the next busy period behaves in the same way, that is, the number in the system is the same for EAS and LAS-DA outside the intervals $(m-, m+)$, $m = 1, 2, \ldots$. The two systems may be different at m, however. In the EAS system, departures (if any) occur in the intervals $(m-, m)$, and they therefore affect the state of the system at time m. On the other hand, arrivals occur in $(m, m+)$, and they do not affect the state at time m. Hence, in EAS, arrivals around m do not affect the state at time m, but departures do. For LAS-DA, on the other hand, departures around m affect the state at time m, but arrivals do not.

In the case of LAS-IA, the first arrival occurs during (A_1-, A_1), and the first departure during $((A_1+S_1-1), (A_1+S_1-1)+)$. Hence, the first departure of the first busy cycle occurs one time unit earlier under LAS-IA than under LAS-DA. It is easy to see that this advances all other departures of the first busy cycle by one time unit as well. The other busy cycles behave the same way. Hence, all departures are advanced by one time unit, that is, departure in LAS-IA occurs one time unit earlier than in LAS-DA. Specifically, the departure occurring in the interval $(m, m+)$ in LAS-DA now occurs in the interval $(m-1, (m-1)+)$. It therefore precedes rather than follows the arrival around m. When compared to LAS-DA, the order of arrivals and departures is therefore reversed. This makes the order in which arrivals and departures occur under LAS-IA similar to the one under EAS. Specifically, the number in the system of LAS-IA in the interval $((m-1)+, m-)$ is identical to the number in the system at m in EAS: In both cases, arrivals occurring around m are not counted, but departures occurring around m in EAS, or $m-1$ in LAS-IA, are.

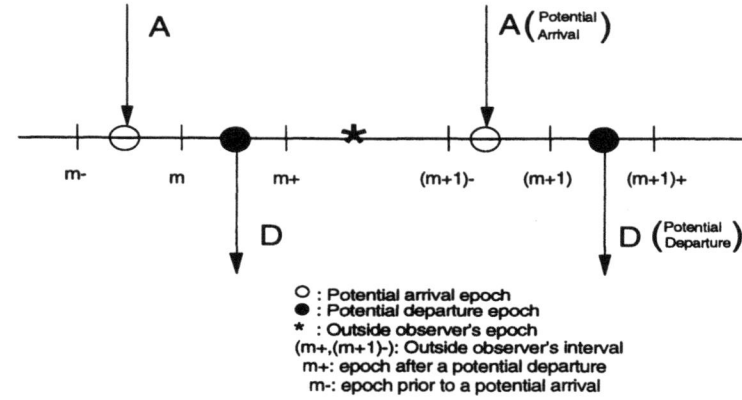

Figure 1 Various epochs in late arrival system with delayed access (LAS-DA)

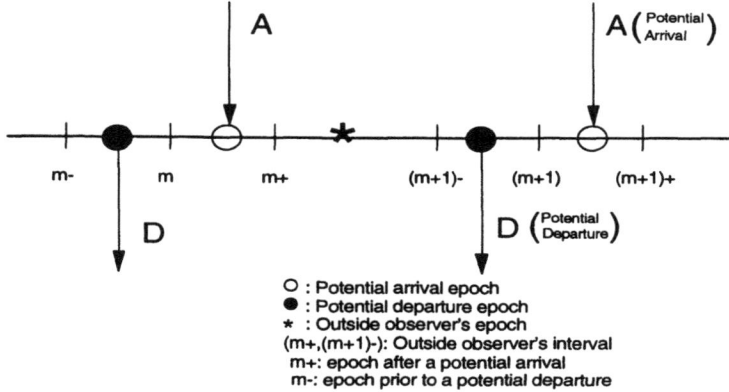

Figure 2 Various epochs in early arrival system (EAS)

Often, one deals with *waiting times, remaining service times* and other entities which express durations. Such durations must also be handled consistently. To do this, it turns out to be convenient to distinguish between slot boundaries, these are the epochs 1, 2, 3, ..., and the intervals between 0 and 1, 1 and 2, ..., the slots. Durations are measured in slots. For instance, there are u slots between the slot boundaries m and $m+u$, and the time between these epochs is therefore u. The same number is obtained by counting slot boundaries, except that one of the slot boundaries, either the one at m or the one at $m + u$ must be dropped. This way of counting must be done carefully because otherwise, basic relations, such as Little's theorem, become invalid.

3 THE GEOM/G/1 QUEUE

3.1 Introduction

In this section, we consider a single-server discrete-time queue Geom/G/1. Specifically, we assume that arrivals can only occur around m, where m is an integer, and that there can only be one arrival around m. This arrival occurs with probability λ. Customers are served by a single server, and the probability that the service time is i is b_i. We consider the state of the system at time $m+$, where m is an integer. The arrangement we assume is LAS-DA. Note, however, that at $m+$, EAS and LAS-DA are un-distinguishable. Generating functions for the number in the system are derived, and numerical work is done by inverting these generating functions.

3.2 General Solution

Consider the state of the system at $m+$, where m is a non-negative integer. At this time, any potential arrival and departure around m has already taken place. The state is described by two variables: the *number of customers* in the system (N_m) and the *remaining service time* of the customer in service (U_m) where for the sake of simplicity m is used for $m+$.
Let

$$\begin{aligned} P_0(m) &= P\{N_m = 0, U_m = 0\} \\ P_n(m, u) &= P\{N_m = n, U_m = u\}, \quad u \geq 1, \quad n \geq 1 \\ P_n(u) &= \lim_{m \to \infty} P_n(m, u), \quad \text{and} \quad P_0 = \lim_{m \to \infty} P_0(m). \end{aligned}$$

Of prime interest is the distribution of N_m in equilibrium, given by

$$P_n = \sum_{u=1}^{\infty} P_n(u).$$

To find P_n, we derive the generating function $P(s) = \sum_{n=0}^{\infty} P_n s^n$.

Note that any arrival around m will increase N_m, and any departure around m will decrease N_m. If there is both an arrival and a departure, N_m stays unchanged. Consider now the remaining service time U_m. U_m will decrease by 1 between m and $m+1$ except when $U_m = 1$. If $U_m = 1$, then there will be a departure before $(m+1)+$, and unless the system empties, a new service time will start at $m+$. This service time is i with probability b_i. In summary, if the system contains at least two elements, and if $U_m = 1$, then $U_{m+1} = i$ with probability b_i, and in the absence of an arrival, N_{m+1} decreases by 1. If $N_m = 1$, the situation is similar, except that in the absence of an arrival, the system becomes empty, in which case $U_{m+1} = 0$. If $N_m = 0$, no departures occur. Considering all these possibilities, with λ being the probability of an arrival, and taking the limit as $m \to \infty$, we have

$$P_0 = (1-\lambda)P_0 + (1-\lambda)P_1(1) \tag{1}$$
$$P_1(u) = (1-\lambda)P_1(u+1) + \lambda b_u P_0 + (1-\lambda)b_u P_2(1) + \lambda b_u P_1(1) \tag{2}$$
$$P_n(u) = (1-\lambda)P_n(u+1) + \lambda P_{n-1}(u+1)$$
$$+ (1-\lambda)b_u P_{n+1}(1) + \lambda b_u P_n(1), \quad n \geq 2. \tag{3}$$

From this set of equations, we will derive the two-dimensional generating function

$$P(z,s) = \sum_{n=1}^{\infty}\left[\sum_{u=1}^{\infty} P_n(u)z^u\right] s^n. \tag{4}$$

Clearly, for $z = 1$, we get

$$P(1,s) = \sum_{n=1}^{\infty}\left[\sum_{u=1}^{\infty} P_n(u)\right] s^n = \sum_{n=1}^{\infty} P_n s^n = P(s) - P_0.$$

Consider first the inner sum of (4) and define

$$P_n^*(z) = \sum_{u=1}^{\infty} P_n(u)z^u, \quad |z| \leq 1. \tag{5}$$

Moreover, let $B(z) = \sum_{i=1}^{\infty} b_i z^i$ be the generating function of the service time, with b being the mean service time. Multiplying (2) and (3) by z^u, summing over $u = 1$ to ∞, we obtain,

$$[1 - (1-\lambda)/z]\, P_1^*(z) = \lambda P_0 B(z) + (1-\lambda)P_2(1)B(z)$$
$$+ P_1(1)[\lambda B(z) - (1-\lambda)] \tag{6}$$
$$[1 - (1-\lambda)/z]\, P_n^*(z) = \lambda P_{n-1}^*(z)/z + (1-\lambda)P_{n+1}(1)B(z)$$
$$+ P_n(1)[\lambda B(z) - (1-\lambda)]$$
$$- \lambda P_{n-1}(1), \quad n \geq 2. \tag{7}$$

From (1), we obtain the following useful relation

$$\lambda P_0 = (1-\lambda)P_1(1). \tag{8}$$

Multiplying (6) and (7) by appropriate powers of s and then summing over $n = 1$ to ∞ yields after some calculation involving also (4) and (5).

$$[1 - (1-\lambda)/z]\,P(z,s) = \lambda s P(z,s)/z + (1-\lambda)B(z)\sum_{n=2}^{\infty} P_n(1)s^{n-1}$$

$$+[\lambda B(z) - (1-\lambda)]\sum_{n=1}^{\infty} P_n(1)s^n$$

$$-\lambda s \sum_{n=1}^{\infty} P_n(1)s^n + \lambda s P_0 B(z).$$

After bringing all terms containing $P(z,s)$ to the left, and using (8), we obtain

$$[1 - (1-\lambda+\lambda s)/z]P(z,s) = \sum_{n=1}^{\infty} P_n(1)s^n\left[\frac{(B(z)-s)(1-\lambda+\lambda s)}{s}\right]$$
$$-\lambda P_0 B(z)(1-s). \quad (9)$$

Unfortunately, this equation expresses $P(z,s)$ in terms of $\sum P_n(1)s^n$. However, by setting $z = 1 - \lambda + \lambda s$, $P(z,s)$ is removed from (9), and we can solve for $\sum P_n(1)s^n$:

$$\sum_{n=1}^{\infty} P_n(1)s^n = \frac{\lambda P_0 B(1-\lambda+\lambda s)(1-s)s}{(1-\lambda+\lambda s)[B(1-\lambda+\lambda s) - s]}. \quad (10)$$

Substituting this value into (9), we obtain, after some simplification,

$$P(z,s) = \frac{z\lambda P_0(1-s)s\{B(z) - B(1-\lambda+\lambda s)\}}{[B(1-\lambda+\lambda s) - s)]\{z - (1-\lambda+\lambda s)\}}. \quad (11)$$

Letting $z = 1$ in the above expression, we can obtain

$$P(1,s) = P(s) - P_0 = \frac{P_0 s(1 - B(1-\lambda+\lambda s))}{B(1-\lambda+\lambda s) - s}. \quad (12)$$

Now

$$\begin{aligned}P(s) &= P_0 + P(1,s)\\ &= P_0 + \frac{P_0 s(1 - B(1-\lambda+\lambda s)}{B(1-\lambda+\lambda s) - s}\\ &= \frac{P_0 B(1-\lambda+\lambda s)(1-s)}{B(1-\lambda+\lambda s) - s}.\end{aligned} \quad (13)$$

To find P_0, let s approach 1. By L'Hospital's rule, one obtains

$$P(1) = -P_0/(b\lambda - 1).$$

Since $P(1) = 1$, this yields

$$P_0 = 1 - \rho, \quad \text{where} \quad \rho = \lambda b. \tag{14}$$

Hence, we get from (13)

$$P(s) = \sum_{n=0}^{\infty} P_n s^n = \frac{(1-\rho)(1-s)B(1-\lambda+\lambda s)}{B(1-\lambda+\lambda s) - s}. \tag{15}$$

For computational aspects of this model, see Section 3.7.

We also have the following interesting relation

$$\sum_{n=1}^{\infty} P_n(1) = \frac{1 - P_0}{b}. \tag{16}$$

To derive (16), add (6), (7) and (8), and set $z = 1$. The left hand side of (16) gives the probability of having a departure, while the $1/b$ is the probability of a departure in a busy system, which means that $\frac{1-P_0}{b}$ is the unconditional departure probability.

3.3 System around Slot Boundaries

In this section, we consider the system state at the points $m-$, m and $m+$. To do this, we change the notation, using N_{m-}, N_m and N_{m+} for the number in the system at $m-$, m and $m+$, and we introduce a similar notation for the remaining service time. Hence, instead of N_m and U_m as in the previous section, we now use N_{m+} and U_{m+}.

There are no events between $(m-1)+$ and $m-$, that is $N_{m-} = N_{(m-1)+}$. In equilibrium, $P\{N_{m+} = n\} = P\{N_{(m-1)+} = n\} = P\{N_{m-} = n\}$. Hence, $P(s)$ in (15) gives the generating function for both N_{m-} and N_{m+}. However, the time advances by 1 from $(m-1)+$ to $m-$, that is $U_{m-} = U_{(m-1)+} - 1$. Hence, in equilibrium,

$$P\{N_{m+} = n, U_{m+} = u\} = P\{N_{m-} = n, U_{m-} = u - 1\}.$$

Under LAS-DA, one has n elements in the system at time m (that is, after a potential arrival, but before a potential departure) if (1) $N_{m-} = n$ and there is no arrival, or (2) if $N_{m-} = n - 1$, and there is an arrival. The latter case occurs only if $n > 0$. Hence, for the point m, one finds, in steady state

$$P\{N_m = 0\} = (1 - \lambda)P_0$$
$$P\{N_m = n\} = (1 - \lambda)P_n + \lambda P_{n-1}, \quad n \geq 1.$$

Here, $P_n = P\{N_{m+} = n\}$.

3.4 Pre-Arrival and Post-Departure Probabilities

Assume that arrivals occur in $(m-, m)$, and departures in $(m, m+)$, that is, assume an LAS-DA policy. We now determine P_n^+, the probability that $N_{m+} = $

n, given there is a departure around m. Collectively, the P_n^+, $n \geq 0$, describe the *post-departure* distribution. We find this distribution for LAS-DA. First, we find the probability of having a departure and having n elements in the system after this departure, and then we divide this probability by the probability of actually having a departure. The former probability is $(1-\lambda)P_1(1)$ if $n = 0$ and $\lambda P_n(1) + (1-\lambda)P_{n+1}(1)$ if $n > 0$. The probability of a departure is $\sum_{n=1}^{\infty} P_n(1)$. Consequently

$$P_0^+ = [(1-\lambda)P_1(1)]/\sum_{n=1}^{\infty} P_n(1) \qquad (17)$$

$$P_n^+ = [\lambda P_n(1) + (1-\lambda)P_{n+1}(1)]/\sum_{n=1}^{\infty} P_n(1), \quad n \geq 1. \qquad (18)$$

Let

$$P^+(s) = \sum_{n=0}^{\infty} P_n^+ s^n \qquad (19)$$

be the probability generating function (p.g.f) of the sequence $\{P_n^+\}_0^{\infty}$. Now following the procedure used in getting (15) (for details see [Chaudhry, 1997]), we can easily get

$$P^+(s) = \frac{(1-\rho)(1-s)B(1-\lambda+\lambda s)}{B(1-\lambda+\lambda s) - s} \qquad (20)$$

Comparing (15) and (20) one can easily see that $P(s) = P^+(s)$ and hence $P_n = P_n^+$. For results given in (15) and (20), one can see several easily available references, the latest being Takagi [Takagi, 1993].

We now obtain the pre-arrival probabilities P_n^-. These are the probabilities that there are n elements in the system at $m-$, given there is an arrival around m. However, since arrivals around m are independent of the N_{m-}, one has

$$P_n^- = P\{N_{m-} = n|\text{ arrival around } m\} = P\{N_{m-} = n\} = P_n.$$

Hence, $P_n^- = P_n$. Since $P_n^+ = P_n$, as was proven above, we have for the LAS-DA policy $P_n = P_n^- = P_n^+$ or equivalently $P(s) = P^-(s) = P^+(s)$, where $P^-(s)$ is the p.g.f. of the sequence $\{P_n^-\}_0^{\infty}$. This result also follows from BASTA (*Bernoulli arrivals see time averages*) or GASTA (geometric arrivals see time averages) property given in [Takagi, 1993, p.6].

3.5 Performance measures

As indicated in Section 3.4, equation (15) or, equivalently, (20) gives simultaneously the p.g.f. for P_n, P_n^- and P_n^+. It can be easily seen by direct differentiation of (20) that

$$E(N) = L = \frac{\lambda^2 B^{(2)}(1)}{2(1-\rho)} + \rho. \qquad (21)$$

Using Little's theorem, we have the mean *waiting time* in the system given by

$$W = \frac{L}{\lambda} = \frac{\lambda B^{(2)}(1)}{2(1-\rho)} + b \qquad (22)$$

where $b = B^{(1)}(1)$. The average number of customers in the queue is given by

$$L_q = L - \rho = \frac{\lambda^2 B^{(2)}(1)}{2(1-\rho)} \qquad (23)$$

and hence the mean waiting time in the queue is

$$W_q = \frac{L_q}{\lambda} = \frac{\lambda B^{(2)}(1)}{2(1-\rho)} \qquad (24)$$

Differentiating (20) twice yields, after some simplification, the following formula for the variance of the number in ths system.

$$\text{Var}(N) = \frac{1}{3(1-\rho)} \left[3\lambda^2 B^{(2)}(1) \left\{ 1 + \frac{\lambda^2 B^{(2)}(1)}{2(1-\rho)} \right\} + \lambda^3 B^{(3)}(1) \right]$$
$$+ L - L^2 \qquad (25)$$

3.6 Queueing Time Distribution

Let the r.v. T_q (T) denote the *queueing time* of a customer in the queue (system). Further, let $W_q(s)$ ($W(s)$) be their p.g.f.'s. Therefore, in LAS-DA under the first-come-first-served (FCFS) queue discipline, we get

$$\begin{aligned}
W_q(s) &= P_0 + \sum_{n=1}^{\infty} P_n^*(s)\{B(s)\}^{n-1} \\
&= P_0 + \frac{\sum_{n=1}^{\infty} P_n^*(s)\{B(s)\}^n}{B(s)} \\
&= P_0 + \frac{\sum_{n=1}^{\infty} \sum_{u=0}^{\infty} P_n(u) s^u \{B(s)\}^n}{B(s)} \\
&= P_0 + \frac{P(s, B(s))}{B(s)} \\
&= \frac{P_0(1-s)}{1 - \lambda - s + \lambda B(s)}. \qquad (26)
\end{aligned}$$

Since $W(s)$ and $W_q(s)$ are related by

$$W(s) = W_q(s) B(s) \qquad (27)$$

we can obtain $W(s)$ using (26) and (27) and it is given by

$$W(s) = \frac{(1-\rho)(1-s) B(s)}{1 - \lambda - s + \lambda B(s)}. \qquad (28)$$

Further, mean waiting times in the queue and in the system can be obtained by direct differentiation of (26) and (28). They are

$$W_q = E(T_q) = W_q^{(1)} = \frac{\lambda B^{(2)}(1)}{2(1-\rho)} \tag{29}$$

and

$$W = E(T) = \frac{\lambda B^{(2)}(1)}{2(1-\rho)} + b. \tag{30}$$

It can be seen that the expression for the mean waiting time in the queue (system) matches the one obtained, using Little's theorem, in the previous section.

To find the variance, use (26) to get

$$W_q^{(2)}(1) = \frac{\lambda B^{(3)}(1)}{3(1-\rho)} + 2\left(E(T_q)\right)^2 \tag{31}$$

and therefore Var(T_q) can be obtained from

$$\text{Var}(T_q) = W_q^{(2)}(1) + W_q^{(1)}(1) - (W_q^{(1)}(1))^2 \tag{32}$$

3.7 Computations and Approximations

We now show how to invert the p.g.f.'s. given in (15) and (26). To do this, we assume that the service-time distribution has a finite support, i.e. let

$$B(s) = \sum_{n=1}^{k} b_n s^n.$$

First, we discuss the inversion of $P(s)$ and some approximations. A similar analysis is then done for $W_q(s)$. In both cases, it is assumed that the roots are distinct. The case with repeated roots can be handled similarly.

Since $P(s)$ is convergent in $|s| \leq 1$, the zeros in $|s| \leq 1$ of the denominator and the numerator of $P(s)$ must coincide. But, using Rouché's theorem, it can be shown that when $\rho < 1$, the equation

$$s - B(1 - \lambda + \lambda s) = 0 \tag{33}$$

has only one root in $|s| \leq 1$, which, in fact, is $s = 1$. The other $(k-1)$ roots, s_i, of this equation are outside the unit circle.

Noting that the degree of the numerator of $P(s)$ in (15) is one higher than that of the denominator and ignoring the cases when $k = 1$ (for more details, see [Hunter, 1983]) or $k = 2$ which can be trivially treated separately, we rewrite (15) as

$$P(s) = (1-\rho)(1-s) + \sum_{i=1}^{k-1} \frac{A_i}{s - s_i} \tag{34}$$

$$= (1-\rho)(1-s) - \sum_{i=1}^{k-1} \sum_{n=0}^{\infty} \frac{A_i}{s_i^{n+1}} s^n \tag{35}$$

with

$$A_i = \lim_{s \to s_i} \frac{(1-\rho)s(s-1)(s-s_i)}{s - B(1-\lambda+\lambda s)}$$

$$= \frac{(1-\rho)s_i(s_i-1)}{1 - \lambda B^{(1)}(1-\lambda+\lambda s_i)} \qquad (36)$$

where $B^{(n)}(s_i)$ denotes the nth derivative of $B(s)$ evaluated at $s = s_i$. The numerical evaluations of the A_i's and s_i's may be checked either through the relation

$$\sum_{i=1}^{k-1} \frac{A_i}{1-s_i} = 1 \qquad (37)$$

or

$$\sum_{i=1}^{k-1} \frac{A_i}{s_i} = 0 \qquad (38)$$

which can be easily derived from (34) or (35) by using the relation $P(1) = 1$ or $P(0) = P_0 = (1-\rho)$, respectively. Thus,

$$P_n = \begin{cases} 1-\rho, & n = 0 \\ -\sum_{i=1}^{k-1} \frac{A_i}{s_i^2} - (1-\rho), & n = 1 \\ -\sum_{i=1}^{k-1} \frac{A_i}{s_i^{n+1}}, & n \geq 2. \end{cases} \qquad (39)$$

The mean and the variance can be obtained from (21), respectively, from (25). Alternatively, one can use the evaluated probabilities for this purpose:

$$E(N) = \sum_{n=0}^{size} n P_n \qquad (40)$$

$$\text{Var}(N) = \sum_{n=0}^{size} n^2 P_n - L^2 \qquad (41)$$

where $size$ is defined as the smallest integer such that $\sum_{n=0}^{size} P_n \geq 1-\epsilon$, ϵ being a small positive number. Thus, once the roots s_i ($i = 1, 2, \ldots, k-1$) are known, the exact probability distribution and the mean and variance of the number of customers at a random-time epoch are completely known by equations (39) to (41). The higher order moments can also be obtained from equation (39). The roots s_i such that $|s_i| > 1$ ($i = 1, 2, \ldots, k-1$), are obtained using the *Chaudhry QROOT software package* [Chaudhry, 1992b].

The mean L and $\text{Var}(N)$ can also be obtained (in terms of roots) from (34) by evaluating its derivatives at $s = 1$, and are given, respectively, by

$$L \equiv E(N) = P^{(1)}(1)$$

$$= -(1-\rho) - \sum_{i=1}^{k-1} \frac{A_i}{(1-s_i)^2} \qquad (42)$$

and

$$\text{Var}(N) = P^{(2)}(1) + P^{(1)}(1) - (P^{(1)}(1))^2$$
$$= 2\sum_{i=1}^{k-1} \frac{A_i}{(1-s_i)^3} + E(N) - (E(N))^2. \tag{43}$$

Approximations of Tail Probabilities

In equation (39), we have an exact expression for the probability distribution P_n in terms of outside roots of equation (33). The tail probabilities can be calculated by a single term corresponding to the smallest root in absolute value, say s_1, which based on our computational experience is not only positive and real, but also close to unity. A proof for the existence of this root can be given similar to the case discussed in Theorem 2 of Chaudhry et al. [Chaudhry et al., 1990]. Also, an approximate analytical expression for s_1, denoted as s_0, is given in the Appendix. It should, however, be pointed out that it is only the far tail that can be approximated very accurately by this single root unless ρ is large. To get the tail probabilities, assume

$$P_n \simeq \frac{-A_1}{s_1^{n+1}} = P_{an}^1, \quad n > n_\epsilon^1 \tag{44}$$

where n_ϵ^1 is chosen as the smallest integer such that $|(P_n - P_{an}^1)/P_n| < \epsilon$, i.e., $|1 - P_{an}^1/P_n| < \epsilon$. But, since the probability P_{an}^1 follows a geometric distribution with the common ratio $1/s_1$, it is better to choose n_ϵ such that $|s_1 P_n/P_{n-1} - 1| < \epsilon$. In this paper, the computations are done using this criterion. Moreover, the approximation gets better if more than one root, in ascending order of magnitude, is used. It should, however, be mentioned that those roots that occur in complex-conjugate pairs should be used in pairs. Thus, the tail probabilities using three roots can be approximated by

$$P_n \simeq -\sum_{i=1}^{3} \frac{A_i}{s_i^{n+1}}, \quad n > n_\epsilon^3 \tag{45}$$

where s_i ($i = 1, 2, 3$) are the roots in ascending order of magnitude. Denoting the approximation using three roots by P_{an}^3, choose n_ϵ^3 such that $|1 - P_{an}^3/P_n| < \epsilon$, $n > n_\epsilon^3$. The numerical performances in of all these approximations can be seen in Table 3. It may, however, be mentioned that since the probability P_{an}^3 does not follow a geometric distribution, n_ϵ^3 may be chosen according to this criterion. The approximation of P_n corresponding to the analytically evaluated root s_0 (see Appendix) is denoted by P_{an}^K and is given by

$$P_{an}^K = \frac{-A_0}{s_0^{n+1}}, \quad n > n_\epsilon^K$$

where A_0 and n_ϵ^K are found as A_1 and n_ϵ^1, respectively. It can be observed that the approximation P_{an}^1 (or P_{an}^3) gives better results than the approximation P_{an}^K, as is expected.

To investigate how good the root s_0 is, some numerical results are given in Table 4, page 385. The numerical values indicate that as ρ increases, s_0 approaches s_1. In fact, s_0, which is obtained in terms of the parameters and the first two moments of the distributions involved, serves both as a good initial approximation and an upper bound of the root s_1. Besides, it can be observed in Table 4 that as ρ increases, A_1 approaches $1-s_1$, and A_0 approaches $1-s_0$. Analytically, this can be seen from equation (36), i.e., $\lim_{\substack{\rho \to 1^- \\ s_1 \to 1^+}} (A_1/(1-s_1)) = 1$, or, $A_1 \to (1-s_1)$ as $\rho \to 1^-$. From equation (44), it then follows that, in heavy traffic, the distribution P_n becomes geometric, i.e.,

$$P_n = \left(1 - \frac{1}{s_1}\right)\left(\frac{1}{s_1}\right)^n, \quad \forall n \geq 0. \tag{46}$$

Further, as $\rho \to 1^-$, both $1-s_0$ and $1-s_1$ coalesce and $\to 0^-$. This phenomena can be seen in Table 4 In fact, if $\rho = 1$, $s_1 = 1$. This means that when $\rho = 1$, the root at unity becomes a repeated root of equation (1) of the Appendix. This can also be seen analytically from (1). It may be remarked here that when $\rho = 1$, the distribution P_n degenerates, as it should.

It is obvious that if we know P_{n_ϵ}, then the value of A_1 can also be computed from

$$A_1 = -P_{n_\epsilon^1} s_1^{n_\epsilon^1 + 1}. \tag{47}$$

Using (44) and taking into account the infinite tail after $n = n_\epsilon$, the sum of the probabilities and moments are given below:

$$\text{Sum of Probabilities} = \sum_{n=0}^{n_\epsilon^1} P_n + \frac{K}{s_1 - 1} \tag{48}$$

$$E(N) = \sum_{n=0}^{n_\epsilon^1} n P_n + K\left[\frac{n_\epsilon^1}{s_1 - 1} + \frac{s_1}{(s_1 - 1)^2}\right] \tag{49}$$

$$\text{Var}(N) = \sum_{n=0}^{n_\epsilon^1} n^2 P_n + K\left[\frac{(n_\epsilon^1)^2}{s_1 - 1} + \frac{2 n_\epsilon^1 s_1}{(s_1 - 1)^2} + \frac{s_1(s_1 + 1)}{(s_1 - 1)^3}\right] - (E(N))^2 \tag{50}$$

where

$$K = \frac{-A_1}{s_1^{n_\epsilon^1 + 1}}.$$

Consider now the waiting time T_q and its transform $W_q(s)$. The mean and variance of T_q are given by (29) and (32), respectively. The transform of T_q

is given by (26). It is clear by Rouché's theorem that the denominator of the first expression on the right hand side of (26) has only one zero inside and on the unit circle $|s| = 1$, which, in fact, is $s = 1$, so that the factor $(s - 1)$ in the numerator will cancel with the corresponding factor in the denominator. Therefore, all the $(k-1)$ zeros of the denominator (except for the zero at $s = 1$) are outside the unit circle. Expressing (26) as

$$W_q(s) = \prod_{i=1}^{k-1} \left(\frac{1-s'_i}{s-s'_i}\right), \tag{51}$$

where $s'_i (i = 1, 2, \ldots, k-1)$ are the outside zeros of $s - (1-\lambda) - \lambda B(s)$, we get

$$E(T_q) = W_q^{(1)}(1) = \sum_{i=1}^{k-1} \frac{1}{s'_i - 1} \tag{52}$$

and

$$\text{Var}(T_q) = \sum_{i=1}^{k-1} \frac{s'_i}{(s'_i - 1)^2}. \tag{53}$$

Moreover, since $s = 1$ is a root of both the numerator and the denominator of (26), expressing (26) into partial fractions, we get

$$W_q(s) = \sum_{i=1}^{k-1} \frac{B_i}{s - s'_i} \tag{54}$$

where

$$B_i = \lim_{s \to s'_i} \frac{(1-\rho)(s-1)(s-s'_i)}{s - (1-\lambda) - \lambda B(s)}$$

$$= \frac{(1-\rho)(s'_i - 1)}{1 - \lambda B^{(1)}(s'_i)}. \tag{55}$$

The numerical evaluation of the B_i's may be checked through the relation

$$\sum_{i=1}^{k-1} \frac{B_i}{1 - s'_i} = W_q(1) = 1. \tag{56}$$

The probabilities $w(n) \equiv P(T_q = n)$, mean $E(T_q)$, and variance $\text{Var}(T_q)$ can be derived from equation (54) and are given explicitly in terms of roots as

$$w(n) = -\sum_{i=1}^{k-1} \frac{B_i}{s'^{n+1}_i}, \quad n \geq 0 \tag{57}$$

$$E(T_q) = -\sum_{i=1}^{k-1} \frac{B_i}{(1-s'_i)^2} \tag{58}$$

and

$$\text{Var}(T_q) = 2\sum_{i=1}^{k-1} \frac{B_i}{(1-s_i')^3} + E(T_q) - (E(T_q))^2 \tag{59}$$

respectively. Moreover, from (52) and (58), it follows that

$$-\sum_{i=1}^{k-1} \frac{B_i}{(1-s_i')^2} = \sum_{i=1}^{k-1} \frac{1}{s_i'-1}. \tag{60}$$

This relation works as another check on the B_i's.

It may be noted that s_i''s and B_i's can be expressed in terms of s_i's and A_i's, respectively and are given by

$$s_i' = 1 - \lambda + \lambda s_i \tag{61}$$

and

$$B_i = \frac{\lambda A_i}{s_i}. \tag{62}$$

Once the A_i's are known, equation (62) facilitates the calculations of the B_i's.

The mean and variance of T_q can be obtained in several ways. One can use (58) and (59), or (52) and (53), or (29) and (32). Further, if the probabilities have been evaluated, one finds $E(T_q) = \sum_{n=0}^{size} nw(n)$ and $\text{Var}(T_q) = \sum_{n=0}^{size} n^2 w(n) - (E(T_q))^2$, where $size$ is defined as the smallest integer such that $\sum_{n=0}^{size} w(n) \geq 1 - \epsilon$, ϵ being a small positive number.

Approximations for Queueing-Time Tail

Proceeding on the same lines as before, the tail probabilities can be calculated by a single term corresponding to the smallest root in absolute value, say s_1', which, again based on our computational experience, is not only a positive real root, but also close to unity. An approximate analytical expression for s_1', denoted as s_0', is also given in the Appendix. Thus, the tail probabilities using one root and three roots, denoted by $w_a^1(n)$ and $w_a^3(n)$, respectively, are given by

$$w_a^1(n) = \frac{-B_1}{s_1'^{n+1}}, \quad n > n_\epsilon^1, \tag{63}$$

and

$$w_a^3(n) = -\sum_{i=1}^{3} \frac{B_i}{s_i'^{n+1}}, \quad n > n_\epsilon^3, \tag{64}$$

The approximation of $w(n)$ corresponding to the analytically evaluated root s_0', denoted by $w_a^K(n)$, is given by

$$w_a^K(n) = \frac{-B_0}{s_0'^{n+1}}, \quad n > n_\epsilon'^K \tag{65}$$

where B_0 is found as B_1. It may be remarked that observations and conclusions can be drawn as in the case of arbitrary-epoch tail probabilities for number in system and, in heavy traffic, the distribution $w(n)$ also becomes geometric, i.e.,

$$w(n) = \left(1 - \frac{1}{s_1'}\right)\left(\frac{1}{s_1'}\right)^n, \quad n \geq 0 \tag{66}$$

Remark: Note that equations (48)-(50) hold good for $\sum_{n=0}^{size} w(n)$, $E(T_q)$ and Var(T_q), with P_n, s_1, etc., replaced by $w_q(n)$, s_1', etc., respectively.

3.8 Comments on Computations

Extensive numerical computations along with various checks for the accuracy of the computations have been performed for various arbitrary, geometric, and deterministic service-time distributions. Selected numerical results are given in Tables 1-6. The selection has been done in such a way that by looking at them one gets a feel and an appreciation of the general applicability of the numerical procedure discussed here. These tables show that the results hold across many assumptions as well as for high values of the parameters of the distributions involved. All indications are that the results are very accurate. As the probability distributions and moments of the number in system and queueing-time are calculated in terms of roots, their accuracy depends on how accurately the roots are calculated by the *Chaudhry QROOT software package* [Chaudhry, 1992b]. The probability vector is calculated until the sum of the probabilities exceeds $1 - 10^{-12}$. The calculations were done on a 486 in double precision.

Results in Tables 1 and 2 consist of the mean, standard deviation (SD), probability distribution and cumulative distribution of number in system as well as, the mean, SD, probability distribution and cumulative distribution of queueing time. In these tables, the cumulative probabilities are denoted $Pr(\cdot)$, and $W(\cdot)$. These tables give detailed results for two different examples: in one case the service-time distribution is taken deterministic and in the other arbitrary. The mean and SD of number in system and of the queueing-time for a customer obtained using roots match those which do not use roots. This, in turn, checks the accuracy of all the computations that were performed. It may be mentioned here that some of the tables were prepared using the *Chaudhry QPACK software package* [Chaudhry, 1992a].

The values of s_1, A_1, ϵ and n_ϵ (n_ϵ^1, n_ϵ^3, n_ϵ^K) for the distribution of N and of s_1', B_1, ϵ' and $n_{\epsilon'}$ ($n_{\epsilon'}^1$, $n_{\epsilon'}^3$, $n_{\epsilon'}^K$) for the distribution of T_q are also shown in Tables 1 and 2 In Table 2 for $\rho = 0.9$, the tail probabilities are calculated using equations (44) and (63) after $n_\epsilon^1 = 30$ and $n_{\epsilon'}^1 = 90$, respectively. It may be mentioned here that if a higher value of ϵ is taken, i.e., $\epsilon > 10^{-16}$, then, in general, for high ρ, the distributions of N and T_q entirely become geometric and may be approximated using equations (46) and (66), respectively.

To investigate the performance of our approximations for the distributions of N and T_q, we consider only the system Geom/G/1 of Table 2 for $\rho = 0.5$ and 0.99. Table 3 gives the numerical values of P_n, P_{an}^1, P_{an}^3, and P_{an}^K for several values of n but only when $\rho = 0.5$ and 0.99. The numerical results indicate that while P_{an}^1 can be used for relatively small values of n compared with the approximation P_{an}^K, the approximation P_{an}^3 gives better results for $n \geq 5$. Besides, the performance of the approximations improves as ρ gets larger. Moreover, it may be noted that for both values of ρ, the approximations are no good for initial values of n; this happens because the approximations are valid only for large values of n. The same observations can be drawn from Table 5 which gives the numerical values of $w(n)$, $w_a^1(n)$, $w_a^3(n)$, and $w_a^K(n)$ for the same cases as in Table 3.

While, for several values of ρ, in Table 4 we give the values of s_0, s_1, A_0, and A_1, along with the corresponding values of n_ϵ for different values of ϵ, in Table 6 we give values of s_0', s_1', B_0, and B_1 along with the different values of $n_{\epsilon'}$ for corresponding different ϵ''s. These numerical values indicate that both n_ϵ and $n_{\epsilon'}$ increase as ϵ decreases, as it should.

Finally, we attach graph of roots for the two cases discussed in Table 1, see Figures 3 and 4.

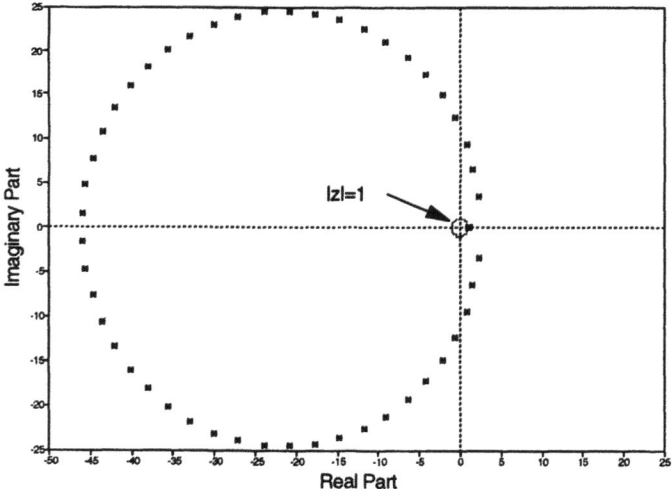

Figure 3 Roots s_i of the equation $s - \lambda B(1 - \lambda + \lambda s)$ for the parameters in Table 1

3.9 Overview

In Section 3.7 we have shown how computations can be done for models such as Geom/G/1. For computational work on models such as GI/Geom/1/N, GI/Geom(n)/1/N and Geom(n)/G(n)/1/N, see [Chaudhry and Gupta, 1996b], Chaudhry et al. [Chaudhry et al., 1996] and Chaudhry and Gupta [Chaudhry and Gupta, 1996a], respectively. Chaudhry and Zhao [Chaudhry and Zhao,

Table 1 Distributions of number in system and queueing time for the Geom/G/1 queue for $\lambda = 0.045$, $b_5 = 0.5$, $b_{25} = 0.3$, $b_{50} = 0.2$, $b = 20$, and $\rho = 0.9$.

Distribution of Number in System		
Probabilities		Cumulative
$P_0 = .1000$	$P_1 = .0953$	$Pr(0) = .1000$
$P_2 = .0898$	$P_3 = .0828$	$Pr(2) = .2851$
$P_4 = .0743$	$P_5 = .0657$	$Pr(4) = .3679$
$P_6 = .0580$	$P_7 = .0512$	$Pr(5) = .4422$
$P_8 = .0398$	$P_9 = .0451$	$Pr(6) = .5079$
$P_{10} = .0351$	$P_{11} = .0310$	$Pr(7) = .6171$
$P_{12} = .0273$	$P_{13} = .0241$	$Pr(9) = .7020$
$P_{14} = .0213$	$P_{15} = .0188$	$Pr(13) = .8195$
$P_{16} = .0165$	$P_{17} = .0146$	$Pr(18) = .9036$
$P_{18} = .0129$	$P_{19} = .0114$	$Pr(228) = 1.000$
MEAN(form.)=7.785		SD(form.)=8.010
MEAN(roots)= 7.785		SD(roots)=8.010
MEAN(probs)= 7.785		SD(probs)=8.010

$\epsilon = 10^{-16}$, $n_\epsilon^1 = 22$, $s_1 = 1.1336$, $A_1 = -0.1395$
Sum of probs=1.000, Number of probs=229

Queueing-Time Distribution		
Probabilities		Cumulative
$w(0) = .1047$	$w(1) = .0049$	$W_q(0) = .1047$
$w(2) = .0052$	$w(3) = .0054$	$W_q(22) = .2002$
$w(4) = .0057$	$w(5) = .0035$	$W_q(46) = .3045$
$w(6) = .0035$	$w(7) = .0036$	$W_q(71) = .4018$
$w(8) = .0036$	$w(9) = .0037$	$W_q(101) = .5002$
$w(10) = .0037$	$w(11) = .0038$	$W_q(139) = .6020$
$w(12) = .0039$	$w(13) = .0040$	$W_q(187) = .7015$
$w(14) = .0041$	$w(15) = .0042$	$W_q(254) = .8002$
$w(16) = .0043$	$w(17) = .0044$	$W_q(370) = .9003$
$w(18) = .0046$	$w(19) = .0047$	$W_q(4172) = 1.000$
MEAN(form.)=153.000		SD(form.)=166.480
MEAN(roots)=153.000		SD(roots)=166.480
MEAN(probs)=153.000		SD(probs)=166.480

$\epsilon' = 10^{-16}$, $n_{\epsilon'}^1 = 400$, $s_1' = 1.0060$, $B_1 = -0.0055$
Sum of probs=1.000, Number of probs=4173

Table 2 Distributions of number in system and queueing time for the Geom/G/1 queue for $\lambda = 0.225$, $b_1 = 0.5$, $b_5 = 0.3$, $b_{10} = 0.2$, $b = 4$, and $\rho = 0.9$.

Distribution of Number in System		
Probabilities		Cumulative
$P_0 = .1000$	$P_1 = .1053$	$Pr(0) = .1000$
$P_2 = .0984$	$P_3 = .0907$	$Pr(1) = .2053$
$P_4 = .0799$	$P_5 = .0695$	$Pr(2) = .3037$
$P_6 = .0603$	$P_7 = .0523$	$Pr(4) = .4744$
$P_8 = .0454$	$P_9 = .0394$	$Pr(5) = .5439$
$P_{10} = .0342$	$P_{11} = .0297$	$Pr(6) = .6041$
$P_{12} = .0258$	$P_{13} = .0223$	$Pr(8) = .7018$
$P_{14} = .0194$	$P_{15} = .0168$	$Pr(11) = .8051$
$P_{16} = .0146$	$P_{17} = .0127$	$Pr(16) = .9040$
$P_{18} = .0110$	$P_{19} = .0096$	$Pr(195) = 1.000$
MEAN(form.)=6.975		SD(form.)=7.102
MEAN(roots)= 6.975		SD(roots)=7.102
MEAN(probs)= 6.975		SD(probs)=7.102
$\epsilon = 10^{-16}$, $n_\epsilon^1 = 30$, $s_1 = 1.1522$, $A_1 = -.1624$		
Sum of probs=1.000, Number of probs=196		

Queueing-Time Distribution		
Probabilities		Cumulative
$w(0) = .1290$	$w(1) = .0187$	$W_q(0) = .1290$
$w(2) = .0214$	$w(3) = .0246$	$W_q(4) = .2219$
$w(4) = .0281$	$w(5) = .0210$	$W_q(8) = .3141$
$w(6) = .0224$	$w(7) = .0238$	$W_q(12) = .4023$
$w(8) = .0251$	$w(9) = .0263$	$W_q(18) = .5116$
$w(10) = .0208$	$w(11) = .0207$	$W_q(24) = .6008$
$w(12) = .0204$	$w(13) = .0198$	$W_q(33) = .7052$
$w(14) = .0187$	$w(15) = .0184$	$W_q(45) = .8032$
$w(16) = .0180$	$w(17) = .0175$	$W_q(66) = .9030$
$w(18) = .0168$	$w(19) = .0161$	$W_q(822) = 1.000$
MEAN(form.)=27.0000		SD(form.)=29.6226
MEAN(roots)= 27.0000		SD(roots)=29.6226
MEAN(probs)= 27.0000		SD(probs)=29.6226
$\epsilon' = 10^{-16}$, $n_{\epsilon'}^1 = 90$, $s_1' = 1.0342$, $B_1 = -.0317$		
Sum of probs=1.000, Number of probs=823		

Table 3 The values of P_n, P_{an}^1, P_{an}^3, and P_{an}^K for the model of Table 2 for $\rho = 0.5$ and 0.99.

n	P_n	P_{an}^1	P_{an}^3	P_{an}^K
		$\rho = 0.5$		
0	.5000	1.0534	1.9117	.4467
5	.0086	.0085	.0086	.0007
10	.0001	.0001	.0001	.0000
20	.0000	.0000	.0000	.0000
		$\rho = 0.99$		
0	.0100	.0134	.0203	.0133
5	.0125	.0125	.0125	.0124
10	.0117	.0117	.0117	.0116
20	.0102	.0102	.0102	.0102
30	.0090	.0090	.0090	.0089
40	.0078	.0078	.0078	.0077
50	.0069	.0069	.0069	.0068
100	.0035	.0035	.0035	.0034
200	.0009	.0009	.0009	.0009
300	.0002	.0002	.0002	.0002
400	.0001	.0001	.0001	.0001
500	.0000	.0000	.0000	.0000

Table 4 The number n_ϵ^1 for the relative error bound ϵ for P_{an}^1 and the values of s_0, s_1, A_0, and A_1 for the model of Table 2 for different values of ρ.

ρ	0.1	0.5	0.9	0.99	0.999
s_0	121.0000	3.6667	1.1646	1.0136	1.0013
s_1	24.6408	2.6201	1.1522	1.0135	1.0013
A_0	-0.9964	-1.6380	-0.1508	-0.0135	-0.0013
A_1	-208.1233	-2.7600	-0.1624	-0.0136	-0.0013
$n_\epsilon^1, \epsilon = 10^{-4}$	15	10	8	8	8
$n_\epsilon^1, \epsilon = 10^{-8}$	22	17	15	15	15
$n_\epsilon^1, \epsilon = 10^{-12}$	41	26	22	20	20
$n_\epsilon^1, \epsilon = 10^{-16}$	54	35	30	27	27

Table 5 The values of $w(n)$, $w_a^1(n)$, $w_a^3(n)$, and $w_a^K(n)$ for the model of Table 2 for $\rho = 0.5$ and 0.99.

n	$w(n)$	$w_a^1(n)$	$w_a^3(n)$	$w_a^K(n)$
		$\rho = 0.5$		
0	.5714	.1095	.2173	.0419
5	.0293	.0435	.0444	.0010
10	.0146	.0173	.0173	.0024
20	.0027	.0027	.0027	.0001
30	.0004	.0004	.0004	.0000
40	.0001	.0001	.0001	.0000
		$\rho = 0.99$		
0	.0133	.0033	.0054	.0033
5	.0027	.0033	.0033	.0033
10	.0031	.0032	.0032	.0032
20	.0031	.0031	.0031	.0031
30	.0030	.0030	.0030	.0030
40	.0029	.0029	.0029	.0029
50	.0028	.0028	.0028	.0028
100	.0024	.0024	.0024	.0023
200	.0017	.0017	.0017	.0017
300	.0012	.0012	.0012	.0012
400	.0009	.0009	.0009	.0009
500	.0006	.0006	.0006	.0006

Table 6 The number $n_{\epsilon'}^1$ for the relative error bound ϵ' for $w_a^1(n)$ and the values of s_0', s_1', B_0, and B_1 for the model of Table 2 for different values of ρ.

ρ	0.1	0.5	0.9	0.99	0.999
s_0'	4.0000	1.3333	1.0370	1.0034	1.0003
s_1'	1.5910	1.2025	1.0342	1.0033	1.0003
B_0	-0.0002	-0.0558	-0.0291	-0.0033	-0.0003
B_1	24.6408	2.6201	1.1522	1.0135	1.0013
$n_{\epsilon'}^1, \epsilon' = 10^{-4}$	71	31	30	30	30
$n_{\epsilon'}^1, \epsilon' = 10^{-8}$	179	91	51	51	51
$n_{\epsilon'}^1, \epsilon' = 10^{-12}$	262	129	90	79	79
$n_{\epsilon'}^1, \epsilon' = 10^{-16}$	378	178	121	113	107

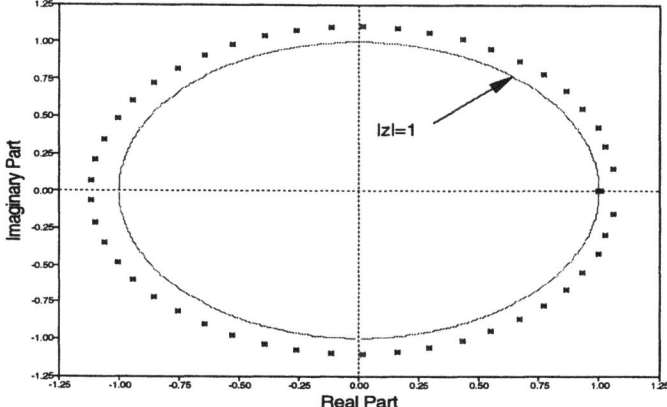

Figure 4 Roots s'_i of the equation $s - (1 - \lambda) - \lambda B(s)$ for the parameters in Table 1.

1994] discuss the first passage time and busy period distributions of discrete-time Markovian queue Geom(n)/Geom(n)/1/N. For many variations and extensions of the model Geom/G/1 and Geom/G/1/N, one may consult Takagi [Takagi, 1993]. In his book, he gives transforms of various results. Some of these transforms can be inverted by using the methods discussed here. For the special case of Geom/G/1 and Geom/G/1/N. viz., Geom/D/1 and Geom/D/1/N, see et al. [Gravey et al., 1990]. For cases when interarrival times are not identically distributed, or the arrival process has some sort of correlation, or applications of discrete-time queues in ATM, see [Bruneel and Kim, 1993] and references within. In this connection, see also [Woodward, 1994]. For a more generalized version of Geom/G/1/N, viz. MMBP/G/1/N in which arrivals follow Markov modulated batch Bernoulli process, see [Tsuchiya and Takahashi, 1993]. This model is more general than the one discussed in Bruneel [Bruneel, 1993]. In this connection, see also [Blondia, 1993].

Using a combinatorial approach, Bhat [Bhat, 1968] discusses, among other things, the joint distribution of the length of and the number of customers served in a busy period for the queues Geom/G/1 and GI/Geom/1. Yang and Li [Yang and Li, 1995] discuss the steady-state queue-size distribution of the Geom/G/1 queue with repeated customers. For a departure process in an m-MMBP/Geom/1/N queue, see [Park and Perros, 1994]. Here, the arrival process is an m-state Markov modulated Bernoulli process. Alfa and Chakravarthy [Alfa and Chakravarthy, 1994] consider a discrete-time queue with Markovian arrival process and phase-type primary and secondary services. Gouweleeuw [Gouweleeuw, 1994] considers the loss probability in finite-buffer queue with batch arrivals and complete rejection.

4 THE GI/G/1 QUEUE

4.1 Introduction

The queueing system GI/G/1 has been discussed by many authors, see, e.g., [Konheim, 1975] and [Grassmann and Jain, 1989]. For other references, see Section 4.7. Here we give closed-form expressions, in terms of roots of certain equations, for the distributions of the waiting time in queue, W_q, in the steady-state for the discrete-time queue GI/G/1. Essentially, this is done by finding roots of the denominator of the p.g.f. of W_q and then resolving the generating function into partial fractions. Numerical examples are given showing the use of the required roots, even when there is a large number of them. The method discussed in this section uses both closed- and non-closed forms of interarrival- and service-time distributions. Approximations of the tail probabilities in terms of one or three roots taken in ascending order are also discussed.

4.2 General solution

The basic assumptions are

(i) Customers arrive singly at a service station at epochs $\{T_i : 0 \leq i < \infty\}$ ($T_0 = 0$) and are serviced by a single server in the order of arrivals. The interarrival time $\{A_i = T_i - T_{i-1} : 1 \leq i < \infty\}$ are i.i.d.r.v.'s distributed as A with probability distribution

$$P(A = k) = a_k, \quad 0 \leq k \leq K_1 < \infty$$

(ii) The service times $\{S_i : 0 \leq i < \infty\}$ are i.i.d.r.v.'s distributed as S with probability distribution

$$P(S = k) = b_k, \quad 0 \leq k \leq K_2 < \infty$$

(iii) Denote $X = S - A$ and assume that $E(X) = E(S) - E(A) < 0$. The utilization factor $\rho = E(S)/E(A)$ is then less than 1 so that the queueing process is positive recurrent. The r.v. X can only assume the integral values $-K_1, -(K_1 - 1), ..., -2, -1, 0, 1, 2, ..., (K_2 - 1), K_2$. Further, A and S are independent r.v.'s.

(iv) Let W_q be the stationary queueing time of a customer. Further, let $w_q(i)$ be the steady-state probability that a customer waits for i units of time in the queue before getting service.

Under the above assumptions and notations, we have, according to [Konheim, 1975]

$$W_q(s) = \frac{s^{-K_1} M(s)(1-s)}{1 - A(1/s)B(s)} \tag{67}$$

where $W_q(s)$, $A(s)$ and $B(s)$ are the p.g.f.'s of the stationary queueing time W_q, interarrival time A, and service time S, respectively, and are convergent in $|s| \leq 1$, i.e.,

$$W_q(s) = \sum_{i=0}^{\infty} w_q(i) s^i \qquad A(s) = \sum_{i=0}^{K_1} a_i s^i \qquad B(s) = \sum_{i=0}^{K_2} b_i s^i$$

with $M(s)$ being a polynomial of degree $K_1 - 1$ [Konheim, 1975], which can be determined uniquely due to the analyticity of $W_q(s)$.

Rewrite (67) as

$$W_q(s) = \frac{M(s)(1-s)}{s^{K_1} - A^*(s) B(s)} \tag{68}$$

where $A^*(s) = \sum_{i=0}^{K_1} a_i s^{K_1 - i}$.

Clearly, the degree of the denominator in (68) is $K_1 + K_2$. Since $W_q(s)$ is convergent in $|s| \leq 1$, the zeros, in $|s| \leq 1$, of the denominator and the numerator of $W_q(s)$ must coincide. But by Rouché's theorem it can be shown that the equation

$$s^{K_1} - A^*(s) B(s) \tag{69}$$

has K_1 roots in $|s| \leq 1$, and hence K_2 roots outside the unit circle, if $\rho < 1$. Further, since the degree of $M(s)$ is known to be $K_1 - 1$, $W_q(s)$ can be written as

$$W_q(s) = \frac{K}{\prod_{i=1}^{K_2} (s - s_i)} \tag{70}$$

where $|s_i|$ ($i = 1, 2, \ldots, K_2$) are the zeros, assumed distinct, outside the unit circle, and the constant K is determined from $W_q(1) = 1$. This implies that

$$W_q(s) = \prod_{i=1}^{K_2} \left(\frac{1 - s_i}{s - s_i} \right). \tag{71}$$

Resolving (71) into partial fractions, we get

$$W_q(s) = \sum_{i=1}^{K_2} \frac{C_i}{s - s_i} \tag{72}$$

where

$$C_i = (1 - s_i) \prod_{\substack{j=1 \\ j \neq i}}^{K_2} \left(\frac{1 - s_i}{s_i - s_j} \right). \tag{73}$$

Set s in (71) equal to zero to obtain

$$w_q(0) \equiv P(W_q = 0) = \prod_{i=1}^{K_2}(1 - 1/s_i). \qquad (74)$$

Expand (72) in powers of s. The coefficient of s^n then yields

$$w_q(n) \equiv P(W_q = n) = -\sum_{i=1}^{K_2} C_i/s_i^{n+1}, \quad n \geq 1. \qquad (75)$$

By taking the derivatives of (71) we get the mean and variance of W_q in the normal way

$$E(W_q) = -\sum_{i=1}^{K_2} \frac{1}{1-s_i} \qquad (76)$$

and

$$\text{Var}(W_q) = \sum_{i=1}^{K_2} \frac{1}{(1-s_i)^2} + E(W_q). \qquad (77)$$

Equations (74) and (75) give probabilities in closed form, in terms of the roots s_i, where s_i are the K_2 roots of (69) outside the unit circle. Once the roots s_i ($i = 1, 2, ..., K_2$) are known, $w_q(n)$ and the moments of W_q are completely determined. The higher-order moments may be found from (72) or (75). To find these outside roots, use the method discussed in the *Chaudhry QROOT software package* [Chaudhry, 1992b], which determines roots inside and on the unit circle. Thus, first find the K_2 roots inside the unit circle of the equation

$$s^{K_2} - A(s)B^*(s) \qquad (78)$$

where $B^*(s) = \sum_{i=0}^{K_2} b_i s^{K_2-i}$.

Equation (78) is obtained by changing s to $1/s$ in (69). After getting the inside roots of (78), the outside roots of (69) are obtained by taking the reciprocals of the inside roots of (78).

4.3 Interarrival and service times with rational transforms

In many cases, $A(1/s)$ and/or $B(s)$ are rational functions. In this case, one could truncate the series, but this may lead to large values of K_1 and K_2. Though this may not be a major problem, it is preferable to use an alternative approach.

Case 1. $B(s)$ rational

Let $B(s)$ be a rational function, say $f(s)/g(s)$. Assume that $f(s)$ and $g(s)$ have no common factors. Suppose further that $g(s)$ is of degree m_2 and $f(s)$

of degree m_1. Since $g(s)$ is a generating function, it is convergent for $|s| \leq 1$. This implies that all the m_2 zeros of $g(s)$ must lie outside the unit circle, unless $g(s)$ is constant in which case it has no zeros. Thus, following the procedure of getting (71), (68) reduces to

$$W_q(s) = \frac{g(s)}{g(1)} \prod_{i=1}^{m} \left(\frac{1-s_i}{s-s_i} \right) \tag{79}$$

where m is the number of roots of (69) outside the unit circle and it is m_2 or m_1, according as $m_1 \leq m_2$ or $m_1 > m_2$. Inverting (79) gives

$$w_q(0) = \frac{g(0)}{g(1)} \prod_{i=1}^{m} (1 - 1/s_i)$$

and

$$w_q(n) = -\sum_{i=1}^{m} D_i/s_i^{n+1}, \quad n \geq 1$$

where

$$D_i = \frac{g(s_i)}{g(1)} C_i.$$

The mean and variance of W_q may be obtained from (79) and are given, respectively, by

$$E(W_q) = \frac{1}{g(1)} \left[g^{(1)}(1) + \sum_{i=1}^{m} \frac{1}{s_i - 1} \right] \tag{80}$$

and

$$\text{Var}(W_q) = \frac{1}{g(1)} \left[g^{(2)}(1) + g^{(1)}(1) - (g^{(1)}(1))^2 + \sum_{i=1}^{m} \frac{s_i}{(s_i - 1)^2} \right]. \tag{81}$$

Case 2. $A(1/s)$ *rational*

Let $A(1/s)$ be a rational function, say $u(s)/v(s)$, where $u(s)$ and $v(s)$ have no common factors. Suppose further that $v(s)$ is of degree n_2 and $u(s)$ of degree n_1. As $A(1/s)$ is convergent in $|s| \geq 1$, it implies that all the n_2 zeros of $v(s)$ must lie inside the unit circle. Therefore (67) reduces to

$$W_q(s) = \prod_{i=1}^{n} \left(\frac{1-s_i}{s-s_i} \right)$$

which is of the same form as that given in (71), where n is the number of roots of (69) outside the unit circle and which are determined, in each case, by the degrees of various functions involved in the numerator and denominator of (67).

Case 3. $A(1/s)$ and $B(s)$ rational

Let both $A(1/s)$ and $B(s)$ be rational functions, then from cases (i) and (ii) above, it follows that $W_q(s)$ is of the same form as (79), where m is the number of roots of (69) outside the unit circle and these roots are determined, once again, depending on the degrees of various functions involved in the numerator and denominator of (67).

4.4 Approximations for tail probabilities

In (75), we have an exact expression for the probability distribution $w_q(n)$ in terms of outside roots. The tail probabilities (see e. g. [Feller, 1968, pp.276-277]) can be approximated by a single term corresponding to the root which is smallest in absolute value, say s_1 which is not only positive and real, but close to unity. A proof for the existence of this root is given in Theorem 3, Chaudhry et al. [Chaudhry et al., 1990]. It should, however, be pointed out here that it is only the far tail that can be approximated very accurately by this single term s_1. To obtain the tail probabilities, assume

$$w_q(n) \simeq \frac{-C_1}{s_1^{n+1}} = w_{aq}^1(n), \quad n \geq n_\epsilon^1 \gg K_2, \quad n = kN, \quad k \in [1, 2, ...] \quad (82)$$

where n_ϵ^1 is chosen as the smallest integer such that $|(w_q(n) - w_{aq}^1(n))/w_q(n)| < \epsilon$, i.e. $|1 - w_{aq}^1(n)/w_q(n)| < \epsilon$ and N is the greatest common divisor of the interarrival and service times. However, since the probability $w_{aq}^1(n)$ follows a geometric distribution with the common ratio of non-zero probabilities $1/s_1^N$, it is better to choose n_ϵ^1 such that $|s_1^N w_q(n)/w_q(n-N) - 1| < \epsilon$. In this section, though no tables are being presented computations were done using the latter criterion. Moreover, the approximation gets better if more than one root, in ascending order of magnitude, is used. It should, however, be mentioned that those roots that occur in complex-conjugate pairs should be used in pairs, Thus, the tail probabilities can be approximated, using three roots, by

$$w_{aq}^3(n) \simeq -\sum_{i=1}^{3} \frac{C_i}{s_i^{n+1}}, \quad n \geq n_\epsilon^3 \gg K_2, \quad n = kN, \quad k \in [1, 2, ...]$$

where s_i ($i = 1, 2, 3$) are roots in ascending order of magnitude. Denoting the approximation using three roots by $w_{aq}^3(n)$, choose n_ϵ^3 such that $|1 - w_{aq}^3(n)/w_q(n)| < \epsilon$. It may, however, be mentioned that $w_{aq}^3(n)$ does not follow a geometric distribution, n_ϵ^3 may be chosen according to this criterion.

For the discrete-time queue GI/G/1, Kobayashi [Kobayashi, 1983] shows that the queueing-time distribution of a customer is geometric for $\rho \simeq 1$, i.e.

$$w_q(n) \simeq \left(1 - \frac{1}{s_0^N}\right)\left(\frac{1}{s_0}\right)^n, \quad n = 0, N, 2N, ... \quad (83)$$

where

$$s_0 = 1 + \frac{2a(1-\rho)}{\sigma_a^2 + \sigma_b^2}, \tag{84}$$

σ_a^2 and σ_b^2 being the variances of the interarrival and service times, respectively, and $a = E(A)$ the mean of the interarrival time. In the case of Kobayashi [Kobayashi, 1983], $N = 1$. Result (83) corresponds to the continuous-time result for GI/G/1.

The derivation of (83) given by Kobayashi is very involved. We give here a simpler derivation. Let

$$w_q(n) = \frac{-C_n}{s_0^{n+1}}, \quad n = 0, N, 2N, \ldots$$

where C_0 is the value of C_1 corresponding to the root s_0. One can see from (84) that as $\rho \to 1^-$, $s_0 \to 1^+$. In view of this, identifying C_1 as C_0 and s_1 as s_0 we can see from (73) that $C_0 \to C(1-s_0)$, $C_i \to 0, i \geq 2$, where C is obtained from normalization. Let this approximation be denoted by $w_{aq}^K(n)$.

Some comments on how good the approximations are, using s_1 versus s_0, seem in order. In the derivations of the numerator of (84) on the assumption that $(1-\rho) \simeq 0$, two positive terms are ignored. If this assumption is not made, then the root, say s_1', will be smaller than the root s_0. Using the original equation (69), we can refine the root s_1' to obtain root s_1. In view of these comments, we can say that the root, which can be easily obtained from (84), serves as an upper bound of the root s_1. Further, the result in equation (82) which holds for all ρ should be considered as a *refinement* over the result in (83) which holds only for high ρ. In fact, $w_{aq}^3(n)$, in general, gives better results than $w_{aq}^1(n)$, which, in turn, gives better results than $w_{aq}^K(n)$.

If (75) and (82) are used, the results obtained are extremely accurate. Using (82), the sum of the probabilities and moments, taking into account the infinite tail after $n = M + 1$, are given below:

$$\text{Sum of Probabilities} = \sum_{n=0}^{n_\epsilon^1} w_q(i) + w_q(n_\epsilon^1)\left[\frac{1}{s_1^N - 1}\right] \tag{85}$$

$$E(W_q) = \sum_{i=0}^{n_\epsilon^1} i w_q(i) + w_q(n_\epsilon^1)\left[\frac{n_\epsilon^1}{s_1^N - 1} + \frac{N s_1^N}{(s_1^N - 1)^2}\right] \tag{86}$$

$$\text{Var}(W_q) = \sum_{i=0}^{n_\epsilon^1} i^2 w_q(i) +$$

$$w_q(n_\epsilon^1)\left[\frac{(n_\epsilon^1)^2}{s_1^N - 1} + \frac{2n_\epsilon^1 N s_1^N}{(s_1^N - 1)^2} + \frac{N^2 s_1^N(s_1^N + 1)}{(s_1^N - 1)^3}\right]$$
$$- E^2(W_q). \tag{87}$$

4.5 Special analytical results

In order to show the unification which can be achieved by using the method described here, we consider two special cases for which the solutions are available. Besides, it may be remarked that when geometric distributions are assumed such as $q^i p$, $i \geq 0$ and $q^{i-1}p$, $i \geq 1$ for interarrival or service times, the former allows batches whereas the latter does not.

Case 1. Geom/G/1

It is interesting to note that in this case $W_q(s)$ can be written in a form that does not involve roots. Let the interarrival-time distribution be geometric and given by

$$a_i = q^i p, \quad i \geq 0 \quad \text{with} \quad p + q = 1.$$

Therefore

$$A(1/s) = \frac{ps}{s-q}, \quad |q/s| < 1.$$

Equation (69) now reduces to

$$s - q - psB(s) = 0$$

which has K_2 roots outside and one on the unit circle. This implies that

$$s - q - psB(s) \equiv -pb_{K_2}(s-1)\prod_{i=1}^{K_2}(s-s_i).$$

Therefore, from (70), we get

$$\begin{aligned} W_q(s) &= \frac{-Kpb_{K_2}(s-1)}{s-q-psB(s)} \\ &= \frac{(1-\rho)(s-1)q}{s-q-psB(s)} \end{aligned} \quad (88)$$

where $(1-\rho)q$ is the normalization constant with the utilization factor ρ being $pE(S)/q$. Now, from (88) we get

$$w_q(0) = 1 - \rho \quad (89)$$

which is the expression for $B(0)$ given in Kobayashi [Kobayashi, 1983, p.60].

Case 2. GI/Geom/1

Let the service-time distribution be

$$b_j = \beta^{j-1}\alpha, \quad j \geq 1 \quad \text{with} \quad \alpha + \beta = 1$$

so that
$$B(s) = \frac{\alpha s}{1 - \beta s}, \quad |s| < 1/\beta.$$
Therefore, from (79),
$$W_q(s) = \frac{(1-\beta s)(1-s_1)}{(1-\beta)(s-s_1)}$$
where s_1 is the unique root outside the unit circle of the equation
$$(1-\beta s) - \alpha s A(1/s) = 0.$$
From $W_q(s)$, we get
$$w_q(0) = \frac{1}{\alpha}\left(1 - \frac{1}{s_1}\right)$$
and
$$w_q(n) = \frac{1}{\alpha}\left(1 - \frac{1}{s_1}\right)\left(\frac{1-\beta s_1}{s_1^n}\right), \quad n \geq 1.$$
These results match equations (4.131) and (4.132) of Kobayashi [Kobayashi, 1983].

4.6 Numerical examples

Case 1.

First, we consider a simple numerical example. Consider example 7.1 of Gross and Harris [Gross and Harris, 1985, p.413]. Let
$$a_{10} = 2/5 \quad a_{15} = 3/5 \quad b_9 = 2/3 \quad b_{12} = 1/3$$
so that
$$W_q(s) = \frac{15M(s)(1-s)}{15s^{15} - 2s^{17} - 4s^{14} - 3s^{12} - 6s^9}.$$
In this case, the characteristic equation (CE) is
$$s^9(2s^8 - 15s^6 + 4s^5 + 3s^3 + 6) = 0$$
which has 15 roots for s inside and on the unit circle as it should, and two roots outside. If we change s to $1/s$ in (89) and ignore the term s^9, we get the roots for $1/s$ 1.0000, 0.3885, -0.3481, -1.2681, $0.6353 \pm 1.0274i$, and $-0.5215 \pm 1.0296i$. Using (74) and (75), the following probabilities were obtained

$w_q(0)$	$w_q(1)$	$w_q(2)$	$w_q(3)$	$w_q(4)$	$w_q(5)$	$w_q(6)$
0.8243	0.0334	0.1128	0.0091	0.0156	0.0018	0.0022

...	$w_q(8)$...	$w_q(15)$	...	$w_q(28)$
	0.0003	2.497×10^{-7}		1.444×10^{-12}

with $E(W_q) \doteq 0.377189$ and standard deviation= $\sqrt{\text{Var}(W_q)} = 0.847574$.

Case 2.

Second, we consider an example requiring a large number of roots. Let the interarrival and service time distributions be uniform, i.e., $a_i = 1/900$, $i = 1, ..., 900$ and $b_i = 1/1000$, $i = 1, ..., 1000$ so that the CE (78) is

$$s^{1000} - \sum_{i=1}^{900} \frac{s^i}{900} \sum_{i=1}^{1000} \frac{s^{1000-i}}{1000} = 0.$$

In this case, the CE has 1000 roots for s inside and on the unit circle, and none outside. Using (74) and (75), the following probabilities were obtained

$w_q(0)$	$w_q(1)$	$w_q(8)$	$w_q(15)$	$w_q(18)$
0.16810	0.3909×10^{-3}	0.3917×10^{-3}	0.3930×10^{-3}	0.3936×10^{-3}

with $E(W_q) \doteq 1289.4696$ and standard deviation= $\sqrt{\text{Var}(W_q)} = 1494.7566$.

4.7 Overview

Van Ommeren [van Ommeren, 1991] considers the waiting-time distribution of the discrete-time queueing system GI/G/1 in which a customer arriving at an idle channel has a modified service time. Murata and Miyahara [Murata and Miyahara, 1991] consider the waiting-time distribution of $GI^X/G/1$ in discrete time. In their analysis, they assume that the p.g.f.'s of both interarrival- and service-time distributions be rational functions. Wolf [Wolf, 1982] deals with approximation and bounds for the waiting-time distribution and its moments. Yang and Chaudhry [Yang and Chaudhry, 1996] discuss queue-size distributions for the discrete-time model GI/G/1. In this connection, see also [Hasslinger, 1995] who discusses a polynomial factorization approach to the discrete-time queue-size distribution for the systems GI/G/1 and GI/G/1/N.

5 OVERVIEW AND CONCLUDING REMARKS

In this section, we briefly review some of the other models that have been discussed in the literature. As discussed earlier, for lack of space, we give only a brief account of some of these models.

Chaudhry and Gupta [Chaudhry and Gupta, 1997c, Chaudhry and Gupta, 1998] provide solutions to both the GI^X/Geom/1 queue with infinite buffer and the GI^X/Geom/1/N queue with finite buffer N. As arrivals are in batches and buffer size is finite, two cases arise according as (i) the batch-size is larger than the available buffer, in which case part of the batch is accepted and the rest is rejected, or (ii) the batch size is larger than the available buffer, in which case the whole batch is rejected. In brief, case (i) is referred to as partial rejection and case (ii) as total rejection. For more details, see Chaudhry and Gupta's papers referred above.

The $Geom^X/G/1$ queue has been studied extensively by Dafermos *et al.* [Dafermos et al., 1971], Kobayashi [Kobayashi, 1983], Hunter [Hunter, 1983],

Bruneel and Kim [Bruneel and Kim, 1993], Takagi [Takagi, 1993], Tsuchiya and Takahashi [Tsuchiya and Takahashi, 1993], and others. Generally, Geom^X/G/1 queue has been discussed under the assumption of LAS-DA, but Takagi also presents results for EAS. Almost all the authors with the exception of Chaudhry [Chaudhry, 1997] present the analytic results for these models in terms of p.g.f.'s of associated r.v.'s. Numerical aspects of models such as Geom^X/G/1 are now included in the *Chaudhry QPACK software package* [Chaudhry, 1992a]. For several variations of the model Geom^X/G/1, see [Bruneel, 1993] and [Takagi, 1993]. The numerical work on Geom^X/G/1/N can be done in a manner similar to the methods discussed earlier for some of the models.

The results for the queueing system Geom/G^B/1 (using different methods) are available in many places, see, e.g. [Chaudhry and Templeton, 1983]. For relations among state probabilities at various epochs in Geom/G^B/1, see [Chaudhry et al., 1997]. Some other aspects of this and related models are currently being worked out. An interested reader may write to the author. The distributions of number in the system D^X/D^m/1 has been discussed by Zhao and Campbell [Zhao and Campbell, 1996]. For specialized and related cases, see references in [Zhao and Campbell, 1996]. In this connection, see also [Madill et al., 1985] who discuss the model D/D^{a,b}/1.

The multiserver model GI/Geom/c (EAS) has been discussed by Chan and Maa [Chan and Maa, 1978] using the IMC, but the model GI/Geom/c/c (EAS and LAS-DA) has been discussed by Chaudhry and Gupta [Chaudhry and Gupta, 1997d]. Chan and Maa [Chan and Maa, 1978] do limited computations on the model GI/Geom/c. They present some graphs when the interarrival times are geometrically distributed. For more general cases such as deterministic input, the method proposed by Chan and Maa creates computational problems, particularly for high values of ρ and c. This is because there are alternating signs in the computations of sums. This problem is being looked at by the author and will be reported later.

In the special case of GI/Geom/c/c viz. Geom/Geom/c/c, Chan and Maa show how to get results for the continuous-time model M/M/c. For a model of the type Geom^X/D/c, see [Bruneel and Wuyts, 1994]. For numerical aspects of finding the loss probabilities of GI/G/2, see [Atkinson, 1995]. For some infinite-server discrete-time queues, see [Ushakumari, 1995] and references therein.

Acknowledgments

I would like to thank Professors U.C. Gupta, Department of Mathematics, Indian Institute of Technology, Kharagpur, and J.G.C. Templeton, Department of Mechanical and Industrial Engineering, University of Toronto for reading the first draft and making many suggestions that lead to significant improvements of the chapter. Special thanks are due to W. K. Grassmann of the Department of Computer Science, University of Saskatchewan, for proposing several changes that lead to significant improvements of the chapter. I am also grateful to Mr. Haynes Lee, Research Assistant, RMC, for his help in the preparation of this chapter. This research was supported, in part, by ARP.

Appendix

Case 1: The outside root s_1, with the smallest magnitude, of the characteristic equation (CE) (see denominator of (15)),

$$s - B(1 - \lambda + \lambda s) = 0 \tag{1}$$

can be approximated by

$$s_0 = 1 + \frac{2(1-\rho)}{\lambda^2 B^{(2)}(1)}$$

(see results after equation (45)).

Proof: Using the Taylor series expansion for $B(1 - \lambda - \lambda s)$ in s, we have

$$\begin{aligned} B(1 - \lambda + \lambda s) &= B(1) + (s-1)\lambda B^{(1)}(1) + \frac{(s-1)^2}{2!}\lambda^2 B^{(2)}(1) + o((s-1)^2) \\ &= 1 + (s-1)\rho + \frac{(s-1)^2}{2!}\lambda^2 B^{(2)}(1) + o((s-1)^2). \end{aligned} \tag{2}$$

Using this expansion in (2), we have

$$\begin{aligned} s - B(1 - \lambda + \lambda s) &= (s-1)(1-\rho) - \frac{(s-1)^2}{2!}\lambda^2 B^{(2)}(1) \\ &\quad + o((s-1)^2). \end{aligned} \tag{3}$$

Dropping $o((s-1)^2)$ from (3), we find that whereas one root of (1) is $s = 1$, the other root say s_0, is given by

$$s_0 = 1 + \frac{2(1-\rho)}{\lambda^2 B^{(2)}(1)}. \tag{4}$$

This root is positive, real, and > 1. Since we have neglected $o((s-1)^2)$, this root is the smallest of the $k-1$ roots of CE (1).

Case 2: Proceeding along the same lines as in Case 1 above, the outside root s_1', with the smallest magnitude, of the CE (given as denominator in (26))

$$1 - \lambda - s + \lambda B(s) = 0 \tag{5}$$

can be approximated by

$$s_0' = 1 + \frac{2(1-\rho)}{\lambda B^{(2)}(1)} \tag{6}$$

which is also positive, real, and > 1 (see equation (65)).

References

[Akar and Arikan, 1996] Akar, N. and Arikan, E. (1996). A numerically efficient method for the MAP/D/1/K queue via rational approximations. *Queueing Systems*, 22:97–120.

[Alfa, 1982] Alfa, A. (1982). Time-inhomogeneous bulk server queue in discrete time: A transportation type problem. *Oper. Res.*, 30:650–658.

[Alfa, 1995a] Alfa, A. (1995a). A discrete MAP/PH/1 queue with vacations and exhaustive time-limited service. *Oper. Res. Lett.*, 18:31–40.

[Alfa, 1995b] Alfa, A. (1995b). Modelling traffic queues at a signalized intersection with vehicle-actuated control and Markovian arrival processes. *Comput. Math. Applic.*, 30:105–119. Errata: (1996), 31: 137.

[Alfa, 1998] Alfa, A. (1998). Matrix-geometric solution of discrete time MAP/PH/1 priority queue. *Naval Res. Logist.*, 45:23–50.

[Alfa and Chakravarthy, 1994] Alfa, A. and Chakravarthy, S. (1994). A discrete queue with the Markovian arrival process and phase type primary and secondary services. *Commun. Statist. Stochastic Models*, 10:437–451.

[Alfa et al., 1995] Alfa, A., Dolhun, K., and Chakravarthy, S. (1995). A discrete single server queue with Markovian arrivals and phase type group services. *J. of Applied Math. and Stochastic Analysis*, 8:151–176.

[Alfa and Neuts, 1995] Alfa, A. and Neuts, M. (1995). Modelling vehicular traffic using the discrete time Markovian arrival process. *Transportation Science*, 29:109–117.

[Alfa and Frigui, 1996] Alfa, A. and Frigui, I. (1996). Discrete NT-policy single server queue with Markovian arrival process and phase type service. *European J. Opl. Res.*, 88:599–613.

[Asmussen, 1987] Asmussen, S. (1987). *Applied Probability and Queues*. John Wiley, NY.

[Atkinson, 1995] Atkinson, J. (1995). The general two-server queueing loss system: Discrete-time analysis and numerical approximation of continuous-time systems. *J. Opl. Res. Soc.*, 46:386–397.

[Bacelli and Bremaud, 1994] Bacelli, F. and Bremaud, P. (1994). *Elements of Queueing Theory: Palm-Martingale Calculus and Stochastic Recurrences*. Springer-Verlag, NY.

[Bhat, 1968] Bhat, N. (1968). A study of the queueing systems M/G/1 and GI/M/1. In Beckmann, M. and Künzi, H., editors, *Lecture Notes in Operations Research and Mathematical Economics*, volume 2. Springer-Verlag, NY.

[Bhat and Basawa, 1992] Bhat, U. and Basawa, I. (1992). *Queueing and Related Models*. Clarendon, Oxford.

[Blondia, 1989] Blondia, C. (1989). The $N/G/1$ finite capacity queue. *Commun. Statist.- Stochastic Models*, 5:273–294.

[Blondia, 1993] Blondia, C. (1993). A discrete-time batch Markovian arrival process as B-ISDN traffic model. *Belgian Journal of Operations Research, Statistics and Computer Science*, 32(3, 4):3–23.

[Blondia and Theimer, 1989] Blondia, C. and Theimer, T. (1989). A discrete-time model for ATM traffic. working paper. Philips Research Laboratory, Brussels, Belgium.

[Böhm et al., 1997] Böhm, W., Krinik, A., and Mohanty, S. (1997). The combinatorics of birth-death processes and applications to queues. *Queueing Systems*, 26:255–267.

[Böhm and Mohanty, 1993] Böhm, W. and Mohanty, S. (1993). The transient solution of M/M/1 queues under (m,n)-policy. A combinatorial approach. *Journal of Statistical Planning and Inference*, 34:23–33.

[Böhm and Mohanty, 1994a] Böhm, W. and Mohanty, S. (1994a). On discrete-time Markovian n-policy queues involving batches. *Sankhyā*, 56, Ser. A:144–163.

[Böhm and Mohanty, 1994b] Böhm, W. and Mohanty, S. (1994b). On random walks with barriers and their applications to queues. *Studia Sci. Math. Hungar.*, 29:397–413.

[Böhm and Mohanty, 1994c] Böhm, W. and Mohanty, S. (1994c). Transient analysis of M/M/1 queues in discrete time with general server vacations. *J. Appl. Probab.*, 31A:115–129. In *Studies in Probability (Essays in honour of Lajos Takács)*.

[Böhm and Mohanty, 1994d] Böhm, W. and Mohanty, S. (1994d). Transient analysis of queues with heterogeneous arrivals. *Queueing Systems*, 18:27–46.

[Boucherie and van Dijk, 1991] Boucherie, R. and van Dijk, N. (1991). Product forms for queueing networks with state-dependent multiple job transitions. *Adv. in Appl. Probab.*, 23:152–187.

[Boxma and Syski, 1988] Boxma, O. and Syski, R. (1988). *Queueing Theory and its Applications Liber Amicorum for J.W. Cohen*. North-Holland, Amsterdam.

[Bruneel, 1986] Bruneel, H. (1986). A general treatment of discrete-time buffers with one randomly interrupted output line. *European J. Opl. Res.*, 27:67–81.

[Bruneel, 1993] Bruneel, H. (1993). Performance of discrete-time queueing systems. *Comput. Oper. Res.*, 20:303–320.

[Bruneel and Kim, 1993] Bruneel, H. and Kim, B. (1993). *Discrete-Time Models for Communication Systems Including ATM*. Kluwer, Boston.

[Bruneel and Wuyts, 1994] Bruneel, H. and Wuyts, I. (1994). Analysis of discrete-time multiserver queueing models with constant service times. *Oper. Res. Lett.*, 15:231–236.

[Bunday, 1986] Bunday, B. (1986). *Basic Queuing Theory*. Edward Arnold, London.

[Bunday, 1996] Bunday, B. (1996). *An Introduction To Queueing Theory.* Edward Arnold, London.

[Chan and Maa, 1978] Chan, W. and Maa, D. (1978). The GI/Geom/N queue in discrete time. *INFOR*, 16:232-252.

[Chang and Chang, 1984] Chang, J. and Chang, R. (1984). The application of the residue theorem to the study of a finite queue with batch Poisson arrivals and synchronous servers. *SIAM J. Appl. Math.*, 44:646-656.

[Chaudhry, 1965] Chaudhry, M. (1965). Correlated queueing. *J. of the Canad. Operat. Res. Soc.*, 3:142-151.

[Chaudhry, 1992a] Chaudhry, M. (1992c). *QPACK Software Package.* A&A Publications, 395 Carrie Crescent, Kingston, Ontario K7M 5X7, Canada.

[Chaudhry, 1992b] Chaudhry, M. (1992b). *QROOT Software Package.* A&A Publications, 395 Carrie Crescent, Kingston, Ontario K7M 5X7, Canada.

[Chaudhry, 1997] Chaudhry, M. (1997). On the discrete-time bulk-arrival queues with finite- and infinite-waiting spaces. Working Paper. Dept. of Math. & Computer Science, Royal Military College of Canada, Kingston, Ontario, Canada.

[Chaudhry and Gupta, 1996a] Chaudhry, M. and Gupta, U. (1996a). On the analysis of the discrete-time Geom(n)/G(n)/1/N queue. *Probab. Engrg. Inform. Sci.*, 10:59-68.

[Chaudhry and Gupta, 1996b] Chaudhry, M. and Gupta, U. (1996b). Performance analysis of the discrete-time GI/Geom/1/N queue. *J. Appl. Probab.*, 33:239-255.

[Chaudhry and Gupta, 1997c] Chaudhry, M. and Gupta, U. (1997c). Queue-length and waiting-time distributions of discrete-time GI^X/Geom/1 queueing systems with early and late arrivals. *Queueing Systems*, 25:307-334.

[Chaudhry and Gupta, 1997d] Chaudhry, M. and Gupta, U. (1997d). Algorithmic discussions of distributions of numbers of busy channels for GI/Geom/m/m queues. Working Paper. Dept. of Math. & Computer Science, Royal Military College of Canada, Kingston, Ontario, Canada.

[Chaudhry and Gupta, 1998] Chaudhry, M. and Gupta, U. (1998). Performance analysis of discrete-time finite-buffer batch-arrival GI^X/Geom/1/N queues. *Discrete Event Dynamic Systems*, 8:55-70.

[Chaudhry et al., 1997] Chaudhry, M., Gupta, U. and Goswami, V. (1997). Relations among state probabilities in discrete-time bulk-service queues with finite- and infinite-waiting spaces. Working Paper. Dept. of Math. & Computer Science, Royal Military College of Canada, Kingston, Ontario, Canada.

[Chaudhry et al., 1990] Chaudhry, M., Harris, C., and Marchal, W. (1990). Robustness of rootfinding in single-server queueing models. *ORSA J. Computing*, 2:273-286.

[Chaudhry and Templeton, 1968] Chaudhry, M. and Templeton, J. (1968). On the discrete-time queue length distribution in a bulk service considering correlated arrivals. *Canad. Operat. Res. Soc.*, 2:79-88.

[Chaudhry and Templeton, 1983] Chaudhry, M. and Templeton, J. (1983). *A First Course on Bulk Queues*. NY: John Wiley.

[Chaudhry et al., 1996] Chaudhry, M., Templeton, J. and Gupta, U. (1996). Analysis of the discrete-time GI/Geom(n)/1/N queue. *Comput. Math. Applic.*, 31:59–68.

[Chaudhry et al., 1992] Chaudhry, M., Templeton, J. and Medhi, J. (1992). Computational analysis of multiserver bulk-arrival queues with constant service time $M^X/D/c$. *ORSA*, 40 (Supp. 2):229–238.

[Chaudhry and van Ommeren, 1997] Chaudhry, M. and van Ommeren, J. (1997). Analytically explicit results for the transient solutions of one-sided skip-free finite Markov chains. Working Paper. Dept. of Math. & Computer Science, Royal Military College of Canada, Kingston, Ontario, Canada.

[Chaudhry and Zhao, 1994] Chaudhry, M. and Zhao, Y. (1994). First-passage-time and busy period distributions of discrete-time Markovian queues: Geom(n)/Geom(n)/1/N. *Queueing Systems*, 18:5–26.

[Chaudhry and Zhao, 1997] Chaudhry, M. and Zhao, Y. (1997). Transient solutions for Markov chains having lower Hessenberg transition probability matrices and their applications. Working Paper. Dept. of Math. & Computer Science, Royal Military College of Canada, Kingston, Ontario, Canada.

[Choneyko et al., 1993] Choneyko, I., Mohanty, S., and Sen, K. (1993). Enumeration of restricted three-dimensional lattice paths with fixed number of turns and an application. *J. Statistical Planning and Inference*, 34:57–62.

[Choukri, 1993] Choukri, T. (1993). The transient blocking probabilities in $M/M/N$ systems via large deviations. *Advances in Applied Probab.*, 25:483–486.

[Cohen, 1982] Cohen, J. (1982). *The Single Server Queue*. North-Holland, Amsterdam, second edition.

[Cooper, 1981] Cooper, R. (1981). *Introduction to Queueing Theory*. North-Holland, Amsterdam, second edition.

[Daduna, 1996] Daduna, H. (1996). Discrete time queueing networks: Recent developments. *Tutorial Lecture Notes, Performance '96*, 8:1–42. Lausanne.

[Dafermos et al., 1971] Dafermos, S., Stella, C., and Neuts, M. (1971). A single server queue in discrete time. *Cahiers du Centre de Rech. Opér*, 13:23–40.

[Daigle, 1992] Daigle, J. (1992). *Queueing Theory for Telecommunications*. Addison-Wesley, NY.

[Feller, 1968] Feller, W. (1968). *An Introduction to Probability Theory and Its Applications*, volume 1. John Wiley, NY, third edition.

[Frigui et al., 1997] Frigui, I., Alfa, A, and Xu, X. (1997). Algorithms for computing waiting time distributions under different queue disciplines for the D-BMAP/PH/1. *Naval Res. Logist.*, 44:559–576.

[Garcia and Casals, 1989] Garcia, J. and Casals, O. (1989). A discrete time queueing model to evaluate cell delay variation in an ATM network. working paper. Computer Architecture Dept., Polytechnic University of Catalonia, Spain.

[Gnedenko and Kovalenko, 1989] Gnedenko, B. and Kovalenko, I. (1989). *Introduction to Queueing Theory*. Birkhäuser, Boston, MA.

[Gopinath and Morrison, 1977] Gopinath, R. and Morrison, J. (1977). Discrete-time single server queues with correlated inputs. In Chandy, K. and Reiser, M., editors, *Computer Performance*, pages 263–278. North-Holland, Amsterdam.

[Gouweleeuw, 1994] Gouweleeuw, F. (1994). The loss probability in finite-buffer queues with batch arrivals and complete rejection. *Probab. Engrg. Inform. Sci*, 8:221–227.

[Grassmann and Jain, 1989] Grassmann, W. and Jain, J. (1989). Numerical solutions of the waiting time distribution and idle time distribution of the arithmetic GI/G/1 queue. *Oper. Res.*, 37:141–150.

[Gravey and Hébuterne, 1992] Gravey, A. and Hébuterne, G. (1992). Simultaneity in discrete time single server queues with Bernoulli inputs. *Perf. Eval.*, 14:123–131.

[Gravey et al., 1990] Gravey, A., Louvion, J., and Boyer, P. (1990). On the Geo/D/1 and Geo/D/1/n queues. *Perf. Eval.*, 11:117–125.

[Gross and Harris, 1985] Gross, D. and Harris, C. (1985). *Fundamentals of Queueing Theory*. John Wiley, NY, second edition.

[Hashida et al., 1991] Hashida, O., Takahashi, Y., and Shimogawa, S. (1991). Switched batch Bernoulli process (SBBP) and the discrete-time SBBP/G/1 queue with applications to statistical multiplexer performance. *IEEE J. Selected Areas Commun.*, 9:394–401.

[Hasslinger, 1995] Hasslinger, G. (1995). A polynomial factorization approach to the discrete time $GI/G/1/(N)$ queue size distribution. *Perf. Eval.*, 23:217–240.

[Hasslinger and Rieger, 1996] Hasslinger, G. and Rieger, E. (1996). Analysis of open discrete time queueing networks: A refined decomposition approach. *J. Opl. Res. Soc.*, 47:640–653.

[Heimann and Neuts, 1973] Heimann, D. and Neuts, M. (1973). The single server queue in discrete time. Numerical analysis IV. *Nav. Res. Logist.*, 20:753–766.

[Henderson et al., 1995] Henderson, W., Pearce, C., Taylor, P., and van Dijk, N. (1995). Insensitivity in discrete-time generalized semi-Markov processes allowing multiple events and probabilistic service scheduling. *Ann. Appl. Probab.*, 5:78–96.

[Heyman and Sobel, 1982] Heyman, D. and Sobel, M.J. (1982). *Stochastic Models in Operations Research*, volume 1. McGraw-Hill, NY.

[Hunter, 1983] Hunter, J. (1983). *Discrete Time Models: Techniques and Applications*, volume 2 of *Mathematical Techniques of Applied Probability*. Academic Press, NY.

[Ishizaki et al., 1994a] Ishizaki, F., Takine, T., and Hasegawa, T. (1994a). Analysis of a discrete-time queue with a gate. In Labetoulle, J. and Roberts, J., editors, *The Fundamental Role of Teletraffic in the Evolution of Telecommunications Networks*, pages 169–178. Elsevier, Amsterdam.

[Ishizaki et al., 1994b] Ishizaki, F., Takine, T., Takahashi, Y., and Hasegawa, T. (1994b). A generalized SBBP/G/1 queue and its applications. *Perf. Eval.*, 21:163–181.

[Johnson and Narayana, 1996] Johnson, M. and Narayana, S. (1996). Descriptors of arrival-process burstiness with applications to the discrete Markovian arrival process. *Queueing Systems*, 23:107–130.

[Kalashnikov, 1994] Kalashnikov, V. (1994). *Mathematical Methods in Queuing Theory*. Kluwer, Amsterdam.

[Kashyap and Chaudhry, 1988] Kashyap, B. and Chaudhry, M. (1988). *An Introduction to Queueing Theory*. A&A Publications, Kingston, Ontario.

[Kinney, 1962] Kinney, J. (1962). A transient discrete time queue with finite storage. *Ann. Math. Stat.*, 33:130–106.

[Kleinrock, 1975] Kleinrock, L. (1975). *Queueing Systems: Theory*, volume 1. John Wiley, NY.

[Klimko and Neuts, 1973] Klimko, E. and Neuts, M. (1973). The single server queue in discrete time. Numerical analysis II. *Nav. Res. Logist.*, 20:305–319.

[Knessl, 1990] Knessl, C. (1990). On the transient behavior of the M/M/m/m loss model. *Communications in Statistics- Stochastic Models*, 6:749–776.

[Kobayashi, 1983] Kobayashi, H. (1983). Discrete-time queueing systems. In Louchard, G. and Latouche, G., editors, *Probability Theory and Computer Science*, pages 53–121. Academic Press, NY.

[Konheim, 1975] Konheim, A. (1975). An elementary solution of the queueing system G/G/1. *SIAM J. Comput.*, 4:540–545.

[Kouvatsos, 1994] Kouvatsos, D. (1994). Entropy maximisation and queueing network models. *Ann. Oper. Res.*, 48:63–126.

[Kouvatsos, 1995] Kouvatsos, D. (1995). *Performance Modelling and Evaluation of ATM Networks*, volume 1. Chapman & Hall, London.

[Kouvatsos, 1996] Kouvatsos, D. (1996). *Performance Modelling and Evaluation of ATM Networks*, volume 2. Chapman & Hall, London.

[Kouvatsos and Fretwell, 1995] Kouvatsos, D. and Fretwell, R. (1995). Closed form performance distributions of a discrete time $GI^G/D/1/N$ queue with correlated traffic. In *Proceedings of the Sixth IFIP WG6.3 Conference of Computer Networks*, pages 141–163, Istanbul, Turkey.

[Kuehn, 1979] Kuehn, P. (1979). Approximate analysis of general queueing networks by decomposition. *IEEE Trans. Commun.*, 27:113–126.

[Lipsky, 1992] Lipsky, L. (1992). *Queueing Theory*. Macmillan, NY.

[Liu and Mouftah, 1997] Liu, X. and Mouftah, H. (1997). Queueing performance of copy networks with dynamic cell splitting for multicast ATM switching. *IEEE Trans. Selected Areas Commun.*, 45:464–472.

[Liu and Neuts, 1994] Liu, D. and Neuts, M. (1994). A queueing model for an ATM control scheme. *Telecommunication Systems*, 2:321–348.

[Lucantoni, 1991] Lucantoni, D. (1991). New results on the single server queue with a batch Markovian arrival process. *Commun. Statist.– Stochastic Models*, 7:1–46.

[Madill et al., 1985] Madill, B., Chaudhry, M., and Buckholtz, P. (1985). On the diophantine queueing system, $D/D^{a,b}/1$. *J. Opl. Res. Soc.*, 36:531–535.

[Medhi, 1991] Medhi, J. (1991). *Stochastic Models in Queueing Theory*. Academic Press, NY.

[Meisling, 1958] Meisling, T. (1958). Discrete-time queueing theory. *Oper. Res.*, 6: 96-105.

[Mitra and Weiss, 1988] Mitra, D. and Weiss, A. (1988). On the transient behavior in Erlang's model for large trunk groups and various traffic conditions. In *Proceedings of the 12th International Teletraffic Congress*, volume 5.1B4.1–5.1B4.8. Torino, Italy.

[Miyazawa and Takagi, 1994] Miyazawa, M. and Takagi, H. (1994). Advances in discrete time queues. *Queueing Systems*, 18(1-2).

[Mohanty, 1991] Mohanty, S. (1991). On the transient behavior of a finite discrete time birth-death process. *Assam Statistical Review*, 5:1–7.

[Mohanty and Panny, 1989] Mohanty, S. and Panny, W. (1989). A discrete-time analogue of the M/M/1 queue and the transient solution. An analytic approach. In Revesz, P., editor, *Coll. Math. Soc. János Bolyai*, volume 57 of *Limit Theorems in Probab. and Statist.*, pages 417–424.

[Mohanty and Panny, 1990] Mohanty, S. and Panny, W. (1990). A discrete-time analogue of the M/M/1 queue and the transient solution: A geometric approach. *Sankhyā*, 52(Ser. A):364–370.

[Murata and Miyahara, 1991] Murata, H. and Miyahara, H. (1991). An analytic solution of the waiting time distribution for the discrete-time $GI/G/1$ queue. *Perf. Eval.*, 13:87–95.

[Neuts, 1973] Neuts, M. (1973). The single server queue in discrete time. Numerical analysis I. *Nav. Res. Logist.*, 20:297–304.

[Neuts and Klimko, 1973] Neuts, M. and Klimko, E. (1973). The single server queue in discrete time. Numerical analysis III. *Nav. Res. Logist.*, 20:557–567.

[Park and Perros, 1994] Park, D. and Perros, H. (1994). m-MMBP characterization of the departure process of an m-MMBP/Geo/1/K queue. *ITC*, 14:75–84.

[Park et al., 1994] Park, D., Perros, H., and Yamahita, H. (1994). Approximate analysis of discrete-time tandem queueing networks with bursty and correlated input and customer loss. *Oper. Res.*, 15:95–104.

[Pestien and Ramakrishnan, 1994] Pestien, V. and Ramakrishnan, S. (1994). Features of some discrete-time cyclical queueing networks. *Queueing Systems*, 18:117–132.

[Prabhu, 1965] Prabhu, N. (1965). *Queues and Inventories: A Study of Their Basic Stochastic Processes*. John Wiley, NY.

[Pujolle et al., 1986] Pujolle, G., Claude, J., and Seret, D. (1986). A discrete queueing system with a product form solution. *Computer Networking and Performance Evaluation*, pages 139–147.

[Robertazzi, 1990] Robertazzi, T. (1990). *Computer Networks and Systems: Queueing Theory and Performance Evaluation*. Springer-Verlag, NY.

[Sharma, 1990] Sharma, O. (1990). *Markovian Queues*. Ellis Harwood, NY.

[Srivastava and Kashyap, 1982] Srivastava, H. and Kashyap, B. (1982). *Special Functions in Queueing Theory and Related Stochastic Processes*. Academic Press, NY.

[Takács, 1971] Takács, L. (1971). Discrete queues with one server. *J. Appl. Probab.*, 8:691–707.

[Takagi, 1993] Takagi, H. (1993). *Queueing Analysis – A Foundation of Performance Evaluation: Discrete-Time Systems*, volume 3. North-Holland, Amsterdam.

[Tijms, 1986] Tijms, H. (1986). *Stochastic Modelling and Analysis: A Computational Approach*. John Wiley, NY.

[Tjhie and Rzehak, 1996] Tjhie, D. and Rzehak, H. (1996). Analysis of discrete-time TES/G/1 and TES/D/1-K queueing systems. *Perf. Eval.*, 27&28:367–390.

[Tran-Gia et al., 1995] Tran-Gia, P., Blondia, C., and Towsley, D. (1995). Discrete-time models and analysis methods. *Perf. Eval.*, 21(1-2).

[Tsuchiya and Takahashi, 1993] Tsuchiya, T. and Takahashi, Y. (1993). On discrete-time single-server queues with Markov modulated batch arrival Bernoulli input and finite capacity. *J. Oper. Res. Soc. Japan*, 36:22–45.

[Ushakumari, 1995] Ushakumari, P. (1995). *Analysis of Some Infinite-Server Queues and Related Optimization Problems*. PhD thesis, The Cochin University of Science and Technology, Cochin - 682 022, India.

[van Dijk, 1990] van Dijk, N. (1990). An insensitive product form for discrete-time communication networks. *Performance '90*, pages 77–89.

[van Doorn and Schrijner, 1996] van Doorn, E. and Schrijner, P. (1996). Limit theorems for discrete-time Markov chains of the nonnegative integers conditioned on recurrence to zero. *Communi Statists.- Stochastic Models*, 12:77–102.

[van Ommeren, 1991] van Ommeren, J. (1991). The discrete-time single-server queue. *Queueing Systems*, 8:279–294.

[Wally and Viterbi, 1996] Wally, S. and Viterbi, A. (1996). A tandem of discrete-time queues with arrivals and departures at each stage. *Queueing Systems*, 23:157–176.

[White et al., 1975] White, J., Schmidt, J., and Bennett, C. (1975). *Analysis of Queueing Systems*. Academic Press, NY.

[Wolf, 1982] Wolf, D. (1982). Approximating the stationary waiting time distribution function of $GI/G/1$-queues with arithmetic interarrival time and service time distribution function. *Oper. Res. Spektrum*, 4:135–148.

[Woodward, 1994] Woodward, M. (1994). *Communication and Computer Networks: Modelling with Discrete-Time Queues*. California IEEE Computer Society Press, Los Alamitos, CA.

[Xie and Knessl, 1993] Xie, M. and Knessl, C. (1993). On the transient behavior of the Erlang loss model: Heavy usage asymptotics. *SIAM J. Appl. Math.*, 53:555–599.

[Yang and Chaudhry, 1996] Yang, T. and Chaudhry, M. (1996). On the steady-state queue size distributions of the discrete-time GI/G/1 queue. *Adv. Appl. Probab.*, 28:1177–1200.

[Yang and Li, 1995] Yang, T. and Li, H. (1995). On the steady-state queue size distribution of the discrete-time Geo/G/1 queue with repeated customers. *Queueing Systems*, 21:199–215.

[Yoshimoto et al., 1993] Yoshimoto, M., Takine, T., Takahashi, Y., and Hasegawa, T. (1993). Waiting time and queue length distributions for go-back-n and selective-repeat ARQ protocols. *IEEE Trans. Commun.*, 41:1687–1693.

[Zhao and Campbell, 1996] Zhao, Y. and Campbell, L. (1996). Equilibrium probability calculations for a discrete-time bulk queue model. *Queueing Systems*, 22:189–198.

Mohan L. Chaudhry joined Kurukshetra University as a lecturer in 1963. He left India in 1966 to join as a Senior Post-doctoral fellow in the Department of Industrial Engineering, University of Toronto, Toronto. In 1967, Dr. Chaudhry joined the Department of Mathematics and Computer Science at the Royal Military College of Canada, Kingston, Ontario, where he currently holds a professorship. He has served the college in various capacities including as Head of the Department during 1990-91. In 1995-96, he became a Distinguished Professor of the Department. He has also served and continues to serve in various professional societies in different capacities such as member and/or Associate Editor of Editorial Boards. He has won several awards and grants and has held appointments at several universities in India, Canada and USA. Currently, he is an adjunct professor at DalTech, Nova Scotia, Queen's University, Kingston and University of Toronto, Toronto. Having published over 5 dozen articles in numerous international Operations Research and Statistics journals, he has coauthored two books: (i) A First Course in Bulk Queues, 1983, Wiley, New York and (ii) An Introduction to Queueing Theory, 1988, A & A Publications, 395 Carrie Crescent, Kingston, Ontario, K7M 5X7 (phone/fax: 613-389-7697). In addition, he has edited or coedited special issues of several journals and has prepared software packages for several queueing models.

11 THE PRODUCT FORM TOOL FOR QUEUEING NETWORKS

Nico M. van Dijk[1] and Winfried Grassmann[2]

[1]Department of Econometrics and Operations Research
University of Amsterdam
The Netherlands
nivd@fee.uva.nl

[2]Department of Computer Science
University of Saskatchewan
Saskatoon, Sask S7N 5A9
Canada
grassman@cs.usask.ca

1 INTRODUCTION

Queueing networks are used widely as modelling and evaluation tools in manufacturing, telecommunications, computer networking, and related areas. Much of the research effort has been devoted to so-called Jackson networks, that is, networks with Poisson arrivals, exponential service times and routing independent of the state of the system and the history of the customer. The steady-state distribution of Jackson networks can be expressed in a so-called product form. This computationally attractive form will be shown to be directly related to the principle of balance per station. This principle will be used to provide practical insights concerning the following questions

1. When can a product form be expected?

2. Why is this product form often violated in practice?

3. How can one restore a product form to obtain simple bounds?

These three aspects will be illustrated, in turn, as follows: first, standard assembly line queueing systems and Jacksonian networks are treated to demonstrate the derivation and characterization of product forms. More realistic cases with blocking, overflow and breakdowns are then introduced to demonstrate when and why the product form is violated, and how it can be restored. Finally, modifications are introduced to obtain simple performance bounds. The quality of these bounds will then be illustrated by numerical examples for a number of simple generic networks. The extension to more complicated networks will also be addressed briefly and illustrated by numerical examples.

Queueing networks are of particular importance in the following three major application areas:

1. In production or manufacturing systems, parts and tools are transported along workstations in order to be processed at each station. This creates interdependencies between the workstations. In particular, a saturation of a station with finite storage capacity may hold up several upstream stations.

2. In computer networking, jobs or programs from different users have to be executed, which may require several successive stages like: the input of data, the retrieval of files, the actual execution, and the storage or printing of output. These stages are usually highly interactive and different jobs (users) may have to contend for limited resources or capacities (e.g. a processor, a read/write device, a disk or printer).

3. In telecommunications, communication channels are set up between devices such as telephones, processors, personal computers, fax-machines, terminals, or satellites, using media such as copper wires, optical fibers or radio frequencies. As the exchanges and channels have limited capacities, resource contentions may arise.

The three application areas described above can be formulated as queueing networks with two essential characteristics:

1. Dependencies between the various intermediate stages (stations) and/or users (jobs).

2. Uncertainties (randomness) of service times, such as task, process, execution, call, or transmission times.

There are a number of typical performance measures for such networks, including the *throughput*, the *utilization*, the mean *delay* or *waiting time*, the mean *sojourn* or *response time*, measures which can be applied to the individual queues, to parts of the network, or to the entire network. Of importance is also the *blocking probability* of the individual queues.

Our aim is to provide computational tools to calculate these performance measures. To this end, we generally need to focus on the long-run or steady state distributions of the joint queue lengths. Because of the curse of dimensionality, numerical methods rapidly become prohibitively expensive if not infeasible for realistically sized systems. Our effort is therefore aimed at methods which are not subject to an exponential increase in computational complexity as the number of dimensions increases. Specifically, we will concentrate on product form solutions. More precisely, our aim is to provide insights as to when product form expressions can be derived, to find what features destroy the product form, and how one can use product forms to find performance bounds and approximations in cases where no product form exists.

Accordingly, the chapter establishes conditions for the existence of product form expressions, and, if they do not exist, it provides simple performance bounds based on insights gained from analyzing product forms.

For simplicity, our discussion will be restricted to simple networks, involving only a single class of customers and standard-type multi-server stations. On the other hand, we include complicating practical features, most notably finite capacity constraints (buffers), which can be regarded as generic for more complex realistic systems. This sets the stage for more complicated settings which will be discussed briefly. The analysis is done for both open and closed networks.

The outline of this chapter is as follows. The starting point of our discussion are the equilibrium equations of the continuous-time Markov chain describing the queueing network. These equations are called *global balance equations*. The global balance equations give rise to the *partial balance equations*, which in turn lead to the key concept of *station balance*. If the station balance equations are satisfied, then a so-called *product form* solution can be obtained for the joint steady state distribution of the station queue length. This will be discussed in detail in Section 2.

In Section 3, it will be shown by counter examples that the conditions resulting in product form expressions are often violated because of features such as blocking, dynamic routing and breakdowns. However, as will be explained in Section 4, the insights gained from Section 3 can be exploited to provide simple performance bounds for such systems.

It is impossible to do justice to all aspects of queueing networks in one short chapter. For instance, we were unable to include the very interesting and useful results provided by mean-value analysis [Reiser, 1980]. Readers intersted in further informtion about queueing networks should consult [Walrand, 1988, Robertazzi, 1990, Van Dijk, 1993, de Souza e Silva and Muntz, 1990, Kelly, 1979], among others. Furthermore, Chapter 8 contains a discussion of queueing networks by using a somewhat different perspective.

2 STATION BALANCE AND PRODUCT FORMS

2.1 Global and Partial Balance

In this section, global and partial equations are derived for both open and closed networks. In an open network, arrivals to and departures from the network are possible, whereas in a closed network, no elements arrive or depart. The number of elements in a closed network is therefore constant, say M.

We assume that all arrivals to the network are Poisson, and all service times are exponential. Besides its mathematical convenience, this can be justified by the fact that if only the arrival rates and service rates are known, the exponential distribution minimizes the entropy [Guiasu, 1986]. Moreover, the pooled output of several renewal processes converges to the Poisson process [Heyman and Sobel, 1982], a fact which is of particular relevance in queueing networks where the inputs of the stations often consist of the output of several other stations. Last, but not least, exponentiality formally justifies the usage of rate or flow equations, as they are generally used in practical engineering. These equations will play a crucial role throughout this chapter.

Exponential queueing networks can be modeled as continuous-time Markov chains (CTMCs) (see Chapter 2), characterized by a state description i and transition rates q_{ij}. Here, a state i may be identified with a multi-dimensional state description (a vector), representing information about the number of jobs and the types of jobs at the various service stations. This CTMC-formulation will justify the usage of balance principles which will form the basis for our analysis. The presentation that will be given below is informal and focused purely on steady-state probabilities. It merely aims to give the fundamental insight for justifying the use of global and partial balance questions.

The equilibrium equations of a CTMC with state space S can be written as (see Chapter 2)

$$\pi_j \sum_{\substack{k \in S \\ k \neq j}} q_{jk} = \sum_{\substack{k \in S \\ k \neq j}} \pi_k q_{kj}, \quad j \in S. \tag{1}$$

This system of equations expresses the fact that the total rate or probability flow out of any state j is equal to the total rate or probability flow into that state j, where "out of" and "into" mean into or out of a state other than j. Hence, the equilibrium equations balance the flow, and they are therefore called *balance*.

An important point to make here is the fact that the summations in the global balance equations can be taken either with both summations over all k, or over just $k \neq j$ as a transition from state j into itself $(k = j)$ will contribute equally to both sides. This observation is mathematically trivial but can be convenient for verification and modeling purposes.

The probability π_i is not only the probability to be in state i at a point far enough in the future, it is also the long run proportion of time the system spends in state i, and this proportion is in turn equal to the probability to be in state i at a randomly selected moment within a long enough observation interval.

It is in general difficult to solve the equilibirum equations directly for queueing networks of reasonable size. A solution method that works for some queueing networks is as follows. For each state j, the balance equations given by (1), called *global balance equations*, is divided into a number of independent *partial balance equations* which are easier to solve. Of course, this usually does not work because there are more partial balance equations than global ones, and the system is now overdetermined. Hence, in general, the system given by all partial balance equations does not have a solution. By happy coincidence, it turns out that the partial balance equations do have a solution for some important queueing networks, and this solution can be used to solve the global balance equations.

To formulate the partial balance equations mathematically, let $S(j)$ denote the set of states such that $q_{ji} > 0$, and let $R(j)$ similarly denote the set of states with $q_{kj} > 0$, and partition $S(j)$ into subsets $S_1(j), S_2(j), \ldots, S_{N_j}(j)$, and $R(j)$ into sets $R_1(j), R_2(j), \ldots, R_{N_j}(j)$. This partition generates the following partial equations from (1):

$$\sum_{k \in S_i(j)} \pi_j q_{jk} = \sum_{k \in R_i(j)} \pi_k q_{kj}, \quad i = 1, 2, \ldots, N_j, \quad j \in \mathcal{S}. \qquad (2)$$

The sum of these partial equations over i yields (1), and any solution π_i, $i \in \mathcal{S}$ of (2) is therefore a solution of (1). However, (1) may have solutions which are not solutions of (2).

2.2 Station Balance

The remaining question is how to partition the rates, that is, how to form the $S_i(j)$ and $R_i(j)$ such that the partial balance equations can be satisfied. The idea is to include in $S_i(j)$ all rates out of state j due to a departure from station i, and to include in $R_i(j)$ all rates into state j due to an arrival at station i. In addition, there is a set of rates, say $S_0(j)$, which includes all rates due to arrivals from the outside resulting in state j, and there is similarly a set $R_0(j)$, including all rates out of j due to departures. In summary, the partial balance equations can be formed by following the following balance equations, which must hold for each state j and each station i.

Station Balance

The rate out of any state j due to a departure at any station i
or due to an arrival to the network
=
The rate into that state j due to an arrival at that station i
or due to a departure from the network.

The formulation of the station balance principle simplifies if the outside is considered as a station, say station 0. Arrivals to the network then become departures from station 0, and departures from the network become arrivals to station 0. In this sense, one can say that the rates into state j by an arrival to station i must be balanced with the rates out of state j due to a departure from station i. With this convention, there is a station balance equation for each state and for each station. Hence, if there are N stations in the network, each global equation generates N partial equations by means of the station balance principle. This implies that there is no guarantee that a solution satisfying all station balances can be found. However, if a solution satisfying all station

2.3 An Instructive Two-Stage Tandem Line

The model under study here involves a system with two service stations in series with a Poisson input at the first station.

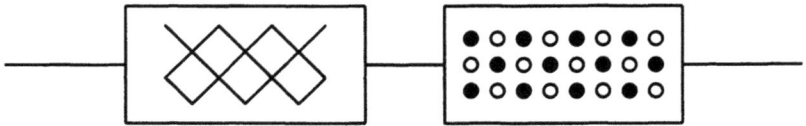

Specifically, jobs arrive at station 1 at a rate λ and are served at station 1 at a rate μ_1 (single-server station). Upon completion at station 1, a job instantly precede to station 2 which serves jobs at a rate μ_2 (single-server station). The state with n_1 jobs in station 1 and n_2 in station 2 is denoted by (n_1, n_2), and its steady state probability is denoted by $\pi(n_1, n_2)$. Both n_1 and n_2 can be arbitrarily large. The global balance equations are:

$$\begin{array}{c}\textbf{Total outrate} \\ \left\{\begin{array}{l} \pi(n_1,n_2)\mu_1 \mathcal{I}\{n_1>0\}+ \\ \pi(n_1,n_2)\mu_2 \mathcal{I}\{n_2>0\}+ \\ \pi(n_1,n_2)\lambda \end{array}\right\} \end{array} = \begin{array}{c}\textbf{Total inrate} \\ \left\{\begin{array}{l} \pi(n_1-1,n_2)\lambda \mathcal{I}\{n_1>0\}+ \\ \pi(n_1+1,n_2-1)\mu_1 \mathcal{I}\{n_2>0\}+ \\ \pi(n_1,n_2+1)\mu_2 \end{array}\right\}. \end{array} \tag{3}$$

In these equations, we made use of the indicator function $\mathcal{I}\{B\}$, which is equal to 1 if the logical expression B is true, and 0 otherwise.

We now form the partial balance equations. According to the station balance principle, we match the *outrate* due to an arrival to station 1 with the *inrate* due to a departure from station 1, that is, we have to equate the first terms on the the left and right hand side of (3), which yields

$$\pi(n_1, n_2)\mu_1 = \pi(n_1 - 1, n_2)\lambda, \quad n_1 \geq 0. \tag{4}$$

Note that if $n_1 = 0$, both sides of this partial equation are zero. Similarly, we have to match the outrate due to a departure from station 2 with the inrate due to an arrival to station 2, as expressed the second term in (3).

$$\pi(n_1, n_2)\mu_2 = \pi(n_1 + 1, n_2 - 1)\mu_1, \quad n_2 \geq 0. \tag{5}$$

Again, when $n_2 = 0$, both sides are zero. Finally, we have to match the arrivals to the system with the departures from the system, which yields

$$\pi(n_1, n_2)\lambda = \pi(n_1, n_2 + 1)\mu_2. \tag{6}$$

Equations (4) and (6) are easily solved. Repeated application of (4) yields

$$\pi(n_1, n_2) = \frac{\lambda}{\mu_1} \pi(n_1 - 1, n_2) = \left(\frac{\lambda}{\mu_1}\right)^2 \pi(n_1 - 2, n_2) = \ldots$$

Continuing this way, one finds

$$\pi(n_1, n_2) = \pi(0, n_2) \left(\frac{\lambda}{\mu_1}\right)^{n_1}.$$

The solutions for relations (6) is similarly

$$\pi(n_1, n_2) = \pi(n_1, n_2 - 1) \left(\frac{\lambda}{\mu_2}\right) = \ldots = \pi(n_1, 0) \left(\frac{\lambda}{\mu_2}\right)^{n_2}.$$

Combination of these two relations yields:

$$\pi(n_1, n_2) = \pi(0, 0) \left(\frac{\lambda}{\mu_1}\right)^{n_1} \left(\frac{\lambda}{\mu_2}\right)^{n_2}.$$

In order to be acceptable as a solution, this expression must also satisfy (5). This is the case as one can easily verify. Since all the partial balance equations hold, so do the global balance equations given by (1). To solve the problem, we still need to determine $\pi(0,0)$, using the condition that all probabilities must sum to one. By using the geometric series, we find that

$$\pi(0,0) = \left(1 - \frac{\lambda}{\mu_1}\right)\left(1 - \frac{\lambda}{\mu_2}\right).$$

The solution is therefore

$$\pi(n_1, n_2) = \left[\left(1 - \frac{\lambda}{\mu_1}\right)\left(\frac{\lambda}{\mu_1}\right)^{n_1}\right]\left[\left(1 - \frac{\lambda}{\mu_2}\right)\left(\frac{\lambda}{\mu_2}\right)^{n_2}\right].$$

Note that this solution yields a product form, with the first factor corresponding to station 1, and the second one to station 2.

2.4 Multi-Station Assembly Lines

The product form results obtained above can easily be generalized to systems with more than two stations. Specifically, consider N single server stations in sequence, which may be open or closed, as indicated in the following figure

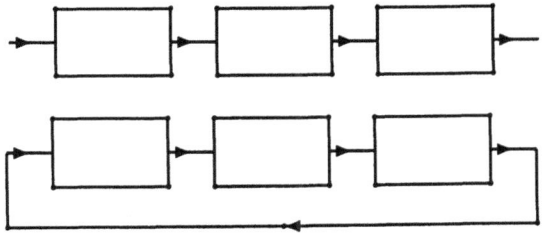

Closed networks are frequently encountered in manufacturing, where the parts are put onto pallets or into containers, and as soon as the part leaves the system, the same pallet or container returns to station 1. Station i, $i = 1, 2, \ldots, N$, serves at a rate μ_i whenever jobs are present. Upon service completion at station i, a job instantly moves to the next station $i + 1$ for $i < N$, while for $i = N$, it moves to station 1 in the closed case or leaves the system in the open case. In the closed case the total number of jobs in the system is fixed (say M). In the open case this number is unlimited and jobs arrive at station 1 at a rate λ. For the purpose of formulating the station balance equations, let $\overline{n}$ be the state with n_i jobs in station i, $i = 1, 2, \ldots, N$. Hence

$$\overline{n} = (n_1, n_2, \ldots, n_N).$$

Also, let $e_i = (0, \ldots, 0, 1, 0, \ldots, 0)$ denote the vector with N components with the i-th component equal to 1 and all other equal to 0, so that a transition from $\overline{n}$ to $\overline{n} + e_i$ denotes an arrival to station i, a transition from state $\overline{n}$ to state $\overline{n} - e_i$ denotes a departure from station i, and a transition from state $\overline{n}$ to state $\overline{n} - e_i + e_j$ occurs if a job moves from i to j. With these conventions, the station balance relations become:

Closed case:

$$\pi(\overline{n})\mu_1 = \pi(\overline{n} - e_1 + e_N)\mu_N, \quad n_1 > 0$$
$$\pi(\overline{n})\mu_j = \pi(\overline{n} + e_{j-1} - e_j)\mu_{j-1}, \quad n_j > 0,\ j = 2, \ldots, N.$$

Open case:

$$\pi(\overline{n})\mu_1 = \pi(\overline{n} - e_1)\lambda, \quad n_1 > 0$$
$$\pi(\overline{n})\mu_j = \pi(\overline{n} + e_{j-1} - e_j)\mu_{j-1}, \quad n_j > 0,\ j = 2, \ldots, N.$$

By substitution (or in fact by construction) one can again directly verify the following product form solution, where c is a normalization constant.

$$\pi(\overline{n}) = \frac{1}{c} \prod_{k=1}^{N} \left(\frac{1}{\mu_k}\right)^{n_k} : \text{ Closed Case} \qquad (7)$$

$$\pi(\overline{n}) = \frac{1}{c} \prod_{k=1}^{N} \left(\frac{\lambda}{\mu_k}\right)^{n_k} : \text{ Open Case.} \qquad (8)$$

The constant c must be determined in such a way that the sum of the probabilities of all admissible states is equal to one, which means that c must be equal to the sum of all products on the right. The constant c in the open case is easy to find. In fact,

$$\pi(\overline{n}) = \prod_{k=1}^{N} \left(1 - \frac{\lambda}{\mu_k}\right)\left(\frac{\lambda}{\mu_k}\right)^{n_k}. \qquad (9)$$

In the closed case, the factor c is more difficult to obtain, as will be shown later.

In (7) and (8), the equilibrium probabilities are expressed as products, hence the name *product form solution*. It is interesting to note that each station generates exactly one factor of this product. Of particular interest is (9): if all the stations had been isolated (and therefore independent) M/M/1 queueing systems, exactly the same result would have been obtained.

2.5 Jackson Networks

In the situations that have been presented so far, the routing from one station to another was deterministic. Below we will show that even with randomized routing, the station balance remains valid for all states and stations, and the solution is still of product form.

Specifically, consider closed or open queueing networks with N service stations such as the ones below:

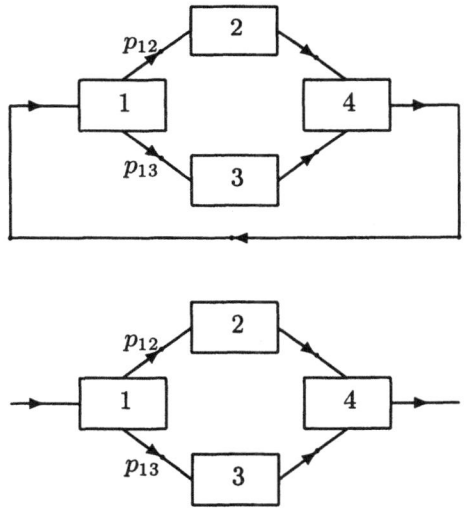

In the closed case, the total number of jobs is fixed at M. In the open case this number is unlimited and jobs enter the system at station j at a Poisson rate γ_j. Station j completes jobs at a rate $\mu_j, j = 1, \ldots, N$. The routing is random but state independent as follows. After completion of a service at station i a job instantly goes to station j with fixed probability p_{ij}, or in the open case, leaves the system with probability $p_{i0} = 1 - \sum_j p_{ij}$. We also define p_{0j} as $\gamma_j / \sum_i \gamma_i$. The solution is done in two steps. First, certain values λ_i are obtained, which can be interpreted as arrival rates at station i as if in isolation. In the open case, the $\lambda_i, i = 1, 2, \ldots N$ are uniquely determined by the external arrival rates $\gamma_j, j = 1, 2, \ldots N$. In the closed case, the λ_i can only be determined up to a factor. In fact, if λ_1 is set to 1, then the λ_i give the number of visits to station i before returning to station 1. The equations used for determining the λ_i are called *traffic equations*. They are obtained by equating the flow into station

i with the flow out of station i, $i = 1, 2, \ldots, N$, which leads to the following balance equations.

$$\begin{array}{ll} \lambda_j = \sum_i \lambda_i p_{ij}, & j = 1, \ldots, N : \text{ Closed case} \\ \lambda_j = \gamma_j + \sum_i \lambda_i p_{ij}, & j = 1, \ldots, N : \text{ Open case.} \end{array} \quad (10)$$

In the closed case, the traffic equations have a solutions which is unique up to a factor if the matrix $P = [p_{ij}]$ forms a non-decomposable stochastic matrix. In the open case, the solution is unique if P is a non-decomposable substochastic matrix.

The station balance equations can be obtained as before in that the flow out of state $\bar{n}$ because of a departure from station j, which is $\pi(\bar{n})\mu_j$, must be equal to the flow into state $\bar{n}$ at station j through an arrival from the outside or from another station i, which is $\pi(\bar{n} + e_i - e_j)\mu_i p_{ij}$. To have one set of balance equations for both the open and the closed case, we define $\gamma_j = 0$ for all j and $p_{i0} = 0$ for all i when the network is closed. The balance equations can now be written as

$$\pi(\bar{n})\mu_j = \sum_i \pi(\bar{n} + e_i - e_j)\mu_i p_{ij} + \pi(\bar{n} - e_j)\gamma_j, \quad (11)$$

$$\pi(\bar{n}) \sum_j \gamma_j = \sum_i \pi(\bar{n} + e_i)\mu_i p_{i0}. \quad (12)$$

It is required that these equations hold for any state $\bar{n}$ and any station j with $n_j > 0$. We aim to show that the following expression satisfies the station balance equations for the closed and the open case

$$\pi(\bar{n}) = \frac{1}{c} \prod_{k=1}^{N} \left(\frac{\lambda_k}{\mu_k}\right)^{n_k}. \quad (13)$$

This equation implies

$$\pi(\bar{n} + e_i) = \pi(\bar{n})\frac{\lambda_i}{\mu_i}$$

$$\pi(\bar{n} - e_i) = \pi(\bar{n})\frac{\mu_i}{\lambda_i}$$

$$\pi(\bar{n} + e_i - e_j) = \pi(\bar{n})\frac{\lambda_i}{\mu_i}\frac{\mu_j}{\lambda_j}.$$

Substituting these relations into (11) yields

$$\pi(\bar{n})\mu_j = \sum_i \pi(\bar{n})\frac{\lambda_i}{\mu_i}\frac{\mu_j}{\lambda_j}\mu_i p_{ij} + \pi(\bar{n})\frac{\mu_j}{\lambda_j}\gamma_j.$$

This is obviously equivalent to the traffic equation (10), which shows that the solution given by (13) satisfies (11). Similarly, substituting (13) into (12) yields

$$\pi(\bar{n}) \sum_j \gamma_j = \sum_i \pi(\bar{n})\frac{\lambda_i}{\mu_i}\mu_i p_{i0}$$

which simplifies to

$$\sum_j \gamma_j = \sum_i \lambda_i p_{i0}.$$

This last relation is true because the arrivals to the network ($\sum_j \gamma_j$) must be equal to the departures from the network ($\sum_i \lambda_i p_{i0}$).

Note that (13) is essentially identical to (8). Also, the normalization constant is found the same way as in the case of (8). It follows that (9), the expression derived for the open case, holds unchanged. Hence, for the open case, the problem is completely resolved. However, in the closed case, the normalization constant c must still be determined. This will be done next.

2.6 Normalization Constants in Closed Single Server Networks

If S is the state space, the normalization constant c can be found as

$$c = \sum_{\bar{n} \in S} \prod_{k=1}^{N} \rho_k^{n_k}.$$

Here, $\rho_k = \lambda_k/\mu_k$. If M and N are large, the curse of dimensionality makes itself felt, and the number of different states is huge. This makes the direct calculation of c impractical. However, there is an efficient method to find c. We define

$$c_j(m) = \sum_{n_1+n_2+\ldots+n_j=m} \prod_{k=1}^{j} \rho_k^{n_k}. \tag{14}$$

Clearly, $c = c_N(M)$. The following relations can be used to calculate $c_j(m)$ recursively:

$$c_j(0) = 1, \quad j = 1, 2, \ldots, N \tag{15}$$
$$c_1(n) = \rho_1^n, \quad n = 0, 1, \ldots, M \tag{16}$$
$$c_j(n) = \rho_j c_j(n-1) + c_{j-1}(n), \quad j = 1, 2, \ldots, M, \; n = 0, 1, \ldots, N. \tag{17}$$

Equations (15) and (16) follow directly from (14), provided a product over an empty set is assigned the value 1. Equation (17) is proven as follows

$$\rho_j c_j(n-1) + c_{j-1}(n)$$

$$= \rho_j \sum_{n_1+\cdots+n_j=n-1} \prod_{k=1}^{j} \rho_k^{n_k} + \sum_{n_1+\cdots+n_{j-1}=n} \prod_{k=1}^{j-1} \rho_k^{n_k}$$

$$= \sum_{n_1+\cdots+n_j=n-1} \left(\prod_{k=1}^{j-1} \rho_k^{n_k}\right) \rho_j^{n_j+1} + \sum_{n_1+\cdots+n_{j-1}=n} \prod_{k=1}^{j-1} \rho_k^{n_k}$$

$$= \sum_{\substack{n_1+\cdots+n_j=n \\ n_j \geq 1}} \prod_{k=1}^{j} \rho_k^{n_k} + \sum_{\substack{n_1+\cdots+n_j=n \\ n_j=0}} \prod_{k=1}^{j} \rho_k^{n_k}$$

$$= \sum_{n_1+\cdots n_j=n} \prod_{k=1}^{j} \rho_k^{n_k}. \qquad = c_j(n).$$

Equations (15) to (17) allow one to find all $c_j(n)$ in $O(MN)$ operations, that is, in polynomial time. Besides allowing one to determine c efficiently, they can also be used to find many important parameters. In particular, the marginal distribution of the number in station j, say X_j, can be found as

$$P\{X_j \geq n\} = \rho_j^n c_N(M-n)/c. \qquad (18)$$

This is proven as follows

$$cP\{X_j \geq n\} = \sum_{\substack{n_1+\cdots n_N=M \\ n_j \geq n}} \prod_{k=1}^{N} \rho_k^{n_k}$$

$$= \rho_j^n \sum_{n_1+\cdots+n_N=M-n} \prod_{k=1}^{N} \rho_k^{n_k}$$

$$= \rho_j^n c_N(M-n).$$

A station is busy as long as it contains one or more elements. The utilization U_j can therefore be found from (18) with $n = 1$.

$$U_j = \rho_j c_N(M-1)/c.$$

As long as station j is busy, its output is μ_j. Hence, the unconditional output of station j, which must be equal to its input and therefore its throughput, is given as

$$T_j = U_j \mu_j = \lambda_j c_N(M-1)/c.$$

Hence, T_j is proportional to λ_j, with $c_N(M-1)/c$ acting as the factor of proportionality.

Given (18), the average number in the system is

$$L_j = \sum_{n=1}^{M} P\{X_j \geq n\} = \sum_{n=1}^{M} \rho_j^n c_N(M-n)/c.$$

The time spent by a job in station j can now be found by applying Little's formula:

$$W_j = L_j/T_j = \sum_{n=1}^{M} \frac{\rho_j^n c_N(M-n)}{\lambda_j c_N(M-1)}.$$

Hence, besides allowing one to find the normalization constant c, the $c_j(n)$ allow one to find many other measures of interest without great difficulty.

As was mentioned earlier, the λ_j can only be determined up to a factor. Hence, if λ_j, $j = 1, 2, \ldots, N$ solves the traffic equations, so does $b\lambda_j$. Mathematically, the factor b makes no difference. However, it can make a huge difference numerically, and ill chosen factors b will almost certainly lead to exponent overflows or underflows. To avoid this, one must choose b such that no $b\lambda_j$ is very large or very small.

2.7 A Generalized View of Product Form

Expression (13) is naturally called "product form" as it factorizes to expressions for the individual stations corresponding to their steady-state distributions as if they were in isolation. In particular, in the open case, these expressions correspond to the marginal steady-state distributions $\pi_i(n_i)$ of each individual station i. It is from this factorization, as first revealed by Jackson [Jackson, 1957, Jackson, 1964] that the term "product form" originates.

However, in more complicated networks with state-independent routing and service speeds, explicit steady-state expressions might still be obtainable based on partial balance principles such as station balance, but these expressions may no longer factorize to expressions for individual stations. As a common factor, however, they all reveal another form of factorization, which also justifies the term "product form". To illustrate this, note that equation (13) can be rewritten as:

$$\pi(\overline{n}) = \frac{1}{c} \left[\prod_i \lambda_i^{n_i} \right] \left[\prod_i \left(\frac{1}{\mu_i} \right)^{n_i} \right].$$

Here the first term between brackets only concerns the routing probability without regard of how long jobs stay at the various stations. The second term between brackets, in contrast, only concerns the service speeds at the various stations which determine the actual residence or sojourn times of jobs at these stations. More generally, partial balance principles appear to be inherent with explicit solutions of the more general product form:

$$\pi(\overline{n}) = \frac{1}{c} R(\overline{n}) S(\overline{n}), \qquad (19)$$

where $R(\overline{n})$ is the routing part, independent of service times and disciplines, and $S(\overline{n})$ is the service part independent of the routing.

As an illustration of this more general concept of a product form, let us extend Jackson's classical model to allow for load dependent service speeds. Hence, assume that station i serves jobs with a service capacity $f_i(n_i)$ when n_i jobs are present, while the service requirement of a single job is exponential with parameter μ_i at station i.

The station balance relation (11) (the closed case being included with $\gamma_j = 0$ for all j) then becomes:

$$\pi(\overline{n})f_j(n_j)\mu_j = \sum_i \pi(\overline{n} + e_i - e_j)f_i(n_i + 1)\mu_i p_{ij} + \pi(\overline{n} - e_j)\gamma_j.$$

One can verify that these partial station balance equations are satisfied by (19) with

$$S(\overline{n}) = \prod_i \prod_{k=1}^{n_i} \frac{1}{\mu_i f_i(k)} \qquad (20)$$

$$R(\overline{n}) = \prod_i \lambda_i^{n_i}. \qquad (21)$$

Insensitivity. Another topic of practical importance is the *insensitivity property*. The insensitivity property for a service station means that the steady state distribution $\pi(\overline{n})$ still applies when the exponential service assumption at station i is dropped, provided the mean service time is kept at $1/\mu_i$. It could even apply to deterministic service requirements. Clearly, as exponentiality assumptions are hard to check in the first place and quite unrealistic in the second, this property is of great practical value. The insensitivity property applies to:

- Single-server stations ($f(n) = 1$), but only under special service protocols, most notably, a processor sharing protocol (all jobs present get an equal capacity fraction) or a last-come first-served preemptive discipline.

- Stations with an unlimited number of servers ($f(n) = n$) under the natural assumption that an arriving job is instantaneously assigned a free server.

On the other hand, the insensitivity property will necessarily fail for any station in which an arriving job may have to wait for service.

The insensitivity property is directly related to the notion of partial balance, this time formulated not only at the level of the stations, but also at the level of the individual jobs. More precisely, for each job, the rate out of any state due to the job should be balanced by the rate into that state due to that job. Clearly, when this more detailed balance holds, the station balance also holds, but the reverse is not true. This notion is referred to in literature as *local balance* [Schassberger, 1978], or more naturally as *job-local-balance* [Hordijk and Van Dijk, 1982, Hordijk and Van Dijk, 1983a]. This is the reason that the insensitivity property requires instantaneous service attention, which is absent if jobs have to wait. As the more precise description of this most detailed form of partial balance requires a more detailed state description, and, in addition to this, results from renewal theory, we refer to [Schassberger, 1978, Hordijk and Van Dijk, 1982, Hordijk and Van Dijk, 1983b, Hordijk and Van Dijk, 1983a, Van Dijk and Tijms, 1986] for details.

Multiple job-types. Finally, we should mention here that product form solutions are also available for networks with several types of customers. In this case, the partial balance equations must not only take total number of jobs at each station into account, but also the number of job-types at each station. When service times are stochastically equal for all job types, the station balance principle and the station balance equations still apply and lead to similar product forms [Kelly, 1979, Basket et al., 1975]. Different service characteristics for different job types, however, require a more specialized form of partial balance, viz. a balance per job type at each station. Under this more restrictive form of station balance, to be called *job-class balance*, similar product forms are still obtainable [van Dijk, 1990a, Van Dijk and Akylildiz, 1990, Hordijk and Van Dijk, 1982, Hordijk and Van Dijk, 1983b, Basket et al., 1975].

2.8 The Convolution Algorithm

If the service rate is load dependent, equation (17) no longer applies, and other methods, such as the *convolution algorithm* described here, have to be used to find c. We write

$$\pi(\overline{n})c = \prod_{i=1}^{N} r_i^{(n_i)}, \qquad (22)$$

with

$$r_i^{(n_i)} = \prod_{k=1}^{n_i} \frac{\lambda_i}{\mu_i f_i(k)}.$$

Let

$$c_j(m) = \sum_{n_1+\cdots+n_j=m} \prod_{k=1}^{j} r_k^{(n_k)}. \qquad (23)$$

Clearly, $c = c_N(M)$. To find the $c_j(n)$, one uses

$$c_j(0) = 1 \qquad (24)$$

$$c_1(n) = r_1^{(n_i)} \qquad (25)$$

$$c_j(n) = \sum_{m=0}^{n} c_{j-1}(n-m) r_j^{(m)}. \qquad (26)$$

Here, j ranges from 1 to N, and n from 0 to M. Since (24) and (25) are obvious, we only need to prove (26). One has

$$c_j(n) = \sum_{n_1+\cdots+n_j=n} \prod_{k=1}^{j} r_k^{(n_k)}$$

$$= \sum_{m=0}^{n} \left(\sum_{n_1+\cdots+n_{j-1}=n-m} \prod_{k=1}^{j-1} r_k^{(n_k)} \right) r_j^{(m)}$$

$$= \sum_{m=0}^{n} c_{j-1}(n-m) r_j^{(m)}.$$

The number of multiplications needed to find $c_j(n)$ for $n = 0, 1, \ldots, M$ is $M(M+1)/2$ as is easily verified. Since this must be done N times, the complexity is $O(NM^2)$. This is polynomial, and we have avoided the combinatorial explosion.

If X_N is the number of jobs at station N, one finds, using (23) and (22)

$$P\{X_N = n\} = c_{N-1}(M-n) r_N^{(n)}/c.$$

We can now calculate the throughput of station N as follows

$$T_N = \sum_{m=0}^{M} P\{X_N = m\} \mu_N f(m).$$

For all stations i, $i = 1, 2, \ldots, N$, λ_i must be proportional to T_i. If b is the factor of proportionality, $T_N = \lambda_N b$, that is, $b = T_N/\lambda_N$, and

$$T_i = \lambda_i b = T_N \lambda_i / \lambda_N.$$

Hence, all throughputs are easily calculated.

In some applications, it is necessary to find the distribution of X_i, the number of jobs in station i, $i \neq N$. To do this, define

$$c_-^i(n) = \sum_{n_1 + \cdots + n_{i-1} + n_{i+1} + \cdots + n_j = n} \prod_{\substack{k=1 \\ k \neq i}}^{j} r_k^{(n_k)}.$$

The $c_-^i(n)$ can be determined from the following equations, which are derived like (26):

$$c_N(n) = \sum_{m=0}^{n} c_-^i(n-m) r_i^{(m)}.$$

This equation can be solved for $c_-^i(n)$, yielding

$$c_-^i(n) = \frac{1}{r_i^{(0)}} \left(c_N(n) - \sum_{m=1}^{m} c_-^i(n-m) r_i^{(m)} \right)$$

and

$$P\{X_i = n\} = c_-^i(M-n) r_i^{(n)}/c.$$

3 WHY PRODUCT FORM SOLUTIONS TYPICALLY FAIL

3.1 Introduction

We now discuss when and why station balance fails so frequently in realistic situations, so that the product form is no longer available. One approach to deal with such cases is to use the solution methods discussed in Chapters 4 through 8, even though they are often computationally expensive. Here, we use a different approach, and we show that station balances can give practical insights, which may be used to test the limitations of the product form. These investigations will pave the way for finding useful bounds, which is the topic of Section 4.

3.2 Situations where Station Balance Fails

First, we discuss some cases where the station balances cannot be satisfied. We concentrate on cases where in some states, it is either impossible to leave a station through a departure, or cases in which it is impossible to enter a station through an arrival.

The prime example where a state cannot be left through a departure at a given station i is when departures are blocked. Consider, for instance, a two-station tandem queue with the state description (n_1, n_2) as in the Figure 1 below The dotted lines in this figure indicate blocking. As soon as n_2 reaches N_2, the

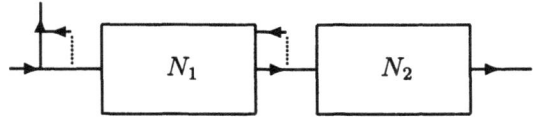

Figure 1 Tandem queue with blocking

first station is blocked, that is, it cannot continue working. Equivalently, one could assume that once finished, the job returns to station 1. Moreover, the input is blocked as soon as n_1 reaches N_1. In this case, arriving jobs are lost. For this model, one has

$$\text{Station 1, state } (n_1, N_2): \quad \text{In} > 0 \quad \text{Out} = 0 \qquad (27)$$
$$\text{Station 2, state } (N_1, n_2): \quad \text{In} = 0 \quad \text{Out} > 0 \qquad (28)$$

In state (n_1, N_2) it is impossible to have a departure from station 1 because it is blocked. Yet this same state can be entered through an arrival to station 1 while the system is in the state $(n_1 - 1, N_2)$. Hence, the rate of leaving state (n_1, N_2) through a departure from station 1 is zero, while the rate of entering this state by an arrival to station 1 is greater than zero. This means that in state (n_1, N_2), the station balance for station 1 cannot be satisfied as long as arrivals are allowed in state (n_1, N_2). Station balance is violated at station 2 in states (N_1, n_2) with $n_2 > 0$: The rate of leaving this state by a departure from station 2 is greater than zero. However, the rate of entering the state due to an arrival at station 2 is zero as state $(N_1 + 1, n_2 - 1)$ is not feasible.

Another generic situation in which states cannot be entered by arrivals at a particular station involves state dependent routing, including overflow. As an example, consider an overflow model with two stations 1 and 2, with jobs arriving at station 1 as shown in Figure 2 below. Station 1 can accommodate

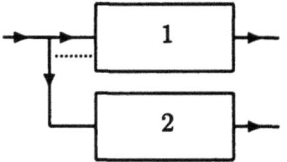

Figure 2 Overflow model

at most N_1 jobs. When saturated, arriving jobs are rerouted to station 2, say with an infinite capacity. Once rerouted, a job will have to complete its service at station 2. One has, if (n_1, n_2) is again the state description:

$$\text{Station 2, state } (n_1, n_2), \ n_1 < N: \quad \text{In} = 0 \quad \text{Out} > 0 \tag{29}$$

State (n_1, n_2) cannot be entered through an arrival at station 2 if $n_1 < N_1$, because in this case, any arrival would go to the first station. Yet, state (n_1, n_2) can be left through a departure at station 2. Indeed, this will happen whenever there is a departure from station 2.

Besides blocking and state-dependent routing, the station balance can also be violated by other state dependent features occuring frequently in practical applications, such as diminished service rates or service breakdowns. Specifically, if one station no longer operates, as in Figure 3 below, its output through departures vanishes. However, there may still be input from other stations.

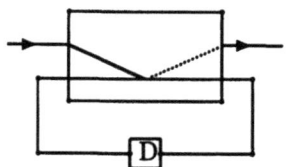

Figure 3 Breakdown model

The station balance is therefore lost.

$$\text{Station down:} \quad \text{In} > 0 \quad \text{Out} = 0 \tag{30}$$

Finally, queueing systems which require additional variables to describe the state of the system, or the entities therein, typically violate the product form.

Examples of such systems include job-class priorities, shortest-job-first disciplines, synchronizations, such as the fork and join queue (see Chapter 2) and rendez-vous networks [Woodside, 1988]. We should also mention here that if a job requires several resources, such as processors and memory, then no product form can be expected. And except in cases where insensitivity applies, no product form can be expected if the service times are not exponential.

3.3 Situations Where Product Form Still Applies

As illustrated above, blocking, state dependent routing and breakdowns will generally destroy the station balance. There are, however, exceptions to this rule. Consider blocking, in which case it is impossible to leave certain states through departures at a certain station i. If it is also impossible to enter the same states through an arrival at i, then the station balance will still hold. As an example, consider the figure below, which represents a closed two-station model with capacity constraints N_1 and N_2 for the number of jobs at station 1 and 2, respectively, where N_1 and/or N_2 can be smaller than M, the total number of jobs.

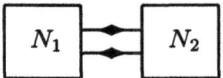

When a station is saturated, the other station stops serving. The service rates of the stations are, respectively, μ_1 and μ_2. Although it is sufficient to specify only the number of jobs at one of the two stations, we use the state description (n_1, n_2) to make the station balance characterization more explicit. The station balance equations for station 1 and 2 then become, provided n_1 and n_2 are greater than zero:

$$\begin{array}{rcl}\textbf{Out} & & \textbf{In} \\ \pi(n_1,n_2)\mu_1\mathcal{I}\{n_2 < N_2\} & = & \pi(n_1-1, n_2+1)\mu_2\mathcal{I}\{n_2 < N_2\} \\ \pi(n_1,n_2)\mu_1\mathcal{I}\{n_1 < N_1\} & = & \pi(n_1-1, n_2+1)\mu_2\mathcal{I}\{n_1 < N_1\}.\end{array}$$

Since the same indicator functions appear in both the right and the left hand sides, the station balance equation holds, and the solution has product form. In fact, we have

$$\pi(n_1, n_2) = \frac{1}{c}\left(\frac{1}{\mu_1}\right)^{n_1}\left(\frac{1}{\mu_2}\right)^{n_2}, \quad n_1 \leq N_1,\ n_2 \leq N_2.$$

Here, c is to be determined such that the sum of all probabilities equals 1.

As a second example, consider an arbitrary closed Jackson network as analyzed in Section 2.5, but in addition assume that there is a finite capacity (buffer) constraint of size N_j in the total number of jobs at some or all stations j, and assume the following protocol: when station j is saturated ($n_j = N_j$), transitions from any other station, other than station j itself, are blocked, that is, all stations except station j stop serving. We now show why the station balances are satisfied in all states.

If the network has no full buffer, it is identical to the Jackson network discussed in Section 2.5, and the station balances for $n_k < N_K$, $k = 1, 2, \ldots, N$ are given by

$$\pi(\overline{n})\mu_j = \sum_k \pi(\overline{n} + e_k - e_j)\mu_k p_{kj}. \qquad (31)$$

Consider now a state $\overline{n}$ in which only the buffer of station j is full, that is, $n_j = N_j$, $n_i < N_i, i \neq j$. We denote this state by $\overline{N}_j$. According to the protocol, the entire network freezes, except for station j. Consequently, there is no way that any other buffer fills, unless, of course, there is a departure from station j, at which time station j can no longer block other stations. It is easily seen that there can be at most one full buffer at a time. Moreover, the rate of leaving $\overline{N}_j$ through a departure at station $i \neq j$ is zero because station i is frozen. If follows that the rate of leaving state $\overline{N}_j$ through a departure at state $i \neq j$ is zero. Since the only way n_j can become N_j is through an arrival at station j, $\overline{N}_j$ cannot be entered through an arrival at $i \neq j$. Hence, for all stations $i \neq j$, both the rate of entering $\overline{N}_j$ and the rate of leaving $\overline{N}_j$ are zero. This shows that the station balance holds for every station except j, no matter which state the system is in. However, from the discussion above, it follows that the station balance equation for station j is given by (31) even when station j is saturated $(n_j = N_j)$.

The product form solution

$$\pi(\overline{n}) = \frac{1}{c} \prod_{k=1}^{N} \left(\frac{\lambda_k}{\mu_k}\right)^{n_k} \quad n_j \leq N_j, \ j = 1, \ldots, N \qquad (32)$$

thus satisfies (31) in all cases, even for station j while $n_j = N_j$. It remains to find c, which is the sum of all the products given by (32). This can be done by appropriate modifications of the convolution algorithm.

4 HOW PRODUCT FORMS CAN STILL BE USEFUL

4.1 Motivation and Bounding Methodology

Motivation. As shown in the previous section, exact product-form expressions cannot generally be expected in realistic situations involving features such as blocking due to finite buffers, breakdowns or state dependent routing. Nevertheless, the station balance insights could well remain useful.

The simplest way to deal with situations violating station balance is to simply ignore features that destroy it. This "approximate use of product forms" might well be justified depending on the situation. For example, when

- buffer sizes are relatively large
- down times are rather low

so that the violation of station balance will only take place in states having a very small probability.

But clearly, this is impossible if such features play an essential role. A more thoughtful alternative would be to modify the system such that the station balances (and product form) are restored. Such a modification will generally not lead to a very accurate approximation, but it can still be practically useful to provide:

- A quick approximation of orders of magnitude
- Simple secure lower and/or upper bounds.

This may be useful to gain qualitative and quantitative insights, to optimize a design, or to do sensitivity analysis. As a special advantage of such an approach, we also note that product-form expressions and the bounds obtained from them can be independent of the service time distributions as long as the sensitivity principle is justified (see Section 2.7). This principle covers, among others, processor sharing, last-come first served single server stations, and, most notably, pure multiserver queues with immediate access to the server.

Bounding methodology. The bounding methodology suggested here can be summarized as follows. Whenever one is faced with a system which does not satisfy the required partial balance, in the setting of this chapter the station balance, we must modify the original system so that:

1. The concept of station balance is guaranteed, and
2. Bounds for a performance measure can be expected.

We emphasize that the modifications are not meant to be made to the original system in reality but solely as a theoretical modification to justify simple computations. The following are performance measures that may be of interest in practical applications:

- A system throughput.
- A blocking or loss probability.
- Station utilization or efficiency.
- A steady-state probability of a special event.

To make this approach practical, three questions must be addressed.

1. How can the modifications be found?
2. How can the modifications be used to obtain bounds?
3. How accurate or inaccurate, or, rather, how useful are these bounds?

These questions, and their answers, will be illustrated by numerical examples in the next section. Nevertheless, let us first provide some general response.

In the examples discussed in Section 3.2, the station balance is violated if in a certain state, the outrate from that state through a departure from a certain station is zero, while the inrate into that state through an arrival at this station is greater than zero. In this case, one can modify the system such that the inrate due to an arrival at this station also becomes zero. A similar method applies if the inrate into a certain state through an arrival at a certain station is equal to zero, while the outrate due to a departure from this station is greater than zero, as is the case in the overflow model. Such a modification may lead to a product-form expression that can be verified formally by direct substitution in the station balance equations.

The second question deals with finding bounds. We normally use intuitively appealing principles. For instance, more arrivals from the outside will increase traffic, and hence increase throughput and queue lengths. A formal guarantee, in contrast, is usually much harder to establish, despite this intuitive support. Nevertheless, by using a rather technical comparison technique, rigorous proofs have been given that the "intuitively obvious bounds" to be presented are indeed bounds. (See [van Dijk, 1991, van Dijk, 1989b, van Dijk, 1989a, van Dijk, 1990b, van Dijk, 1990c, van Dijk et al., 1988, van Dijk and van der Wal, 1989, van Dijk and Lamond, 1988]).

Last, but not least, the numerically practical significance of the bounds must be determined. Clearly, one cannot expect accurate results when modifying transition rates in situations occurring frequently. Nevertheless, as will be illustrated by numerical examples, the bounds can provide reasonable quick and secure indicators of orders of magnitude in situations where otherwise no simple results are available. Further, as will also be shown, the bounds may provide qualitative insights useful for optimization purposes. In addition, for so-called insensitive disciplines, the bounds will also be insensitive, that is, be independent of service time distributions.

4.2 Simple Product Form Bounds For Finite Networks

In this section we illustrate the bounding methodology by a number of generic examples. Although these examples are very simple in structure, each one represents a feature that will destroy product form. The examples can be used as building blocks for dealing with more complicated non-product-form networks. We will elaborate on this in Section 4.4

In these examples, we exploit the ideas of Section 4.1, by proposing product-form bounds for which it is intuitively obvious that they are really bounds. Formal proofs that this is the case, however, are not trivial. Nevertheless, for each of the examples, formal proofs have been provided so that the bounds are 100% secure. Generally, the first intuition thus seems to support the bounding methodology as a practical tool.

Finite tandem model. Reconsider the two-station assembly line model of Figure 1 from Section 3.2, with finite buffer sizes N_1 and N_2 for the total number of jobs at station 1 and station 2 respectively. Jobs enter the first station at a rate λ and are served at a rate μ_1 at station 1 and μ_2 at station 2, provided blocking does not take place. However, due to the capacity constraints, blocking arises. When the first station is saturated ($n_1 = N_1$), arriving jobs are blocked and lost. When the second station is saturated ($n_2 = N_2$), jobs which complete a service at station 1 are recycled to undergo a new service at station 1, so that the effective output rate of station 1 becomes 0 as if serving at station 1 had been stopped. As indicated by (27), the station balance fails for station 1 in any state (n_1, N_2) with $n_1 > 0$ because in this state, no departures occur at station 1, yet the state can be entered through arrivals. Equation (28) indicates that the station balance also fails at station 2 in any state (N_1, n_2) with $n_2 > 0$ because the only possibility to enter this state through an arrival to station 1 is from the state $(N_1 + 1, n_2 - 1)$, and this state is not admissible. Hence, the inrate to this state through an arrival to station 2 is zero, while the outrate from this state through a departure from station 2 is greater than zero. The following modification to restore station balance can therefore be suggested:

1. When station 2 is saturated ($n_2 = N_2$), not only departures but also arrivals at station 1 are to be blocked, that is, arrivals are to be rejected. In this way, the inrate of (27) is reduced to 0.

2. When station 1 is saturated ($n_1 = N_1$), not only arrivals but also departures from the system are to be blocked, reducing in this way the outrate of (28) to zero. To this end, system departures should be recycled to undergo a new service at station 2 or, equivalently, service at station 2 is stopped whenever station 1 is full.

Intuitively, station balance seems hereby restored so that a product form can be expected. Indeed, for the modified finite tandem queue, one can check the validity of the product form:

$$\pi(n_1, n_2) = \frac{1}{c} \left(\frac{\lambda}{\mu_1}\right)^{n_1} \left(\frac{\lambda}{\mu_2}\right)^{n_2} \tag{33}$$

by substituting these expressions in the station balance relation for stations 1 and 2:

$$\pi(n_1, n_2)\mu_1 \mathcal{I}\{(n_2 < N_2)\} = \pi(n_1 - 1, n_2)\lambda \mathcal{I}\{(n_2 < N_2)\}, \quad n_1 > 0$$
$$\pi(n_1, n_2)\mu_2 \mathcal{I}\{(n_1 < N_1)\} = \pi(n_1 + 1, n_2 - 1)\lambda \mathcal{I}\{(n_1 < N_1)\}, \quad n_2 > 0.$$

The normalization constant c has to be determined such that the sum of all probabilities over the state space R_U is 1, where R_U is given as

$$R_U = \{(n_1, n_2) | (n_1, n_2) \neq (N_1, N_2), n_1 \leq N_1, n_2 \leq N_2\}$$

Clearly, in this modified system arrivals will be rejected more frequently, as if arriving customers anticipate congestions ahead and change their mind, while in addition, service at station 2 is sometimes interrupted so that the system will remain more congested. The fact that the arrival loss probability is increased suggests an upper bound B_U for the loss probability B. This bound can be computed based on the product form above.

To find a lower bound, we overcome the failures of station balances of the two stations by never letting jobs be blocked once having been accepted into the system. The finite capacity constraints N_1 and N_2 can nevertheless be taken into account by allowing an arriving job to be rejected only if the total number of jobs already present exceeds the total capacity, that is, if $n_1 + n_2 = N_1 + N_2$. Furthermore, service always continues and jobs are never blocked from entering station 2 or leaving the system. Each station can thus contain up to $N_1 + N_2$ jobs. The system is hereby reduced to a standard product form system: the global balance equations are satisfied by the the product form given by (33), but with the set of admissible states R_U replaced by:

$$R_L = \{(n_1, n_2) | n_1 + n_2 \leq N_1 + N_2\}.$$

Furthermore, as the total capacity $N_1 + N_2$ to accommodate jobs is now pooled and used in a dynamic manner wherever needed, either at station 1 or station 2, the system becomes more efficient and the arrival loss probability is reduced, leading thus to a *lower-bound* B_L.

We can thus easily compute a lower and upper bound B_L and B_U for the original loss probability B by:

$$\pi_L(n_1, n_2) = c_L \left(\frac{\lambda}{\mu_1}\right)^{n_1} \left(\frac{\lambda}{\mu_2}\right)^{n_2}, \quad (n_1, n_2) \in R_L$$

$$\pi_U(n_1, n_2) = c_U \left(\frac{\lambda}{\mu_1}\right)^{n_1} \left(\frac{\lambda}{\mu_2}\right)^{n_2}, \quad (n_1, n_2) \in R_U$$

and

$$B_L = \sum_{n_1+n_2=N_1+N_2} \pi_L(n_1 + n_2)$$

$$B_U = \sum_{n_2} \pi_U(N_1, n_2) + \sum_{n_1} \pi_U(n_1, N_2).$$

Furthermore, by providing a lower and an upper bound for the loss probability B, we also obtain a lower and an upper bound for the throughput by the relation

$$T = (1 - \lambda)B.$$

The bounds B_L and B_U have been compared with the numerical values of B for a wide range of system parameters. Without a loss of generality, we use $\lambda = 1$. The results are given in Table 1. Since the worst performance appears when $\lambda = \mu_1 = \mu_2$, various examples with $\lambda = \mu_1 = \mu_2 = 1$ are included. The results can be summarized as follows. For systems with heavy congestion, the bounds are quite narrow. For the more realistic case of light congestions, they remain good indicators of the order of magnitude. Also, qualitative results are accurately indicated by the bounds, such as the dominance of the smallest service rate or the smallest buffer size. If approximations are needed, one can use the average value $(B_L + B_U)/2$. This value is also given in Table 1.

THE PRODUCT FORM TOOL FOR QUEUEING NETWORKS   433

Table 1   Bounds for the tandem line

N_1	N_2	μ_1	μ_2	B_L	B_U	B	$\frac{B_L+B_U}{2}$
1	1	1	1	0.500	0.667	0.600	0.583
2	2	1	1	0.333	0.500	0.422	0.417
2	2	0.2	0.2	0.842	0.893	0.867	0.867
2	2	0.2	0.2	0.900	0.912	0.903	0.906
3	3	0.5	0.1	0.250	0.400	0.325	0.325
5	5	1	1	0.200	0.348	0.280	0.274
5	5	1	1	0.166	0.286	0.222	0.226
10	10	1	1	0.090	0.167	0.124	0.128
20	20	1	1	0.047	0.091	0.064	0.069
40	40	1	1	0.003	0.007	0.003	0.005
5	5	0.1	0.1	0.909	0.948	0.917	0.928
5	5	0.5	0.5	0.900	0.910	0.900	0.904
5	5	0.5	0.5	0.539	0.674	0.584	0.611
5	5	0.5	2.0	0.500	0.512	0.508	0.506
5	5	2.0	2.0	0.002	0.032	0.018	0.017
5	10	0.1	0.1	0.906	0.948	0.909	0.926
5	10	0.5	0.5	0.533	0.671	0.547	0.602
5	10	2.0	2.0	0.000	0.017	0.009	0.009

Overflow model. State-dependent or dynamic routing due to saturated or blocked stations, service priorities or minimal utilization requirements is a common feature in practical queueing network applications. Examples are widely found in flexible manufacturing, the allocation of a public or private communications network, the skill based routing in call-centers or alternate routing in telecommunications, or low- and high-speed processors that are alternately used in computer networks, depending on the actual workload.

As a prototypical situation, consider an overflow model with two stations given in Figure 2. Station 1 represents a regular service facility with s regular servers and jobs arriving at rate λ. Each of these servers has a unit service speed. When all regular servers are busy, an arriving job is rerouted to a group of m overflow servers. If this group is also saturated, arriving jobs are rejected and lost. The overflow servers have a service speed f per server, where, typically, one may think of $f > 1$ (faster but more expensive service during overflow). Moreover, N_1 and N_2 represents the buffer size of the regular, respectively, the overflow servers.

As indicated by (29) in Section 3.2, station balance is destroyed by a positive service rate at the overflow station when a regular server is free. One way to regain the station balance is as follows: jobs that complete service at an overflow server have to repeat service at this server, so that service of overflow servers is

Table 2 Bounds for the overflow model ($\mu = 1$)

s	m	λ	f	B_L	B_U
3	1	6	2	0.395	0.443
			4	0.297	0.354
		3	2	0.160	0.208
			4	0.104	0.148
		2	2	0.074	0.105
			4	0.046	0.070
3	2	6	2	0.239	0.312
		3	4	0.060	0.107
5	1	6	2	0.211	0.270
		3	2	0.037	0.066
5	2	6	2	0.117	0.191
		3	2	0.010	0.034

effectively stopped whenever a regular server is free. This reduces the outrate in (29) to zero.

With (n_1, n_2) denoting that n_1 regular and n_2 overflow servers are busy, one might thus expect a product-form expression. For example, with $N_1 = s$, $N_2 = m$ (thus no waiting rooms), arrival rate λ and exponential service parameter $\mu = 1$, one easily verifies:

$$\pi(n_1, n_2) = c \frac{1}{n_1!} \frac{1}{n_2!} \lambda^{n_1} \left(\frac{\lambda}{f}\right)^{n_2}, \quad n_1 \leq s, n_2 \leq m$$

by substitution in the station balance equations for the regular (with $n_1 > 0$) and overflow stations, respectively, which are given by:

$$\pi(n_1, n_2) = \pi(n_1 - 1, n_2)\lambda$$
$$\pi(n_1, n_2)n_2 f \mathcal{I}\{(n_1 = s)\} = \pi(n_1, n_2 - 1)\lambda \mathcal{I}\{(n_1 = s)\}.$$

Clearly, as the overflow servers are kept busier, this modification suggests an upper bound B_U for the loss probability. Conversely, by assuming that arriving jobs are randomly assigned to any of the remaining free servers, whether regular or overflow, while assuming that $f \geq 1$, a lower bound B_L is easily obtained by Erlang's classical loss formula. The results given in Table 2 show that these bounds are quite reasonable first indicators for small overflow systems.

Breakdown model. With present-day computer communications technology increasingly expanding to interconnected local, metropolitan and wide area networks, with manufacturing systems becoming more and more computerized, and last but not least with telecommunications becoming more dependent on central links (e.g. for Internet) and central switches, the effect of breakdowns or failures will generally become more significant and can have critical consequences for performance. Combining reliability and performance analysis has therefore become increasingly important over the last few years (see also Chapter 12) and seems to have become a significant research topic, under the name of *performability*, for both theoretical and practical purposes. As a simple example consider the single server queueing system described by Figure 3. Suppose that this system is subject to breakdowns at a constant rate γ_1 independent of the number of jobs present. A breakdonw renders the system inoperative for an exponential repair time with parameter γ_0. Here, a distinction is made between the active case in which a breakdown only takes place when the system works, and the independent case in which the breakdown can arise any time.

The service station is defined as station 1, and the up or down status as station 2. As indicated by (30), the station balance is violated when the system is down, in both the active and independent case, as departures from station 1 are stopped, yet the arrivals to station 1 are still possible. An obvious modification to overcome this rate imbalance and to restore station balance is to reject all arrivals once the system is down.

To analyze the system, let n be the number of jobs present and let θ express whether the system is up ($\theta = 1$) or down ($\theta = 0$). The state (n, θ) might then satisfy a product-form expression. For example, for a finite system (of capacity N) and no waiting room, arrival rate λ and exponential service rate μ per server, one can easily verify that the modified system has the solution

$$\pi(n,\theta) = [c\gamma_\theta]^{-1} \left[\frac{1}{n!} \left(\frac{\lambda}{\mu}\right)^n \right], \quad n \le N, \ \theta = 0, 1, \ldots$$

by substituting this expression into the station balance relations:

$$\pi(n,1)n\mu = \pi(n-1,1)\lambda, \quad n \ge 1$$
$$\pi(n,1)\gamma_1 = \pi(n,0)\gamma_0, \begin{cases} n \ge 1 & \text{active case} \\ n \ge 0 & \text{independent case.} \end{cases}$$

Here, the first equation represents the service station, and the second one the up and down status of the service station.

Clearly, the rate at which jobs enter the system decreases if arrivals are rejected once the system is down, which suggests that this modification provides an *upper bound* B_U for the loss probability. Conversely, by assuming that the system can never break down ($\gamma_1 = 0$), a *lower bound* on the loss probability can be expected. In this case, $\pi(n, 0) = 0$, that is, the second equation becomes redundant. Table 3 provides numerical results that seem to indicate reasonably accurate bounds for realistic situations, in which the proportion of down time is small, say 5% or less. In these tables, M is the total capacity (also the number of servers in the multi-server case), $\rho = \lambda/\mu$ and $\tau = \gamma_1/\gamma_0$ (roughly the fraction of time that the system is down). Because of the insensitivity results mentioned in Section 2.7, the bounds obtained for the pure multi-server case also apply for non-exponential service and down times.

Table 3 Bounds for breakdown model

Independent case: pure multi-server

M	ρ	τ	B_L	B_U
30	25	0.05	0.052	0.098
		0.01	0.052	0.062
20	15	0.05	0.045	0.091
		0.01	0.045	0.065
		0.005	0.045	0.055
10	5	0.01	0.018	0.028
		0.005	0.018	0.024
		0.001	0.018	0.020
10	1	0.01	0.000	0.010
		0.005	0.000	0.005
		0.001	0.000	0.001

Active case: single server

M	ρ	τ	B_L	B_U
4	1	0.2	0.20	0.31
		0.1	0.20	0.26
		0.05	0.20	0.23
		0.02	0.20	0.21
8	2	0.20	0.50	0.58
	1	0.10	0.12	0.19
		0.05	0.12	0.16
		0.02	0.12	0.14

4.3 An Optimal Design Application

To illustrate that the bounding methodology is of potential practical use, despite the inaccuracy of the bounds, reconsider the assembly-line system of Section 4.2 with two successive single-server stations with buffer sizes N_1 and N_2, respectively, where the numbers N_1 and N_2 are to be determined so as to minimize the average cost rate:

$$C = 1000B + (N_1 + N_2)^2.$$

Here, B is the loss probability of an arriving job. By using the lower and upper bounds B_L and B_U for B as obtained in Section 4.2, we also obtain lower and upper bounds C_L and C_U for C. These values are graphically illustrated in Figure 4 as a function of $N_1 + N_2$ with $N_1 = N_2$. The computation of these curves was based on the simple expressions derived above. Although the lower and upper bound cost curves are widely apart, they lead to more or less the same optimal numbers 8 or 10 when we assume that $N_1 = N_2$. Presumably, the real optimum is at or between these two numbers. This information may be helpful for a more detailed study, which may have to be done by using simulation.

Given one has upper and lower bounds for the costs, one can provide a bound for the minimum cost. Indeed, the minimum cost cannot be reached for any $N_1 + N_2$ where the lower bound exceeds the minimum of the upper bound. In Figure 4, this means that the optimum must be between 4 and 17.

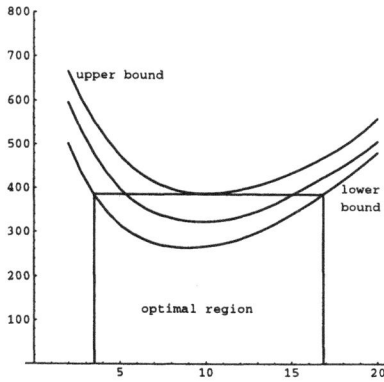

Figure 4 Optimization of a two-station tandem queue

4.4 Constructing Larger Models

In this section, we briefly discuss and illustrate possible extensions of the foregoing that allow one to analyze somewhat larger models. In particular, one can add more stations, combine different features, and one can break down an entire network into subnetworks.

Adding additional stations. The bounding methods used can definitely be scaled up to allow for more than two stations. As more stations are added, the bounds will become substantially less accurate but, nevertheless, they still remain reasonable indicators of orders of magnitude. This is illustrated in Table 4 for a tandem line with three stations, as illustrated by the graph at the bottom of the table. Similar statements hold for multiple overflow stations or for multiple breakdowns.

Combinations of bounding methods. In the two-station assembly line example in Section 4.2, each of the stations 1 and 2 can also be subject to breakdowns. To deal with this case, we combine the modifications derived for finite buffers with the ones used for breakdowns. The results of such a combination are shown in Table 5, with a graph of the setup included at the bottomof the table. In this table, γ_{i0} and γ_{i1} are the repair and breakdown rates for station i, $i = 1, 2$.

Components treated as sub-networks. In manufacturing situations one frequently encounters storage buffers or total capacity constraints for a whole group of workstations. (This typically applies to so-called "push and pull" manufacturing systems.) By regarding the whole group of workstations between such buffers as one aggregate station, the same modification approach can be applied.

438 COMPUTATIONAL PROBABILITY

Table 4 Bound for three station tandem line

N_1	N_2	N_3	μ_1	μ_2	μ_3	B_L	B_U	B	$\frac{B_L+B_U}{2}$
1	1	1	1	1	1	0.50	0.86	0.64	0.68
1	1	1	10	1	1	0.40	0.77	0.54	0.58
2	2	2	10	1	1	0.25	0.56	0.36	0.41
5	5	5	1	1	1	0.16	0.42	0.25	0.29
5	5	5	10	1	1	0.12	0.30	0.18	0.21
1	5	5	10	10	1	0.08	0.24	0.13	0.16
1	5	5	100	100	1	0.08	0.17	0.09	0.13
1	1	5	10	10	10	0.13	0.31	0.17	0.21
1	1	5	100	100	1	0.12	0.18	0.13	0.15
1	5	10	10	10	10	0.00	0.18	0.10	0.09
1	1	5	100	100	100	0.00	0.02	0.01	0.01

Table 5 Bounds for tandem line with individual breakdowns ($\lambda = 1$)

N_1	N_2	μ_1	μ_2	γ_{10}	γ_{11}	γ_{20}	γ_{21}	B_L	B_U
4	4	1	1	100	1	100	1	0.20	0.34
5	5	100	1	50	1	50	1	0.09	0.20
10	10	1	1	100	1	100	1	0.09	0.18
10	10	100	1	50	1	50	1	0.04	0.13
10	10	10	10	50	1	50	1	0	0.04
10	10	10	10	100	1	100	1	0	0.02

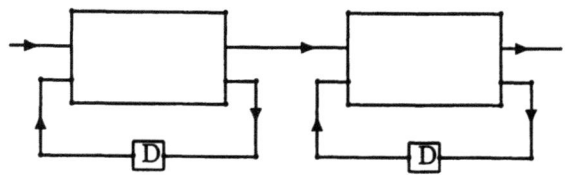

Table 6 Bound for tandem line with breakdowns ($\lambda = 1$)

T	μ_1	μ_2	μ_3	μ_4	μ_5	μ_6	γ_0	γ_1	B_L	B_U
20	1	1	1	1	1	1	25	1	0.23	0.26
20	1	1	1	1	1	1	50	1	0.23	0.25
20	1	1	1	1	1	1	100	1	0.23	0.24
20	10	1	10	1	10	1	50	1	0.13	0.15
20	10	1	10	1	10	1	100	1	0.13	0.14
20	10	10	10	10	10	10	50	1	0.001	0.02
10	10	10	10	10	10	10	100	1	0	0.01

As a first example, the breakdown system of Section 4.2 can be replaced by an entire Jackson network, for example an assembly line of arbitrary length or a computer network system, for which a common resource, controller or central computer is subject to breakdowns which renders the whole system inoperative. The same modification now applies if the whole network is treated like a single station. In Table 6, we show an assembly line of six stations with a total capacity constraint for no more than N jobs and subject to breakdowns.

As a second example, consider a four-station tandem queue, in which the total number of the first two stations is restricted to M_1, and the number in the second two stations to M_2 (see figure at the bottom of Table 7). Arrivals are lost when the first cluster is at capacity, and if the second cluster is at capacity, blocking occurs. A similar modification as in the two station case can now be applied at cluster level to enforce simple product form bounds. Table 7 provides some numerical results, with B_L and B_U lower and upper bounds for the blocking probability, $B = (B_L + B_U)/2$ and B obtained by numerical computation. This also establishes bounds on the throughput T because $T = \lambda(1 - B)$.

Decomposition. One of the most commonly employed approaches for approximate computations of complicated networks is that of decomposition. The total network is decomposed into subcomponents constituting small subnetworks or even individual stations and approximate results for these subcomponents are combined in an appropriate manner. By providing simple bounds for decomposed subcomponents, one can obtain simple bounds or approximations for the entire network.

For example, suppose that one is able to provide a lower and upper bound for the throughout (output) of subnetwork 1 in isolation with an input λ_1, and of subnetwork 2 in isolation with an input λ_2. Then, for given λ_1, using the throughput bounds of subnetwork 1 as extreme input values for λ_2 and using the throughput lower and upper bounds of subnetwork 2 for this given λ_2 value, a secure lower and upper bound for the total system throughput (output) λ_3 is also attained. Clearly, the inaccuracy of the bounds for subnetwork 1

Table 7 Bounds for finite two cluster tandem line

μ_1	μ_2	μ_3	μ_4	T_1	T_2	B_L	B	B_U	$\frac{B_L+B_U}{2B}$ in %
1	1	1	1	3	5	.33	.42	.52	101.2%
1	1	1	1	6	6	.25	.30	.40	107.6%
1	1	1	1	8	8	.20	.24	.33	109.3%
2	2	1	1	10	10	.054	.084	.101	112.6%
1	2	3	2	10	10	.10	.12	.17	92.2%
1	2	3	2	10	10	.021	.049	.065	87.3%

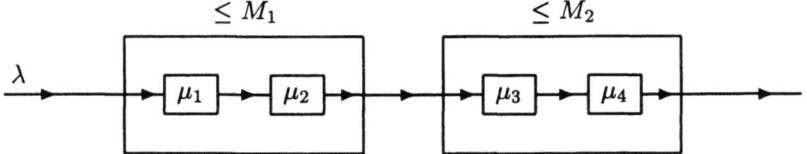

and subnetwork 2 accumulate, but the bounds will still be of the right order of magnitude. If one needs approximations, one can take the average of the lower and upper throughput bounds of subnetwork 1 as an approximate input λ_2^* value for subnetwork 2. The average of the lower and upper throughput bounds of subnetwork 2 with this given λ_2^* value can then be used to obtain an approximate throughput estimate λ_3^*.

5 CONCLUSIONS

This chapter dealt with queueing networks in which arrivals are Poisson and all service times are exponential. If the routing is independent of the state of the network, then the principle of *station balance* applies for every station i, that is, in every state, the rate of entering the state through an arrival to station i is equal to the rate of leaving the state through a departure from station i. When station balance holds, then the solution is of product form, which simplifies the computational complexity for finding the solution considerably.

In practical applications, the assumption that routing is independent of the state of the network is often violated because of finite buffers, state dependent routing, or service breakdowns. In such cases, the station balance is often violated because either the inrate is zero while the out-rate is greater than zero, or because the outrate is greater than zero while the inrate is equal to zero. Based upon this insight, one can modify the system such that both rates are zero, and recover the station balance in this way. This may lead to easy to compute bounds for specific performance measures such as a loss probability or throughput. A number of generic cases is studied in detail. Also, numerical support is provided that shows that the bounds obtained in this fashion are reasonable for many practical applications.

Acknowledgments

The first author would like to acknowledge the Operations Research Center at the Massachusetts Institute of Technology (MIT) for facilitating a visit during which this chapter is written. The authors are also grateful to the Natural Sciences and Research Council of Canada for the financial support to realize this chapter.

References

[Basket et al., 1975] Basket, F., Chandy, K. M., Muntz, R. R., and Palacios, F. (1975). Open, closed and mixed networks with queues with different classes of customers. *Journal of the ACM*, 22:248–260.

[de Souza e Silva and Muntz, 1990] de Souza e Silva, E. and Muntz, R. R. (1990). Queueing networks: Solutions and applications. In Takagi, H., editor, *Stochastic Analysis of Computer and Communication Systems*, pages 319–399. North-Holland, Amsterdam.

[Guiasu, 1986] Guiasu, S. (1986). Maximum entropy condition in queueing theory. *J. Opl. Res. Soc.*, 37:293–301.

[Heyman and Sobel, 1982] Heyman, D. P. and Sobel, M. J. (1982). *Stochastic Models in Operations Research*, volume 1. McGraw-Hill, New York.

[Hordijk and Van Dijk, 1982] Hordijk, A. and Van Dijk, N. M. (1982). Stationary probabilities for networks of queues. In *Applied Probability, Computer Science, Vol II, The Interface*, pages 423–451. Birkhäuser, Boston.

[Hordijk and Van Dijk, 1983a] Hordijk, A. and Van Dijk, N. M. (1983a). Adjoint processes, job local balance and insensitivity for stochastic networks. *Bull 44th Session Int. Stat. Inst.*, 50:776–788.

[Hordijk and Van Dijk, 1983b] Hordijk, A. and Van Dijk, N. M. (1983b). *Networkds of Queues: Part I: Job-Balance and the Adjoint Process; Part II: General Routing and Service Characteristics*. Lecture Notes in Control and Informations Sciences, (Eds Baccelli F. and Fayelle, G.). Springer Verlag.

[Jackson, 1957] Jackson, J. R. (1957). Networks of waiting lines. *Operations Research*, 5:518–521.

[Jackson, 1964] Jackson, J. R. (1964). Job shop like queueing systems. *Management Science*, 10(1):131–142.

[Kelly, 1979] Kelly, F. P. (1979). *Reversibility and Stochastic Networks*. John Wiley and Sons, New York.

[Reiser, 1980] Reiser, M. (1980). Mean-value analysis of closed multichain queuing networks. *JACM*, 27(2):313–322.

[Robertazzi, 1990] Robertazzi, T. G. (1990). *Computer Networks and Systems*. Springer-Verlag, New York, NY.

[Schassberger, 1978] Schassberger, R. (1978). Insensitivity of steady-state distributions of generalized semi-Markov processes with speeds. *Adv. Appl. Probability*, 10:836–851.

[van Dijk, 1989a] van Dijk, N. M. (1989a). A simple throughput bound for large closed queueing networks with finite capacities. *Performance Evaluation*, 10:153–167.

[van Dijk, 1989b] van Dijk, N. M. (1989b). "Stop = Recirculate" for exponential product form queueing networks with departure blocking. *Oper. Res. Letters*.

[van Dijk, 1990a] van Dijk, N. M. (1990a). A discrete-time product form for random access protocols. In *Proceedings of IEEE INFOCOM'90*, pages 1071–1077, San Francisco, California.

[van Dijk, 1990b] van Dijk, N. M. (1990b). Mixed parallel processors with interdependent blocking. *Applied Stochastic Models and Data Analysis*, 6:85–100.

[van Dijk, 1990c] van Dijk, N. M. (1990c). Queueing systems with restricted workload: an explicit solution. *J. Appl. Prob.*, 27:393–400.

[van Dijk, 1991] van Dijk, N. M. (1991). On "stop = repeat" servicing for nonexponential queueing networks with blocking. *J. Appl. Prob.*, 28:159–173.

[Van Dijk, 1993] Van Dijk, N. M. (1993). *Queuing Networks and Product Forms*. Wiley, New York, NY.

[Van Dijk and Akylildiz, 1990] Van Dijk, N. M. and Akylildiz, I. F. (1990). Networks with mixed processor sharing parallel queues and common pools. In *Performance 90*, Amsterdam. North Holland.

[van Dijk and Lamond, 1988] van Dijk, N. M. and Lamond, B. F. (1988). Simple bounds for finite single-server exponential tandem queues. *Operations Research*, 36:470–477.

[Van Dijk and Tijms, 1986] Van Dijk, N. M. and Tijms, H. C. (1986). Intensitivity in two-mode blocking models with applications. In Boxma, O. J. and Cohen, W. J., editors, *Teletraffic Analysis and Computer Performance Evaluation*. North Holland, Amsterdam.

[van Dijk et al., 1988] van Dijk, N. M., Tsoucas, P., and Walrand, J. (1988). Simple bounds and monotonicity of the call congestion of finite multiserver delay systems. *Probability in the Engineering and Informational Sciences*, 2:129–138.

[van Dijk and van der Wal, 1989] van Dijk, N. M. and van der Wal, J. (1989). Simple bounds and monotonicity results for finite multi-server exponential tandem queues. *Queueing Systems*, 4:1–16.

[Walrand, 1988] Walrand, J. (1988). *An Introduction to Queueing Theory*. Prentice-Hall, Englewood Cliffs, NJ.

[Woodside, 1988] Woodside, C. M. (1988). Throughput calculations for basic stochastic rendezvous networks. *Performance Evaluation*, 9(9):143–160.

Nico van Dijk is director of the Operations Research and Management program at the University of Amsterdam, The Netherlands. His theoretical research interests include exact analytic expressions and simple performance estimates and bounds for queueing networks, as well as analytic error bounds and comparison results for approximating queueing systems. His interest in practical applications have lead him to investigate hybrid analytic and simulation methods for practical queueing problems.

Generally, his aim is to advocate the role of analytic and quantitative methods for actual applications in fields such as telecommunications, service industries and public transportation. As such he has become strongly involved in the analysis of a variety of real life queueing problems in postal offices, airports, railways and presently, call centers. Accordingly, he has written several popularizing articles in daily newspapers, magazins and journals.

Winfried Grassmann got his education in economics in Zurich, Switzerland. After his Masters, he joined the Operations Research Department of Swissair, the Swiss flag carrier. There, he developed a system for inventory control, and a system for dealing with rotating parts. Both systems were implemented with great success. While at Swissair, he also finished his Ph.D., which he defended in 1968 with summa cum laude. He then joined the Computer Science Department of the University of Saskatchewan, where he taught Operations Research and Computer Science. Dr. Grassmann was on the editorial boards of Naval Research Logistics and Operations Research, and he is presently associate editor of the INFORMS Journal on Computing. He has written a book on stochastic processes and, more recently, a book on logic and discrete mathematics. His main areas of research are queueing theory and Markov modeling, two areas in which he has published widely. His papers have appeared in Operations Research, Journal of Applied Probability, Interfaces, Naval Research Logistics, INFOR and other journals. For his lifetime achievements, he received the 1999 merit award of the Canadian Operational Research Society.

12 TECHNIQUES FOR SYSTEM DEPENDABILITY EVALUATION

Jogesh K. Muppala[1], Ricardo M. Fricks[2] and Kishor S. Trivedi[3]

[1]Dept. of Computer Science
The Hong Kong University of Science and Technology
Clear Water Bay Kowloon
Hong Kong
muppala@cs.ust.hk

[2]SIMEPAR - The Meteorological System of Paraná
Paraná State Power Company
Curitiba, PR 80420-170, Brazil
fricks@simepar.br

[3]Dept. of Electrical and Computer Engineering
Duke University
Durham, NC 27706, USA
kst@ee.duke.edu

1 INTRODUCTION

A major application area for the probabilistic and numerical techniques explored in the earlier chapters is in characterizing the behavior of complex computer and communication systems. While system *performance* has received a lot of attention in the past, increasingly system *dependability* is gaining importance. The proliferation of computer and computer-based communication systems has contributed to this in no small measure. This chapter is thus a step in the direction of summarizing the techniques, tools and recent developments in the field of system dependability evaluation.

Laprie [Laprie, 1985] defines system dependability as the ability of a system or a product to deliver its intended level of service to its users, especially in the light of failures or other incidents that impinge on its level of service.

Dependability can thus be viewed as an all-encompassing term that considers the system reliability, availability, safety and maintainability. Depending on the application environment, one or more of these characteristics is an appropriate reflection of the system behavior. For example, in a management information system environment, the proportion of time that the system is able to deliver its intended level of service (system availability) is an important measure. In an aircraft flight-control system, system failures may be catastrophic. Thus the ability of the system to continue delivering its service without a catastrophic failure (system reliability) is of greater importance.

Heimann et al [Heimann et al., 1990] list three different reasons for using dependability in system evaluation:

1. *Dependability allows comparisons with cost and performance.* Along with cost and performance, dependability is the third critical criterion based on which system related decisions are made.

2. *Dependability provides a proper focus for product-improvement efforts.* Dependability evaluation enables us to identify which aspect of the system behavior viz. subsystem dependability, fault coverage, or maintenance strategy plays a crucial role in determining the overall system dependability.

3. *Dependability can take into account safety and risk issues.* Dependability evaluation enables us to identify unsafe situations and the inherent risks involved in the system being unable to deliver its intended level of service.

Traditionally, reliability block diagrams and fault trees were commonly used for system reliability and availability analysis [Shooman, 1970]. These model types allow a concise description of the system under study and can be evaluated efficiently, but they cannot easily represent dependencies occurring in real systems [Sahner and Trivedi, 1987a]. Markov models, on the other hand, are capable of capturing various kinds of dependencies that occur in reliability/availability models [Dugan et al., 1986, Goyal et al., 1986, Trivedi, 1982].

In this chapter we give an overview of dependability evaluation techniques and tools. We start by defining the various measures of dependability in Section 2. Then we introduce the different modeling techniques commonly used in representing system dependability behavior in Section 3. Thereafter, we review the computational methods used for evaluating the dependability measures from the dependability models in Section 4. Finally we mention some important issues encountered in dependability modeling in Section 5.

2 MEASURES OF DEPENDABILITY

Dependability measures can be viewed as belonging to two major categories: system reliability, and system availability. Depending on the specific situation under investigation, one or more of these measures may be appropriate. System *reliability* measures are typically relevant in situations where the systems are highly sensitive to occurrences of system failures or interruptions. For example, in aircraft flight control and spacecraft mission control, the system is expected to provide uninterrupted service. System *availability* measures are more suitable for systems where short interruptions can be tolerated. Most of the commercial applications of computer systems, for example airline reservations systems, automated banking systems, fall into this category.

Other specialized measures may be defined for specific systems, which better reflect the abilities of the systems under consideration. Heimann et al. [Heimann et al., 1990] provide a good discussion on this topic.

2.1 Reliability Measures

System *reliability* is one of the most commonly used measure for evaluating mission critical systems.

Definition 12.1 *The* **reliability** $R(t)$ *of a system at time t is the probability that the system failure has not occurred in the interval* $[0,t)$. *If X is a random variable that represents the time to occurrence of system failure, then* $R(t) = P(X > t)$.

We can compute the system *unreliability* as $1 - R(t)$, a more appropriate measure for highly-reliable systems given the finite precision of numbers in digital computers. Another important and often used measure of interest is the *mean time to failure* of the system.

Definition 12.2 *The* **mean time to failure** *MTTF of a system is the expected time until the occurrence of the (first) system failure. If X is a random variable that represents the time to occurrence of system failure, then* $MTTF = E[X]$.

Given the system reliability $R(t)$, the MTTF can be computed as,

$$\text{MTTF} = \int_0^\infty R(t)dt.$$

2.2 Availability Measures

System *availability* measures are especially relevant in repairable systems, where brief interruptions in service are tolerated. Depending on the time horizon of interest, system availability measures can be expressed in three different forms.

Definition 12.3 *The* **instantaneous availability** $A(t)$ *of a system at time t is the probability that the system is functioning correctly at time t.*

Like the reliability measure, in some applications it is better to compute the system *unavailability* $U(t) = 1 - A(t)$.

Definition 12.4 *The* **interval availability** $\overline{A}(t)$ *of a system during the interval* $[0,t)$ *is the expected proportion of time within the interval the system was functioning correctly.*

It must be noted that

$$\overline{A}(t) = \frac{1}{t} \int_0^t A(x)dx.$$

Definition 12.5 *The* **steady-state availability** A_{SS} *represents the long-term probability that the system is available:*

$$A_{SS} = \lim_{t \to \infty} A(t).$$

The *steady-state unavailability* of the system is given by $U_{SS} = 1 - A_{SS}$.

Definition 12.6 *The* **limiting interval availability** $\overline{A}$ *is the expected fraction of time the system is operating:*

$$\overline{A} = \lim_{t \to \infty} \overline{A}(t).$$

If $\lim_{t \to \infty} A(t)$ exists, then the steady-state availability and the limiting interval availability are equal [Barlow and Proschan, 1975, Leemis, 1995], i.e.,

$$\overline{A} = \lim_{t \to \infty} \frac{1}{t} \int_0^t A(x)dx = A_{SS}.$$

Other possible measures of system availability are [Goyal et al., 1987, Heimann et al., 1990]: (cumulative) availability distribution, tolerance availability, capacity oriented availability, tolerance capacity oriented availability, and degraded capacity time.

3 DEPENDABILITY MODEL TYPES

Several model types are used in the representation of the dependability behavior of complex systems and obtaining various measures of dependability. These methods can be broadly classified into two different types: (1) non-state space models, and (2) state space models, depending on the nature of their constitutive elements and solution techniques. Non-state space models do not require the enumeration of systems states, while state space modeling techniques demand the collection of variables the values of which define the state of the system at a given point. Non-state space models allow a concise description of the system under study and can be evaluated efficiently, but they cannot represent system dependencies occurring in real systems [Sahner and Trivedi, 1987a] unlike state space based methods [Dugan et al., 1986, Goyal et al., 1986, Muppala and Trivedi, 1995, Trivedi, 1982].

State space models may be deterministic or stochastic in nature. Models are said to be deterministic if their elements are sufficiently specified so that the model behavior is exactly determined. Stochastic models, on the other hand, have probabilistic nature and do not determine numerical values for the variables as the deterministic models do. They normally determine probabilities and/or moments associated with system state and output variables. The state probabilities are then calculated and, subsequently, the dependability measures of interest are determined.

Stochastic models are usually the method of choice when modeling dependability of computer systems since phenomena involving significant uncertainties and unpredictable variability (inherent in the system or in its inputs) frequently need to be represented. Through the probabilistic approach, the repercussions of such uncertainties in the model solution can be taken into consideration.

In the practice of reliability engineering, stochastic models are further classified as Markovian or non-Markovian. This distinction is based on the joint distribution of the underlying stochastic process. Markov models have often been used for hardware and software dependability assessment of computer systems. Reasons for the popularity of Markov models include the ability to capture various dependencies, the equal ease with which steady-state, transient, and cumulative transient measures can be computed, and the extension to Markov reward models useful in performability analysis [Trivedi et al., 1992]. A wide range of dependability modeling problems fall in the domain of Markov models (both homogeneous and non-homogeneous). However, some important aspects of system behavior in stochastic models cannot be easily captured through a Markov model. The common characteristic these problems share is that the *Markov property* [Trivedi, 1982] is not valid (if valid at all) at all time instants. This category of problems is jointly referred as non-Markovian models.

Non-Markovian models can be analyzed introducing supplementary variables [Cox, 1955a] or phase-type expansions [Botta et al., 1987, Neuts, 1978] in the state space of the model. In some circumstances, however, it is possible to analyze non-Markovian models by considering some appropriately chosen embedded epochs in the process evolution where the Markov property applies [Çinlar, 1975, Kulkarni, 1995]. Several well-known classes of stochastic processes such as regenerative, semi-Markov, and Markov regenerative processes are based on this concept. Unlike the supplementary variable and phase-type expansion approaches, embedding techniques do not increase the cardinality of the state space of the model, and this is their utmost advantage over the other non-Markovian techniques mentioned.

In the remainder of this subsection, we briefly review several model types that are commonly employed in dependability evaluation. To better illustrate the application of the reviewed techniques we use the following example:

Example 1 *Consider an internetworking system consisting of two local area network (LAN) segments connected together by a bridge as shown in Figure 1. Each LAN segment has its own local file server and supports the data needs of the workstations on the segment. The data on each file server is replicated on the other file server so that the workstations can continue to access the data even if one of the file servers has failed. Data consistency is maintained between the two servers by requiring that any updates to data on one server be immediately reflected on the other server. The bridge is a critical component in maintaining the data consistency. Thus the system is operational as long*

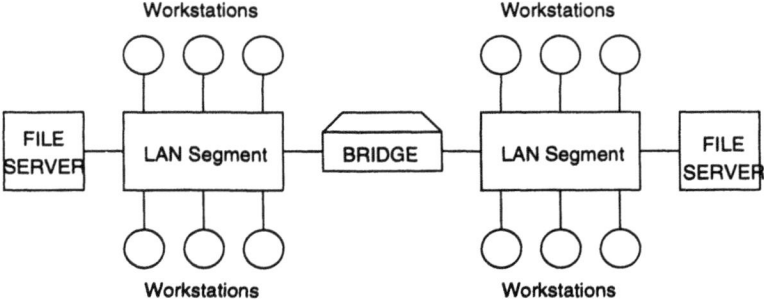

Figure 1 The internetworking system

as one of the file servers, and the bridge are operational. The failures of the workstations, which serve as the access points to the network, are ignored.

3.1 Non State Space Models

Reliability Block Diagrams (RBDs). In a reliability block diagram model, the components/subsystems are represented as blocks. The interconnections among the blocks reflect the operational dependency of the system on its constituent components/subsystems. Those components, which are all required to be functioning for the (sub)system to be operational, are connected in series.

Component blocks are connected in parallel iff the failure of all of them will result in the failure of the (sub)system. In this model-type the failures of the individual components are assumed to be independent, in that the failure of a component will not influence the failure behavior of any other component. RBDs are viewed as the probability of success approach to systems modeling [Shooman, 1970].

kOFn block structures are also possible in RBDs [Sahner and Trivedi, 1987b]. In a kOFn structure, the block with n components is operational if at least k of its components are operational. Series and parallel block connections represent special cases of kOFn blocks with $k = n$ and $k = 1$, respectively. Other researchers have also considered RBDs with repeated blocks [Buzacott, 1970].

The RBD model for the internetworking system in Example 1 is shown in Figure 2.

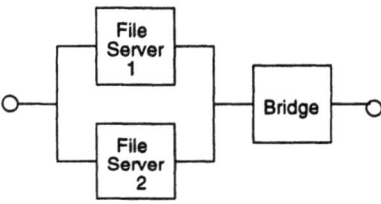

Figure 2 Reliability block diagram model of the internetworking system.

Fault Trees (FTs). Fault trees, unlike RBDs, represent the probability of failure approach to systems modeling [Shooman, 1970]. Fault trees use boolean gates (e.g., AND and OR gates) to represent the operational dependency of the system on its components.

Components are represented as nodes in the tree, which form inputs to the gates. When the component fails, the corresponding input to a gate becomes TRUE. If any input to an OR gate becomes TRUE, then its output also becomes TRUE. The inputs to an OR gate are those components which are all required to be functioning for the (sub)system to be functioning. The inputs to an AND gate, on the other hand, are those components all of which should fail for the (sub)system to fail. Whenever the output of the *topmost* gate becomes TRUE, then the system is considered failed.

The fault-tree corresponding to the internetworking system in Example 1 is shown in Figure 3.

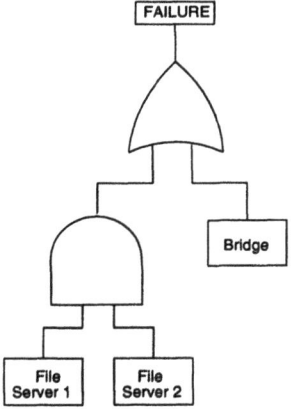

Figure 3 Fault-tree model of the internetworking system.

Several extensions to FTs have been considered including the use of NOT, EXOR, Priority AND, kOFn gates [Dugan et al., 1992]. Another extension considered is fault trees with repeated events (FTREs). In this case different gates are allowed to share inputs.

Example 2 *We modify the internetworking system in Example 1 as follows. We assume that if the bridge fails, the LAN segments can continue to operate as stand-alone segments as long as the local file server is up. Strict data consistency requirements between the two servers are relaxed in these circumstances.*

The system in Example 2 can be modeled using fault trees with repeated events. The corresponding FTRE model is shown in Figure 4. As long as the bridge is up, the system is considered failed only when both the file-servers are failed. However, if the bridge is failed, then the failure of even one file server causes the system to fail. Note that in this example, a similar result

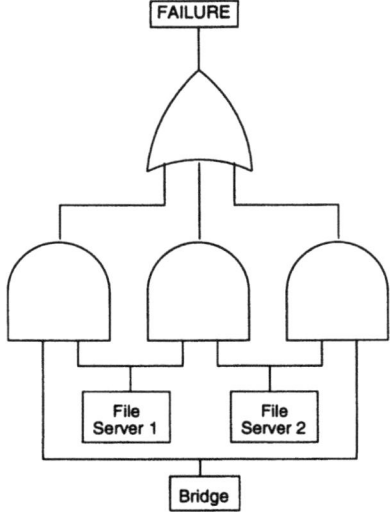

Figure 4 FTRE model of the internetworking system.

can be obtained by using a single kOFn gate allowing dissimilar components, with $k = 2$, $n = 3$, and inputs corresponding to the state of all three modeled components.

Reliability Graphs (RGs). A reliability graph $G = (U, V)$ is a special type of an acyclic digraph where U represents the set of nodes, and V the set of edges in the graph. A component is represented by a directed edge in the graph connecting two nodes. The failure of the component is represented as the deletion of the edge from the graph. Some special edges called ∞-edges represent a component that do not fail. Two special nodes in the graph are designated as the *source* and the *sink* nodes. The source node has no incoming edges, and the sink node has no outgoing edges. The system is considered operational as long as at least one directed path exists from the source node to the sink node.

The reliability graph corresponding to the internetworking system in Example 1 is shown in Figure 5.

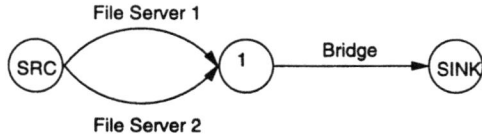

Figure 5 Reliability graph model of the internetworking system.

3.2 State Space Models

One major drawback of the non state space methods presented in the earlier section is that these methods assume stochastic independence between system components. Many intricate system dependencies that arise in modern complex systems cannot be adequately represented by these methods. We are then forced to use state space based methods like continuous time Markov chains, which can handle many of these dependencies. In a later subsection, we will elaborate on various kinds of system dependencies that can be represented using state space models.

Example 3 *We introduce repairs in our internetworking system Example 1 and suppose that the bridge has repair priority over the file servers. Furthermore, whenever the system is down, no further failures can take place. Hence, when the bridge is down, the file servers cannot fail. Similarly when both the file servers are down, the bridge does not fail.*

Here we have introduced failure and repair dependencies among the components of the system. None of the non state space models can handle these interdependencies.

Continuous Time Markov Chains (CTMCs). We have already seen in the earlier chapters the definition and the properties of continuous time Markov chains. Here, we briefly review the use of CTMCs in the modeling of system dependability behavior. The state description of a CTMC can be used to explicitly keep track of the state of the components and subsystems comprising the system. Transitions among the states represent failure/repair events that occur in the system and result in a state change. In using the (homogeneous) CTMC, we are implicitly assuming that the times to occurrence of events (failures, repairs) in the system are all exponentially distributed. This assumption can be relaxed in situations where the distributions can be represented by phase-type approximations [Trivedi, 1982], while still using the CTMCs. Alternatively, non-homogeneous CTMC can be used to allow globally time dependent failure rates [Trivedi, 1982].

Let $\{Z(t), t \geq 0\}$ represent a homogeneous finite-state continuous time Markov chain (CTMC) with state space Ω. that is $Z(t) \in \Omega$. We assume that Ω is a countable set, and without loss of generality, we will assume that $\Omega = \{1, 2, \ldots, n\}$. The infinitesimal generator matrix is given by $\mathbf{Q} = [q_{ij}]$ where $q_{ij}, (i \neq j)$ represents the transition rate from state i to state j, and the diagonal elements are $q_{ii} = -q_i = -\sum_{j \neq i} q_{ij}$. Further, let $q = \max_i |q_{ii}|$ and let η be the number of non-zero entries in $\mathbf{Q}$. Of great interest is the distribution of $Z(t)$ for some fixed t. Hence, let $\pi_i(t) = P\{Z(t) = i\}$.

The behavior of the internetworking system with the failure and repair dependencies in Example 3 can be represented by the continuous-time Markov chain shown in Figure 6. In this figure the label (i, j) of each state is interpreted as follows: i represents the number of file servers that are still functioning and j is 1 or 0 depending on whether the bridge is up or down respectively. The

454   COMPUTATIONAL PROBABILITY

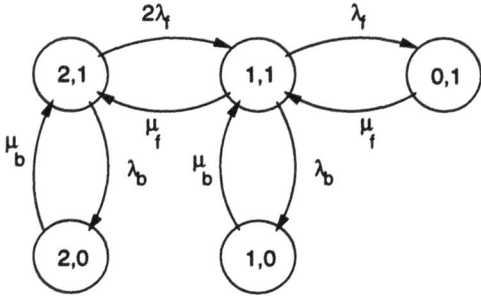

Figure 6 CTMC for the internetworking system.

Table 1 Component parameters.

rates		
failure	repair	meaning
λ_f	μ_f	constant rates of file servers
λ_b	μ_b	constant rates of the bridge

failure and repair rate of all modeled component types are assumed to be time independent with rates listed in Table 1.

Example 4 *We can modify Example 3 further by assuming that the internetworking system does not recover whenever both file servers fail, or whenever the bridge fails. When this situation occurs, no further repairs are carried out on the system.*

We obtain the CTMC corresponding to Example 4 by turning the states $(0,1)$, $(1,0)$, and $(2,0)$ of the CTMC in Figure 6 into absorbing states. The corresponding Markov chain is shown in Figure 7.

Markov Reward Models (MRMs). Markov reward models are extensions to Markov chains obtained by assigning transformation functions $r[X(.)]$ (called *reward rate functions*) to map elements from the state space Ω of a Markov chain into real numbers This approach is similar to the one discussed in Chapter 3. One can also associate reward rates with transitions, as shown in Chapter 3 and [Howard, 1971], but we will not do this here. If the MRM spends τ_i time units in state i during the interval $[0,t)$, then a reward $r_i\tau_i$ is accumulated up to time t. In other words, in state i, the reward rate is r_i.

Rewards can be used in many different ways [Ciardo et al., 1992a, Ciardo et al., 1992b, Heimann et al., 1990, Muppala et al., 1996, Muppala et al., 1993]. Here, we show how to express the availability in terms of rewards. We divide the state space Ω into upstates $\mathcal{U}$ and downstates $\mathcal{D}$. For each $i \in \mathcal{U}$, let $r_i = 1$,

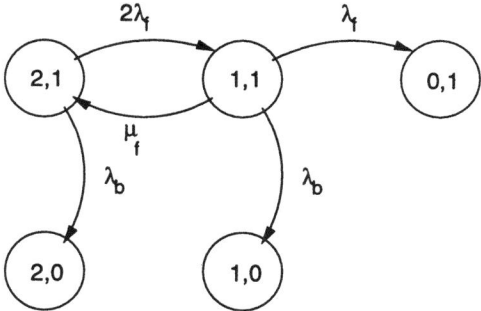

Figure 7 CTMC with absorbing states

and for each $i \in \mathcal{D}$, let $r_i = 0$. Given this reward assignment, the steady-state availability A_{SS} is the expected reward rate in steady-state:

$$A_{SS} = \sum_{i \in \Omega} r_i \pi_i = \sum_{i \in \mathcal{U}} \pi_i.$$

With the same reward assignment, the instantaneous and interval availability can also be computed. The instantaneous availability $A(t)$ is the expected instantaneous reward rate at time t:

$$A(t) = \sum_{i \in \Omega} r_i \pi_i(t).$$

The interval availability $\overline{A}(t)$ is the expected time-averaged accumulated reward over the interval $[0, t)$:

$$\overline{A}(t) = \frac{1}{t} \sum_{i \in \Omega} r_i \int_0^t \pi_i(x) dx.$$

Distinct reward rate functions associated with a given Markov chain produce distinct MRMs. Another property is that the definition of reward rates is orthogonal to the analysis type that is used. Thus, for instance, with the same reward definition we can compute the steady-state availability, as well as instantaneous availability, and interval availability in a dependability model.

Non-Markovian Models. The modeling framework presented so far allows the solution of stochastic problems enjoying the **Markov property**: *the probability of any particular future behavior of the process, when its current state is known exactly, is not altered by additional knowledge concerning its past behavior* [Trivedi, 1982]. If the past history of the process is completely summarized in the current state and is independent of the current time, then the process is said to be *(time-) homogeneous*. Otherwise, the exact characterization of the present state needs the associated time information, and the process is said to be *non-homogeneous*. A wide range of real problems fall in the class of Markov models (both homogeneous and non-homogeneous). However, some important aspects of system behavior in a dependability model cannot be easily captured in a Markov model. The common characteristic these problems share is that the Markov property is not valid (if valid at all) at all time instants. This category of problems is jointly referred to as *non-Markovian* models and can be analyzed using several approaches. Among them we summarize three of the major options: (i) associating supplementary variables to non-exponential random variables; (ii) replacing non-exponential distributions by arrangements of exponential distributions, also called phase-type expansions; and (iii) searching for embedded epochs in the system evolution where the Markov property can be applied.

Supplementary Variables. This method, originally discussed in [Cox, 1955a], allows for the solution of dependability models when the lifetime and/or repair distributions of network components are non-exponential. It is the most direct method of solving the modeling problem and is based on the inclusion of sufficient supplementary variables in the specification of the state of the system to make the whole process Markovian. In dependability models the supplementary variables are the times expended in repairs and ages of network components. The purpose of the added supplementary variables is to include all necessary information about the history of the stochastic process. The resulting Markov process is in continuous time and has a state space which is multidimensional of mixed type, partly discrete and partly continuous.

Since, after the inclusion of the supplementary variables, the stochastic process describing the system behavior satisfies the Markov property, it is possible to derive the Chapman-Kolmogorov equations describing the dynamic behavior for such a process. The resultant set of ordinary or partial differential equations can be defined together with boundary conditions and analyzed. Several non-Markovian dependability models solved using the supplementary variables technique have been reported in recent literature [Cao, 1994, Chen et al., 1993, Dhillon and Anude, 1993, Dhillon and Anude, 1994, Dhillon and Yang, 1993, Gopalan and Dinesh Kumar, 1996, Wu et al., 1994].

Phase Type Expansions. The use of phase type distributions dates back to the pioneering work of Erlang on congestion in telephone systems at the beginning of this century [Brokemeyer et al., 1948]. His approach (named *method of stages*), although simple, was very effective in dealing with non-exponential distributions and has been considerably generalized since then. The age (repair time) of a component is assumed to consist of a combination of stages each of which is exponentially distributed. The division into stages is an operational device and may not necessarily have any physical significance in the process being modeled (component failure or repair activity). The whole stochastic process becomes Markovian provided that the description of the state of the system contains the information as to which stage of the component state duration has been reached.

The major advantage of the phase type expansions is that once a proper stage combination has been found to represent or approximate a distribution, solutions can be obtained for the resulting Markov chain even with fairly complex models [Singh et al., 1977].

The basic phase type expansion techniques approximate a non-exponential distribution by connecting dummy stages with independent and exponential sojourn time distribution in series or parallel (or combination of both). A process with sequential phases gives rise to hypoexponential or an Erlang distribution, depending upon whether or not the phases have identical parameters. Instead, if a process consists of alternate phases (parallel connection) then the overall distribution is hyperexponential. The basic instrument when selecting one of these distributions to represent a non-exponential interval is given by the *coefficient of variation*. The coefficient of variation, C_X, of a random variable is a measure of deviation from the exponential distribution and is given by

$$C_X = \frac{\sigma_X}{E[X]},$$

where σ_X is the standard deviation of the random variable and $E[X]$ is its expectation. This coefficient varies as follows according to the selected distribution:

C_X	Distribution
> 1	Hyperexponential
1	Exponential
< 1	Hypoexponential Erlang
0	Deterministic

Important generalizations of the basic stage devices are the Coxian distributions [Cox, 1955b], Phase Type [Neuts, 1978], and Generalized Hyperexponential [Botta and Harris, 1986]. An alternative representation of the Coxian distribution with computational advantages is the exponential polynomial or exponomial form [Sahner et al., 1995].

Markov Renewal Theory. A set of techniques that proved very powerful for the solution of non-Markovian models of dependability is based on concepts grouped under the umbrella of Markov renewal theory [Çinlar, 1975, Kulkarni, 1995], a collective name that includes **Markov renewal sequences** (MRSs), and two other important classes of stochastic processes with embedded MRSs, named **semi-Markov processes** (SMPs) and **Markov regenerative processes** (MRGPs). Mathematical definitions for these stochastic processes are given now.

We first introduce the *renewal process*. Suppose we are interested in a single event related with the system (e.g., when system components fail). Additionally, assume the times between successive occurrences of this type of event are *independent and identically distributed* (*iid*) random variables. Let $S_0 < S_1 < S_2 < \ldots$ be the time instants of successive events to occur (as shown in Figure 8). The sequence of non-negative *iid* random variables, $\mathbf{S} = \{S_{n+1} - S_n; n = 0, 1, 2, \ldots\}$ is a *renewal process*. Otherwise, if we do not start observing the system at the exact moment an event has occurred (i.e., $S_0 \neq 0$) the stochastic process $\mathbf{S}$ is a *delayed renewal process*.

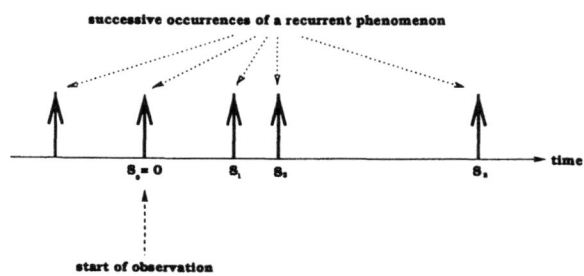

Figure 8 A sample realization of a renewal process.

In reliability, one frequently has more than one event. For example, in a system, one may observe the two events "system fails" and "system comes up again". Hence, at time S_1, there may be a failure, at S_2 the system comes up again, at S_3 it fails, and so on. In this case, there are two system states, namely "up" and "down" which alternate. The times $S_1, S_2, \ldots$ are the state changes, and the convention is that the state at these times is the state entered. Hence, if there is a failure at time S_1, then the system is in the state "down" at S_1. The distribution of $S_{n+1} - S_n$ are now the up-times, respectively, the down times, depending on the state of the system, and these times typically have different distributions.

The idea of having the times $S_{n+1} - S_n$ depend on a state can be generalized. Indeed, we can assume that there is a set of states Ω, which can be thought of as the set $0, 1, \ldots$ as before, and the state at S_n is given by $X_n \in \Omega$. The X_n now form a process on their own. In particular, they may form a discrete-time Markov chain (DTMC). The points S_n are called *Markov regeneration epochs*, or *Markov renewal moments*, and together with the states of the *embedded Markov chain* (EMC) X_n, they define a Markov renewal sequence. The Markov renewal epochs S_n are not renewal moments as described in renewal theory, since the distributions of the time interval between consecutive moments are not necessarily iid.

Definition 12.7 *The bivariate stochastic process* $(\mathbf{X}, \mathbf{S}) = \{X_n, S_n; n = 0, 1, 2, \ldots\}$ *is a* **Markov renewal sequence (MRS)** *provided that*

$$\Pr\{X_{n+1} = j, S_{n+1} - S_n \leq t \mid X_0, \ldots, X_n; S_0, \ldots, S_n\}$$

$$= \Pr\{X_{n+1} = j, S_{n+1} - S_n \leq t \mid X_n\}, \; n = 0, 1, 2, \ldots, \; j \in \Omega, \; t \geq 0.$$

The random variables S_n are the regeneration epochs, and the X_n are the states at these epochs.

Thus $(\mathbf{X}, \mathbf{S})$ is a special case of bivariate Markov process in which the increments $S_1 - S_0, S_2 - S_1, \ldots$ are all non-negative and are conditionally independent given $X_0, X_1, \ldots$. These increments are called the *sojourn times*; if $X_n = j$, then $S_{n+1} - S_n$ is called the sojourn time in state j or the nth sojourn time.

We will always assume *time-homogeneous* MRSs; that is, the conditional transition probabilities $K_{i,j}(t)$, where

$$K_{i,j}(t) = \Pr\{X_{n+1} = j, S_{n+1} - S_n \leq t \mid X_n = i\}$$

are independent of n for any $i, j \in \Omega$, $t \geq 0$. Therefore, we can always write

$$K_{i,j}(t) = \Pr\{X_1 = j, S_1 \leq t \mid X_0 = i\}, \quad i, j \in \Omega, \; t \geq 0.$$

The matrix of transition probabilities $\mathbf{K}(t) = [K_{i,j}(t)]$ is called the *kernel* of the MRS.

There are no restrictions regarding the structure of the EMC on a MRS. For instance, there is no imposition that $\{X_n; n = 0, 1, 2, \ldots\}$ should be irreducible. Therefore, we can start at time S_0 in a state that will not be reached again at any other Markov regeneration epoch in the future evolution of the process.

The semi-Markov process is a generalization of the DTMC. In a DTMC, $S_{n+1} - S_n$ is always 1. The CTMC is also a semi-Markov process, with the distributions of $S_{n+1} - S_n$ having exponential distributions with rate q_i, if $X_n = i$.

Given an MRS $(\mathbf{X}, \mathbf{S})$ with state space Ω and kernel $\mathbf{K}(t)$, we can introduce the counting process

$$\mathbf{N}(t) = \sup\{n : S_n \leq t\}, \quad t \geq 0,$$

to count the number of Markov regeneration epochs up to time t, but not considering the one at zero. Using the counting process just defined, we introduce the definition of SMP.

Definition 12.8 *A* **semi-Markov process (SMP)** *is a process* $\mathbf{Y} = \{Y_t; t \geq 0\}$ *defined by*

$$Y_t = X_{\mathbf{N}(t)}, \quad t \geq 0.$$

It follows from this definition that Y_t is equal to X_n if t is between S_n and S_{n+1}.

An SMP (for a sample realization see Figure 9) is a stochastic process which moves from one state to another, with the successive states visited forming a discrete-time Markov chain, and that the time required for each successive

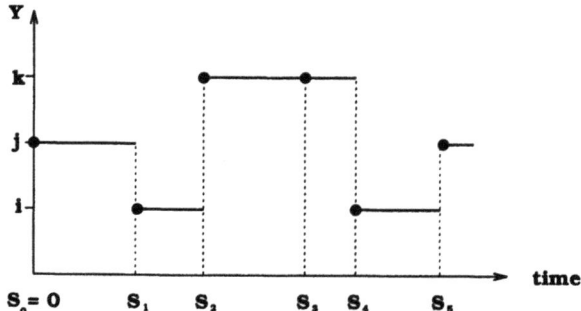

Figure 9 A sample realization of a semi-Markov process.

move is a random variable whose distribution may depend on the two states between which the move is being made. The SMP definition implies that the process only changes state (possibly back to the same state as shown in Figure 9) at the Markov regeneration epochs S_n.

The name "semi-Markov" comes from the somewhat limited Markov property which **Y** has: the future of **Y** is independent of its past provided the present is a Markov regeneration epoch. Note that since we consider $S_0 = 0$, the initial condition $Y_0 = i$ always means that the SMP has just entered state i at time zero.

A stochastic process $\mathbf{Z} = \{Z_t; t \geq 0\}$ with state space Ω is called *regenerative* if there exist time points at which the process probabilistically restarts itself. Such random times when the future of **Z** becomes a probabilistic replica of itself are named *times of regeneration* for **Z**. This concept may be weakened by letting the future after a time of regeneration depend also on the state of an MRS at that time. We then say that **Z** is a Markov regenerative process:

Definition 12.9 A Markov regenerative process (MRGP) *is a stochastic process* $\{Z_t; t \geq 0\}$, $Z_t \in \Omega$ *with an embedded MRSs* $(\mathbf{X},\mathbf{S})$, $X_n \in \mathcal{F}$, *which has the additional property that all conditional finite distributions of* $\{Z_{S_n+t}; t \geq 0\}$, *given* $\{Z_u; 0 \leq u \leq S_n, X_n = i\}$, *are the same as those of* $\{Z_t, t \geq 0\}$, *given* $X_0 = i$.

As a special case, the definition implies that for $i \in \mathcal{F}$, $j \in \Omega$,

$$\Pr\{Z_{S_n+t} = j \mid Z_u, 0 \leq u \leq S_n, X_n = i\} = \Pr\{Z_t = j \mid X_0 = i\}.$$

This means that the MRGP $\{Z_t; t \geq 0\}$ does not have the Markov property in general, but there is a sequence of embedded time points $S_0, S_1, ..., S_n, ...$ such that the states $X_0, X_1, ..., X_n, ...$ realized at these points satisfy the Markov property. It also implies that the future of the process **Z** from $t = S_n$ onwards depends on the past $\{Z_u, 0 \leq u \leq S_n\}$ only through X_n.

In contrast to SMPs, state changes in MRGPs may occur between two consecutive Markov regeneration epochs (see Figure 10). Such changes do not

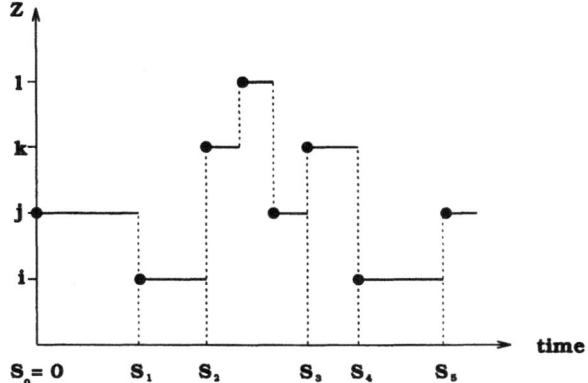

Figure 10 A sample realization of a Markov regenerative process.

imply regeneration. The states that can be visited at Markov regeneration epochs may form a set different from Ω, the state space of Z_t. For this reason, we assumed in the above definition that $X_n \in \mathcal{F}$. Of course, $\mathcal{F} \subseteq \Omega$.

The stochastic process between the consecutive Markov regeneration epochs, usually referred to as *subordinated process*, can be any continuous-time discrete-state stochastic process over the same probability space. Recently published examples considered subordinated homogeneous CTMCs [Choi et al., 1994, German and Lindemann, 1994], non-homogeneous CTMCs [Garg et al., 1997], SMPs [Bobbio and Telek, 1995, Ciardo et al., 1994], MRGPs [Telek, 1994], or a more general stochastic reward process.

4 COMPUTATIONAL METHODS

4.1 Non State Space Models

Reliability Block Diagrams (RBD). To compute the reliability of a system represented as a reliability block diagram, we normally break it down into its serial and parallel parts and compute the reliabilities of these parts and then compose the solution to obtain the reliability of the whole system. Given a system consisting of N components with $F_i(t)$ representing the failure time distribution of component i, we know that $F(t)$, the distribution of the failure time of a subsystem with N components is given by [Sahner et al., 1995]:

$$F(t) = \begin{cases} 1 - \prod_{i=1}^{N}(1 - F_i(t)) & \text{for a series structure,} \\ \prod_{i=1}^{N} F_i(t) & \text{for a parallel structure.} \end{cases}$$

In the RBD model of the internetworking system in Example 1, suppose $R_f(t)$ and $R_b(t)$ represent the individual reliabilities of the file server and the bridge respectively, then the overall system reliability $R(t)$ is given by,

$$R(t) = [1 - (1 - R_f(t))^2] R_b(t). \tag{1}$$

Note that this expression is independent of the distribution associated with the time to failure of each component. In the present example, we assumed that the times to failure of each component are exponentially distributed. Hence, the reliability is given by,

$$R(t) = [1 - (1 - e^{-\lambda_f t})^2] e^{-\lambda_b t}.$$

The system mean time to failure (MTTF) is given by:

$$\text{MTTF} = \int_0^\infty R(t) dt = \int_0^\infty (2e^{-(\lambda_f+\lambda_b)t} - e^{-(2\lambda_f+\lambda_b)t}) dt = \frac{2}{\lambda_f + \lambda_b} - \frac{1}{2\lambda_f + \lambda_b}.$$

We can also use the reliability block diagrams to compute the system unavailability if we assume that the failure and repair time distributions are all independent. This situation occurs when the system has enough repair resources to repair all the failed components simultaneously. Given the component instantaneous unavailability $U_i(t)$, the subsystem unavailability is computed as,

$$U(t) = \begin{cases} 1 - \prod_{i=1}^N (1 - U_i(t)) & \text{for a series structure,} \\ \prod_{i=1}^N U_i(t) & \text{for a parallel structure.} \end{cases}$$

The same formula can be extended to the steady-state unavailability.

For the internetworking system in Example 1 suppose we further assume that failed components can be repaired. The system unavailability $U(t)$ is given by,

$$U(t) = 1 - (1 - U_b(t))(1 - U_f^2(t)).$$

The instantaneous system availability $A(t) = 1 - U(t)$ is then,

$$A(t) = [1 - (1 - A_f(t))^2] A_b(t).$$

Let the time to repair a file server and the time to repair the bridge be exponentially distributed with the parameters μ_f and μ_b respectively. Hence, the instantaneous availability is given by

$$A(t) = \left\{ 1 - \left[\frac{\lambda_f}{\lambda_f + \mu_f} \left(1 - e^{-(\lambda_f+\mu_f)t} \right) \right]^2 \right\} \left(\frac{\mu_b}{\lambda_b + \mu_b} + \frac{\lambda_b}{\lambda_b + \mu_b} e^{-(\lambda_b+\mu_b)t} \right)$$

Fault Trees. Analysis of a fault tree is similar to the reliability block diagram analysis. Given the failure time distributions of component i, $F_i(t)$, the failure time distribution $F_G(t)$ for a gate with n inputs is computed as [Sahner et al., 1995]:

$$F_G(t) = \begin{cases} \prod_{i=1}^n F_i(t) & \text{and gate,} \\ 1 - \prod_{i=1}^N (1 - F_i(t)) & \text{or gate.} \end{cases}$$

For the internetworking system in Example 1, the reliability expression obtained from the fault-tree will also be the same as in Equation (1) above. Even in case of the fault-trees, the expression is independent of the distribution associated with the time to failure of the component.

If we consider a kOFn gate, where the output becomes 1 if k of the n components fail, then the corresponding distribution of failure time for the gate is given by:

$$F_G(t) = \begin{cases} \sum_{i=k}^{n} \binom{n}{i} F(t)^i (1-F(t))^{n-i}, & \text{identically distributed inputs,} \\ \sum_{|J| \geq k} \left(\prod_{j \in J} F_j(t) \right) \left(\prod_{j \notin J} (1 - F_j(t)) \right), & \text{non-identically distributed inputs.} \end{cases}$$

If we use fault trees with repeated components, then one of the approaches we can employ in solving the model is the *factoring* or *decomposition* method [Sahner et al., 1995]. The basic idea is to select a component and break the model down into two cases, the first assuming the component has failed, and the second assuming that it has not failed. For each case, a similar fault tree is obtained. The process may need to be repeated several times until fault trees are reached that do no contain any repeated component.

Revisiting Example 2, the fault-tree in Figure 4 can be factored into two cases, conditioned on whether the bridge is up or down as shown in Figure 11.

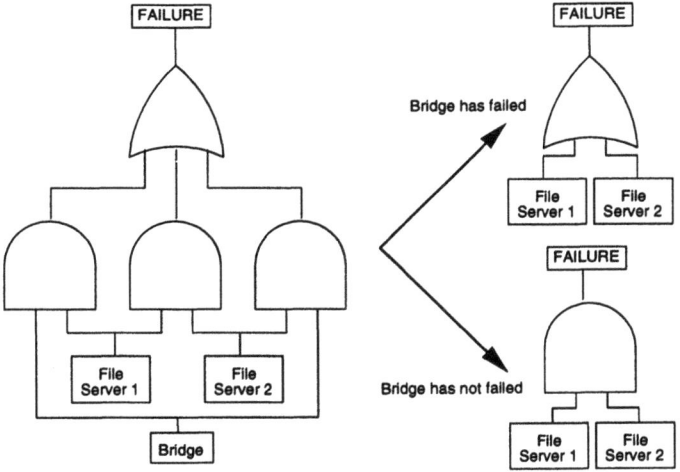

Figure 11 Decomposition of the FTRE model of the internetworking system.

In this case, the distribution of time to failure $F(t)$ is given by,

$$F(t) = F_f^2(t)(1 - F_b(t)) + F_b(t)(1 - (1 - F_f(t))^2)$$

and the corresponding reliability $R(t) = 1 - F(t)$ is given by,

$$R(t) = [1 - (1 - R_f(t))^2] R_b(t) + (1 - R_b(t)) R_f^2(t)$$

Reliability Graphs. Reliability graphs can be analyzed using the factoring algorithm illustrated in the earlier section. However for general network reliability problems or fault tree efficient algorithms are available [Rai et al., 1995, Veeraraghavan and Trivedi, 1991]. These algorithms typically use the following two steps [Sahner et al., 1995]:

1. First, generate a set of minimal paths (or minimal cuts). A *path* is a set of components all of which have to be up for the system to be up. A path is considered *minimal* if it has no proper subpaths.

2. Second, manipulate the paths to obtain the system reliability. Generally either the *inclusion-exclusion* technique or the *sum of disjoint products* technique is used. If the paths are already disjoint, then the path reliability can be computed easily by multiplying the reliabilities of the components comprising the path, and the system unreliability is obtained by multiplying the path unreliabilities.

4.2 Markovian State Space Models

In this section, we discuss state space models that are formulated as continuous time Markov chains. Both transient and steady state analysis is considered. For transient analysis, see also Chapter 3 for further details. Non-Markovian models are described in the next section.

Instantaneous transient analysis. Let $\pi_i(t) = Pr\{Z(t) = i\}$ be the unconditional probability of the CTMC being in state i at time t. Then the row vector $\boldsymbol{\pi}(t) = [\pi_1(t), \pi_2(t), \ldots, \pi_n(t)]$ represents the *transient state probability vector* of the CTMC. The behavior of the CTMC can be described by the following Kolmogorov differential equation:

$$\frac{d}{dt}\boldsymbol{\pi}(t) = \boldsymbol{\pi}(t)\mathbf{Q}, \quad \text{given} \quad \boldsymbol{\pi}(0), \qquad (2)$$

where $\boldsymbol{\pi}(0)$ represents the initial probability vector (at time $t=0$) of the CTMC.

Cumulative Transient Analysis. Define $\boldsymbol{L}(t) = \int_0^t \boldsymbol{\pi}(u)du$. Then $L_i(t)$ is the expected total time spent by the CTMC in state i during the interval $[0, t)$. $\boldsymbol{L}(t)$ satisfies the differential equation:

$$\frac{d}{dt}\boldsymbol{L}(t) = \boldsymbol{L}(t)\mathbf{Q} + \boldsymbol{\pi}(0), \qquad \boldsymbol{L}(0) = \mathbf{0}, \qquad (3)$$

which is obtained by integrating Equation (2). When doing this, use

$$\int_0^t \frac{d}{dt}\boldsymbol{\pi}(t) = \boldsymbol{\pi}(t) - \boldsymbol{\pi}(0)$$

and

$$\frac{d}{dt}\boldsymbol{L}(t) = \boldsymbol{\pi}(t).$$

Steady-State Analysis. Let π_i be the steady-state probability of state i of the CTMC, and let $\pi = \lim_{t\to\infty} \pi(t)$ be the steady-state probability vector. We know that in the steady state $\frac{d}{dt}\pi(t) = 0$. By substituting this into Equation (2) we can derive the following equation for the steady state probabilities:

$$\pi\mathbf{Q} = 0, \quad \sum_{i\in\Omega} \pi_i = 1. \tag{4}$$

Let us return to the internetworking system of Example 3. Since the system is repairable, it is meaningful in this case to compute the availability of the system. We note that the system is available as long as it is in states (2,1) and (1,1). Hence the instantaneous availability of the system $A(t)$, which is the probability that the system is operational at time t is given by,

$$A(t) = \pi_{(2,1)}(t) + \pi_{(1,1)}(t).$$

If we consider interval availability $A_I(t)$, the expected fraction of time during the interval $[0, t)$ that the system is available, then it can be computed as,

$$A_I(t) = \frac{L_{(2,1)}(t) + L_{(1,1)}(t)}{t}.$$

The steady-state availability A_{SS} is given by,

$$A_{SS} = \pi_{(2,1)} + \pi_{(1,1)}.$$

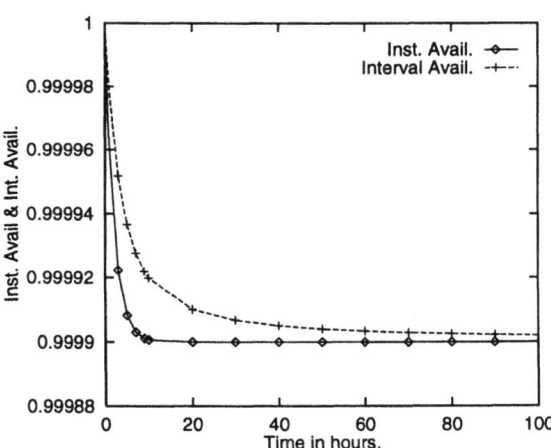

Figure 12 Availability for the internetworking system.

Availabilities for this system, as functions of time, are plotted in Figure 12. For the plots, we assume that $\lambda_f = 0.0001$ hr^{-1}, $\lambda_b = 0.00005$ hr^{-1}, $\mu_f = 1.0$ hr^{-1}, and $\mu_b = 0.5$ hr^{-1}. We notice that both the instantaneous and interval availability decrease as expected and reach the steady-state value of the availability A_{SS} which is 0.9999.

Up-to-Absorption Analysis. Let A represent the set of absorbing states (a state is considered an absorbing state if there are no outgoing transitions from that state, i.e., an absorbing state i has $q_{ij} = 0, \forall j, (j \neq i)$). Let B ($= \Omega - A$) be the set of the transient states in the CTMC. From the matrix $\mathbf{Q}$ a new matrix can be constructed by restricting $\mathbf{Q}$ to states in B only: $\mathbf{Q}_B$ is of size $|B| \times |B|$, where $|B|$ is the cardinality of the set B.

Let $z_i = \int_0^\infty \pi_i(\tau)d\tau$, $i \in B$, the mean time spent by the CTMC in state i until absorption. The row vector $z = [z_i]$ satisfies the following equation:

$$z\mathbf{Q}_B = -\pi_B(0), \qquad (5)$$

where $\pi_B(0)$ is the vector $\pi(0)$ restricted to the states in the set B. The above equation can be obtained by taking the limit as

$t \to \infty$ of Equation (3), with $z = L_B(\infty)$ and noting that $\frac{d}{dt}L_B(\infty) = 0$. The mean time to absorption (MTTA), of the CTMC into an absorbing state is computed as

$$\text{MTTA} = \sum_{i \in B} z_i .$$

The mean time to failure MTTF of the internetworking system in Example 4, which is the same as the mean time to absorption for the Markov chain given in Figure 7, is obtained as

$$\text{MTTF} = z_{(2,1)} + z_{(1,1)}.$$

Assuming that $\lambda_f = 0.0001$ hr^{-1}, $\lambda_b = 0.00005$ hr^{-1}, and $\mu_f = 1.0$ hr^{-1} we obtain the mean time to failure as 19992 hours.

Furthermore, since this Markov chain has absorbing states, we can also compute the reliability of the system. Since all system failure states are absorbing, it follows that if the system is functioning at time t, it must be functioning throughout the interval $[0, t)$. Thus,

$$R(t) = \pi_{(2,1)}(t) + \pi_{(1,1)}(t).$$

The reliability for the example computer system is plotted in Figure 13. Also plotted is the system reliability considering that file servers cannot be repaired when the system is operational. The state transition diagram of the CTMC modeling this case is similar to the one presented in Figure 7, except that there is no transition from state (1,1) back to state (2,1).

4.3 Non-Markovian Models

Let $\mathbf{Z} = \{Z_t; t \geq 0\}$ be a stochastic process with discrete state space Ω and embedded MRS $(\mathbf{X}, \mathbf{S}) = \{X_n, S_n; n = 0, 1, 2, \ldots\}$ with kernel matrix $\mathbf{K}(t)$. For such a process we can define a matrix of conditional transition probabilities as:

$$V_{i,j}(t) = \Pr\{Z_t = j \mid Z_0 = i\}, \quad i \in \mathcal{F}, \, j \in \Omega, \, t \geq 0.$$

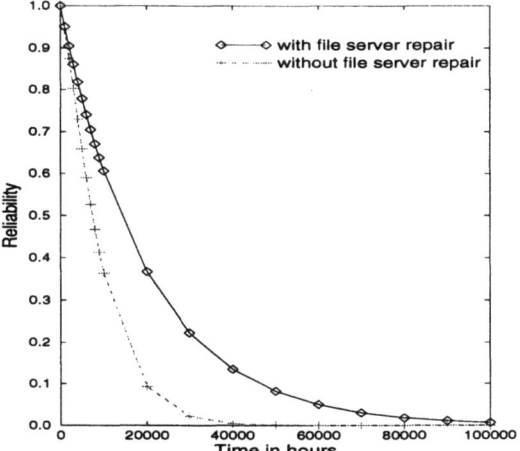

Figure 13   Reliability for the computer system

In many problems involving Markov renewal processes, our primary concern is finding ways to effectively compute $V_{i,j}(t)$ since several measures of interest (e.g., reliability and availability) are related to the conditional transition probabilities of the stochastic process.

At any instant t, the conditional transition probabilities $V_{i,j}(t)$ of $\mathbf{Z}$ can be written as [Çinlar, 1975, Kulkarni, 1995]:

$$V_{i,j}(t) = \Pr\{Z_t = j, S_1 > t \mid Z_0 = i\} + \sum_{k \in \mathcal{E}} \int_0^t dK_{i,k}(u) V_{k,j}(t-u).$$

This equation is obtained by conditioning on the state at the first transition, respectively, if there is none, on the fact that $S_1 > t$.

If we construct a matrix $\mathbf{E}(t) = [E_{i,j}(t)]$ with

$$E_{i,j}(t) = \Pr\{Z_t = j, S_1 > t \mid Z_0 = i\}, \quad i \in \mathcal{F},\ j \in \Omega,\ t \geq 0.$$

Then, the set of integral equations $V_{i,j}(t)$ defines a **Markov renewal equation**, and can be expressed in matrix form as

$$\mathbf{V}(t) = \mathbf{E}(t) + \int_0^t d\mathbf{K}(u)\mathbf{V}(t-u),$$

where the Lebesgue-Stieltjes integral,

$$\int_0^t d\mathbf{K}(u)\mathbf{V}(t-u) = \int_0^t \mathbf{k}(u)\mathbf{V}(t-u)du,$$

is taken term by term. If the stochastic process $\mathbf{Z}$ is a SMP then $\mathbf{E}(t)$ is a diagonal matrix with elements

$$E_{i,i}(t) = 1 - K_i(t).$$

The Markov renewal equation represents a set of coupled *Volterra integral equations of the second kind* [Fröberg, 1969] and can be solved in time-domain or in Laplace-Stieltjes domain. For a discussion of approaches to solve these equations see [German et al., 1995, Telek et al., 1995]. References for the application of Markov renewal theory in the solution of performance and reliability/availability models, see [Bobbio et al., 1995, Fricks et al., 1997, Garg et al., 1997, German et al., 1995, Logothetis and Trivedi, 1995, Logothetis and Trivedi, 1997, Mainkar and Trivedi, 1993].

To better distinguish the roles of matrices $\mathbf{E}(t)$ and $\mathbf{K}(t)$ in the description of the MRGP we use the following terminology:

- We call matrix $\mathbf{E}(t)$ the **local kernel** of the MRGP. It not only describes the situation from 0 to S_1, but, because of the Markovian property, it also describes the state probabilities of the process during the interval between any successive Markov regeneration epochs.

- Since matrix $\mathbf{K}(t)$ describes the evolution of the process from the Markov regeneration epoch perspective, without describing what happens in between these moments we call it the **global kernel** of the MRGP. The matrix is the joint conditional probability of the time to the next Markov regeneration and the state right after the regeneration given the state at the current Markov regeneration.

The equilibrium probabilities for Z_t can be obtained by arguments involving time proportions. These time proportions are equal to the long run probabilities if the times between events has a density on some interval [Ross, 1970, Theorem 5.8].

Let Z_t be defined as before, and let $(\mathbf{X}, \mathbf{S})$ be the embedded MRS of Z_t. Also, let $Z_t \in \Omega$ and $X_n \in \mathcal{F}$. The steady-state probability that X_n is equal i is denoted by π_i, $i \in \mathcal{F}$. Of course, π_i is also the proportion of epochs n with $X_n = i$. Similarly, let V_j be the proportion of time during which $Z_t = j$. Clearly, the expected time spent from 0 to S_1 in state j is given as

$$\int_0^\infty E_{ij}(t)dt.$$

Because the process can be restarted at S_n, this is also the expected time between S_n and S_{n+1} the process spends in state $Z_t = j$. After these initial remarks, we state the following theorem [Ross, 1970, page 108]

Theorem 1 *Let $V_j = \lim_{t \to \infty} V_{ij}(t)$. If the process Z_t is positive recurrent, and if the times between S_n and S_{n+1} are continuous, one has*

$$V_j = \frac{\sum_{i \in \mathcal{F}} \pi_i \int_0^\infty E_{ij}(t)dt}{\sum_{k \in \mathcal{F}} \pi_k \eta_k}$$

with

$$\eta_k = \int_0^\infty P\{S_1 > t | Z_0 = k\}dt = E(S_1 | Z_0 = k).$$

To see why the theorem holds, consider the process from 0 to S_m, where m is suitably large, and find the proportion of time the process Z_t spends in state j within this interval. Since π_i is the expected proportion of Markov renewal epochs with $X_n = i$, the expected proportion of time the system is in state j becomes

$$m \sum_{i \in \mathcal{F}} \pi_i \int_0^\infty E_{ij}(t)dt. \qquad (6)$$

Adding all these times yields the total length of the interval for 0 to S_m, provided integration and summation can be interchanged:

$$\sum_{j \in \Omega} m \sum_{i \in \mathcal{F}} \pi_i \int_0^\infty E_{ij}(t)dt = m \sum_{i \in \mathcal{F}} \pi_i \int_0^\infty \sum_{j \in \Omega} E_{ij}(t)dt$$

$$= m \sum_{i \in \mathcal{F}} \pi_i \int_0^\infty P\{S_1 > t\}dt = m \sum_{i \in \mathcal{F}} \pi_i \eta_i. \qquad (7)$$

The theorem now follows by dividing (6) by (7).

The result given above simplifies in the case of a SMP. In this case, $E_{ij}(t)$ is 0 for $i \neq j$, and $P\{S_1 > t | X_0 = i\}$ if $i = j$. Hence

$$V_i = \frac{\pi_i \eta_i}{\sum_{k \in \Omega} \pi_k \eta_k}. \qquad (8)$$

For instance, if the expected up-time of a machine is m_U, and the expected down-time is m_D, then half of the events are failures, and half repair completions, that is, $\pi_U = 0.5$, $\pi_D = 0.5$, where U and D denote the states "up" and "down". Hence, (8) yields

$$V_U = m_U/(m_U + m_D).$$

The equation for V_D is similar.

5 ISSUES IN DEPENDABILITY ANALYSIS

5.1 Power Hierarchy of Model Types

Malhotra and Trivedi [Malhotra and Trivedi, 1994] give a classification of the various dependability model types based on their modeling power. They establish this hierarchy through a proof procedure that relies on the algorithms for conversion from one model type to another. Among the non state space models, they show that fault-trees with repeated events are the most powerful, followed by reliability graphs, and at the lowest level of the hierarchy are reliability block diagrams and fault trees without repeated events. Among the state space based methods, Markov chains are less general than Markov regenerative processes, but are more powerful than all the non state space models, including fault-trees with repeated nodes. Figure 14 summarizes the power hierarchy among the model types.

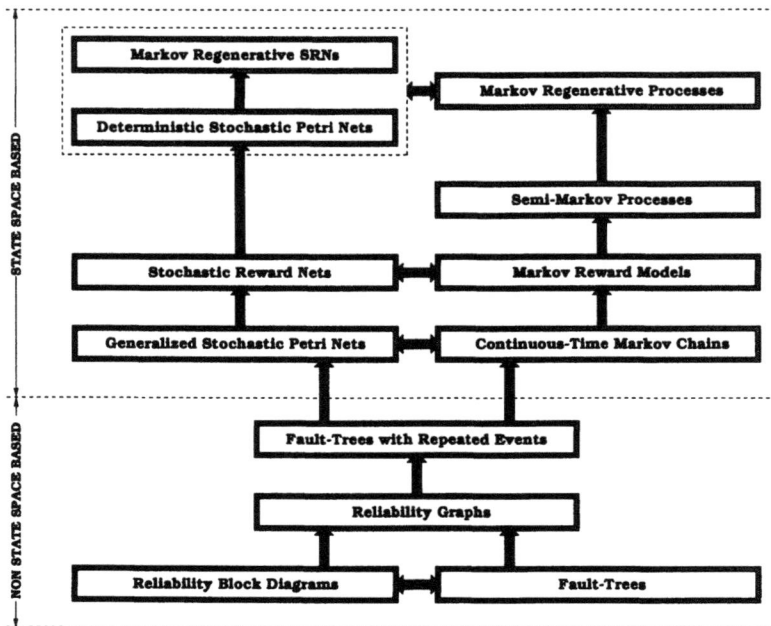

Figure 14 Power Hierarchy Among Model Types

5.2 Largeness of the State Space

While the use of automated generation methods for the Markov models is helpful, the size of the underlying state space still poses difficulties in the generation, storage, and solution of the models.

Several methods for reducing the effect of the state-space explosion have been proposed in the literature. These *(largeness avoidance)* methods include state truncation [Muppala et al., 1992], structural [Ciardo and Trivedi, 1993], time scale [Ammar and Islam, 1989] and behavioral decomposition [Dugan et al., 1986], model composition [Sahner and Trivedi, 1987b], and fixed point iteration [Mainkar and Trivedi, 1996, Tomek and Trivedi, 1991].

In the following subsections, we briefly review various largeness avoidance methods that have been successfully used in dependability modeling.

State Truncation. State truncation eliminates those system states with low probability of occurrence from further consideration in the system model. Typically in highly reliable systems, the occurrence of many failures is of low probability, and the system spends most of its time in a small subset of highly probable states. Since it is possible to identify states to be truncated before generating them, this technique is widely used [Muppala et al., 1992]. Error incurred due to state truncation has been studied [Li and Silvester, 1984, Muntz et al., 1989].

Model Composition. If an overall model can be thought of as consisting of submodels, these submodels can be solved in isolation and parameters of the solution passed from one submodel to another as required. Such model composition is available in the SHARPE package [Sahner et al., 1995]. For additional references see [Malhotra and Trivedi, 1993].

Model Decomposition. Time scale decomposition aims at separating the activities in the system based on the rates of their occurrence into fast and slow activities. Complex systems have several concurrent fast and slow activities that can be decomposed using the time scale. The overall system model is decomposed into a hierarchical sequence of aggregated subnets each of which is valid under a certain time scale [Ammar and Islam, 1989]. In particular, for behavioral decomposition [Dugan et al., 1986], the system can be decomposed into the slow time scale failure occurrence model (FORM) and the underlying fast time scale fault/error handling model (FEHM). The FORM model examines the dynamics of the occurrence of failures in the system. Upon occurrence of the failure, the FEHM model represents the details of how the faults are detected and handled appropriately. The FEHM model enables us to classify the faults into

transient, intermittent and permanent faults. Detailed fault/error handling models can be developed [Dugan et al., 1986] and solved. The results are then used in computing system dependability measures using the FORM model. Bobbio and Trivedi studied this problem in the context of CTMCs [Bobbio and Trivedi, 1986].

Fixed Point Iteration. Fixed point iteration is applicable in systems which cannot be "cleanly" decomposed into hierarchical submodels, because the interactions among the submodels cannot be ordered [Mainkar and Trivedi, 1996, Tomek and Trivedi, 1991]. The submodels have weak interactions among them, whereby some parameters in a submodel can be defined to be dependent on other submodels. In such cases, starting with an initial estimate for the parameters, the submodels are iteratively solved until convergence of the solution is attained.

5.3 Model Stiffness

Stiffness is another undesirable characteristic of many practical Markov dependability models, which adversely affects the computational efficiency of numerical solution techniques. Stiffness arises if the model solution has rates of change that differ widely. The linear system of differential Equations is considered stiff if for $i = 2, ..., m$, $Re(\lambda_i) < 0$ and

$$\max_i |Re(\lambda_i)| >> \min_i |Re(\lambda_i)|, \tag{9}$$

where λ_i are the eigenvalues of $\mathbf{Q}$. (Note that since $\mathbf{Q}$ is a singular matrix, one of its eigenvalues, say λ_1, is zero.) However, the above equation misses the point that the rate of change of a solution component is directly influenced by the length of the solution interval. To overcome that shortcoming, Miranker [Miranker, 1981] defined stiffness as follows: "A system of differential equations is said to be stiff on the interval [0,t) if there exists a solution component of the system that has variation on that interval that is large compared to $1/t$".

Stiffness of a Markov model could cause severe instability problems in the solution methods if the methods are not designed to handle stiffness. The two basic approaches to overcome stiffness are:

- *Stiffness-avoidance*: In this case, stiffness is eliminated from a model by solving a set of non-stiff models. One such technique based on aggregation is described in [Bobbio and Trivedi, 1986].

- *Stiffness-tolerance*: This approach employs solution methods that remain stable for stiff models.

Let us consider the source of stiffness in Markov chains. In a dependability model, repair rates are several orders of magnitude (sometimes 10^6 times) larger than failure rates. Failure rates could also be much larger than the reciprocal of mission time (which is the length of the solution interval). Such Markov chains have events (failures or repairs) occurring at widely different time scales. This results in the largest eigenvalue of $\mathbf{Q}$ being much larger than the inverse of mission time [Clarotti, 1986]; consequently the system of differential equations (Equation (2)) is stiff. According to the Gerschegorin circle theorem [Golub and van Loan, 1989], the magnitude of the largest eigenvalue is bounded above by twice the absolute value of the largest entry in the generator matrix. In a Markov chain, this entry corresponds to the largest total exit rate from any of the states. Therefore, the stiffness index of a Markov chain can be defined as qt, the product of the largest total exit rate from a state, q, and the length of the solution interval t [Reibman and Trivedi, 1988]. The above discussion suggests that stiffness can be arbitrarily increased by increasing q or t. The largest rate q can be increased by increasing model parameters. However, this increase changes the eigen-structure of matrix $\mathbf{Q}$. In some models it results in an increase in the magnitude of the smallest non-zero eigenvalue of the matrix. This implies that those models reach steady-state faster.

Numerical ODE solution methods which are not designed to handle stiffness, become computationally expensive for stiff problems. The solution of a stiff model entails very small time steps, which increases the total number of time steps required and the total computation time manifolds. The original version of Jensen's method does not handle stiffness well either [Reibman and Trivedi, 1988]. Recently, hybrid methods have been proposed [Malhotra, 1996] which combine stiff and non-stiff ODE methods to yield efficient solutions of stiff Markov models.

6 CONCLUSIONS

We have briefly reviewed the techniques and tools and recent developments in the field of system dependability evaluation. Commonly used measures of system dependability were defined. Several modeling techniques, both non-state-space and state-space based, are then briefly presented. Simple examples illustrating the use of the modeling techniques were also presented. Computational methods for obtaining the dependability measures from the system models were then reviewed. Some important issues that arise in the dependability evaluation of complex systems were then briefly discussed.

References

[Ammar and Islam, 1989] Ammar, H. H. and Islam, S. M. R. (1989). Time scale decomposition of a class of generalized stochastic Petri net models. *IEEE Transactions on Software Engineering*, 15(6):809–820.

[Barlow and Proschan, 1975] Barlow, R. E. and Proschan, F. (1975). *Statistical Theory of Reliability and Life Testing - Probability Models*. Holt, Rinehart and Winston, New York, NY, USA.

[Bobbio et al., 1995] Bobbio, A., Kulkarni, V. G., Puliafito, A., Telek, M., and Trivedi, K. S. (1995). Preemptive repeat identical transitions in Markov regenerative stochastic petri nets. In *Proceedings of the 6th International Workshop on Petri Nets and Performance Models - PNPM'95*, pages 113–122, Durham, NC, USA.

[Bobbio and Telek, 1995] Bobbio, A. and Telek, M. (1995). Markov regenerative spn with non-overlapping activity cycles. In *Proc. of the International Computer, Performance and Dependability Symposium - IPDS'95*.

[Bobbio and Trivedi, 1986] Bobbio, A. and Trivedi, K. S. (1986). An aggregation technique for the transient analysis of stiff Markov chains. *IEEE Transactions on Computers*, C-35(9):803–814.

[Botta and Harris, 1986] Botta, R. F. and Harris, C. M. (1986). Generalized hyperexponential distributions: Weak convergence results. *Queueing Systems - Theory and Applications*, 1(2):169–190.

[Botta et al., 1987] Botta, R. F., Harris, C. M., and Marchal, W. G. (1987). Characterizations of generalized hyperexponential distribution functions. *Communications in Statistics - Stochastic Models*, 3(1):115–148.

[Brokemeyer et al., 1948] Brokemeyer, F., Halstron, H. S., and Jensen, A. (1948). The life and works of A. K. Erlang. *Transactions of the Danish Academy of Technical Sciences*, 2.

[Buzacott, 1970] Buzacott, J. (1970). Network approaches to finding the reliability of repairable systems. *IEEE Transactions on Reliability*, R-19(4):140–146.

[Cao, 1994] Cao, J. (1994). Reliability analysis of M/G/1 queueing system with repairable service station of reliability series structure. *Microelectronics and Reliability*, 34(4):721–725.

[Çinlar, 1975] Çinlar, E. (1975). *Introduction to Stochastic Processes*. Prentice-Hall, Englewood Cliffs, NJ, USA.

[Chen et al., 1993] Chen, Y.-M., Fujisawa, T., and Osawa, H. (1993). Availability of the system with general repair time distributions and shut-off rules. *Microelectronics and Reliability*, 33(1):13–19.

[Choi et al., 1994] Choi, H., Kulkarni, V. G., and Trivedi, K. S. (1994). Markov regenerative stochastic Petri nets. *Performance Evaluation*, 20:337–357.

[Ciardo et al., 1992a] Ciardo, G., Blakemore, A., Chimento, P., Muppala, J., and Trivedi, K. (1992a). Automatic generation and analysis of Markov reward models using stochastic reward nets. In *Linear Algebra, Markov Chains, and Queueing Models, IMA Volumes in Mathematics and its Applications*, volume 48, Heidelberg, Germany. Springer-Verlag.

[Ciardo et al., 1994] Ciardo, G., German, R., and Lindemann, C. (1994). A characterization of the stochastic process underlying a stochastic Petri net. *IEEE Transactions on Software Engineering*, 20:506–515.

[Ciardo et al., 1992b] Ciardo, G., Muppala, J. K., and Trivedi, K. S. (1992b). Analyzing concurrent and fault-tolerant software using stochastic reward nets. *Journal of Parallel and Distributed Computing*, 15:255–269.

[Ciardo and Trivedi, 1993] Ciardo, G. and Trivedi, K. S. (1993). A decomposition approach for stochastic Petri net models. *Performance Evaluation*, 18(1):37–59.

[Clarotti, 1986] Clarotti, C. (1986). The Markov approach to calculating system reliability: Computational problems. In Serra, A. and Barlow, R., editors, *Proceedings of the International School of Physics, Course XCIV*, pages 55–66. North-Holland.

[Cox, 1955a] Cox, D. R. (1955a). The analysis of non-Markovian stochastic processes by the inclusion of supplementary variables. *Proc. Camb. Philos. Soc.*, 51(3):433–441.

[Cox, 1955b] Cox, D. R. (1955b). Use of complex probabilities in the theory of stochastic processes. *Proc. Camb. Philos. Soc.*, 51:313–318.

[Dhillon and Anude, 1993] Dhillon, B. S. and Anude, O. C. (1993). Common-cause failure analysis of a non-identical unit parallel system with arbitrarily distributed repair times. *Microelectronics and Reliability*, 33(1):87–103.

[Dhillon and Anude, 1994] Dhillon, B. S. and Anude, O. C. (1994). Income optimization of repairable and redundant system. *Microelectronics and Reliability*, 34(11):1709–1720.

[Dhillon and Yang, 1993] Dhillon, B. S. and Yang, N. (1993). Availability of a man-machine system with critical and non-critical human error. *Microelectronics and Reliability*, 33(10):1511–1521.

[Dugan et al., 1992] Dugan, J. B., Bavuso, S., and Boyd, M. (1992). Dynamic fault-tree models for fault-tolerant computer systems. *IEEE Transactions on Reliability*, R-41(9):363–377.

[Dugan et al., 1986] Dugan, J. B., Trivedi, K. S., Smotherman, M. K., and Geist, R. M. (1986). The Hybrid Automated Reliability Predictor. *AIAA Journal of Guidance, Control and Dynamics*, 9(3):319–331.

[Fricks et al., 1997] Fricks, R., Telek, M., Puliafito, A., and Trivedi, K. S. (1997). Markov renewal theory applied to performability evaluation. In Bagchi, K. and Zobrist, G., editors, *State-of-the Art in Performance Modeling and Simulation. Modeling and Simulation of Advanced Computer Systems: Applications and Systems*, pages 193–236, Newark, NJ, EUA. Gordon and Breach Publishers.

[Fröberg, 1969] Fröberg, C. (1969). *Introduction to Numerical Analysis, 2nd. ed.* Addison-Wesley, Reading, MA, USA.

[Garg et al., 1997] Garg, S., Puliafito, A., M. T., and Trivedi, K. (1997). Analysis of preventive maintenance in transactions based software systems. *submitted for publication.*

[German and Lindemann, 1994] German, R. and Lindemann, C. (1994). Analysis of deterministic and stochastic Petri nets by the method of supplementary variables. *Performance Evaluation*, 20(1-3):317–335.

[German et al., 1995] German, R., Logothetis, D., and Trivedi, K. S. (1995). Transient analysis of Markov regenerative stochastic Petri nets: A comparison of approaches. In *Proceedings of the 6th International Workshop on Petri Nets and Performance Models - PNPM'95*, pages 103–111, Durham, NC, USA.

[Golub and van Loan, 1989] Golub, G. and van Loan, C. F. (1989). *Matrix Computations*. Mathematical Sciences. Johns Hopkins University Press, Baltimore, MD, 2nd edition.

[Gopalan and Dinesh Kumar, 1996] Gopalan, M. N. and Dinesh Kumar (1996). On the transient behaviour of a repairable system with a warm standby. *Microelectronics and Reliability*, 36(4):525–532.

[Goyal et al., 1986] Goyal, A., Carter, W. C., de Souza e Silva, E., Lavenberg, S. S., and Trivedi, K. S. (1986). The system availability estimator. In *Proceedings of the Sixteenth International Symposium on Fault-Tolerant Computing*, pages 84–89, Los Alamitos, CA. IEEE Computer Society Press.

[Goyal et al., 1987] Goyal, A., Lavenberg, S. S., and Trivedi, K. S. (1987). Probabilistic modeling of computer system availability. *Annals of Operations Research*, 8:285–306.

[Heimann et al., 1990] Heimann, D., Mittal, N., and Trivedi, K. S. (1990). Availability and reliability modeling of computer systems. In Yovits, M., editor, *Advances in Computers*, volume 31, pages 176–233. Academic Press, San Diego, CA.

[Howard, 1971] Howard, R. A. (1971). *Dynamic Probabilistic Systems, Vol II: Semi-Markov and Decision Processes*. John Wiley and Sons, New York, NY, USA.

[Kulkarni, 1995] Kulkarni, V. G. (1995). *Modeling and Analysis of Stochastic Systems*. Chapman Hall.

[Laprie, 1985] Laprie, J. C. (1985). Dependable computing and fault-tolerance: Concepts and terminology. In *Proceedings of the Fifteenth International Sym-

posium on Fault-Tolerant Computing, pages 2–7, Los Alamitos, CA. IEEE Computer Society Press.

[Leemis, 1995] Leemis, L. M. (1995). *Reliability: Probability Models and Statistical Methods*. Prentice-Hall, Englewood Cliffs, NJ, USA.

[Li and Silvester, 1984] Li, V. and Silvester, J. (1984). Performance analysis of networks with unreliable components. *IEEE Transactions on Commun.*, COM-32(10):1105–1110.

[Logothetis and Trivedi, 1995] Logothetis, D. and Trivedi, K. (1995). Time-dependent behavior of redundant systems with deterministic repair. In Stewart, W. J., editor, *Computations with Markov Chains*. Kluwer Academic Publishers.

[Logothetis and Trivedi, 1997] Logothetis, D. and Trivedi, K. S. (1997). The effect of detection and restoration times for error recovery in communication networks. *to appear in the Journal of Network and Systems Management*.

[Mainkar and Trivedi, 1993] Mainkar, V. and Trivedi, K. S. (1993). Approximate analysis of priority scheduling systems using stochastic reward nets. In *Proceedings of the 13th International Conference on Distributed Computing Systems - ICDCS'93*, pages 466–473, Pittsburgh, PA, USA.

[Mainkar and Trivedi, 1996] Mainkar, V. and Trivedi, K. S. (1996). Sufficient conditions for the existence of a fixed point in stochastic reward net-based iterative models. *IEEE Transactions on Software Engineering*, 22(9):640–653.

[Malhotra, 1996] Malhotra, M. (1996). A computationally efficient technique for transient analysis of repairable Markovian systems. *Performance Evaluation*, 24(4):311–331.

[Malhotra and Trivedi, 1993] Malhotra, M. and Trivedi, K. S. (1993). A methodology for formal expression of hierarchy in model solution. In *Proceedings of the Fifth International Workshop of Petri Nets and Performance Models, PNPM'93*, pages 258–267, Toulouse, France.

[Malhotra and Trivedi, 1994] Malhotra, M. and Trivedi, K. S. (1994). Power-hierarchy of dependability-model types. *IEEE Transactions on Reliability*, R-43(3):493–502.

[Miranker, 1981] Miranker, W. (1981). *Numerical Methods for Stiff Equations and Singular Perturbation Problems*. D. Reidel, Dordrecht, Holland.

[Muntz et al., 1989] Muntz, R. R., de Souza e Silva, E., and Goyal, A. (1989). Bounding availability of repairable computer systems. *IEEE Transactions on Computers*, C-38(12):1714–1723.

[Muppala et al., 1996] Muppala, J. K., Malhotra, M., and Trivedi, K. S. (1996). Markov dependability models of complex systems: Analysis techniques. In Özekici, S., editor, *Reliability and Maintenance of Complex Systems*, pages 442–486, Berlin, Germany. Springer.

[Muppala et al., 1992] Muppala, J. K., Sathaye, A., Howe, R., and Trivedi, K. (1992). Dependability modeling of a heterogeneous VAXcluster system using stochastic reward nets. In Avresky, D., editor, *Hardware and Software Fault Tolerance in Parallel Computing Systems*, pages 33–59. Ellis Horwood Ltd.

[Muppala and Trivedi, 1995] Muppala, J. K. and Trivedi, K. S. (1995). System dependencies in Markov dependability modelling. In Mittal, R., Muthukrishnan, C. R., and Bhatkar, V. P., editors, *Fault-Tolerant Systems and Software, Proceedings of FTS-95*, pages 38–47. Narosa Publishing House, New Delhi, India.

[Muppala et al., 1993] Muppala, J. K., Woolet, S. P., and Trivedi, K. S. (1993). On modeling performance of real-time systems in the presence of failures. In *Readings in Real-Time Systems*, pages 219–239, Los Alamitos, CA, USA.

[Neuts, 1978] Neuts, M. F. (1978). Renewal process of phase type. *Naval Research Logistics Quarterly*, 25(3):445–454.

[Rai et al., 1995] Rai, S., Veeraraghavan, M., and Trivedi, K. S. (1995). A survey of efficient reliability computation using disjoint products approach. *Networks*, 25:147–163.

[Reibman and Trivedi, 1988] Reibman, A. L. and Trivedi, K. S. (1988). Numerical transient analysis of Markov models. *Computers and Operations Research*, 15(1):19–36.

[Ross, 1970] Ross, S. M. (1970). *Applied Probability Models with Optimization Applications*. Holden-Day, San Francisco.

[Sahner and Trivedi, 1987a] Sahner, R. A. and Trivedi, K. S. (1987a). Performance and reliability analysis using directed acyclic graphs. *IEEE Transactions on Software Engineering*, SE-13(10):1105–1114.

[Sahner and Trivedi, 1987b] Sahner, R. A. and Trivedi, K. S. (1987b). Reliability modeling using SHARPE. *IEEE Transactions on Reliability*, R-36(2):186–193.

[Sahner et al., 1995] Sahner, R. A., Trivedi, K. S., and Puliafito, A. (1995). *Performance and Reliability Analysis of Computer Systems: An Example-Based Approach Using the SHARPE Software Package*. Kluwer Academic Publishers, Dordrecht, The Netherlands.

[Shooman, 1970] Shooman, M. L. (1970). The equivalence of reliability diagram and fault-tree analysis. *IEEE Transactions on Reliability*, R-19(5):74–75.

[Singh et al., 1977] Singh, C., Billington, R., and Lee, S. Y. (1977). The method of stages for non-Markovian models. *IEEE Transactions on Reliability*, 26(2):135–137.

[Telek, 1994] Telek, M. (1994). *Some Advanced Reliability Modeling Techniques*. PhD thesis, Technical University of Budapest, Department of Telecomunications, Budapest, Hungary.

[Telek et al., 1995] Telek, M., Bobbio, A., Jereb, L., and Trivedi, K. (1995). Steady state analysis of Markov regenerative spn with age memory policy. In *Proceedings of the International Conference on Performance Tools and MMB '95*, Heidelberg, Germany.

[Tomek and Trivedi, 1991] Tomek, L. A. and Trivedi, K. S. (1991). Fixed point iteration in availability modeling. In Cin, M. D. and Hohl, W., editors, *Proceedings of the 5th International GI/ITG/GMA Conference on Fault-Tolerant Computing Systems*, pages 229–240, Berlin. Springer-Verlag.

[Trivedi, 1982] Trivedi, K. S. (1982). *Probability & Statistics with Reliability, Queueing, and Computer Science Applications*. Prentice-Hall, Englewood Cliffs, NJ, USA.

[Trivedi et al., 1992] Trivedi, K. S., Muppala, J. K., Woolet, S. P., and Haverkort, B. R. (1992). Composite performance and dependability analysis. *Performance Evaluation*, 14(3 & 4):197–216.

[Veeraraghavan and Trivedi, 1991] Veeraraghavan, M. and Trivedi, K. S. (1991). An improved algorithm for the symbolic reliability analysis of networks. *IEEE Transactions on Reliability*, R-40(3):347–358.

[Wu et al., 1994] Wu, S., Huang, R., and Wan, D. (1994). Reliability analysis of a repairable system without being repaired "as good as new". *Microelectronics and Reliability*, 34(2):357–360.

Jogesh K. Muppala Jogesh K. Muppala received the Ph. D. degree in Electrical Engineering from Duke University, Durham, NC in 1991, the M. S. degree in Computer Engineering from The Center for Advanced Computer Studies, University of Southwestern Louisiana, Lafayette, LA in 1987 and the B. E. degree in Electronics and Communication Engineering from Osmania University, Hyderabad, India in 1985. He is currently an associate professor in the Department of Computer Science, The Hong Kong University of Science and Technology (HKUST), Clearwater Bay, Kowloon, Hong Kong. He was previously with Software Productivity Consortium, Herndon, VA, USA. His research interests include performance and dependability modeling, high speed networking, distributed systems, and stochastic Petri nets. He has published over 30 refereed journal and conference papers in these areas. He is the program co-chair for the 1999 Pacific Rim International Symposium on Dependable Computing. He has also served on program committees of international conferences. He was recently given the Teaching Excellence Appreciation Award by the Dean of Engineering at HKUST. Dr. Muppala is a member of IEEE and IEEE computer society.

Ricardo M. Fricks Ricardo M. Fricks received his Ph.D. degree in electrical and computer engineering from Duke University (Durham, NC, USA) in 1997. He also holds a M.Sc. degree in computer science from the Universidade Federal do Rio Grande do Sul (Porto Alegre, Brazil), and a diploma in electrical engineering from the Universidade Federal do Paraná (Curitiba, Brazil). Ricardo is currently a senior researcher at SIMEPAR - an atmospheric and earth sciences research laboratory jointly maintained by the Paraná State Power Company and the Agronomical Institute of Paraná - and an associate professor at the graduate school of informatics at the Pontificia Universidade Católica do Paraná. Ricardo's research interests include power systems reliability modeling and application of point processes to performability modeling of computer systems. He is a member of the IEEE and the IEEE Computer Society.

Kishor S. Trivedi Kishor S. Trivedi received the B.Tech. degree from the Indian Institute of Technology (Bombay), and M.S. and Ph.D. degrees in computer science from the University of Illinois, Urbana-Champaign. He is the author of a well known text entitled, Probability and Statistics with Reliability, Queueing and Computer Science Applications, published by Prentice-Hall. He has recently published another book entitled, Performance and Reliability Analysis of Computer Systems, published by Kluwer Academic Publishers. His research interests are in reliability and performance assessment of computer and communication systems. He has published over 200 articles and lectured extensively on these topics. He has supervised 30 Ph.D. dissertations. He is a Fellow of the Institute of Electrical and Electronics Engineers. He is a Golden Core Member of IEEE Computer Society.

He is a Professor in the Department of Electrical and Computer Engineering at Duke University, Durham, NC. He also holds a joint appointment in the Department of Computer Science at Duke. He is the Duke-Site Director of an NSF Industry-University Cooperative Research Center between NC State University and Duke University for carrying out applied research in computing and communications. He has been on the Duke faculty since 1975. He has served as an editor of the IEEE Transactions on Computers from 1983-1987. He is a co-designer of HARP, SAVE, SHARPE and SPNP modeling packages. These packages have been widely circulated.

Index

Absorbing state
  in Markov decision process, 330
Absorption, 466
Accelerating convergence, 273
Acceleration, 268, 270
Action space, 326
Adjoint, 215, 230
Agarwal, 289, 298
Aggregate station, 437
Aggregation, 57
Aggregation matrix, 102
Aggregation step, 103
Aliased sequence, 277
Aliasing, 266
  See also Error, aliasing
Alternating series, 268
Analytic, 206, 208, 210, 216, 235, 238
Analyticity condition, 294
And gate, 462
Approximation, 375, 377–378, 380–381, 388, 392–393, 396
  of transition matrix, 180, 357
Arbitrary-epoch tail probabilities, 381
Arbitrary service time, 381
Argument principle, 207, 239
Arnoldi's method, 53, 83, 97
Arrival-first, 366
Asmussen, 293, 299, 301
Assembly line, 415
Asymptotic behavior, 243
Asymptotic formulas, 282
Asymptotic parameter, 294
Asymptotics, 303
Automata: *see* Stochastic automata
Availability, 46, 465
  instantaneous, 447
  interval, 447
  limiting interval, 448
  point availability, 69
  steady-state, 448

Average-cost criterion, 326, 360
Back-substitution, 182–183
Bailey's bulk queue, 209, 247–248
Balance equations, 412
  global, 130, 411–412, 414
  job-class, 423
  local, 130, 422
  partial, 411–414, 418, 422
  station, 411, 413–414, 418, 422
    and blocking, 427
    failure of, 425
    restored, 427, 429, 435
Barthez, 272
BASTA, 373
Beneš, 282, 284, 287
Bernoulli arrivals, 373
Bernoulli process, 387
Bertozzi, 304
Binomial average, 270
Binomial distribution, 270
Block elimination, 175, 179, 183
  and paradigms of Neuts, 187–189
Block iterative methods, 93
Block Jacobi, 94
Block SOR, 94
Block-splitting, 93
Blocking, 158, 425–427
Blocking probability, 410
  Erlang, 283
  time dependent, 280, 282
Borovkov, 281
Boundary probabilities, 206, 210, 218
Bounding methodology, 428–429
Bounds, 351, 429, 432, 434–435, 437
  and optimal design, 436
  for waiting time, 396
  intuitively obvious, 430
  on loss probability, 429, 432, 434
  on minimal costs, 436
  on throughput, 429, 432, 439

on utilization, 429
Branch point, 238
Breakdown, 426, 434–437, 439
Bremmer, 260
Bromwich inversion integral, 263
Buffer, 411, 437–438
Busy cycle, 291, 367
Busy period, 261, 367, 387
  in M/G/1 queue, 280
Call center, 194, 433
  bilingual, 199
Capacity constraint, 411, 427
Cauchy contour integral, 276
Censored chain, 214, 221, 230
Censoring, 173, 175, 177, 180, 182, 189–190
Central limit theorem, 259
Characteristic function, 258
Chaudhry, 289, 298
  QPACK software package, 381, 397
  QROOT software package, 376, 381, 390
Chebyshev-Subspace iteration, 97
Closed queueing network: see Queueing network
Coefficient of variation, 457
Cohen, 257, 290, 301
Combinatorial explosion, 424
Communication systems, 365
Compact storage schemes, 82, 85
Companion matrix, 222
Complementary cdf, 262
Complementation
  of transition matrix, 180
Complex-conjugate, 377, 392
Complex Fourier series, 266
Complex-valued functions, 262
Complex variables, 263
Computation time, 114
Computational complexity, 7, 139, 309
Computer networking, 410
Computer systems, 365
Conditional decomposition, 309
Congestion, 432
Conservation of probability, 206, 209–210, 212, 218, 226, 228, 231, 249
Continuous, 210, 216, 238, 243
Continuous-time control, 359
Continuous-time Markov chain, 154, 219, 232, 412, 453
Continuous-time Markov decision chain, 358–359
Contour integral, 263, 289
Contour, 276
Contraction mapping, 342, 349, 358
Control
  continuous-time, 359
Convergence
  pointwise, 358

rate of, 89, 358
Convex, 243–244
Convolution algorithm, 304, 423, 428
Convolution, 206, 259, 285, 315
  smoothing, 272
  n-fold, 261
Conway, 304
Costs, 325, 328
  terminal, 329
  time-homogeneous, 335
  uniformly bounded, 357
Coupling matrix, 102
Covariance function, 281, 287
Cramér-Lundberg approximation, 294
Crout method, 184, 191
CTMC, 154, 412
  See also Continuous-time Markov chain
Cumulants, 299
Cumulative distribution function, 262
Cumulative reward, 45, 48, 60–61
  cumulative impulse reward, 46
  cumulative operational time, 46, 69
  cumulative rate reward, 45, 64, 71
Curse of dimensionality, 7, 410
Cutset, 141

DA (Delayed Access), 367
Damped sequence, 277
Damping, 266
Darling, 286
Davies, 272
Davis, 263, 265
Decision, 325
Decision rule, 326
  greedy, 353
Decomposable, 157, 174
Decomposition
  for dimension reduction, 309–311
  of a network, 439
Decompositional methods, 97
Decreasing failure rate, 27
Delay, 410
  average, 197–198
Delayed access, 367
Departure-first, 366
Dependability, 65, 445–446, 469–473
Depth-first search, 105
DES: see Discrete event system
Descriptor, 123
Determinant, 216–218, 238, 240, 282
Deterministic arrivals, 397
Deterministic service time, 381, 422
Diagonally dominant, 240
  irreducibly, 240
  strictly, 240
Difference equations, 225
Difference operator, 264
Diffusion process, 281

INDEX   483

Dimension reduction
  by decomposition, 309–311
Directed graph, 105
Direct methods: *see* Solution method
Disaggregation step, 103
Discount factor, 330
Discounted-cost problem, 330
Discrete event
  simulation, 3
  system, 3, 14
Discrete-time Markov chain, 12, 14, 154–155, 207
Discretization error, 265–266, 268
Distribution
  binomial, 270
  gamma, 260, 297–298
  geometric, 397
  infinitely divisible, 247
  marginal, 420–421, 424
  normal, 281
  phase-type, 5, 27
  stationary, 81, 206, 214, 219, 229
  steady-state, 281
Divide and conquer, 99
Doetsch, 260, 262
Drift, 212, 219
DTMC, 12
  See also Discrete-time Markov chain
Dubner, 272
Duffield, 280
Durbin, 272
Dynamic programming, 332, 336

EAS (early arrival system), 367–368, 397
Eigenvalues, 51, 207, 217, 222, 234, 238, 243, 282
  distinct, 239
  multiplicity of, 239
Eigenvectors, 207, 217, 243, 282
Elimination of non-optimal actions, 355
Emptiness
  probability of, 29
Entropy, 412
Erlang, 365
Erlang blocking probability, 281, 283
Erlang loss model, 279
Error, 265
  absolute, 293
  aliasing, 266–267, 269–272, 277–279, 303, 311–312
  bound, 265, 272, 275
  discretization, 265–268, 276, 312
    balancing with roundoff error, 271
  estimate, 270
  in Euler summation, 272, 274
  relative, 293
  roundoff, 265, 268, 271–272, 278–279, 312
    control of, 268, 279

  truncation, 265, 270, 274
Euler summation, 268, 270, 273, 278, 314–315
Euler's constant, 301
Event, 3, 16
  immediate, 15, 29–33
  schedule of, 3, 14
Exchangeability, 60, 64
Expected level, 193
Exponential damping, 300
Exponential polynomial, 457
Exponential service time, 412

Fault tree, 451, 462–463
  with repeated events, 451
Feldmann, 301
Feller, 259, 268
Fill-in, 82, 85, 92
Finite differences, 264
Finite-horizon problem, 329, 332, 334
Finite-state approximations, 357
First passage time, 47, 284, 387
Fixed point, 341, 349
  iteration, 91, 471
Flannery, 296
Flow equations, 412
Fluid flow models, 71
Fork and join queue, 427
Fourier
  coefficients, 266
  series, 265–266
  method, 266
  transform, 168, 258
    discrete, 277
FT, 451
  See also Fault tree
FTRE, 451
  See also Fault tree with repeating events
Functional dependency cycle, 140
Functional rates, 135
Functional transitions, 118, 123

G matrix, 221
Gamma distribution, 260, 297–298
GASTA, 373
Gaussian elimination, 86–87, 340
Gauss-Seidel, 91
  backward, 82, 89
  forward, 82, 89
  ordering of states, 90
Gaver, 258, 272, 301
Gaver-Stehfest procedure, 264
Generalized eigenvector, 222
Generalized Erlang, 292
Generalized tensor product, 135, 137–139
Generating function, 206, 208, 275, 303, 307
Generator, 219, 232

and synchronizing events, 119
Geometric arrivals, 373
Geometric, 160, 366, 377–378, 381, 392, 394
Geometric distribution, 397
Georganas, 304
Giffin, 260
GI/G/1 paradigm, 156, 191–193
GI/G/1 queue: see Queue, GI/G/1
GI/G/1 type process, 156–158
GI/M/1 paradigm, 168, 187–189
GI/M/1 type process, 155–157, 159, 164
GI/M/1 queue: see Queue, GI/M/1
Global balance, 130, 411–412, 418, 422
Global generator, 118
G/M/1 type process, 206, 223, 228–229, 232–233
   See also GI/M/1 type process
G/M/1 queue: see Queue, GI/M/1
GMRES, 98, 115
Graph, 310
GSPN (Generalized Stochastic Petri Net) 17–22
   colored 22
GTH advantage, 86, 182
GTP: see Generalized tensor product

Harmonic, 217, 249
Heavy-traffic, 281
   approximation, 293
Hessenberg matrix, 98
   lower, 155
   upper, 156
High-level formalism, 14
History, 326
Homogeneous in time, 455–456
Horners's method, 169
Hosono, 272
Hyperexponential, 291
Hypoexponential, 291

IA (Immediate Access), 367
Ideal decomposition, 309
Idle time, 291
Ill-conditioned, 87–88
ILU: see Incomplete LU factorization
ILU preconditioning, 90
ILU(0), 92
ILUK, 93
ILUTH, 92
Imbedded Markov chain, 159, 173, 180, 365
Immediate access, 367
Incomplete LU factorization, 92
Increasing failure rate, 27
Indicator function, 6
Infinite state space, 58
Infinite-horizon discounted problem, 325
Infinite-server center, 306
Infinitely divisible distribution, 247

Infinitesimal generator, 81
Initial approximation, 83
Initial value theorem, 269
Insensitivity, 422
   of bounds, 430, 435
Instantaneous state, 31
   preservation, 31
Interarrival time
   rational transform, 290
Interdependence graph, 310
Inverse of a matrix, 174–175, 185
Inversion dimension, 310
Inversion of generating functions, 375
Irreducible, 207, 210, 217, 226
Iterative aggregation/disaggregation, 103
Iterative methods, 82

Jacobs, 301
Jackson, 421
Jackson networks, 24, 417
Jagerman, 264, 272, 281–282, 284
Jagerman-Stehfest procedure, 263
Jensen's method, 54, 233
   See also Randomization, Uniformization
Job, 305, 422–423
   class, 305
      job-class balance, 423
      job-local balance, 422
      priority, 427
      type, 423
Johnsonbaugh, 273
Jordan chain, 217, 222, 234, 246
Jordan form, 222, 234
Jump chain, 220, 232

Kao, 259
Keilson, 283, 286
Kendall functional equation, 261
Kernel of the MRS, 459
Kingman, 299
Kleinrock, 260
Knessl, 282
Kobayashi, 259, 304
Kolmogorov's equations, 44, 48, 52, 56
Krylov subspace, 98
Krylov subspace method, 53, 55
Kwok, 272

Laguerre
   coefficients, 264
   functions, 264
   generating functions, 264
   polynomials, 264
   series, 264
   series algorithm, 272, 279
   series inversion formula, 264
   series representation, 272
Lam, 310–311, 315
Laplace curtain, 258

INDEX

Laplace transform, 52, 55, 66, 262, 289
Laplace-Stieltjes transform, 262
Largeness avoidance, 470
LAS (Late Arrival System), 367
LAS-DA, 367, 372–374, 397
LAS-IA, 367–368
Last-come first-served
  preemptive-resume, 306
  preemptive, 422
Late arrival system, 367
Latouche, 169–170, 178
Lavenberg, 304, 308
LCFSPR, 306
  See also Last-come first-served
  preemptive-resume
Least squares, 98
Leontief substitution system, 346
Leung, 303, 308
Level, 155, 214, 218–219, 221, 229, 233
Levy-Desplanques theorem, 240
Lien, 310–311, 315
Lifetime, 47
Linear independence, 248
Linear programming, 345
  dual problem, 346
  in Markov decision process, 345
Little's theorem, 369, 374–375, 420
Load dependent service speeds, 421
Local balance: see balance equations
Local transition, 118–119, 121, 141
Long-run dynamics, 100
Long-tail service-time, 299
Loss of significance, 86
Loss probability
  bound on, 434
LU decomposition, 83–84
Lucantoni, 262, 272, 279–280
Lumping of automata, 114

Manufacturing, 410
Marginal distribution, 420–421, 424
Marking, 18
  tangible, 19
  vanishing, 19
Markov chain
  continuous-time, 154, 219, 232, 412, 453
  discrete-time, 12, 14, 154–155, 207
Markov decision
  chain, 326
  model, 325
  process, 325
Markov policy, 327
Markov property, 455
Markov regenerative
  epoch, 458
  process, 449, 457, 460, 468
    kernel of, 468
Markov renewal

  equation, 467
  moments, 458
  sequence, 457–458
  theory, 457
Markov reward models, 60–61, 454
Martin, 272
Master/slave, 118
Matrix analytic methods, 154, 206, 213, 221, 229
Matrix equation, 166
Matrix exponential, 50, 55
Matrix geometric, 161, 168, 179, 195, 197, 229, 233
  methods, 258
  solution, 161
Maximal connected subsets, 310
McKenna, 304
MDC, 326
  See also Markov decision chain
Mean time to absorption, 466
Mean time to failure, 447, 462, 466
Mean-value analysis, 411
Memory requirement, 114
Meromorphic, 242
Method of phases, 5
M/G/1 paradigm, 164, 170, 190–191
M/G/1 queue: see Queue, M/G/1
M/G/1 type process, 155–157, 206–207, 213, 218, 221
Mitra, 282
M/M/1 bulk service, 246
Modeling power of a formalism, 25
Moment generating function, 294
Moments, 309
Monotone convergence theorem, 169
MRGP, 457
  See also Markov regenerative process
MRM, 454
  See also Markov reward models
MRS, 457–458
  See also Markov renewal sequence
MTTA, 466
  See also Mean time to absorption
MTTF, 462
  See also Mean time to failure
Multi-dimensional generating functions, 304
Multi-dimensional transforms, 262, 303
Multichain closed queueing networks, 305–316
Multiplicity
  of factors in denominator, 307
  of roots, 246
  of zeros, 210, 216, 225, 231

NCD, 82, 100
  See also
    Nearly-completely-decomposable
Nearly alternating, 269

Nearly alternating series, 265
Nearly-completely-decomposable, 82, 100–101
Nearly decomposable system, 89
Nearly separable, 100
Nearly uncoupled, 100
Nerman, 301
Network: see Queueing network, Stochastic automata network, GSPN
Neuts, 155–156, 159–160, 258, 298
Newton-Raphson root finding procedure, 293–294, 313–314
Non-decomposable matrix, 418
Non-homogeneous Markov chain, 55, 59, 62
Non-Markovian models, 71
Non-product-form networks, 430
Non-skip-free, 209, 213, 222, 224, 229, 234
Nonlinear equations, 313
Nonsingular, 240
Norm, 348
Normal distribution, 281
Normal form, 218, 232, 242
Normalization constant, 303, 416, 419, 423, 431
Normalizing constant, 132, 160
Norming constant, 193
Null recurrent, 212, 218, 228, 232, 237
Null vector, 207, 217, 222
Numerical computations, 381
Numerical differentiation, 276
Numerical example, 395
Numerical integration, 263, 265, 296
Numerical inversion algorithms, 261
Numerical performance, 377
Numerical transform inversion, 257

O'Cinneide, 273, 275
ODE, 48, 55, 59, 70
   Euler's method, 48
   multi-step method, 49
   Runge-Kutta methods, 50
   single step method, 49
   trapezoid method, 49
Ollson, 301
One-stage problem, 336
Optimal
   ϵ-optimal, 329, 342
Optimal admission policy, 357
Optimal control of queues, 326
Optimal decision rule, 336
Optimal policy, 329, 336
Optimal stopping problem, 330, 332
Optimal value function, 325, 329, 332, 334–336
Optimality equation, 325, 332, 334, 336–342, 345, 355, 359–361
Or gate, 462
Ordinary differential equation, 282

   See also ODE
Ordinary tensor product, 135
Ornstein-Uhlenbeck process, 281
   reflected, 281
Orthogonal projection process, 96
OTP, 135
   See also Ordinary tensor product
OU process, 281
   See also Ornstein-Uhlenbeck process
Overflow, 425–426, 433–434
Overrelaxation: see SOR

Pareto mixture of exponentials, 301
Partial balance, 411
Partial fractions expansion, 245
Partial fractions, 291
Partial generating function, 305, 308
Partition function, 304, 307
Performability, 45, 62, 69, 449
   point performability, 60
Performance analysis tool
   Q^2, 273
Performance analysis, 435
Performance measures, 410
Periodic function, 265
Periodic sequence, 277
Perron root, 101
Perturbation, 239
Petri net, 17, 114–115
   See also GSPN
Phase, 155, 214, 219, 229
Phase-type distribution, 5, 27
Phase-type expansion, 449, 456
Pivoting, 85
Poisson arrivals, 412
Poisson-Charlier polynomials, 282
Poisson summation formula, 266–267, 277
Pole, 243, 245
Policy, 326
   deterministic, 327
   improvement, 344
   iteration, 340, 343
      modified, 349–350
   Markov, 327
   optimal, 329, 336
   stationary, 327
Pollaczek Contour Integral, 294
Pollaczek, 289, 294
Pollaczek-Khintchine formula, 290, 301
Polling, 72, 280
Polynomial, 225, 245
Positive recurrent, 207, 209, 212, 218, 224, 228, 232
Post-Widder formula
   discrete analog of, 264
Post-Widder inversion formula, 264
Poularikas, 260
Power method, 82, 86–87

INDEX   487

Power series coefficients, 211, 214, 235, 243
Precision, 271, 279
Preconditioned power iterations, 82, 88–89
Preconditioning, 91, 99
  matrix, 91
Prefactor, 268–269, 279
Present value, 330
Press, 296
Priority
  job class, 427
  shortest-job first, 427
Probability density function, 262
Probability generating function, 165, 210
Probability of emptiness, 291
Processor sharing, 306, 422
Product form, 115, 129–134, 409, 411, 415, 417, 421
  as approximation, 428
  generalized view, 421
  restored, 429
Production, 410
Project completion time, 259
Projection, 206
  method, 53, 147
  process, 95–96, 99
  techniques, 82
Prolog, 22
Puterman, 325, 349

QBD process: *see* quasi birth and death process
Quadrature, 263
Quasi birth and death process, 155–157, 170–171, 177, 189, 194
Queue
  BMAP/G/1, 280
  $D^X/D^m/1$, 397
  $E_k/E_k/1$, 298
  $\Gamma_\alpha/\Gamma_\beta/1$, 297
  Geom/G/1, 366, 394
  Geom(n)/G(n)/1/N, 382
  $Geom^X/G/1$, 396
  $Geom/G^B/1$, 397
  GI/G/1, 156, 158, 366, 392
    discrete-time, 388
    moment generating function, 294
  GI/Geom/1, 382, 394
  GI/Geom/1/N, 382
  GI/Geom/c/c, 397
  GI/M/1, 159
  $H_2/G/1$, 293
  $K_2/G/1$, 292
  $K_m/G/1$, 290
  M/D/c, 365
  M/G/1, 164, 280
  M/M/∞, 287
  MMBP/G/1/N, 387
  $M/M^X/1$, 158
  $M_t/G_t/1$, 280
  $M^X/M/1$, 158
  $M^X/M^Y/1$, 158
Queueing network, 113, 409–441
  closed, 303–309, 311–316, 415–417
    multichain, 305–316
    single server, 419
  Jackson, 24, 417
  non-product form, 430
  open, 415–417
  with blocking, 124
Queueing time, 374, 388

R matrix, 233
  See also Rate matrix
Rabinowitz, 263, 265
Radius of convergence, 236, 243–244, 276
Ramaswami, 170
Ramaswami method, 221
Random variable, 262
Randomization method, 54, 233
  See also Jensen's method, Uniformization
Randomly changing environment, 158, 160
Rank, 216
Rate equations, 412
Rate matrix, 161, 179
Rational generating function, 245
Rational transforms, 390–392
RBD, 450, 461
  See also Reliability block diagram
Reachable state space
  reduction of, 145
Real analytic, 243
Recurrent, 157
  See also Positive recurrent
Recursive relation, 283
Reducible, 218, 232, 242
Reference function, 358
Regular, 249
Regular function, 217
Regular splitting, 348
Reif, 304
Reiser, 304
Relative traffic intensities, 306
Relative-cost function, 361
Reliability, 47, 69, 435, 447
Reliability block diagram, 461, 450
Reliability graph, 452
Reliability models
  composition, 471
  decomposition, 471
  non-Markovian, 466
  non-state space, 448
  state space, 448
Remaining service time, 369
Removal
  of functional terms, 144

488 COMPUTATIONAL PROBABILITY

of synchronizing events, 144
Rendez-vous, 118
  networks, 427
Renewal function, 261
Renewal process, 261, 457
Repeating rows
  method of, 213, 229
Residual vector, 96
Resource sharing, 124
Response time, 410
Reversible process, 283
Reward, 6, 328
  impulse reward, 26, 46, 63
  instantaneous reward, 60
  rate and impulse based, 66
  rate functions, 454
  rate reward, 26, 45, 63
  unbounded, 357
RG, 452
  See also Reliability graph
Riemann-Lebesgue lemma, 268
Riordan, 282
Romberg integration, 296
Root vector, 206, 215, 223, 225
Roots, 160, 205, 229, 289, 366, 375–377,
    379–382, 388–392, 394–396, 398
  distinct, 375
  multiple, 246
  of unity, 237, 242
  repeated, 375
Ross, 286
ROU process, 281
  See also Ornstein-Uhlenbeck process,
    reflected
Rouché's theorem, 207, 235, 238, 375, 379,
    389
roundoff error: see Error, roundoff
Routing
  matrix, 306
  random, 417
  state dependent, 425

Sample path, 183
SAN, 113, 117
  See also Stochastic automata network
SAVE, 23
Scalar elimination, 181
Scaled generating function, 312
Scaling, 269, 279, 281, 284
  in closed queueing networks, 311–314
Scheduled maintenance, 72
Schur vector, 96
Search algorithm, 296
Semi-Markov decision chain, 358–359
Semi-Markov decision process, 326
Semi-Markov process, 280, 449, 457, 459,
    467
Sensitivity analysis, 70

Series
  alternating, 268
  Fourier, 265–266
    See also Fourier, series
  complex Fourier series, 266
  Laguerre
    series, 264
    See also Laguerre series
  nearly alternating, 265
  power series coefficients, 211, 214, 235, 243
Service centers, 305
Service-time distribution
  arbitrary, 381
  deterministic, 381
  exponential, 412
  geometric, 381
SHARPE package, 471
Short-run dynamics, 100
Siegert, 286
Sigma function, 284
Significance
  loss of, 86
Similarity transformation, 51
Simulation, 11, 15, 114
  discrete event, 3
  Monte Carlo, 3
Singularity, 207, 263, 295
  critical, 292
  dominant, 292
Skip-free
  to the left, 208–209, 213
  to the right, 223–224, 229
Slot, 365–366, 369
  boundaries, 369, 372
Smoothness, 268
SMP: see Semi-Markov process
Sojourn time, 410
Sojourn
  above n, 161–163, 178
  in E', 174
Solution method
  Arnoldi, 83
  combined direct-iterative, 90
  decompositional, 82, 97
  direct, 82–83
  iterative, 82
SOR (Successive OverRelaxation), 82
Spans, 351
Sparse matrix format, 34
Sparseness, 310
  of routing chain, 310
Sparsity, 139
Spectral expansion, 282
Spectral radius, 89, 218, 241, 243
Splitting, 348
Squared coefficient of variation, 291
Srikant, 281

INDEX 489

SSOR (Symmetric SOR), 90
Stable algorithm, 87
Stages, 326, 330
  method of, 456
State reduction, 181–184
State space, 16, 326
  explosion, 114, 143
  generation of, 12, 33
  potential, 16
  storage, 33
State truncation, 470
State variables, 3
State-action frequencies, 347
States
  recurrent, 157
  tangible, 19
  transient, 157
  vanishing, 19
Station balance, 411, 413–414, 418, 422
  and blocking, 427
  and failure of, 425
  restored, 427, 429, 435
Stationary
  distribution, 81, 206, 214, 219, 229
    See also Stationary, probability vector
  policy, 327
  probability vector, 207, 220, 224, 81
    See also Stationary, distribution
Stationary-excess cdf, 290
Steady-state
  detection, 57
  analysis, 465
  distribution, 281
  probability, 81, 412
  solutions, 5
    See also Solution method
Stehfest, 272
Step-by-step nested inversion, 305
Stiff models, 56, 67, 69, 471
Stiffness, 472
Stochastic activity network, 20
Stochastic automata, 114, 119
  grouping of, 143
  ordering of, 143
  superposition, 115
  synchronization, 114, 123, 126
Stochastic automata network, 113, 118–123
  interacting, 118–119
  non-interacting, 117–118
Stochastic complement, 213, 221
Stochastic confusion, 30
Stochastic loss network, 303
Stochastic matrix, 81
Stochastic Petri net, 114–115
  See also GSPN
Stochastic process algebra, 114
Stochastic reward net, 20

Stochastic system, 1
Stochastically smaller, 300
Stopping criterion, 341
Stroot, 272
Sturm sequence, 247
Subdominant eigenvalue, 89–90
Sub-level, 155, 158, 160
Subnetwork, 437, 439
Subspace iteration, 96–97, 147
Substochastic, 179–180
  matrix, 105, 174
Successive approximation
  in policy space, 340
  in function space, 340
Successive substitution, 169
Summable, 224–225, 230, 234
Superharmonic, 250
Superposition of automata, 115
Superregular, 250
Supplementary variables, 4, 449
Supremum norm, 358
Symbolic mathematical software, 276
Symmetric Gauss Seidel, 90
Synchronization of automata, 114, 123, 126
Synchronized transitions, 118–119, 121, 123
  functional, 119
  constant, 119
Synchronizing event, 118–123, 126–127, 141–142
  owner of, 142
System dependencies, 453

Tail probabilities, 377, 380–381, 388, 392
  arbitrary epoch, 381
Takács, 257
Takács theorem, 241
Tandem queue, 414–415, 425, 430–431
Tangible state, 19
TANGRAM, 22–23
Telecommunication, 410, 433
Telephone operator, 194
  bilingual, 194
Telephony, 365
Telescoping property, 308
Templeton, 289, 298
Tensor, 115
  addition, 116
  algebra, 115, 117
  product, 116–119, 121
  sum, 116–117, 123
Terminal cost, 329
Teukolsky, 296
Throughput, 307, 410, 420, 424, 429, 432, 439
Time to achieve a reward level, 46, 71
Time-dependent
  blocking probability, 280
  mean, 280

Toeplitz matrix, 168, 206–208, 215
Tolerance criterion, 83
Traffic equation, 417–418
Traffic rate equation, 306
Transform method, 206–207, 213, 248
Transient, 212, 218, 228, 232
  analysis
    cumulative, 464
    instantaneous, 464
  state, 157
Transient solutions, 5, 43-72, 280   of
  Erlang loss model, 279–288
Transition
  enabled, 18
  immediate, 18
  priority, 19
  timed, 18
Transition function, 44
Transition matrix, 5
  approximation, 357
  storage, 34
Translate, 266
Trapezoidal rule, 265, 276
Tree algorithm, 311, 315
Truncation
  error, 265, 270, 274
  of state space, 131
  of transition matrix, 180
Types of jobs, 423

UDL factorization, 186–187
UL factorization, 193
Uniform convergence, 358
Uniformization, 54–55, 58, 66, 70, 221, 233
  See also Jensen's method,
    Randomization

adaptive uniformization, 57
Jensen's method, 54
randomization technique, 54
regenerative randomization, 58
uniformization power method, 56, 68
Uniformly bounded costs, 357
Upper-bound extrapolation, 354
Utilization, 410, 420, 429

Value determination, 344
Value function, 328
Value iteration, 340
Vandermonde matrix, 248–249
Van Der Pol, 260
Vanishing state, 19
Vector subspaces, 95
Vector-descriptor multiplication, 134
Vetterling, 296
Visit ratio, 306
Volterra integral equation, 468

Waiting time, 158, 369, 374, 388, 410
  in GI/G/1 queue, 288
Weakly coupled, 309
Weeks, 264
Weeks' algorithm, 264
Weighted supremum norms, 358
Weiss, 272, 282
Well-defined model, 30
Wiener-Hopf, 206, 223
Wimp, 268, 273
Winding number, 239

Zero, 205, 238, 242, 291
  See also Roots
  multiple, 246
  multiplicity of, 210, 216, 225, 231
  of generating function, 375

GPSR Compliance
The European Union's (EU) General Product Safety Regulation (GPSR) is a set of rules that requires consumer products to be safe and our obligations to ensure this.

If you have any concerns about our products, you can contact us on

ProductSafety@springernature.com

In case Publisher is established outside the EU, the EU authorized representative is:

Springer Nature Customer Service Center GmbH
Europaplatz 3
69115 Heidelberg, Germany

www.ingramcontent.com/pod-product-compliance
Ingram Content Group UK Ltd.
Pitfield, Milton Keynes, MK11 3LW, UK
UKHW022229230426

12048UKWH00016BA/1154